STATISTICS IN MEDICINE

Second Edition

STATISTICS IN MEDICINE

Second Edition

ROBERT H. RIFFENBURGH

Clinical Investigation Department
Naval Medical Center, San Diego
San Diego, California

ELSEVIER
ACADEMIC
PRESS

AMSTERDAM • BOSTON • HEIDELBERG • LONDON
NEW YORK • OXFORD • PARIS • SAN DIEGO
SAN FRANCISCO • SINGAPORE • SYDNEY • TOKYO

Senior Acquisitions Editor	Nancy Maragioglio
Project Manager	A. B. McGee
Associate Acquistions Editor	Kelly Sonnack
Marketing Manager	Linda Beattie
Cover Design	Cate Barr
Composition	Cepha Imaging Pvt. Ltd.
Cover Printer	Phoenix Color Corp.
Interior Printer	Courier Companies, Inc.

Elsevier Academic Press
30 Corporate Drive, Suite 400, Burlington, MA 01803, USA
525 B Street, Suite 1900, San Diego, California 92101-4495, USA
84 Theobald's Road, London WC1X 8RR, UK

This book is printed on acid-free paper.

Library of Congress Cataloging-in-Publication Data
Application submitted

British Library Cataloguing in Publication Data
A catalogue record for this book is available from the British Library

ISBN-13: 978-0-12-088770-5
ISBN-10: 0-12-088770-3

For all information on all Elsevier Academic Press publications
visit our Web site at www.books.elsevier.com

Printed in the United States of America
08 09 10 11   9 8 7 6 5 4 3 2

This book is dedicated to
Gerrye, my love and best friend for more than five decades;
Robin and Rick;
Scott, Ellen, Alex, and Jasmine;
the memory of Marc;
Karen, Matt, Bryce, Aaron, and Sophia;
and Doug, Tricia, Nichole, and Ryan.

When told, "I'm too busy treating patients to do research," I answer:

When you treat a patient, you have treated a patient.
When you do research, you have treated ten thousand patients.

Contents

Part I   A Study Course of Fundamentals

Chapter 1   Data, Notation, and Some Basic Terms   3

Chapter 2   Distributions   15

Chapter 3   Summary Statistics   37

Chapter Summaries   531

References and Data Sources   581

Tables of Probability Distributions   585

Symbol Index   603

Subject Index   607

Foreword to the Second Edition

THE NEED FOR CLARITY

To refer to the last half-century as revolutionary in the statistical world would be a gross understatement. I believe the effects of this revolution are more profound in the world of medicine and other health-related sciences than in most scientific disciplines. Most studies in health sciences today involve formal comparisons, requiring the use of appropriate statistical input on study design and analysis. Some have become very sophisticated, as the FDA requires complex and often large clinical trials, and most epidemiologic studies are more elaborate than heretofore.

As a consequence, for those who aspire to engage in research activities, formal training in statistics is a *sine qua non*. This is true even though most medical research today is, by and large, collaborative, involving a team of scientists, including at least one trained statistician. As with any endeavor, communication among team members is crucial to success and requires a common language to bridge the gaps between their disciplines. Thus, although it is certainly not necessary or expedient for most research scientists to become statisticians or for most statisticians to become medical doctors, it is essential that they develop these avenues of communication.

In fact, all health care practitioners who even occasionally glance at their voluminous literature encounter statistical jargon, concepts, notation, and reasoning. Of necessity, therefore, the training of most such practitioners involves some exposure to statistics in order that clear communication and understanding result.

Those statisticians who collaborate with medical researchers struggle mightily with this problem of communicating with their colleagues. Often this involves writing texts or reference books designed not to turn medical researchers into statisticians but rather to train them in a common language in order to be good consumers of our statistical products and good collaborators with their statistical colleagues.

PROVIDING THE TOOLS

Dr. Riffenburgh addressed just this problem in the first edition of his successful book on statistics in medicine. He presented statistics in language accessible to health care practitioners so that they could effectively understand and communicate statistical information. To the second edition he brings new material, all the while using the formula so successful in the first edition. He has added three completely new chapters on multifactorial studies, equivalence testing, and more uncommon tools. In addition, he has expanded several of the original chapters with new sections on issues of study design, ethics, statistical graphics, bias, and *post hoc* power. These additions, all presented in the same spirit as in the first edition, make this an even more comprehensive resource.

This book is useful to medical researchers who have some training in statistics and need only a reminder of some of the basic statistical tools and concepts. Herein they will find, unencumbered by unnecessary formulas, equations, symbols, or jargon, ready access to most of the fundamental tools useful in medical research. Such access allows one to quickly find understandable explanations of most of the material found in the literature, thereby enabling understanding of the statistical analyses reported and facilitating a proper interpretation of the results.

This book is also useful as a text. It is designed so that, in the hands of a knowledgeable instructor, it can be used to provide students in most health-related fields with the statistical background necessary for them to function as researchers and/or practitioners in today's increasingly quantitative world.

I congratulate Dr. Riffenburgh for his career as a medical statistician and for this useful text/reference book, which I commend to all who teach statistics to students in the health-related fields and to all who are engaged in or are consumers of health-related research.

W. M. (Mike) O'Fallon, PhD*

*Dr. O'Fallon, Professor Emeritus of Biostatistics at the Mayo Clinic, has served as the Head of the Division of Biostatistics and Chair of the Department of Health Science Research at Mayo. He was president of the American Statistical Association in 2000 and served on its Board of Directors for 6 years, He has participated in hundreds of medical research endeavors and is a co-author on more than 400 peer-reviewed manuscripts in medical literature. He chaired an NIH Study Section and has served on an FDA advisory committee.

Foreword to the First Edition

In my five years as Commander of Naval Medical Center San Diego, which has been termed the world's largest and most technologically advanced military health care complex, I was always humbled by the vast spectrum of clinical challenges successfully met by the staff. In my current position, I have come to realize that all of modern medicine deals with a similar depth and variety of cases as it transitions to a managed-care environment. We have a critical obligation not only to care for our patients, but also to assess thoroughly our decisions on therapy, treatment effects, and outcomes. Scientifically rigorous medical studies including sound statistical analyses are central to this process.

Each year, the demand for sound statistics in medical research is increasing. In recent years, statistical analysis has become the hallmark of studies appearing in the leading journals. It is common for authors to have their manuscripts returned for rework on their statistics. The demand for quality statistics puts intense pressure on medical education to include more statistical training, on practitioners to evaluate critically the medical literature, and on clinical staff to design and analyze their investigations with rigor.

Currently, the field lacks a book that can be used efficiently as both text and reference. Clinical investigators and health care students continuously complain that traditional texts are notoriously difficult for the nonmathematically inclined whose time is saturated with clinical duties. This book addresses that problem. The explanations, step-by-step procedures, and many clinical examples render it a less daunting text than the usual statistics book. This book represents a practical opportunity for health care trainees and staff not only to acquaint themselves with statistics in medicine and understand statistical analysis in the medical literature, but also to be guided in the application of planning and analysis in their own medical studies.

Dr. Riffenburgh, a former professor of statistics, has consulted on medical statistics for decades. He was a linchpin in hundreds of the nearly 1000 studies current at NMCSD during my tenure. Investigators from all disciplines benefited from his collegial advice and rigorous analysis of their projects. His understanding of the unique needs in the training of health care professionals was one motivation for this book. I hope you will find it a valuable addition to your professional library.

Vice Admiral Richard A. Nelson, M.D., USN*
Surgeon General of the Navy Chief,
Bureau of Medicine and Surgery

---

*A Navy physician for more than 30 years, Admiral Nelson has served in clinical practice and has directed a number of medical departments and agencies. He has been certified by the American Board of Preventive Medicine and has received numerous medals and awards for his service. In the past decade, he has been Commanding Officer, Naval Hospital Bremerton, WA; Fleet Surgeon, U.S. Atlantic Fleet; Command Surgeon, U.S. Atlantic Command; Medical Advisor, NATO's Supreme Allied Command Atlantic (SACLANT); and Commander, Naval Medical Center San Diego.

Acknowledgments

With special thanks to all those acknowledged in the first edition. Although all of you have moved on to new challenges, your contributions to this book continue to be recognized and appreciated.

The author is grateful to Dr. Michael O'Fallon for writing the Foreword to this second edition. Dr. O'Fallon, who directed biostatistics at the Mayo Clinic for many years and has been president of the American Statistical Association, has been an esteemed colleague, and his continued support is much appreciated. Editing and proofreading were assisted by Kevin Cummins and Benjamin Ely. Thank you all for your efforts and attention.

Databases

Databases are abbreviated as *DBs,* followed by the DB number. The super-script number at the end of each main heading refers to source information listed in References and Data Sources. Databases may be downloaded in Microsoft Excel format from the Elsevier Web site (access to information is available via the Internet at http://books.elsevier.com/companions/0120887703).

PURPOSE FOR REPEATED USE OF THESE DATA SETS

In this text, these data are used for primary illustration of many of the statistical methods presented. Once familiar with a data set, the reader may concentrate on the statistical aspects being discussed rather than focusing on the medical aspects of the data. These data sets were selected as much for data reasons as for medical reasons: A mix of data types illustrates different methods, adequate sample size, and other features.

DB1. INDICATORS OF PROSTATE BIOPSY RESULTS[33]

BACKGROUND

The data from DB1 were obtained from 301 male patients examined in the Urology Department at the Naval Medical Center San Diego who exhibited one of several reasons to suspect the declining health of their prostate glands. These data were recorded for the patients in the order they presented. These data have the advantage that they include biopsy outcomes regardless of the results of preliminary examinations; thus, false-negative results (negative biopsy outcome when a tumor is present) and false-positive results (positive biopsy outcome when no tumor is present) can be examined.

DATA

For each patient, recorded data (units indicated in parentheses) are as follows:

Age (years, >0, integers)
Digital rectal examination result
Transurethral ultrasound result

Prostate-specific antigen level (ng/ml, >0, rounded to one decimal place)
Volume of prostate (ml, >0, rounded to one decimal place)
Prostate-specific antigen density level (prostate-specific antigen level/volume of prostate, >0, rounded to two decimal places)
Biopsy result.

There are too many data for the reader to do calculations without a computer. Some summary statistics are given in the following table for the full set of 301 patients, of whom 296 patients are at risk for prostate cancer (data were excluded for 5 patients with severe benign prostatic hypertrophy):

Data set	Age (years)	PSA (ng/ml)	Volume (ml)	PSAD
301 patients				
Mean	66.76	8.79	36.47	0.269
Standard deviation	8.10	16.91	18.04	0.488
296 patients				
Mean	66.75	8.79	35.46	0.272
Standard deviation	8.13	17.05	16.35	0.491
Patients 11–301				
Mean	66.82	8.85	36.60	0.270
Standard deviation	8.14	17.19	18.12	0.496

PSA, prostate-specific antigen level; PSAD, prostate-specific antigen density level.

Data for the first 10 patients are given in Table DB1.1 for use in calculations. Note that all data are numeric. Most analysis software packages require this, and it is much easier to transfer data from one package to another (e.g., from a spreadsheet to a statistical package) with data in numeric form.

Table DB1.1
Prostate Test Data from the First 10 of the 301 Patients in the Data Set

Patient no.	Age (years)	Digital rectal examination	Transurethral ultrasound	PSA (ng/ml)	Volume (ml)	PSAD	Biopsy result
1	75	0	1	7.6	32.3	0.24	0
2	68	1	0	4.1	27.0	0.15	0
3	54	0	0	5.9	16.2	0.36	1
4	62	1	1	9.0	33.0	0.27	1
5	61	0	0	6.8	30.9	0.22	1
6	61	0	1	8.0	73.7	0.11	0
7	62	0	0	7.7	30.5	0.25	0
8	61	1	1	4.4	30.5	0.14	0
9	73	1	1	6.1	36.8	0.17	0
10	74	1	1	7.9	16.4	0.48	0

0, negative; 1, positive; PSA, prostate-specific antigen level; PSAD, prostate-specific antigen density level.

DB2. EFFECTIVENESS OF A DRUG IN REDUCING NAUSEA AFTER GALLBLADDER REMOVAL[31]

BACKGROUND

A sample of 81 patients undergoing gallbladder removal by laparoscope was randomized into two groups to receive preoperative ondansetron hydrochloride or a placebo. (Data were edited slightly for the exercise and do not reflect exact original results.) Patients rated their nausea from 1 (no nausea) to 5 (unbearable nausea) 2 hours after the end of surgery.

DATA

	Nausea scores					
	1	2	3	4	5	Total
Drug	31	7	2	0	0	40
Placebo	25	6	5	2	3	41
Total	56	13	7	2	3	81

To illustrate some analytic techniques, rating scores of 2 to 5 may be combined to form a 2×2 contingency table of nausea or no nausea.

	No nausea (score 1)	Nausea (score 2–5)	Total
Drug	34	6	40
Placebo	22	19	41
Total	56	25	81

DB3. EFFECT OF AZITHROMYCIN ON SERUM THEOPHYLLINE LEVELS OF PATIENTS WITH EMPHYSEMA[12]

BACKGROUND

Theophylline dilates airways in patients with emphysema. When azithromycin is indicated for an infection in such a patient, does it alter the serum theophylline level? Serum theophylline levels were measured in 16 patients with emphysema at baseline (just before the start of azithromycin administration), 5 days (at the end of the course of antibiotics), and at 10 days. Clinically, it is anticipated that the antibiotic will increase the theophylline level.

DATA

Patient no.	Age (years)	Sex	Baseline[a]	5 days[a]	10 days[a]
1	61	F	14.1	2.3	10.3
2	70	F	7.2	5.4	7.3
3	65	M	14.2	11.9	11.3
4	65	M	10.3	10.7	13.8
5	64	F	15.4	15.2	13.6
6	76	M	5.2	6.8	4.2
7	72	M	10.4	14.6	14.1
8	69	F	10.5	7.2	5.4
9	66	M	5.0	5.0	5.1
10	62	M	8.6	8.1	7.4
11	65	F	16.6	14.9	13.0
12	71	M	16.4	18.6	17.1
13	51	F	12.2	11.0	12.3
14	71	M	6.6	3.7	4.5
15	—[b]	M	9.9	10.7	11.7
16	50	M	10.2	10.8	11.2

[a] Data are serum theophylline levels (mg/dl) in 16 patients with emphysema.
[b] Age for Patient 15 is not available.
F, female; M, male.

DB4. EFFECT OF PROTEASE INHIBITORS ON PULMONARY ADMISSIONS[70]

BACKGROUND

Protease inhibitors (PIs) are used in the treatment of infection by human immunodeficiency virus (HIV). A military hospital provides unlimited access to a full spectrum of HIV medications. Has this access reduced admissions caused by secondary infections? In particular, has unlimited access to PIs reduced the annual admission rate of patients with pulmonary complications? The number of admissions for HIV, separated into patients with and those without pulmonary complications, was obtained for the 4 years (1992–1995) before access to PIs and for the 2 years (1996–1997) after PIs became available.

DATA

	Patients with pulmonary complications	Patients without pulmonary complications	Total
Before PIs were available	194	291	485
After PIs were available	25	67	92
Total	219	358	577

PI, protease inhibitor.

DB5. EFFECT OF SILICONE IMPLANTS ON PLASMA SILICON[57]

BACKGROUND

Silicone implants in women's breasts occasionally rupture, releasing silicone into the body. A large number of women have received silicone breast implants, and some investigators believe that the presence of the implants increases plasma silicon levels, leading to side effects. A study was begun to test this belief. A method was developed for accurately measuring the silicon level in blood. For each of 30 women studied, plasma silicon levels (mcg/g dry weight) were taken before surgical placement of the implants. A postsurgical washout period was allowed, and plasma silicon levels were retaken.

DATA

Patient no.	Preoperative level	Postoperative level
1	0.15	0.21
2	0.13	0.24
3	0.39	0.10
4	0.20	0.12
5	0.39	0.28
6	0.42	0.25
7	0.24	0.22
8	0.18	0.21
9	0.26	0.22
10	0.12	0.23
11	0.10	0.22
12	0.11	0.24
13	0.19	0.45
14	0.15	0.38
15	0.27	0.23
16	0.28	0.22
17	0.11	0.18
18	0.11	0.15
19	0.18	0.04
20	0.18	0.14
21	0.24	0.24
22	0.48	0.20
23	0.27	0.24
24	0.22	0.18
25	0.18	0.19
26	0.19	0.15
27	0.32	0.26
28	0.31	0.30
29	0.19	0.22
30	0.21	0.24

Plasma silicon levels before and after surgery were measured as mcg/g dry weight.

DB6. LASER REMOVAL OF TATTOOS AS RELATED TO TYPE OF INK USED[68]

BACKGROUND

A frequently used component of tattoo ink is titanium. A dermatologist suspected that tattoos applied with titanium ink are more difficult to remove than those applied with other inks. Fifty patients wanting tattoo removal were tested for ink type, and laser removal was attempted. The number of patients responding and the number of patients not responding to the removal attempt are given here according to the type of ink used.

DATA

	Responded	Did not respond	Total
Titanium ink	5	30	35
Nontitanium ink	8	7	15
Total	13	37	50

DB7. RELATION OF BONE DENSITY TO INCIDENCE OF FEMORAL NECK STRESS FRACTURES[54]

BACKGROUND

Femoral neck stress fractures occur in young people engaged in sports, combat training, and other similar activities. Such fractures have a relatively high rate of poor healing, which, in many cases, leaves the patient somewhat impaired. Many such events could be avoided if risk factors (i.e., those characteristics that predict a greater probability of fracture) could be identified. One potential risk factor is bone density. A large DB (several thousand) of bone density measures of healthy people has been compiled at the University of California–San Francisco (UCSF). Norms were established for age and sex. Do people with femoral neck stress fractures tend to have lower bone density measures than healthy individuals? Bone density was measured for 18 young people with femoral neck stress fractures.

DATA

The following data include age, sex, bone density measure, and UCSF norm-value for that age and sex.

Patient no.	Sex	Age (years)	Bone density (mg/ml)	UCSF norm
1	F	22	139.6	185
2	F	24	167.2	188
3	F	39	140.1	181

(Continued)

Patient no.	Sex	Age (years)	Bone density (mg/ml)	UCSF norm
4	M	18	163.6	225
5	M	18	140.0	225
6	M	18	177.4	225
7	M	18	147.0	225
8	M	19	164.9	223
9	M	19	155.8	223
10	M	19	126.0	223
11	M	19	163.4	223
12	M	21	194.9	219
13	M	21	199.0	219
14	M	21	164.6	219
15	M	22	140.8	218
16	M	22	152.4	218
17	M	23	142.0	216
18	M	40	97.3	184

F, female; M, male; UCSF, University of California–San Francisco.

DB8. COMPARING TWO TYPES OF ASSAYS ON THE EFFECT OF GLYCOSAMINOGLYCANS ON THE BLADDER SURFACE[60]

BACKGROUND

Glycosaminoglycans (GAGs) alter the adherence properties of the bladder surface. Four spectrophotometric readings on the GAG levels in tissues of disease-free bladders from cadavers exposed to GAGs were taken by each of two types of assays. (Mean instrument error has been established as 0.002.)

DATA

Assay type	Spectrophotometric readings on GAG levels (mmol/kg)
Type 1	0.74, 0.37, 0.26, 0.20
Type 2	0.51, 0.53, 0.42, 0.29

GAG, glycosaminoglycan.

DB9. PREDICTION OF GROWTH FACTORS BY PLATELET COUNTS[58]

BACKGROUND

There are two types of platelet gel growth factors: platelet-derived growth factor (PDGF), used primarily in cutaneous wound healing, and transforming growth factor (TGF), used primarily to promote bone healing. Users wish to know the level of a growth factor in platelet gel, but the assay is costly and time-consuming. Can a simple platelet count predict the level of a growth factor in platelet gel? Platelet counts and levels of PDGF and TGF in platelet gel were measured for 20 patients. Because the large numbers in the data are a little tedious to calculate, some data summaries are given to assist the student.

DATA

Variable	Mean	Standard deviation
Platelet count	295,000.0	164,900.0
Gel PDGF	31,622.9	19,039.7
Gel TGF	31,340.0	16,767.0

PDGF, platelet-derived growth factor; TGF, transforming growth factor.

Patient no.	Platelet count	Gel PDGF	Gel TGF
1	576,000	61,835	60,736
2	564,000	74,251	61,102
3	507,000	62,022	55,669
4	205,000	10,892	15,366
5	402,000	34,684	41,603
6	516,000	48,689	47,003
7	309,000	26,840	33,400
8	353,000	42,591	36,083
9	65,000	7,970	12,049
10	456,000	49,122	46,575
11	306,000	25,039	35,992
12	281,000	30,810	34,803
13	226,000	19,095	27,304
14	125,000	9,004	10,954
15	213,000	16,398	18,075
16	265,000	25,567	26,695
17	102,000	17,570	13,783
18	159,000	21,813	17,953
19	33,000	16,804	7,732
20	237,000	31,462	23,922

PDGF, platelet-derived growth factor; TGF, transforming growth factor.

DB10. TESTS OF RECOVERY AFTER SURGERY ON HAMSTRINGS OR QUADRICEPS[3]

BACKGROUND

Tests of strength and control following surgery on hamstring or quadriceps muscles or tendons have had varying degrees of success in indicating the quality of recovery. Two proposed tests are the time to perform and the distance covered in a triple hop along a marked line, contrasted between the healthy leg and the leg that underwent surgery. Strength of the operated muscle in the two legs has not been shown to be different for eight postoperative patients. The question is whether or not the proposed tests will detect a difference.

DATA

	Time to perform (seconds)		Distance covered (cm)	
Patient no.	Operated leg	Nonoperated leg	Operated leg	Nonoperated leg
1	2.00	1.89	569	606
2	2.82	2.49	360	450
3	2.62	2.39	385	481
4	2.97	3.13	436	504
5	2.64	2.86	541	553
6	3.77	3.09	319	424
7	2.35	2.30	489	527
8	2.39	2.13	523	568

DB11. SURVIVAL OF MALARIAL RATS TREATED WITH HEMOGLOBIN, RED BLOOD CELLS, OR PLACEBO[16]

BACKGROUND

The malaria parasite affects the transmission of oxygen to the brain. The infusion of red blood cells (RBCs) in a malaria-infected body slows deterioration, giving the body more time to receive the benefit of antimalarial medication. Does the infusion of hemoglobin provide the same benefit? Three groups of 100 rats each were infected with a rodent malaria. One group was given hemoglobin, a second was given RBCs, and the third was given hetastarch (an intravenous fluid with protein). Survival numbers (and, coincidentally, percentage) at the outset (day 0) and during 10 succeeding days were recorded.

DATA

	Rats surviving by treatment type *(n)*		
Day	Hemoglobin	Red blood cells	Hetastarch
0	100	100	100
1	100	100	100
2	77	92	83
3	77	92	67
4	69	83	58
5	69	83	58
6	62	83	50
7	38	83	33
8	38	75	33
9	31	58	33
10	23	58	33

DB12. IDENTIFICATION OF RISK FACTORS FOR DEATH AFTER CARINAL RESECTION[52]

BACKGROUND

Resection of the tracheal carina is a dangerous procedure. From 134 cases, the following data subset is given from the variables that were recorded: age at surgery, prior tracheal surgery, extent of the resection, intubation required at end of surgery, and patient death. Because the data set is large for an exercise, some data summaries are given to assist the student.

DATA

Variable	Mean	Standard deviation	Variable	Mean	Standard deviation
Age (years)	47.84	15.78	Extent (cm)	2.96	1.24
Age, died = 0	48.05	16.01	Extent, died = 0	2.82	1.21
Age, died = 1	46.41	14.46	Extent, died = 1	3.96	1.00

0, no; 1, yes.

Patient no.	Age at surgery (years)	Prior surgery	Extent of resection (cm)	Intubated	Died
1	48	0	1	0	0
2	29	0	1	0	0
3	66	0	1	0	0
4	55	1	1	0	0
5	41	0	1	0	0
6	42	0	1	0	0
7	44	0	1	0	0
8	67	1	1	0	0
9	54	1	1	0	0
10	36	0	1.5	0	0
11	63	0	1.5	0	0
12	46	0	1.5	0	0
13	80	0	1.5	0	0
14	47	0	1.5	0	0
15	41	0	1.5	0	0
16	62	1	1.5	0	0
17	53	0	1.5	0	0
18	62	0	1.5	0	0
19	22	0	1.5	0	0
20	65	0	2	0	0
21	73	0	2	0	0
22	38	0	2	0	0
23	55	0	2	0	0
24	56	0	2	0	0
25	66	0	2	0	0

(Continued)

Patient no.	Age at surgery (years)	Prior surgery	Extent of resection (cm)	Intubated	Died
26	36	0	2	0	0
27	48	0	2	0	0
28	49	0	2	0	0
29	29	1	2	0	0
30	14	1	2	0	0
31	52	0	2	0	0
32	39	0	2	1	0
33	27	1	2	0	0
34	55	0	2	0	0
35	62	1	2	0	0
36	37	1	2	0	0
37	26	1	2	0	1
38	57	0	2	0	0
39	60	0	2	0	0
40	76	0	2	0	0
41	28	1	2	0	0
42	25	0	2	0	0
43	62	0	2	0	0
44	63	1	2	0	0
45	34	0	2	0	0
46	56	0	2.3	0	0
47	51	0	2.5	0	0
48	42	0	2.5	0	0
49	54	1	2.5	0	0
50	62	0	2.5	0	0
51	47	0	2.5	0	0
52	34	1	2.5	0	0
53	52	0	2.5	0	0
54	49	0	2.5	1	0
55	69	0	2.5	0	0
56	32	0	2.5	0	0
57	19	1	2.5	0	0
58	38	0	2.5	0	0
59	55	0	2.5	0	0
60	40	0	2.5	0	0
61	63	0	2.5	0	0
62	51	0	2.5	0	0
63	8	1	2.5	0	0
64	54	0	2.5	0	0
65	58	1	2.5	1	1
66	48	0	2.5	0	0
67	19	0	2.5	0	0
68	25	0	2.5	1	0
69	49	0	3	0	0
70	27	1	3	0	0
71	47	1	3	1	0
72	58	0	3	0	0
73	8	1	3	0	0
74	41	1	3	0	0
75	18	1	3	0	0

(Continued)

Patient no.	Age at surgery (years)	Prior surgery	Extent of resection (cm)	Intubated	Died
76	31	0	3	0	0
77	52	0	3	0	0
78	72	0	3	0	0
79	62	0	3	0	0
80	38	0	3	0	0
81	59	0	3	0	0
82	62	1	3	0	0
83	33	1	3	1	1
84	62	0	3	0	0
85	25	0	3	1	0
86	46	0	3.2	0	1
87	52	1	3.5	0	0
88	43	0	3.5	0	0
89	55	0	3.5	0	0
90	66	1	3.5	0	0
91	30	0	3.5	0	0
92	41	0	3.5	0	0
93	53	0	3.5	0	1
94	70	0	3.5	0	0
95	61	1	3.5	1	1
96	28	1	3.6	0	0
97	60	1	4	1	1
98	46	0	4	0	0
99	52	0	4	0	0
100	17	0	4	0	0
101	61	0	4	0	0
102	46	0	4	0	1
103	51	0	4	1	1
104	71	0	4	0	0
105	24	0	4	0	0
106	67	0	4	0	0
107	68	0	4	0	0
108	56	0	4	1	0
109	39	0	4	1	1
110	60	0	4	0	1
111	62	1	4.2	0	0
112	57	1	4.2	0	0
113	60	0	4.5	0	0
114	32	0	4.5	0	0
115	34	0	4.5	0	0
116	48	1	4.5	0	0
117	69	0	4.5	1	0
118	27	1	4.5	1	1
119	66	0	4.5	1	1
120	43	0	4.5	0	1
121	60	0	4.8	0	0
122	66	0	4.8	1	1
123	57	0	5	0	0
124	71	0	5	0	0
125	31	0	5	0	0

(Continued)

Patient no.	Age at surgery (years)	Prior surgery	Extent of resection (cm)	Intubated	Died
126	35	0	5	0	0
127	52	0	5	0	0
128	66	1	5	0	0
129	47	0	5	0	0
130	26	0	5.5	1	1
131	28	0	6	1	1
132	63	0	6	0	0
133	62	0	6	0	0
134	26	0	6	1	0

0, no; 1, yes.

DB13. QUALITY TEST ON WARFARIN INTERNATIONAL NORMALIZED RATIO VALUES[19]

BACKGROUND

Warfarin is a blood anticoagulant used to treat certain cardiac patients. It is measured in International Normalized Ratio units (INRs). Because small doses are ineffective and large doses are quite dangerous (in strong doses, it is used as a rodenticide), maintaining the INR level within range is crucial. In-clinic readings are available immediately, whereas the slower laboratory readings are the gold standard. To test the quality of its tests, a Coumadin Clinic obtained both readings on all 104 patients treated during a certain period. Following is a comparison of these readings. Patients are "In Range" if their INR level is $2 < INR \leq 3$.

DATA

| Patient no. | INR readings | | Difference | Below (−1), in (0), and above (1) range | |
	Clinic	Laboratory		Clinic	Laboratory
1	2.46	2.30	0.16	0	0
2	2.86	3.20	−0.34	0	1
3	2.63	2.40	0.23	0	0
4	2.26	2.00	0.26	0	0
5	2.30	2.20	0.10	0	0
6	2.16	1.80	0.36	0	−1
7	2.24	2.40	−0.16	0	0
8	2.38	1.90	0.48	0	−1
9	1.50	2.00	−0.50	−1	0
10	1.60	1.40	0.20	−1	−1
11	1.30	1.80	−0.50	−1	−1
12	3.18	2.90	0.28	1	0

(Continued)

Patient no.	INR readings		Difference	Below (−1), in (0), and above (1) range	
	Clinic	Laboratory		Clinic	Laboratory
13	2.20	2.00	0.20	0	0
14	1.60	1.50	0.10	−1	−1
15	2.40	2.20	0.20	0	0
16	2.30	1.90	0.40	0	−1
17	2.60	2.30	0.30	0	0
18	1.70	1.30	0.40	−1	−1
19	3.10	3.00	0.10	1	1
20	2.40	2.00	0.40	0	0
21	2.33	1.90	0.43	0	−1
22	4.01	3.70	0.31	1	1
23	2.45	2.30	0.15	0	0
24	2.10	1.80	0.30	0	−1
25	2.86	2.70	0.16	0	0
26	2.93	3.40	−0.47	0	1
27	2.00	1.70	0.30	0	−1
28	1.92	1.60	0.32	−1	−1
29	2.34	2.10	0.24	0	0
30	2.40	2.20	0.20	0	0
31	2.35	1.80	0.55	0	−1
32	2.47	2.20	0.27	0	0
33	2.44	2.50	−0.06	0	0
34	1.29	1.40	−0.11	−1	−1
35	1.71	1.50	0.21	−1	−1
36	1.80	1.70	0.10	−1	−1
37	2.20	2.10	0.10	0	0
38	1.91	1.90	0.01	−1	−1
39	1.29	1.10	0.19	−1	−1
40	2.10	2.60	−0.50	0	0
41	2.68	2.40	0.28	0	0
42	3.41	2.80	0.61	1	0
43	1.92	1.50	0.42	−1	−1
44	2.85	3.30	−0.45	0	1
45	3.41	3.50	−0.09	1	1
46	2.70	2.60	0.10	0	0
47	2.43	2.10	0.33	0	0
48	1.21	1.10	0.11	−1	−1
49	3.14	3.30	−0.16	1	1
50	2.50	2.50	0.00	0	0
51	3.00	3.10	−0.10	1	1
52	2.50	2.50	0.00	0	0
53	1.90	1.70	0.20	−1	−1
54	2.40	2.00	0.40	0	0
55	3.00	3.70	−0.70	1	1
56	2.08	1.90	0.18	0	−1
57	2.66	2.50	0.16	0	0
58	2.00	1.70	0.30	0	−1

(Continued)

Patient no.	INR readings		Difference	Below (−1), in (0), and above (1) range	
	Clinic	Laboratory		Clinic	Laboratory
59	2.38	2.10	0.28	0	0
60	2.69	2.40	0.29	0	0
61	2.28	1.90	0.38	0	−1
62	3.23	3.30	−0.07	1	1
63	2.47	2.20	0.27	0	0
64	2.14	2.10	0.04	0	0
65	1.93	1.90	0.03	−1	−1
66	2.71	2.50	0.21	0	0
67	2.14	1.90	0.24	0	−1
68	1.99	1.90	0.09	−1	−1
69	2.95	2.40	0.55	0	0
70	2.15	2.00	0.15	0	0
71	2.69	3.20	−0.51	0	1
72	2.98	3.40	−0.42	0	1
73	2.01	1.60	0.41	0	−1
74	3.35	3.40	−0.05	1	1
75	2.29	2.00	0.29	0	0
76	3.31	3.10	0.21	1	1
77	3.00	3.40	−0.40	1	1
78	2.20	2.00	0.20	0	0
79	2.15	1.84	0.31	0	−1
80	2.19	2.20	−0.01	0	0
81	2.07	1.70	0.37	0	−1
82	2.09	1.90	0.19	0	−1
83	2.75	2.80	−0.05	0	0
84	3.50	3.50	0.00	1	1
85	2.50	3.10	−0.60	0	1
86	2.23	1.90	0.33	0	−1
87	2.15	1.90	0.25	0	−1
88	2.26	1.90	0.36	0	−1
89	2.94	2.60	0.34	0	0
90	2.92	2.30	0.62	0	0
91	2.86	2.80	0.06	0	0
92	2.28	1.90	0.38	0	−1
93	2.65	2.60	0.05	0	0
94	2.62	2.50	0.12	0	0
95	2.10	2.00	0.10	0	0
96	2.90	3.10	−0.20	0	1
97	1.50	1.30	0.20	−1	−1
98	3.50	3.40	0.10	1	1
99	1.50	1.30	0.20	−1	−1
100	2.50	2.90	−0.40	0	0
101	2.43	2.20	0.23	0	0
102	2.68	2.40	0.28	0	0
103	2.05	1.70	0.35	0	−1
104	1.65	1.40	0.25	−1	−1

INR, International Normalized Ratio.

DB14. EXHALED NITRIC OXIDE AS AN INDICATOR OF EXERCISE-INDUCED BRONCHOCONSTRICTION[46]

BACKGROUND

Exercise-induced bronchoconstriction (EIB) is an indicator of asthma. If exhaled nitric oxide (eNO) can flag EIB, then it may be used in the diagnosis of incipient asthma in a subject before engaging in vigorous exercise. In this experiment, eNO was measured on 38 subjects, who then underwent vigorous exercise. Six were found through a decrease in forced expiratory volume in 1 second (FEV_1) to have EIB (EIB = 1), whereas 32 did not have EIB (EIB = 0). After 5 and 20 minutes, eNO again was measured. Does the change in eNO at 5 minutes or at 20 minutes flag the EIB?

DATA

Patient no.	EIB	eNO relative to exercise			eNO before−after (i.e., difference)		Relative humidity (%)	Age (years)	Sex
		0 Before	5 minutes	20 minutes	5 minutes	20 minutes			
1	0	57.7	36.9	38.5	20.8	19.2	38	31	1
2	0	51.3	53.0	44.8	−1.7	6.5	42	35	1
3	0	21.8	17.8	20.6	4.0	1.2	44	26	2
4	0	43.5	48.9	51.4	−5.4	−7.9	43	27	2
5	0	10.5	10.1	11.8	0.4	−1.3	51	25	1
6	0	11.6	9.2	10.0	2.4	1.6	39	22	1
7	0	15.8	13.4	15.9	2.4	−0.1	28	28	1
8	0	25.0	24.0	37.5	1.0	−12.5	33	26	1
9	1	133.0	127.0	147.0	6.0	−14.0	31	25	1
10	0	31.6	28.3	37.4	3.3	−5.8	34	33	1
11	0	68.0	50.2	53.9	17.8	14.1	44	31	1
12	0	41.2	30.1	38.7	11.1	2.5	51	32	1
13	0	4.8	4.7	4.4	0.1	0.4	49	33	2
14	0	71.0	49.9	55.9	21.1	15.1	33	24	2
15	0	36.4	28.7	42.1	7.7	−5.7	33	21	1
16	0	35.8	28.6	31.9	7.2	3.9	32	20	1
17	0	65.2	49.0	52.9	16.2	12.3	34	28	1
18	0	47.1	30.5	39.0	16.6	8.1	28	38	1
19	0	21.2	20.9	19.6	0.3	1.6	33	37	1
20	0	20.5	24.4	23.8	−3.9	−3.3	37	33	2
21	0	19.4	14.1	16.5	5.3	2.9	33	31	1
22	0	9.0	7.0	4.0	2.0	5.0	34	24	2
23	0	7.9	8.9	9.7	−1.0	−1.8	34	26	1
24	1	21.0	22.0	25.0	−1.0	−4.0	44	23	1
25	1	19.0	11.9	16.2	7.1	2.8	42	21	1
26	0	12.0	13.2	12.0	−1.2	0.0	51	37	1

(Continued)

Patient no.	EIB	eNO relative to exercise			eNO before−after (i.e., difference)		Relative humidity (%)	Age (years)	Sex
		0 Before	5 minutes	20 minutes	5 minutes	20 minutes			
27	0	10.8	9.8	15.2	1.0	−4.4	49	33	2
28	0	54.5	37.0	44.8	17.5	9.7	48	25	2
29	0	24.6	18.0	17.8	6.6	6.8	33	22	1
30	0	8.0	8.3	10.0	−0.3	−2.0	37	21	1
31	0	8.9	6.8	6.6	2.1	2.3	49	39	1
32	1	12.8	16.7	14.7	−3.9	−1.9	50	38	2
33	1	23.0	25.0	27.0	−2.0	−4.0	46	31	1
34	1	29.0	31.0	32.0	−2.0	−3.0	46	27	1
35	0	6.7	6.4	5.5	0.3	1.2	52	19	2
36	0	14.2	14.9	14.1	−0.7	0.1	47	26	1
37	0	10.0	8.4	8.9	1.6	1.1	62	39	1
38	0	8.0	8.3	10.0	−0.3	−2.0	61	28	1

1, male; 2, female; EIB, exercise-induced bronchoconstriction; eNO, exhaled nitric oxide.

DB15. COMPARISON OF KIDNEY COOLING METHODS USED TO PROLONG SURGICAL TIME WINDOW[7]

BACKGROUND

In kidney surgery, the length of operation often causes tissue necrosis unless the kidney is quickly cooled to less than 15°C. The traditional method is to pack the kidney in crushed ice as soon as it is exposed. The question arose: Could the kidney be cooled as well or better by infusing it with cold saline solution? Six anesthetized pigs were opened on both sides. One randomly chosen kidney was cooled internally with cold saline (treatment 1); the other was cooled externally with ice (treatment 2). The temperature, measured at both surface (depth 1) and kidney depth (depth 2), was recorded before cooling (baseline, time 0) and at 5, 10, and 15 minutes after start of cooling.

DATA

Observation number	Pig no.	Time (minutes)	Treatment[a]	Depth[b]	Temperature (°C)
1	1	0	1	1	36.7
2	1	5	1	1	33.2
3	1	10	1	1	24.5
4	1	15	1	1	26.6
5	1	0	2	1	35.9
6	1	5	2	1	12.7

(Continued)

Observation number	Pig no.	Time (minutes)	Treatment[a]	Depth[b]	Temperature (°C)
7	1	10	2	1	6.9
8	1	15	2	1	3.9
9	1	0	1	2	36.8
10	1	5	1	2	33.4
11	1	10	1	2	24.8
12	1	15	1	2	28
13	1	0	2	2	36.1
14	1	5	2	2	21.1
15	1	10	2	2	19.1
16	1	15	2	2	8.7
17	2	0	1	1	34.6
18	2	5	1	1	31.5
19	2	10	1	1	32
20	2	15	1	1	30.7
21	2	0	2	1	37.4
22	2	5	2	1	10.6
23	2	10	2	1	8.5
24	2	15	2	1	3.7
25	2	0	1	2	34.7
26	2	5	1	2	31.7
27	2	10	1	2	32
28	2	15	1	2	27.8
29	2	0	2	2	37.6
30	2	5	2	2	24.6
31	2	10	2	2	14.5
32	2	15	2	2	12.9
33	3	0	1	1	32.6
34	3	5	1	1	32.3
35	3	10	1	1	30.8
36	3	15	1	1	29.9
37	3	0	2	1	35.8
38	3	5	2	1	9.6
39	3	10	2	1	7
40	3	15	2	1	3
41	3	0	1	2	33.7
42	3	5	1	2	32
43	3	10	1	2	28.7
44	3	15	1	2	26.7
45	3	0	2	2	36
46	3	5	2	2	17.4
47	3	10	2	2	14.4
48	3	15	2	2	12.2
49	4	0	1	1	37.1
50	4	5	1	1	35
51	4	10	1	1	33.1
52	4	15	1	1	33.2
53	4	0	2	1	35.7
54	4	5	2	1	14.3
55	4	10	2	1	4.7
56	4	15	2	1	4.7

(Continued)

DB15. Comparison of Kidney Cooling Methods Used to Prolong Surgical Time Window

Observation number	Pig no.	Time (minutes)	Treatment[a]	Depth[b]	Temperature (°C)
57	4	0	1	2	37.3
58	4	5	1	2	34.9
59	4	10	1	2	33.1
60	4	15	1	2	32.6
61	4	0	2	2	35.9
62	4	5	2	2	21.1
63	4	10	2	2	7.2
64	4	15	2	2	10.5
65	5	0	1	1	28
66	5	5	1	1	28.7
67	5	10	1	1	22
68	5	15	1	1	22.7
69	5	0	2	1	37.3
70	5	5	2	1	14.5
71	5	10	2	1	6.7
72	5	15	2	1	4.1
73	5	0	1	2	28.4
74	5	5	1	2	27.5
75	5	10	1	2	24.1
76	5	15	1	2	20.2
77	5	0	2	2	37.4
78	5	5	2	2	25.5
79	5	10	2	2	10.2
80	5	15	2	2	10.6
81	6	0	1	1	36.4
82	6	5	1	1	33.1
83	6	10	1	1	30
84	6	15	1	1	28.6
85	6	0	2	1	34.1
86	6	5	2	1	15.4
87	6	10	2	1	8.8
88	6	15	2	1	6.1
89	6	0	1	2	36.5
90	6	5	1	2	30
91	6	10	1	2	29.3
92	6	15	1	2	28.5
93	6	0	2	2	34.2
94	6	5	2	2	16.6
95	6	10	2	2	16.6
96	6	15	2	2	9.6

[a]Treatment 1 indicates that kidney was cooled internally with cold saline; treatment 2 indicates that kidney was cooled externally with ice.
[b]Depth 1 indicates surface; depth 2 indicates kidney depth.

Part I

A Study Course of Fundamentals

Chapter 1

Data, Notation, and Some Basic Terms

1.1. ABOUT THIS BOOK

A UNIQUE BIOSTATISTICS SET OFFERING BOTH TEXTBOOK AND REFERENCE GUIDE

This two-part source for health scientists offers both an introductory textbook for those first encountering biostatistics and a reference guide for those with basic knowledge who wish to read journal articles or conduct research. Part I is a textbook designed for a first course in biostatistics for students in medicine, dentistry, nursing, pharmacy, and other health care fields. Part II is a reference, guiding practicing clinicians in reading medical literature or conducting occasional medical research. The set's level and tone are designed to be user-friendly by the health care clinician. It is not designed for the statistician. Indeed, some statisticians consider it too far removed from the mode of presentation of traditional statistics textbooks.

ONLY THE MORE IMPORTANT METHODS AND CONCEPTS ARE COVERED

This book does not attempt to cover all biostatistical methods. It is intended for clinical health care professionals and students rather than for full-time research workers. It might be called a "30/90 book," in that it covers 30% of the statistical methods used for 90% of medical studies. In Chapter 3 of Bailar and Moseller[2], Emerson and Colditz analyzed 301 articles appearing in 4 volumes of the *New England Journal of Medicine* for statistical content. Part I covers 82% of the biostatistical methods used in these 301 articles, and Parts I and II combined cover 90%. In the years since this article was published, a few more statistical methods have become more commonly used in medical research, and these methods have been included in this second edition.

The Important Issue Is What the Student Can Remember

The student uses what is remembered 5 years after the course, not what is forgotten. It is better for the student to understand and retain some key essentials than to be exposed to a profusion of methods that will not be retained or a depth that will not be understood. Therefore, the textbook part covers only the basic concepts and the most frequently seen methods in biostatistics. Only one method per concept appears. Part I does not pretend to be a methods book. Students need statistical concepts at this stage in their careers, not methods they are not ready to use.

This Book Starts at the Beginning of Mathematics

This book starts at "ground zero" in mathematics and statistics, but it assumes an intelligent, alert, and motivated student with ability in arithmetic. It is designed to be offered without formal university-level prerequisites. It has been honed by use as a text for several lecture series to staff and residents at Naval Medical Center and other hospitals.

The Role of Part II: Reference Guide

Clinicians reading articles or conducting studies should know the fundamentals as provided in Part I. Part II forms a "how to" guide to the common statistical methods used in medical research. It is designed to aid in assessing the literature and conducting research.

Using a Few Data Sets Repeatedly Allows the User to Focus on the Method

This book opens with a number of databases (DBs) with sample sizes small enough to be given fully. These DBs are drawn from various medical areas, including urology (DB1: prostate cancer), surgery (DB2: gallbladder removal), dermatology (DB6: tattoo removal), orthopedics (DB7: femoral neck fractures), internal medicine (DB11: treating malaria), ENT-HNS (DB12: carinal resection), pulmonary medicine (DB14: bronchoconstriction), and others. Supplementary examples with larger sample sizes from other medical fields are also included, summary statistics being provided for the user. The DB set is available online in Microsoft Excel format at the Elsevier Web site (access to information is available via the Internet at http://books.elsevier.com/companions/0120887703).

Each Paragraph Leads with Its Topic or Gist as a Heading

To facilitate scanning and review, each paragraph in the book begins with a leading phrase providing the topic and, when possible, the gist of the paragraph, as can be noted in this opening section.

4

1.2. STAGES OF SCIENTIFIC KNOWLEDGE

STAGES

We gather data because we want to know something. These data are useful only if they provide information about what we want to know. A scientist usually seeks to develop knowledge in three stages. The first stage is to *describe* a class of scientific events. The second stage is to *explain* these events. The third stage is to *predict* the occurrence of these events. The ability to predict an event implies some level of understanding of the rule of nature governing the event. The ability to predict outcomes of actions allows the scientist to make better decisions about such actions. At best, a general scientific rule may be inferred from repeated events of this type.

THE CAUSATIVE PROCESS IS OF INTEREST, NOT THE DATA

A process, or set of forces, generates data related to an event. It is this process, not the data per se, that interests us. Following is a brief explanation of the three stages of gathering knowledge:

(1) *Description:* the stage in which we seek to describe the data-generating process in cases for which we have data from that process. Description would answer questions such as: What is the range of prostate volumes for a sample of urology patients? What is the difference in average volume between patients with negative biopsy results and those with positive results?

(2) *Explanation:* the stage in which we seek to *infer* characteristics of the (overall) data-generating process when we have only part (usually a small part) of the possible data. Inference would answer questions such as: For a sample of patients with prostate problems, can we expect the average of volumes of patients with positive biopsy results to be less than those of patients with negative biopsy results for all American men with prostate problems? Inferences usually take the form of tests of hypotheses.

(3) *Prediction:* the stage in which we seek to make predictions about a characteristic of the data-generating process on the basis of newly taken related observations. Such a prediction would answer questions such as: On the basis of a patient's negative digital rectal examination, prostate-specific antigen (PSA) of 9, and prostate volume of 30 ml, what is the probability that he has cancer? Such predictions allow us to make decisions on how to treat our patients to change the chances of an event. For example: Should I perform a biopsy on my patient? Predictions usually take the form of a mathematical model of the relationship between the predicted (dependent) variable and the predictor (independent) variables.

PHASE I–IV STUDIES

Stages representing increasing knowledge in medical investigations often are categorized by *phases.* A *Phase I* investigation is devoted to discovering if a treatment is safe and gaining enough understanding of the treatment to design formal studies. For example, a new drug is assessed to learn the level of dosage to study and if this level is safe in its main and side effects. A *Phase II* investigation is a preliminary investigation of the effectiveness of treatment. Is this drug more effective than existing drugs? Is the evidence of effectiveness strong enough to justify further study? *Phase III* is a large-scale verification of the early findings, the step from "some evidence" to "proof." The drug was shown more effective on several Phase II studies of 20 or 30 patients, each over a scatter of subpopulations; now it must be shown to be more effective on, perhaps, 10,000 patients, in a sample comprehensively representing the entire population. In *Phase IV*, an established treatment is monitored to detect any changes in the treatment or population of patients that would affect its use. Long-term toxicities must be detected. It must be determined whether or not a microorganism being killed by a drug can evolve to become partially immune to it.

1.3. QUANTIFICATION AND ACCURACY

CONCEPT OF STATISTICS

Knowledge gained from data is usually more informative and more accurate if the data are *quantitative*. Whereas certain quantities, such as the acceleration of an object in a vacuum, may be expressed with certainty and therefore lie within the realm of (deterministic) mathematics, most quantities that we deal with in life, and particularly in medicine, exhibit some uncertainty and lie within the realm of probability. *Statistics* deals with the development of probabilistic knowledge using observed quantities. These quantities may be as simple as counting the number of patients with flu symptoms or measuring the blood pressure of patients with heart disease, but they should be quantities that shed light on what we want to learn. Statistics as a discipline is interested not in the data per se but rather in the process that has generated these data.

DATA SHOULD BE QUANTIFIED

Often, the information being gathered arises in quantitative form naturally, such as the counts or measures just mentioned. In other cases, the information is given in words or pictures (true–false, "patient is in severe pain," x-ray findings) and must be converted to quantities. A great deal of time and work may be saved, indeed sometimes whole studies salvaged, if the form in which the data are needed is planned in advance and the data are recorded in a form subject to analysis.

QUANTIFYING DATA

Quantities should be clearly defined and should arise from a common measurement base. Counts are straightforward. True–false, presence–absence, or male–female may be converted to 0–1. Verbal descriptions and visual images are more difficult. Cancer stages may be rated as A, B, C, or D and, in turn, converted to 1, 2, 3, or 4; however, to be compared among patients or among raters, they must be based on clearly defined clinical findings. On an x-ray film, the diameter of an arterial stenosis may be measured with a ruler. However, because the scale (enlargement) of radiographs and the size of patients may vary, the ruler-measured stenosis is better given in ratio to a nonstenotic diameter of the artery at a standardized location.

ACCURACY VERSUS PRECISION

Quantification per se is not enough; the quantities must be sufficiently accurate and precise. *Accuracy* refers to how well the data-gathering intent is satisfied, that is, how close the arrow comes to the bull's-eye. *Precision* refers to the consistency of measurement, that is, how tightly the arrows cluster together, unrelated to the distance of the cluster from the bull's-eye.

HOW MUCH PRECISION?

To decide how much precision is enough, identify the smallest unit that provides useful clinical information. (Just to be safe, 1 unit farther might be recorded and then rounded off after analysis.) Body temperature of a patient recorded in 10-degree units is useless. Tenths-of-a-degree units are most useful. Carrying temperature to hundredths-of-a-degree adds no clinical benefit, because too many unknown variables influence such minute temperature differences.

1.4. DATA TYPES

TYPES OF DATA

Quantitative data may be of three major types: continuous (also called interval), rank-order (also called ordinal), or categorical (also called nominal).

(1) *Continuous* data are positions on a scale. (A common ruler is such a scale.) In *continuous* data, these positions may be as close to one another as the user wishes to discern and record. Prostate volumes and PSA level form examples of this type. (See DB1 from the Database collection.) One patient's volume was measured as 32.3 ml. If we needed more accuracy for some reason and had a sufficiently accurate measuring device, we might have recorded it as 32.34 or 32.3387.

Discrete data are a subset of continuous data that are recorded only as distinct values; there is a required distance between adjacent data. Age recorded in integral years is an example of this type. The age could be 62 or 63, but we agree not to record 62.491.

(2) *Rank-order* data are indicators of some ordering characteristic of the subject, such as ranking smallest to largest, most likely to survive to least likely, and so on. There are two classes of data for which ranks are needed. First, ranking is the only data form available for judgment cases in which the investigator cannot measure the variable but can judge the patients' order. An example is triaging the treatment order of patients after a disaster or military battle. Second, ranking forces equal distances between adjacent continuous data where the distances are disparate or carry disparate clinical implications. An example is platelet counts in thrombocytopenia; the clinical implication of a difference from 20,000 to 80,000

is far different from that of 220,000 to 280,000. The continuous data average of counts of 30,000, 90,000, and 400,000 for three patients is 170,000, implying that the group is healthy. Ranking the patients as 1, 2, and 3 removes the misinformation arising from the first minus second difference being 70,000 and the second minus third difference being 310,000.

When continuous data are ranked, *ties* are given the average of the ranks they would have had had they not been tied. For example, let us rank, from smallest to largest, the first eight preoperative serum silicon levels of DB5. The ranks would be as follows: first, 0.13; second, 0.15; third, 0.18; fourth, 0.20; and fifth, 0.24. The sixth and seventh are tied, both being 0.39; their rank would be 6.5. Last, the eighth is 0.42. Notably, ranking retains some, but not all, of the information of continuous data.

(3) *Categorical* data are indicators of type or category and may be thought of as counts. We often see 0–1 (or 1–2) for male–female, false–true, or healthy–diseased. In such cases, the names of the categories may occur in any sequence and are not orderable; nonorderable categorical data are sometimes called *nominal* data. In Table DB1.1, the 0–1 indicators of negative–positive biopsy results were categorical data; seven (a proportion of 70%) of the patients had negative biopsy results and three (30%) had positive biopsy results.

Ratios are occasionally named as a fourth type of data. For the purpose of using data in statistical methods, ratios generally behave like continuous data and thus are not separated out in this book.

DISTINGUISHING BETWEEN TYPES

Sometimes an investigator may choose the data type in which numeric outcomes from a study or experiment are recorded. PSA values are basically continuous data. However, they may also be ranked: smallest, next smallest, …, largest. If PSA is categorized into three groups, <4, 4–10, and >10, the values are still ordered but we have lost a lot of information. So many ties will arise that analysis methods will be sensitive only to three ranks, one for each category. Although we could analyze them as unranked categories A, B, and C, the methods treating them as ranks first, second, and third are still "stronger" methods. When rendering data into categories, note whether or not the categories fall into a natural ordering. If they do, treat them as ranks. For example, categories of ethnic groups do not fall into a natural order, but the pain categories of severe, moderate, small, and absent do.

Note that we can always change data from higher to lower type—that is, continuous to discrete to ranked to categorical—but the reverse is not possible. Thus, it is *always preferable to record data as high up on this sequence as possible*; it can always be dropped lower.

ROUNDING

Rounding, most often a rather benign convenience, can at times change data type. The continuous observations 1/3 and 2/3, if rounded to two decimal places would become 0.33 and 0.67, now discrete observations. If rounded to integers, they would become 0 and 1, which might be mistaken for categorical data.

We should carry full accuracy in calculations and then round to the accuracy that has clinical relevance. For example, if four readings of intraocular pressure (by an applanation tonometer) were 15.96, 17.32, 22.61, and 19.87, then the mean would be 18.94. In clinical use, the portion after the decimal is not used and would only add distracting detail, so the mean would be reported as 19.

RATINGS

Ratings form a class of data all their own, in that they may be any type. There is a great deal of confusion between ratings and rankings in medical literature, even in some published textbooks on biostatistics, and the user must be careful to distinguish between them. In contrast to ranks, which are judgments about one patient or event relative to others, ratings are judgments about a patient or event on that patient's own merits alone, regardless of others, as in the rating of a tumor as one of four cancer stages. In contrast to ranks, which should be all different except perhaps for occasional ties, ratings quite properly could be all the same. Ratings work like continuous data if there are many categories, and they may be analyzed as such if the samples are of fair size. Ratings work like categorical data if there are few categories. Regardless of the number of categories, ratings may be ranked and rank methods used, but there are usually so many ties that they weaken rank methods.

STRING (OR ALPHABETIC) DATA

This book is restricted to quantitative data. Often, data are recorded as words, called *strings* in computer terminology, rather than as numbers, for example, "male–female" or "side effects–no side effects." Most often, these verbal data can be converted to numeric data for statistical analysis. Indeed, if statistical analysis is intended, time and effort will be saved by recording them as numeric data from the outset. If the investigator cannot or will not convert the verbal data to numeric data, some qualitative methods exist to analyze them, but the analyses are more complicated and less effective than quantitative analyses.

1.5. NOTATION (OR SYMBOLS)

An index of symbols appears toward the back of this book. Any reader at ease with formulas and symbols, for example, $\sum$, may not need to review that section.

PURPOSE OF SYMBOLS

Many people have some difficulty with mathematical symbols. They are, after all, a cross between shorthand and a foreign language. They are awkward to use at first, because users must translate before they get a "gut feel" for the significance of the symbols. After some use, a μ or an s^2 takes on an intuitive meaning to the user, and the awkwardness fades. Indeed, symbols are intended to avoid awkwardness, not create it. If we were to write out concepts and their relationships in words, we would soon be so overcome by the verbosity that we would lose track of what we were trying to do.

CATEGORIES OF SYMBOLS

Most symbols arise from one of three categories: names, operators, or relationships.

(1) *Name symbols*, such as x or μ, may be thought of as families. x may denote prostate volume measures all taken together as a family; x is the "family name." μ may denote the mean prostate volume for a population of men. If we need to refer to members of the family, we can affix a "first name" to the family name, usually (but not always) in the form of a *subscript*. x_1 is the name of the first prostate volume listed, x_2 the second, and so on. μ_1 may denote the mean of the population of American men, μ_2 of Japanese men, and so forth. If we think of name symbols in this fashion—for example, y_7 denoting the seventh member of the y family—the mystique of symbols reduces. Most symbols are name symbols.

(2) *Operator (or command) symbols* represent an act rather than a thing. There are not many operator symbols. Operator symbols, such as $\div$ or $\sum$, say, "Do this thing." $\div$ says "Divide the quantity before me by the quantity after me," and $\sum$ says, "Add together all members of the family that follow me." There is no indicator of which symbols are name and which are operator, in the same way that there are none in grammar that distinguishes nouns from verbs. They become clear by context after some use.

(3) *Relationship symbols,* such as $=$ or $>$, express some relationship between two families or two members of a family. We are all familiar with statements such as $2 + 2 = 4$ or $6 > 5$. The characters 2, 4, 5, and 6 are name symbols for members of the family of integers, $+$ is an operator symbol, and $=$ and $>$ are relationship symbols.

FORMULAS

Putting these ideas together, we can see the meaning of the "formula" for a sample mean, in which we add together all n values of observations named x in the sample and divide by n: $(\sum x)/n$. The operator symbol, $\sum$, says, "Add together what follows." The name symbol of what follows is x, the members of the sample of prostate volumes. Thus, we add together all the members of the sample, that is, all prostate volumes. The parentheses say, "We do what is indicated within before taking the next action." The operator symbol, $/$, says, "We divide what comes before $[(\sum x)]$ by what comes after (n)."

BECOMING FAMILIAR WITH SYMBOLS

The foregoing statement is a tortuously long expression of a simple idea. However, the user having difficulty with symbols and formulas who goes through such a mental process on each formula for *only a few times* will soon find that formulas become a natural way to express relationships, much easier than trying to put them into words. The advantage of symbols over words increases as the relationship becomes more complicated.

OUR FIRST FORMULA

Now we can express symbolically the definition of a mean m: $m = (\sum x)/n$.

INDICATOR SYMBOLS

If we want to indicate a member of a family but not wish to specify which member at the moment, we can use what might be called an *indicator symbol*. Any symbol can be used, but the most common in statistics are i, j, and k. If $x_1, x_2, \ldots, x_5$ are the members of the x family of prostate volumes put in order from smallest to largest, we can say $x_i < x_{i+1}$, which is shorthand for the relationship that any member of the x family (any prostate volume) is smaller than the member (volume) to its right.

SYMBOLS FOR RANKS

We need a way to differentiate ranked data from unranked data. In Table DB1.1, x_1 was 32.3; when ranked, the first x is 16.2, the old x_3. A prime ($'$) is a common way to indicate an observation after it has been ranked. Thus x_1' would be the smallest, x_2' the next to smallest, and x_n' the largest. In Table DB1.1, the first three volumes are $x_1 = 32.3$, $x_2 = 27.0$, and $x_3 = 16.2$. If we ranked the 10 values from smallest to largest, we would have $x_1' = 16.2$, $x_2' = 16.4$, and $x_3' = 27.0$.

1.6. SAMPLES, POPULATIONS, AND RANDOMNESS

SAMPLES AND POPULATIONS

By the word *population,* we denote the entire set of subjects about whom we want information. If we were to take our measurements on all patients in the population, descriptive statistics would give us exact information about the population and our analysis would be finished. Because this is generally not possible, we gather information on a portion of the population, known as a *sample*. We use descriptive statistics from the sample to *estimate* the characteristics of the population, and we generalize conclusions about the population on the basis of the sample, a process known as *statistical inference*.

Example

As an example, patients in a hospital would constitute the entire population for a study of infection control in that hospital. However, for a study of infected patients in the nation's hospitals, the same group of patients would be but a tiny sample. The same group can be a sample for one question about its characteristics and a population for another question.

REPRESENTATIVENESS AND BIAS

To make a dependable generalization about certain characteristics, the sample must *represent* the population in those characteristics. For example, men tend to weigh more than women because they tend to be bigger. We could be led into making wrong decisions on the basis of weight if we generalized about the weight of all humans from a sample containing only men. We would say this sample is *biased*. To avoid bias, our sample should contain the same proportion of men as does the human population.

A PICTORIAL EXAMPLE

Let us, for a moment, take the patients in a certain hospital as the population. Suppose we have 250 inpatients during an influenza epidemic. We measure the white blood cell count (WBC in 10^9) of all inpatients and note which arose from 30 patients in the infectious disease wards. Figure 1.1 shows the frequencies of WBCs for the hospital population, with frequencies for those patients from the infectious disease wards appearing darker. It is clear that the distribution of readings from the infectious disease wards (i.e., the sample) is biased and does not represent the distribution for the population (i.e., the entire hospital). If we needed to learn about the characteristics of WBC counts in this hospital, we should ensure a representative sample.

INCREASING REPRESENTATIVENESS BY RANDOM SAMPLES

The attempt to ensure representative samples is a study in itself. One important approach is to choose the sample randomly. A *random sample* is a *sample of elements in which the selection is due to chance alone*, with no influence by any

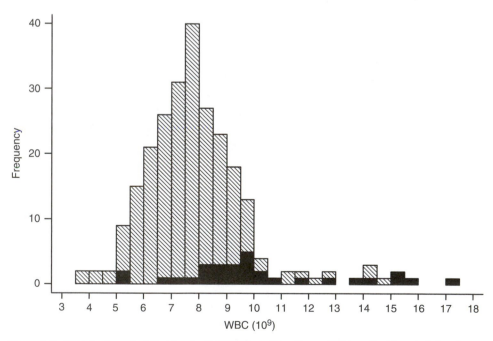

Figure 1.1 Distribution of white blood cell (WBC) readings from a 250-inpatient hospital showing an unrepresentative sample from the 30-patient infectious disease wards (black).

other causal factor. Usually, but not always (e.g., in choosing two control patients per experimental patient), the random sample is chosen such that *any member of the population is as likely to be drawn as any other member*. A sample is not random if we have any advance knowledge at all of what value an element will have. If the effectiveness of two drugs is being compared, the drug allocated to be given to the next arriving patient should be chosen by chance alone, perhaps by the roll of a die, by a flip of a coin, or by using a table of random numbers or a computer-generated random number. A section on how to randomize a sample appears in Chapter 12.

CONVERGENCE WITH SAMPLE SIZE

Another requirement for a dependable generalization about certain characteristics is that the sample must *converge* to the population in the relevant characteristics. Suppose for a moment we thought of our 301 prostate patients as a population. From the data of DB1, we can see that mean PSA is about 8.8, but the mean of the first 10 patients is about 6.8. Table 1.1 shows the mean PSA as the sample size increases. The mean of the first 10 readings is less than the overall mean. It has increased to greater than the mean by 50 readings and continues increase until the 100th reading, after which it decreases, reaching closer and closer to the overall mean. We notice that the pattern of sample values gets closer in nature to the pattern of population values as the sample size gets closer to the population size.

Now think of populations of any size. At some point, enough data are accumulated that the pattern of sample values differs from the pattern of population values only negligibly, and we can treat the results as if they had come from the population; the sample is said to converge to the population. Assume the population on which we are measuring PSA levels representatively is the population of American men. As the (random) sample grows to hundreds and then to thousands, the pattern of PSA values in the sample will differ very little from that of the population.

Table 1.1
Mean Prostate-Specific Antigen (PSA) Level and Its Difference from the Group Mean as Sample Size Grows

Sample size	Mean PSA level	Deviation from group mean
Entire group, $n = 301$	8.8	
10	6.8	−2.0
50	10.7	1.9
100	11.7	2.9
150	10.0	1.2
200	9.5	0.7
250	9.4	0.6
260	9.2	0.4
280	9.0	0.2
290	8.9	0.1

METHODS OF SAMPLING

There are many respectable methods of sampling. Among them are simple random sampling (all members have an equal chance of being chosen) and stratified sampling (dividing the population into segments and sampling a proportionate number randomly from each segment). Section 12.5 addresses how to generate a random sample.

SOURCES OF BIAS

Sources of bias are myriad. Patients who recover may decide not to follow up, whereas ill patients may want continued treatment. Patients with certain characteristics may be noncompliant. Certain diseases may be underreported. Section 10.4 discusses a number of common biases.

CHAPTER EXERCISES

1.1. Give an example of a question from the prostate biopsy database (DB1), the answer for which would fall into (a) the description stage, (b) the explanation stage, or (c) the prediction stage.

1.2. Give an example of a question from the bronchoconstriction database (DB14), the answer for which would fall into (a) the description stage, (b) the explanation stage, or (c) the prediction stage.

1.3. Into what phase-type would the question addressed in the gallbladder database (DB2) fall?

1.4. How could the variable sex in the emphysema database (DB3) be quantified?

1.5. For the platelet counts in the growth factor database (DB9), "normal" is often taken as being in the range from 150,000 to 400,000. How accurately should platelet counts be recorded to avoid interfering with clinical decisions?

1.6. In the femoral neck fracture database (DB7), which variables are (a) categorical and (b) continuous? (c) Can bone density be ranked?

1.7. In the warfarin International Normalized Ratio (INR) database (DB13), which variables are (a) categorical and (b) continuous? (c) Can the differences in INR be ranked?

1.8. Which database contains ratings that could be treated as categorical, ranked, or continuous data?

1.9. In the tattoo removal database (DB6), to what population would the dermatologist prefer to generalize his results? From the information given, is he justified in doing so?

Chapter 2

Distributions

2.1. FREQUENCY DISTRIBUTIONS

The frequency distribution is a concept through which most of the essential elements of medical statistics can be accessed. It is nothing more than the way the data are distributed along the scale (or axis) of the variable of interest. The first 10 prostate volumes from DB1 are given in Table DB1.1. The variable of interest is volume in milliliters (ml). We scan (the entire) DB1 to ascertain the range of volumes, which extends from about 3 ml to about 114 ml. We list 5-ml intervals (choosing intervals is discussed later), starting with 0 to <5 (expressed this way to indicate "up to but not including 5" and done to avoid a possible overlap) and ending with 110 to <115. We go through the data sequentially, making a tick mark for each datum. A space, crossover, or some other indicator might be placed every five tick marks to facilitate counting. Such a tally sheet for the data from Table DB1.1 for 10 patients is shown as Fig. 2.1A. If we rotate Fig. 2.1A a quarter turn counterclockwise and enclose the tick marks into boxes, towers or bars are formed, as shown in the figure. Continuing this procedure for the data for the first 50 patients, we produce Fig. 2.1C. The pattern of the distribution is beginning to emerge. Figure 2.1D represents the distribution of data for 100 patients, Fig. 2.1E represents that for 200 patients, and Fig. 2.1F shows the distribution of data for all 301 patients. Figure 2.1F is the frequency distribution for the volume data.

FREQUENCIES EXPRESSED AS A BAR CHART

A certain number of values has fallen into each defined interval, or bin. This is the *frequency* for that bin. There are 40 patients with volumes between 20 and 25 ml. Bins may be combined; for example, the frequency for volumes less than 30 ml is 122. This type of chart, representing frequencies by the heights of bars, is called a *bar chart.*

RELATIVE FREQUENCIES

The frequency itself does not always tell us what we want to know. Often, of more interest is the *relative frequency*, which is the proportion of the sample in a bin. The proportion of patients with volumes between 20 and 25 ml is

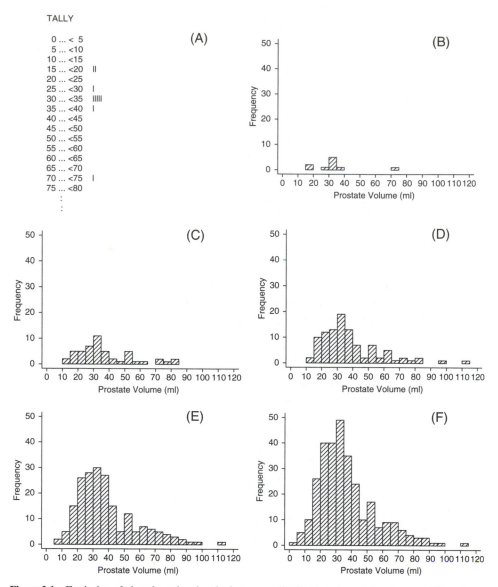

Figure 2.1 Evolution of a bar chart showing the frequency distribution of prostate volume on 301 patients: (A) tally; (B) data for first 10 patients; (C) data for 50 patients; (D) data for 100 patients; (E) data for 200 patients; and (F) data for all 301 patients.

$40/301 = 0.133$. The proportion of patients with volumes less than 30 ml is $122/301 = 0.405$. We have been trained by news reports and such sources to think in terms of parts per hundred, or *percent*, the proportion multiplied by 100. In the preceding sample, 40.5% of the prostate volumes are less than 30 ml.

EFFECT OF INCREASING SAMPLE SIZE

We can observe an interesting phenomenon by comparing Figs. 2.1B–F. The distribution patterns are rather different for the earlier figures but become increasingly stable as we add more and more data. The *increasing stability of statistical behavior with increasing sample size* is an important property of sampling (see later for a more detailed discussion).

Example

Consider the prostate-specific antigen (PSA) data of Table DB1.1 as a sample. What is the range (i.e., largest minus smallest values)? The smallest value is 4.1, and the largest is 9.0. The range is $9.0 - 4.1 = 4.9$. What intervals should we choose for a tally? With only 10 observations, a small number of intervals is appropriate. Interval widths that are easy to plot and easy to interpret are sensible, such as a width of 1.0 for this example. With a range of about 5, we are led to 5 intervals. If we had had a large number of data, we might have chosen 10 or even 15 intervals. Our intervals become >4 to 5, >5 to 6,..., >8 to 9. The student can construct the tally and the bar chart.

2.2. RELATIVE FREQUENCIES AND PROBABILITIES

FREQUENCY DISTRIBUTION VERSUS PROBABILITY DISTRIBUTION

Suppose a prostate volume ≥ 65 ml is taken as suggestive of benign prostatic hypertrophy (BPH). From our sample of 301 prostate patients, 28 readings equal or exceed 65 ml; the proportion $28/301 = 0.093$ shows evidence of BPH. This is the *relative frequency* of BPH suspects in our sample. If *r* patients out of *n* show a certain characteristic, *r/n* is the relative frequency of that characteristic. However, we are less interested in the characteristics of the *sample* of older male patients at our hospital than of the population of older male patients in the United States. We want to know what proportion of the *population* is suspect for BPH. If we could measure the prostate volume of every older man in the United States, we would be able to find the frequency distribution for the population, which we would name the *probability distribution*. We could then find the *probability* that any randomly chosen older man in the United States shows indications of BPH. This probability would be represented as the area under the probability distribution ≥ 65 ml divided by the total area. This calculation is similar to measuring the area under the bars in the sample distribution ≥ 65 ml and dividing by the area under all the bars, yielding 9.3%.

ESTIMATED PROBABILITY: THE TERM *PARAMETER*

Because we cannot measure prostate volume for the entire population, we use what information we can obtain from the sample. The 9.3% relative frequency does not tell us the exact probability of equaling or exceeding 65 ml in the population, but it *estimates* this probability. If we took another similar sample from another hospital, the relative frequency of that sample would be different, but we would expect it to be somewhere in the same vicinity. The term *parameter* is used to distinguish an unchanging characteristic of a population from an equivalent characteristic of a sample, an estimate of this parameter, which varies from sample to sample. How well does the estimate approximate the parameter?

We are able to measure the confidence in this estimate by methods that are discussed later.

2.3. CHARACTERISTICS OF A DISTRIBUTION

The fundamental statistical information is the distribution of data, because it contains all the information we need for our statistical methods. From a sample distribution, we can learn what sample proportion of individuals are part of an interval of interest. From a population distribution, we can learn the probability of a randomly chosen individual being in that interval. We can learn what is typical or characteristic of a distribution and how closely the distribution clusters about this typical value. We can learn about the regularity or "bumpiness" of a distribution and how close it is to symmetric. We can learn about classes or types of distributions. In the next few sections, these characteristics of distributions are addressed conceptually. Formulas, describing characteristics of the distributions in symbols, are given in Chapter 3.

2.4. WHAT IS TYPICAL?

AVERAGES

If we could have only one quantity to give us information about a distribution, it would be the typical, central, or "average" value. There are several averages, three of which are mentioned in this section.

The *mean* is the average used most often. The population mean is denoted μ. We denote the sample mean m for reasons explained in the next section, although some books denote it by the letter representing the variable being averaged with a bar over it (e.g., $\bar{x}$). To obtain the sample mean, we add all values and divide by the number of values added. The mean prostate volume from Table DB1.1 is 32.73 ml. The mean of a probability distribution is the "center of gravity" of that distribution. If the picture of the distribution were cut from a material of uniform density, say, wood, the mean would be the position on the horizontal axis at which the distribution would balance. Thus, values far from the center carry more influence, or "moment," than values near the center.

If we are sampling from two populations, say, x and y, we can use subscripts, which were previously mentioned as the "member-of-the-family" indicators, to keep them straight, such as μ_x and μ_y for the population means, respectively, and m_x and m_y for the sample means, respectively.

The *median* of a sample is the position on the horizontal axis at which half the observations fall on either side. If we order the prostate volumes in Table DB1.1, that is, put them in sequence of increasing value, the median would be the midpoint between the fifth and sixth volumes. The fifth volume is 30.5 ml and the sixth is 30.9 ml, so the median is 30.7 ml. For a distribution with some extreme observations (e.g., income of HMO executives in the distribution of HMO employee incomes, or unusually long-lived patients in the distribution of survival times of cancer patients), the median is a better descriptor of the typical than is the mean. For a probability distribution, the population median is the position on the horizontal axis at which half the area under the curve falls on either side.

The *mode* of a sample is the most frequently occurring value, which is the position on the horizontal axis centered under the tallest bar. The mode of the prostate volumes in Table DB1.1 can be seen in Fig. 2.1B as 32.5. The mode is not very dependable for small samples and should be used only for large samples. The position of the mode depends on the bar width and starting position. Also, the mode is not necessarily unique. An example of a distribution with two modes might be the heights of a mixed sample of men and women, because men, on average, tend to be taller than women. For a probability distribution, the mode is the horizontal axis value under the highest point on the curve. For a multimodal probability distribution, the position of each height on the curve having a lesser height to both left and right is a mode.

CONVERGENCE WITH INCREASING SAMPLE SIZE

In Section 2.1, the increasing stability of a sample with increasing sample size was mentioned. The mean of a sample gets closer to, that is, *converges on*, the population mean as the sample size grows larger. This property is known as the Weak Law of Large Numbers, or the Bienaymé–Tchebycheff Inequality (also Tchebycheff alone, and using various spellings). Although we need not remember the name, this relationship is essential to expressing confidence in our sample estimates of population means and to finding the sample size required for studies and experiments. To illustrate this convergence, sample means of differences between preexercise and 20-minute postexercise nitric oxide levels from database 14 (DB14) are shown in Fig. 2.2 for the first 5, first 10, first 15, ..., first 35, and all 38 patients seen. (They are listed in the order they presented.) The means can be seen to converge on the 1.176 mean of all 38 patients as the sample size increases. Estimates are over, under, under, over, over, over, and over, respectively. Whereas the size of error tends to diminish with each sample size increase, deviation from the mean grows slightly larger at sample sizes 20 and 30 because some patients with greater variability appeared. The convergence occurs in probability, not necessarily with every additional datum.

2.5. THE SPREAD ABOUT THE TYPICAL

TYPES OF SPREAD INDICATORS

After we know the typical value of a distribution, naturally it follows to want to know how closely the values cluster about this average. The *range* (highest minus lowest values) gives us a hint, but it uses the information from only two of our observations, and those are the most unstable values in the sample. As with the average, there are different measures of variability, but the variance and its square root, the standard deviation, are used primarily.

The *variance* of a set of values is the *average of squared deviations from the mean*. Just for computational illustration, consider a population composed of the values 1, 2, and 3, which have the mean 2. First, the mean is subtracted from

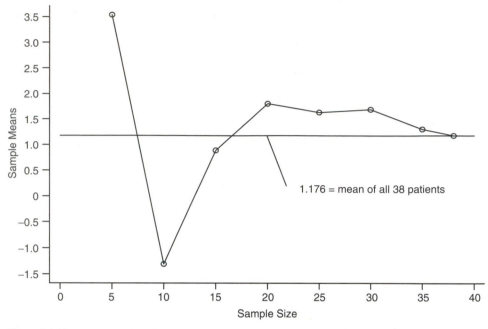

Figure 2.2 From DB14, the difference between preexercise and 20-minute postexercise nitric oxide, sample means of levels are shown for the first 5, first 10, first 15, ..., first 35, and all 38 patients seen. (They are listed in the order they presented.) The means can be seen to converge on the 1.176 mean of all 38 patients as the sample size increases.

each observation, and each difference is squared, yielding $(1 - 2)^2, (2 - 2)^2$, and $(3 - 2)^2$. The squaring makes all values positive, so they do not cancel each other out, and it allows the more deviant observations to carry more weight. Then these squared differences are averaged, that is, added and divided by their number, resulting in the population variance: $(1 + 0 + 1)/3 = 0.666 \ldots$. The population variance is denoted σ^2. A sample variance, commonly denoted s^2, is calculated in the same way if μ is known. However, if μ is estimated by m, it has been found that the divisor must be the sample size less one (as $n - 1$) to converge on the population variance properly.

The standard deviation is the square root of the variance, denoted σ for populations and s for samples. The variance often is used in statistical methods because square roots are difficult to work with mathematically. However, the standard deviation typically is used in describing and interpreting results because it can represent distances along the axis of the variable of interest, whereas the variance is expressed in squared units. The meaning of squared volumes of the prostate would be somewhat obscure.

Example

Let us calculate the variance and standard deviation of prostate volumes from Table DB1.1. Using the sample mean, $m = 32.73$, from the previous section, the observations minus m are $-0.43, -5.73, \ldots, 4.07, -16.33$. Their squares are $0.1849, 32.8329, \ldots, 16.5649, 266.6689$. The sum of these squares is 2281.401. Dividing the sum of squares by 9 gives the variance, often called the *mean square*, as $s^2 = 253.409$. The standard deviation is just the square root of the variance, or $s = 15.92$. Engineers often call the standard deviation the *root-mean-square*, which may help keep its meaning in mind. The computation used here follows the conceptual method of calculation. A simpler computational formula, plus examples, is given in Chapter 3.

GREEK VERSUS ROMAN LETTERS

The beginnings of a commonly used pattern in statistics starts to emerge. It is common among statisticians to use Greek letters to represent population names and Roman letters to represent sample names. Thus, to represent a mean and variance, we use μ and σ^2 for the population and m and s^2 for the sample. Some users of statistics have not adopted this convention, but it helps keep the important distinction between population and sample straight and is used in this text. In my view, $\bar{x}$ to denote the mean is an inconsistent historical remnant.

2.6. THE SHAPE

A distribution may have almost any shape (provided it has areas only above the horizontal axis). Fortunately, only a few shapes are common, especially for large samples and populations where the laws of large numbers hold. Sample distributions, although different from sample to sample, often approximate probability distributions. When they do, this correspondence allows more informative conclusions about the population to be inferred from a sample.

MOST COMMON SHAPES

The two most famous distributions are the uniform and the normal curve, sometimes loosely termed a *bell curve*. The uniform distribution is an equal likelihood case, in which all events (e.g., choosing patients for a certain treatment in a study) have equal chance of occurring, so all probabilities achieve the same height on the graph of the distribution. The bell curve has a graph showing a central hump (mode) with observations tailing off symmetrically, or approximately so, on either side. The normal is a bell curve with certain additional characteristics that are noted in Section 2.8 and Chapter 4.

In describing the shape of a distribution, we look first for number of modes and what might be called smoothness. A "well-behaved" distribution has a clearly seen mode, tails off in not too jagged a fashion on either side, and relates in a meaningful way to the variable of interest. Then, we look for symmetry. If one tail is "dragged out" more than the other, we say that the distribution is *skewed* in that direction. We look for the approximate center of the distribution. In a unimodal unskewed curve, the mode, median, and mean will coincide. If it is right skewed, the median is to the right of the mode and the mean is to the right of both. Finally, we look to see how much it is spread out about its center, that is, its approximate standard deviation. By examining the distribution in this fashion, we can get a much clearer picture of what forces are at work in the generation of the data than through knowing only the mean and the standard deviation.

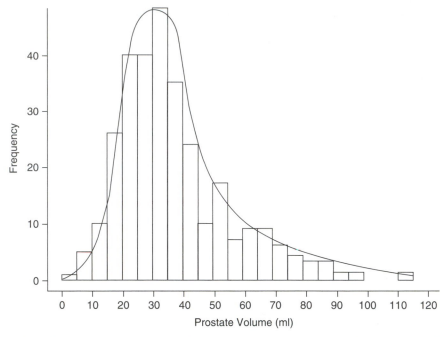

Figure 2.3 The frequency distribution of 301 prostate volumes from Fig. 2.1F, with the population probability distribution approximated by a freehand fit.

Example

Figure 2.3 shows the frequency distribution of 301 prostate volumes from Fig. 2.1F with the population probability distribution approximated by a freehand fit. Note that it is skewed to the right. In agreement with this, the mean of 36.5 ml is to the right of both the median (32.4 ml) and the mode (32.5 ml). Before seeing any data, we might have expected a symmetric distribution. Some of the skew seems to arise from the presence of BPH, which causes a large increase in prostate volume. However, we suspect the main cause is the limitation on the left by zero, whereas there is no limitation to the right. The mean ±1 standard deviation includes the interval on the volume scale of about 21 to 52 ml; probabilistic interpretations of this are better discussed later, but it seems safe to say that we would not consider a prostate volume in this interval to be clinically abnormal.

STANDARDIZING A DISTRIBUTION

Distributions come in various sizes, in particular, having various scales on the variable of interest. However, the scale differences can be overcome by standardizing the distribution. A distribution is *standardized* by subtracting the mean from every observation and dividing the result by the standard deviation. This forces the distribution to have a mean of 0, a standard deviation of 1, and a scale measured in standard deviations. This transformation allows shapes of different distributions to be compared uncluttered by scale differences, and it allows a single table of areas under a curve of a particular type to be made rather than a new table for every scale. The *sample distribution* of a variable, say, x, is standardized when x is transformed into y by using

$$y = \frac{x - m}{s}$$

and the *probability distribution* of x is transformed by using

$$y = \frac{x - \mu}{\sigma}.$$

Probability distributions are transformed further so that their total area under the curve is always 1.

2.7. STATISTICAL INFERENCE

Making inferences about a population on the basis of a sample from that population is a major task in statistics. A statistical inference is a conclusion about a state or process in nature drawn from quantitative, variable evidence in a way that specifies the risk for error about such a conclusion. Although we can never eliminate the risk for error from inference, we can quantify it and limit its probability.

INFERRING A CONFIDENCE INTERVAL

One important form of inference is the generation of confidence intervals. The goal is to measure and control the risk for error about a sample's estimate of a population parameter. Given a sample mean on some variable x, for example, what can be said about the mean of the population? Mathematically, m has been shown to be the best estimate of μ, but how good is it in a particular case? By following a number of steps, we can place a confidence interval about this mean, saying "We are 95% confident that this interval encloses the population mean." (For medical purposes, 95% confidence is used most often, more by habit than because a 5% risk is better than some other risk. We can never be 100% sure.) This procedure *infers* that an interval is likely to contain the mean, reserving a 5% chance that it does not. This chance (5% risk for error) is usually designated as α in statistics.

INFERRING A DIFFERENCE

A decision is a variation on inferring a conclusion. An important form of inference is to decide whether a characteristic, such as a mean, is the same or different between two samples, or between a sample and a population. We state that these two means are the same by means of a *null hypothesis*, so termed because it postulates a null difference between the means. Then, we specify the probability that this hypothesis will be wrong, that is, the chance that we conclude there is no difference when, in fact, there is. This may be thought of as the probability of a false-positive result, named α, frequently taken as 5%. Finally, by using this α, we either do or do not reject the null hypothesis, which is equivalent to inferring that a difference between the means has or has not been shown.

INFERRING EQUIVALENCE

A parallel question, only recently come to be widely used in medical research, is the need to show whether or not two characteristics, such as means, are the same. For example, a new drug is developed that is much cheaper than an existing drug. Does it perform as well? In this case, the error in making a wrong decision may be thought of as a false-negative result, that is, inferring equivalence when, in fact, there is a difference. The null hypothesis we must reject or not becomes the statement that the two mean performances are different. A difficulty not appearing in a difference test is that we must specify how far different the means are, say, a difference of *d* units. The inference becomes the decision that the means are not different by as much as *d* units, and therefore are equivalent for clinical purposes, or that that equivalence has not been shown. The value *d* is usually chosen on clinical grounds so that the decision includes some arbitrary judgment not present in a difference test.

STEPS IN INFERENCE

In making statistical inferences about a sample estimate, we follow several steps: (1) We assume that the data are independent from one another, and often, according to the test some properties of the data distribution involved; for example, "The distribution of the sample mean is normal (bell-shaped)." (2) We arbitrarily choose the probability, often but not necessarily 5%, that the statement to be investigated (confidence interval or null hypothesis about the relationship between this estimate and another value, e.g., a theoretical average) is wrong. (3) We find an interval about our sample estimate on the x-axis designated by those values outside of which lies 5% (or other risk) of the area under the probability distribution (calculated or looked up in tables). (4) If we want a confidence interval, this interval is it. If we are testing a null hypothesis, we reject the hypothesis if the other value in the statement (e.g., the theoretical average) lies outside this interval and fail to reject it if it lies inside. The student can see that there is a logical relationship between confidence intervals and hypothesis testing.

The logic in the preceding paragraph does not follow the way we usually think. It will require intense concentration on initial exposure. Students seldom comprehend it fully at first. Return to it repeatedly, preferably by using it when working through the methods given in later chapters, and give it time to "sink in."

One admonition should be noted. The independence assumption in step (1) is seldom actually stated, but it **is made** whether explicitly or implicitly. Indeed, many statistical inferences are based on *more than one* assumption, for example, that the underlying probability distribution is normal *and* that each sample value drawn cannot be predicted by other sample values (the observations are independent).

EFFECT OF VIOLATED ASSUMPTIONS: ROBUSTNESS

What happens when an assumption is violated? The computations can be made in any case, and there is no flag to alert the user to the violation. When assumptions presume erroneous characteristics of probability distributions, the areas under the curves are computed incorrectly and decisions are made with erroneous confidence. For example, in finding a confidence interval, we may believe that α is 5% when, in fact, it is much greater or smaller. We must give careful attention to assumptions. Fortunately, the commonly used statistical methods often are *robust*, meaning that they are not very sensitive to moderate violations of assumptions.

More details on confidence intervals and hypothesis testing are given in Chapters 4 and 5, respectively, after additional foundation is provided; additional attention to assumptions appears at the end of Section 5.1.

2.8. DISTRIBUTIONS COMMONLY USED IN STATISTICS

Section 2.1 explains how frequency distributions arise from sample data. Section 2.2 demonstrates that the population distribution, arising from sampling the entire population, becomes the probability distribution. Section 2.7 shows that this probability distribution is used in the process of making statistical inferences about population characteristics on the basis of sample information. There are, of course, endless types of possible probability distributions. However, luckily, most statistical methods use only six probability distributions.

The six distributions commonly used in statistics are the normal, t, χ^2 (chi-square), F, binomial, and Poisson. Continuous data depend mostly on the first four distributions. Rank-order methods depend on distributions of ranks rather than continuous data, but several of them can be transformed to calculate probabilities from the normal or chi-square distributions. Categorical data depend mostly on the chi-square, binomial, and Poisson distributions, with larger samples transformed to normal. We need to become familiar with only these six distributions to understand most of the methods given in this text. Figure 2.4 shows examples of these six types of distributions. The following paragraphs describe these distributions and some of their properties needed to use and interpret statistical methods.

NORMAL DISTRIBUTION

The normal distribution, sometimes called Gaussian after the mathematician Carl Friedrich Gauss, is the perfect case of the famous bell curve. We standardize a normal (Gaussian) variable, x, transforming it to z by subtracting the mean and dividing by the standard deviation, for example,

$$z = \frac{x - \mu}{\sigma}.$$

The normal distribution then becomes the *standard normal*, which has mean 0 and standard deviation 1. This transformation usually is made in practice because the probability tables available are usually of the standard normal curve. The statistical "normal" must not be confused with the medical "normal," meaning nonpathologic.

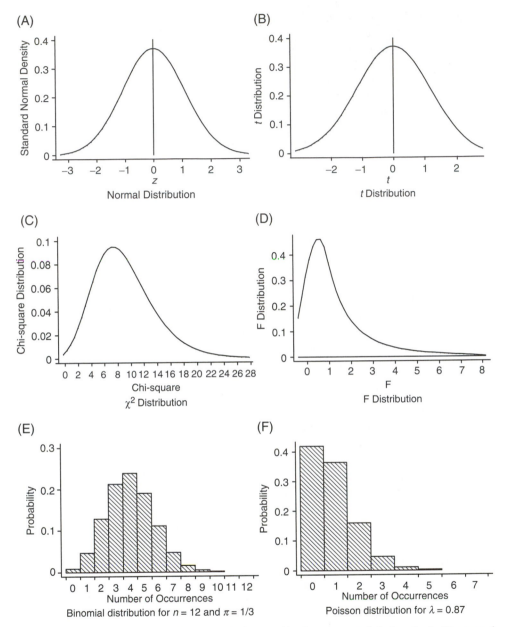

Figure 2.4 The six common probability distributions used in elementary statistical methods. The normal (A) and t (B) are used with inferences about means of continuous data, the χ^2 (C) and F (D) about standard deviations (more exactly, variances) of continuous data, and the binomial (E) about proportions of categorical data. The Poisson (F) and χ^2 are used to approximate the binomial for larger sample sizes.

Shorthand for the Normal

The normal distribution is used so much that a shorthand symbol is helpful. The common symbol is N for "normal" with the mean and variance (square of the standard deviation) following in parentheses. Thus, N(5,4) indicates a normal distribution with mean of 5 and variance of 4 (standard deviation = 2). N(0,1) indicates the standard normal with mean of 0 and standard deviation of 1, used for normal tables.

The Sample Mean Follows Its Own Probability Distribution

Often, we want to make inferences about a population mean based on a sample mean. For example, consider our prostate volume data from 301 patients. Each observation is drawn from a population that has a given probability distribution. If we wanted to make conclusions about a single patient, we would use this distribution. However, suppose we want to use the sample data to make conclusions about the average prostate volume of American men. Because the sample mean is composed of individual volume readings, each of which has a probability distribution, the sample mean must also have a probability distribution, but it will be different.

Sample Means of Continuous Data Form a Normal Distribution

If the sample in question is drawn from a normal population (e.g., if prostate volumes form a normal distribution), the probability distribution of the sample mean is exactly normal. If the sample is drawn from any other distribution (e.g., the prostate volumes are from a skewed distribution), the probability distribution of the sample mean is still approximately normal and converges on normal as the sample size increases. This remarkable result is due to a famous mathematical relationship, named the *Central Limit Theorem*. Because much of our attention to statistical results is focused on means, this theorem is frequently applied.

The Central Limit Theorem is dramatically illustrated in Fig. 2.5. We draw random samples from a far-from-normal distribution, indeed disparate and bimodal. We observe how the frequency distribution of means becomes closer to normal as we take means first of 5 observations, then 10, and finally 20. Figure 2.5A shows 117 PSA readings of patients whose PSA leaves little doubt about a biopsy decision—that is, PSA <4 (no biopsy) or PSA >10 (definite biopsy). A normal curve with the mean and standard deviation of the PSA data is superposed. Figure 2.5B shows the frequency distribution of 200 means of 5 observations each drawn randomly (using a computer random number generator) from the 117 PSA readings, with a normal curve superposed. Figures 2.5C and 2.5D show the same sort of display for means of samples of 10 and 20, respectively. The convergence to the normal can be seen clearly.

t DISTRIBUTION

The *t* distribution answers the same sorts of questions about the mean as does the normal distribution. It arises when we must use the sample standard deviation s to estimate an unknown population standard deviation σ.

The Standard *t*

Standardizing the mean includes dividing by the standard deviation. The known σ is a constant, so the division just changes the scale. However, when σ is unknown, which happens in most cases of clinical studies, we must divide

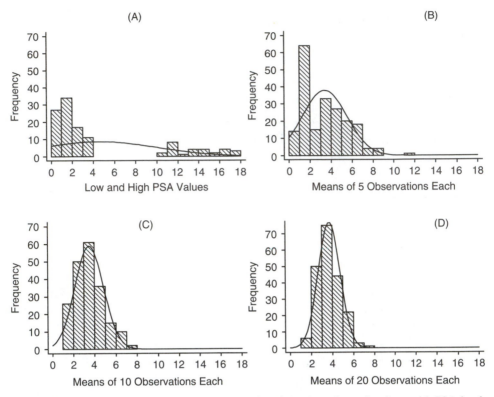

Figure 2.5 A total of 117 prostate-specific antigen (PSA) level readings of patients with PSA levels <4 (no biopsy) or >10 (definite biopsy) (excluding a few high PSAs as benign prostatic hypertrophy). Frequency distribution of 200 means of 5 (B), 10 (C), and 20 (D) observations each drawn randomly from the 117 PSA readings, with a normal curve fitted to the data's mean and standard deviation for each figure. The convergence to the normal can be seen clearly.

by s. Instead of

$$z = \frac{x - \mu}{\sigma}$$

as in the normal, we use

$$t = \frac{x - \mu}{s} \quad \text{or} \quad t = \frac{x - m}{s},$$

depending on whether μ is known or estimated by m. The sample s is not a constant like the population σ, but it is composed of observations and therefore follows a probability distribution. This division by s introduces t as a new variable: one drawn from a normal distribution divided by a variable drawn from a more difficult distribution, that for the root of a sum of squares of normals. The probability distribution for this variable was published in 1908 by W. S. Gossett, who named it t. Sometimes the user hears t referred to as Student's t. Gossett published under the pseudonym "Student" because the policy of this employer, Guinness Brewery, forbade the publication.

The t looks like the normal curve, as seen in Fig. 2.4. However, it is a little fatter because it uses s, which is less accurate than σ. Whereas the normal is a

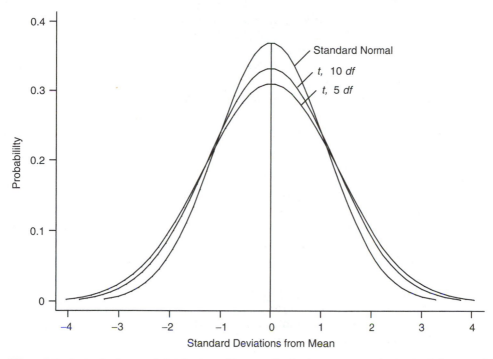

Figure 2.6 A standard normal distribution with two *t* distributions superposed, one for 5 degrees of freedom (*df*) and one for 10 *df*. The mean of the *t* is the same as that of the normal. The spread of the *t* is larger than the normal, much larger for small sample sizes, converging on the normal as the sample size grows larger.

single distribution, *t* is a family of curves. In Fig. 2.6, two standard *t* distributions are superposed on a standard normal. The particular member of the *t* family depends on the sample size, or more exactly, on the degrees of freedom.

DEGREES OF FREEDOM

Degrees of freedom, often abbreviated *df*, is a concept that may be thought of as *that part of the sample size n not otherwise allocated*. This concept relates to quite a number of aspects of statistical methods; thus, *df* may be explained in a number of ways. Some of these aspects are more difficult than others, and even experienced users find some of them challenging. Do not expect to fully understand *df* at once. Comprehension usually starts at a rudimentary level and sophisticates slowly with use. *df* is related to the sample number, usually to the number of observations for continuous data methods and to the number of categories for categorical data methods. It will be enough for a start to conceive of *df* as a sample number adjusted for other sources of information, more specifically, the number of unrestricted and independent data entering into a calculated statistic. In the *t* distribution, we might think informally of *n* "pieces" of information available. Consider the form of *t* in which we use the sample mean *m*. Once we know *m*, we have $n - 1$ pieces of information remaining, which can be selected by the sampling procedure; when we have obtained $n - 1$ observations, the *n*th one may be found by subtraction. (This is, of course, not how the data are obtained, but rather a more abstract mathematical allocation of information.) Because *df* is tied to *n*, the sample values converge to the population values as both *n* and *df* increase. *t* converges on the normal as *df* increases. Figure 2.6 shows the standard normal curve with two *t* curves superposed, one with 10 *df* and the other with 5 *df*. The fewer the *df*, the less accurate an estimate *s* is of σ, thus the greater the standard deviation of *t* (the "fatter" the *t* curve).

CHI-SQUARE (χ^2) DISTRIBUTION

We often want to make decisions, or other inferences, about a standard deviation. For computational ease, we usually make the decision about its square, the variance, because any decision about one implies the equivalent decision about the other. Basically, a variance is a sum of squares of values. If each comes from a normal curve, its mathematical pattern has a right-skewed shape like that in Fig. 2.4C. That distribution is called the chi-square distribution (*ch* pronounced like *k*). It is obtained by multiplying the calculated sample variance s^2 by the constant df/σ^2, where $df = n - 1$, and σ^2 is the population variance. Often, the Greek symbol for chi-square (χ^2) is used. Because all elements are squares, a chi-square cannot be negative. It rises from zero to a mode and then tails off in a skew to the right. As in the normal and the *t*, we use areas under the curve taken from tables or computer calculation routines to make inferences.

Chi-square in a Hypothesis Test: Critical Value

As a further example of hypothesis testing, introduced in Section 2.7, consider the days to heal to a certain standard after a new surgical procedure compared with that after an older established one. Say we know the new procedure is slightly better on average, but could it be so much more variable as to nullify its advantage? The heal-time variance of the old procedure is known from such a large number of cases that it can be taken as having converged to the population variance σ^2. Our estimate of the heal-time variance of the new procedure from a sample of patients is s^2. We are asking if s^2 is probably larger than σ^2. We test the null hypothesis that s^2 is no different from σ^2 by using the ratio

$$\frac{s^2 \times df}{\sigma^2},$$

which can be shown to follow the chi-square distribution. If s^2 is much greater than σ^2, the ratio will be large. We choose our α, the risk of being wrong if we say there is a difference when there is not. This α is associated with a certain value of chi-square, which we term the *critical value*, because it is the value that separates the rejection of the null hypothesis from nonrejection. If the calculated value of $s^2 \times df/\sigma^2$ is greater than the critical value, we reject the null hypothesis and conclude that there is statistical evidence that s^2 is larger. Otherwise, we conclude that s^2 has not been shown to be larger.

F DISTRIBUTION

In the example of the previous paragraph, suppose we were comparing heal times for two new surgical procedures rather than a new one against an established one. The variances for both are estimated from small samples.

We test the two-sample variances s_1^2 and s_2^2 to see if, in the populations being sampled, one is bigger in probability. To do this, we use their ratio, dividing the bigger by the smaller value. (The conclusion may also be extended to standard deviations: Is one standard deviation bigger in probability than the other?) The probability distribution for this ratio is called F, named (by George Snedecor) after Sir Ronald Fisher, the greatest producer of practical theories in the field of statistics. The F distribution looks similar to the chi-square (indeed, it is the ratio of two independent chi-square–distributed variables), as can be seen in Fig. 2.4D. F has one complication not appearing in χ^2: an additional *df* value. Because F involves two samples, the *df* for each sample must be used, making table reference a bit more involved.

RANK-ORDER METHODS

If we ask statistical questions about data recorded as ranks rather than measurements on intervals, we are dealing with fewer assumptions about the distributions of the data being measured. Statistical methods have been developed for rank-type data, often called *nonparametric* methods. In addition, continuous-type data can always be ranked. The advantage of using rank-order methods for continuous data is that we do not need to consider their sampling distributions. In the case of data following well-defined distributions, such as the normal, rank-order methods are not as efficient as methods assuming distributions. However, if the sampling distributions of continuous data are "poorly behaved," assumptions are violated, and results of statistical methods can be wrong. In such cases, rank-order methods give more dependable results and are preferred.

BINOMIAL DISTRIBUTION

When using *categorical data*, we are usually asking questions about counts or proportions, which are based on the same theory, because one can be converted into the other. The binomial distribution has a proportion as its parameter. This proportion, π for a population and p for a sample, and the number in the count, n, fully characterize the binomial. The binomial shown in Fig. 2.4E is the distribution for $\pi = 1/3$ and $n = 12$. If our population were evenly split by sex, for example, the probability that a randomly selected patient is male would be 0.5, the binomial parameter. Suppose we treated 20 patients for some disease and found that 15 were male. Is the disease sex related? If not, this mix of male and female patients would have arisen by chance alone from our binomial with parameter 0.5. We can calculate this probability and find that the chance of 15 or more male patients from an equal population is only 2% (0.0207). We have strong evidence that the disease is more prevalent in males.

Large-Sample Approximations to the Binomial

When we have more than a few observations, the binomial gets difficult to calculate. Fortunately, the binomial can be approximated fairly well by one of two distributions, the normal or the Poisson. If the population proportion (the binomial parameter) is toward the middle of its 0–1 range, say, between

0.1 and 0.9, the normal distribution approximates large-sample binomials fairly well. In this case, the mean of the population being sampled is π and the variance is $\pi(1 - \pi)/n$ (with its square root being the standard deviation). As the population proportion moves farther toward either 0 or 1, the binomial becomes too skewed to use the normal. The binomial distribution shown in Fig. 2.4E has a parameter of 1/3, and already we can see the beginning of a skew.

POISSON DISTRIBUTION

In the case of a population parameter very near 0 or 1, the large sample binomial can be adequately approximated by the Poisson distribution, named after the French mathematician, Siméon Denis Poisson, who published a theory on it in 1837. The Poisson distribution arises in cases in which there are many opportunities for an event to occur, but a very small chance of occurrence on any one trial. Exposure to a common but not very infectious virus should result in an incidence of illness that follows the Poisson. For quite large samples, again there exists a normal approximation. In this case, the mean of the population being sampled remains π but the variance becomes π/n (with its square root being the standard deviation).

2.9. STANDARD ERROR OF THE MEAN

MEASURE OF VARIABILITY IN THE SAMPLE MEAN

Before we meet the concepts of hypothesis testing, it will be convenient to become familiar with the *standard deviation of the sample mean*. This quantity measures the variability of the sample mean m. It has been called the standard error of the mean (SEM) for historical reasons that do not have much meaning today.

POPULATION STANDARD ERROR OF THE MEAN

Let us consider our 301 prostate volumes as a population. The mean of the 301 volumes becomes the population mean $\mu = 36.47$ ml and the standard deviation $\sigma = 18.04$ ml. If we randomly draw prostate volumes, they will generally be somewhat different from one another, and these differences will follow some frequency distribution; σ is the standard deviation of this distribution. Similarly, if we calculate means $m_1, m_2, \ldots$, from repeated samples, they will generally be somewhat different one from another and will follow some frequency distribution; the SEM is the standard deviation of this distribution. It is often symbolized as σ_m. It will be smaller than σ, because it reflects the behavior of 301 observations rather than 1. It turns out that

$$\sigma_m = \frac{\sigma}{\sqrt{n}},$$

the population standard deviation divided by the square root of the population size. In the case of the 301 volumes, $\sigma_m = 18.04/\sqrt{301} = 1.0398$ ml.

Because we know from the Central Limit Theorem that the probability distribution of m is normal, we have the distribution of m as $N(\mu, \sigma_m^2) = N(36.47, 1.0812)$ ($\sigma_m^2 = 1.0398^2$).

SAMPLE STANDARD ERROR OF THE MEAN

In the same way that we can estimate σ by s when we do not know σ, we can estimate σ_m by s_m, using the standard deviation of a sample. The estimated SEM, s_m, is the sample standard deviation divided by the square root of the sample size, or

$$s_m = \frac{s}{\sqrt{n}}.$$

Let us take the 10 prostate volumes given in Table DB1.1 as our sample. We can calculate $m = 32.73$ ml and $s = 15.92$ ml. Then, $s_m = 15.92/\sqrt{10} = 5.3067$ ml.

2.10. JOINT DISTRIBUTIONS OF TWO VARIABLES

EXAMPLE

Suppose we had platelet counts, say, x, on patients with idiopathic thrombocytopenic purpura before treatment and then a second count, say, y, 24 hours after administering immunoglobulin. We wonder whether there is any relationship between x and y; that is, might it be that the lower the pretreatment level, the less effect there is from the immunoglobulin or vice versa? If so, a study might be carried out to learn if we can predict the dosage of immunoglobulin based on the pretreatment platelet count. Or is there no relation?

JOINT FREQUENCY DISTRIBUTION

If we plot the two measures, x on one axis and y on the other, we are likely to find points more concentrated in some areas than others. Imagine drawing a grid on the plot, which is lying flat on a table, and making vertical columns proportional in height to the number of points in each grid square. This *joint frequency distribution* is a three-dimensional analog to a two-dimensional bar chart. Where the points are concentrated, we would find a "hill," sloping off to areas in which the points are sparse. If the hill is symmetrically round (or elliptical with the long axis of the ellipse parallel to one of the axes), any value of x would lead to the same y as the best prediction, that is, the value of y at the top of the hill for that x. We would think that there is no predictive capability. In contrast, if the hill is a ridge with a peak extended from the bottom left to the top right of the graph, the peak y associated with a value of x would be different for each x, and we would think that some predictive relationship exists; for example, y increases as x increases.

33

RELATIONSHIP BETWEEN TWO VARIABLES

There exist statistical methods to assess the relationship between two variables, mostly falling under the topics *correlation* (see Chapter 3) and *regression* (see Chapter 8).

The concept of *covariance* is fundamental to all treatments of the relationship between two variables. It is an extension of the ideas of variance and standard deviation. Let us think of sampling from the two variables x and y; we might keep their standard deviations straight by suffixing an indicator name on the appropriate σ or s, for example, σ_x or s_x. The covariance of two variables measures *how they jointly vary, one with respect to the other*. The common symbol is similar to the standard deviation symbol, but with two subscripts indicating which variables are being included. Thus, the population and sample covariances would be σ_{xy} and s_{xy}, respectively. Formulas for these concepts and examples appear in Section 3.2.

CHAPTER EXERCISES

2.1. For extent of carinal resection in DB12, (a) select intervals for a tally, (b) tally the data, (c) find the median, and (d) convert the tally into a relative frequency distribution. (e) Comment on whether the distribution appears by eye to be near normal. If not, in what characteristics does it differ? (f) Give a visual estimate (without calculation) of the mean and mode. How do these compare with the median found in (c)?

2.2. For patient age in DB12, (a) select intervals for a tally and (b) tally the data. In tallying, (c) record the median when data for 25 patients have been tallied, then for 50, 75, 100, and all patients. Plot the median depending on the number of data sampled similar to the plot of Fig. 2.2. Does the median converge to the final median? (d) Convert the tally into a relative frequency distribution. (e) Comment on whether the distribution appears by eye to be near normal. If not, in what characteristics does it differ? (f) Give a visual estimate (without calculation) of the mean and mode. How do these compare with the final median found in (c)?

2.3. For International Normalized Ratio (INR) readings from the clinic in DB13, (a) select intervals for a tally, (b) tally the data, (c) find the median, and (d) convert the tally into a relative frequency distribution. (e) What does the median compared with the mean $= 2.40$ say about symmetry?

2.4. For INR readings from the laboratory in DB13, (a) select intervals for a tally, (b) tally the data, (c) find the median, and (d) convert the tally into a relative frequency distribution. (e) What does the median compared with the mean $= 2.28$ say about symmetry?

2.5. For computational ease, use the glycosaminoglycans (GAG) levels for type 1 assay in DB8 rounded to 1 decimal place: $0.7, 0.4, 0.3, 0.2$. Calculate (a) the variance and (b) the standard deviation.

2.6. For computational ease, use the glycosaminoglycans (GAG) levels for type 2 assay in DB8 rounded to 1 decimal place: $0.5, 0.5, 0.4, 0.3$. Calculate (a) the variance and (b) the standard deviation.

2.7. As an investigator, you wish to make an inference from DB3 about whether or not the drug affects serum theophylline level (5 days). From Section 2.7, what steps would you go through?

2.8. As an investigator, you wish to make an inference from DB14 about whether or not the difference in exhaled nitric oxide from before exercise to 20 minutes after is zero. From Section 2.7, what steps would you go through?

2.9. The squares of the 60 plasma silicone levels in DB5 are ($\times 10^4$): 225, 169, 1521, 400, 1521, 1764, 576, 324, 144, 676, 100, 121, 225, 361, 729, 784, 121, 121, 324, 324, 576, 2304, 484, 729, 361, 324, 1024, 961, 361, 441, 441, 576, 100, 144, 784, 625, 484, 441, 529, 484, 484, 576, 1444, 2025, 529, 484, 324, 225, 16, 196, 576, 400, 324, 576, 225, 361, 676, 900, 484, 576. Plot a relative frequency distribution of these squares using intervals: 0–0.02, 0.02–0.04, ..., 0.22–0.24. Does it appear more similar to the normal or the chi-square probability distribution in Section 2.8? Why?

2.10. For INR readings from the clinic in DB13, plot the frequency distribution in intervals of 0.25 in width. (This may have been done in Exercise 2.3.) Is it more similar to a normal or a chi-square probability distribution?

2.11. If we were to make an inference about the mean of differences between preoperative and postoperative plasma silicon levels in DB5, what would be the *df*? To make this mean into a *t* statistic, we subtract what value, and then divide by what statistic? To decide if *t* is significantly larger than zero, we choose a "cut point" greater than which *t* is significant and smaller than which *t* is not significant. What do we call this cut point?

2.12. If we were to make an inference about the mean of differences between clinic and laboratory INR readings in DB13, what would be the *df*?

2.13. Assign ranks (small to large) to the platelet counts in DB9.

2.14. Assign ranks (small to large) to the bone density values in DB7. Patient 18 has a bone density reading far less than that for the others. What effect does this have on the distribution of readings? Is this effect also true for the rankings of the readings?

2.15. If we wished to make an inference about the proportion of patients experiencing any nausea in DB2, what distribution would we be using in this inference?

2.16. If we wished to make an inference about the proportion of patients in DB12 who died after 2 cm or less of carinal resection, what distribution would we be using?

2.17. Calculate the sample SEM of the rounded GAG levels (0.7, 0.4, 0.3, 0.2) for type 1 assay in DB8.

2.18. In DB10, the difference in time to perform between legs has mean = 0.16 and standard deviation = 0.2868. Calculate the SEM.

2.19. In DB3, the two related variables baseline and 5-day theophylline levels vary jointly. If we were to calculate a measure of how one varies relative to the other, what would it be called?

Chapter 3

Summary Statistics

3.1. NUMERICAL SUMMARIES, ONE VARIABLE

SECTION FORMAT

Concepts for numerical summary statistics were given in Chapter 2. In this section, formulas for these summary statistics are given so that the user will know just how they are calculated and will have a convenient reference to verify the calculation method used. The symbol representing a concept is given, followed by the formula.

When the number n of observations is indicated, n refers to the sample size when the formula is calculated from a sample, and it refers to the population size when the formula is calculated from a population. The distinction between samples and populations is usually clear by context and is stated if not clear.

QUANTILES

We are all familiar with percentiles. A set of values is divided into 100 parts; the 90th percentile, for example, is the value below which 90% of the data fall. Other quantile types are deciles, in which the set of values is divided into 10 parts, and quartiles, 4 parts. Quantiles are useful descriptors of how a data distribution is shaped. For example, suppose Q_1, Q_2, and Q_3 are the first, second, and third quartiles (one quarter, half, three quarters of the data fall below). It can be seen that, if $Q_2 - Q_1 \approx Q_3 - Q_2$ ($\approx$ means "approximately equal to"), the distribution is approximately symmetric; if $Q_2 - Q_1$ and $Q_3 - Q_2$ are quite different, the distribution is skewed.

MEAN

μ denotes a population's arithmetic mean, and m denotes a sample's arithmetic mean. The two have the same calculation, so the difference is whether n represents the size of a population or a sample. The sample mean m sometimes appears in the literature represented as the variable with an overbar, such as $\bar{x}$. Recall that $\sum$ means "add together all data elements whose symbol follows me." Thus, if the variable x contains the three data elements, 1, 2, and 4, then $\sum x$

implies $1 + 2 + 4$.

$$\mu = \frac{\sum x}{n} \quad \text{or} \quad m = \frac{\sum x}{n}, \tag{3.1}$$

depending on whether n denotes the number in a population or in a sample, respectively. (The usual mean is called arithmetic because the values are added, in contrast to the geometric mean, a form used only in some special cases, in which they are multiplied.)

Example

Consider the mean prostate-specific antigen (PSA) value from Table DB1.1:

$$m = (7.6 + 4.1 + \cdots + 7.9)/10 = 67.5/10 = 6.75$$

MEDIAN

The median, having no standard symbol (sometimes *md* is used, and, occasionally, other representations), may be found for either a population or a sample. The following formula is an algorithm in words rather than in symbols:

Put the *n* observations in order of size.
Median is the middle observation if *n* is odd.
Median is the halfway between the two middle observations if *n* is even.

It will be apparent that the median is also the 2nd quartile (Q_2) and 50th percentile. The median frequently is used to represent the average of survival data, where occasional long survivors skew the data, rendering the mean undescriptive of the typical patient.

Example

Let us put the 10 PSA observations from Table DB1.1 in order of size: 4.1, 4.4, 5.9, 6.1, 6.8, 7.6, 7.7, 7.9, 8.0, 9.0. *n* is even. The two middle observations are 6.8 and 7.6. The median is halfway between them: 7.2.

MODE

The mode, also having no standard symbol (sometimes *mo* is used), may be found for either a population or a sample, provided that *n* is large. The mode can be read from a bar chart but will be approximate, depending on the choice of the bar chart's interval widths and starting point.

Make a bar chart of the data.
Mode is the center value of the highest bar.

Example

We can again use PSA values from Table DB1.1 as an example, although a mode from 10 observations would not be used in practice. The mode requires a large number of observations to have any accuracy and is not often used in

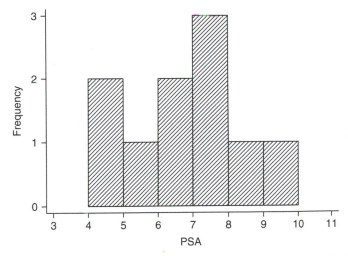

Figure 3.1 Bar chart of prostate-specific antigen values from Table DB1.1. The *mode* is 7.5, the center of the tallest bar.

clinical studies. A bar chart of the observations with a bin width of 1 would appear as in Fig. 3.1. The mode is the center of the tallest bar: 7.5.

VARIANCE

The variance of a population is denoted σ^2. Conceptually, it is the average of the squares of differences between the observations and their mean: $\sigma^2 = \sum(x - \mu)^2/n$. It may be calculated more easily by the equivalent form:

$$\sigma^2 = \frac{\sum x^2 - n\mu^2}{n}. \tag{3.2}$$

The variance of a sample is denoted s^2. It is similar. If m replaces μ, it uses $n - 1$ in place of n to avoid a theoretical bias. The form equivalent to Eq. (3.2) is

$$s^2 = \frac{\sum x^2 - nm^2}{n - 1}. \tag{3.3}$$

The *standard deviations* σ and s are just the square roots of the respective variances.

Example

The mean of the PSA values from Table DB1.1 was $m = 6.75$. The variance may be found by

$$s^2 = (7.6^2 + 4.1^2 + \cdots + 7.9^2 - 10 \times 6.75^2)/9$$
$$= (57.76 + 16.82 + \cdots 62.41 - 10 \times 45.5625)/9$$
$$= (478.890 - 455.625)/9 = 23.265/9 = 2.585.$$

The standard deviation is just the square root of the variance, or

$$s = \sqrt{2.585} = 1.6078.$$

STANDARD ERROR OF THE MEAN

The standard error of the mean (SEM) was introduced in Section 2.9. It is simply the standard deviation of the value m, the mean of n observations. If n is the *population* size, the SEM is symbolized σ_m; if n is the *sample* size, the (estimated) SEM is symbolized s_m. They are calculated using the standard deviations of the observations, σ or s, as appropriate:

$$\sigma_m = \frac{\sigma}{\sqrt{n}} \quad \text{or} \quad s_m = \frac{s}{\sqrt{n}}. \tag{3.4}$$

Example

From the 10 PSA values of Table DB1.1, s was found to be 1.6078. The estimated SEM is $1.6078/\sqrt{10} = 1.6078/3.1623 = 0.5084$.

STANDARD ERROR OF THE MEAN FOR TWO SAMPLES

In many cases, we are faced with two samples that may be drawn from the same population and we want to know if their means are different. To assess this, we need to use the standard deviation of the mean (SEM) using the information from both samples. If the standard deviation of observations from that population is known, say, σ, the SEM is simply σ divided by a sample size term, for example:

$$\sigma_\mu = \sigma \sqrt{\frac{1}{n_1} + \frac{1}{n_2}}. \tag{3.5}$$

If, however, we do not know the population standard deviation, which is usually the case, we must estimate it from s_1 and s_2, the sample standard deviations calculated from the two sets of observations. This requires two steps: (1) use s_1 and s_2 to find the standard deviation of the pooled observations, say, s_p, and (2) then find the SEM, say, s_m, in a form similar to Eq. (3.5). The algebra is worked backward from the formulas for the two sample standard deviations to find s_p, as we would have calculated it if we had pooled all the observations at the outset. It turns out to be

$$s_p = \sqrt{\frac{(n_1 - 1)s_1^2 + (n_2 - 1)s_2^2}{n_1 + n_2 - 2}}. \tag{3.6}$$

Then s_m is

$$s_m = s_p \sqrt{\frac{1}{n_1} + \frac{1}{n_2}}. \tag{3.7}$$

Example

Consider again the PSA values from Table DB1.1. Suppose we had taken two samples: the first four ($n_1 = 4$) and the remaining six ($n_2 = 6$) observations. Their standard deviations are $s_1 = 2.1205$ and $s_2 = 1.3934$. Substituting these values into Eq. (3.6), we obtain

$$s_p = \sqrt{\frac{3 \times 2.1205^2 + 5 \times 1.3934^2}{8}} = \sqrt{2.8997} = 1.7029.$$

The 1.7029 estimate is slightly larger than the 1.6078 we obtained from the original 10 observations because we had to allocate a degree of freedom for each of two means (one appearing in s_1 and the other in s_2) and had to divide by $n - 2$ rather than $n - 1$.

$$s_m = 1.7029 \times 0.6455 = 1.0992$$

3.2. NUMERICAL SUMMARIES, TWO VARIABLES

COVARIANCE

When we consider two measures that vary simultaneously, each one alone can be described by the methods of the preceding section; but we are often interested in how they vary jointly, or "co-vary," one relative to the other. This *covariance,* introduced in Section 2.10 in the context of joint frequency distributions, requires paired recordings—that is, a reading on y for each reading on x. The calculation of the covariance is similar to the variance, except that instead of squaring the $x -$ mean term, we use the $x -$ mean term times its paired $y -$ mean term, as $\sigma_{xy} = \sum (x - \mu_x) \times (y - \mu_y)/n$. These forms, for the population and sample covariances, respectively, become

$$\sigma_{xy} = \frac{xy - n\mu_x\mu_y}{n} \tag{3.8}$$

and

$$s_{xy} = \frac{\sum xy - nm_xm_y}{n - 1}. \tag{3.9}$$

Example

Let age from Table DB1.1 take the x position in Eq. (3.9) and prostate volume take the y position:

$$s_{xy} = (75 \times 32.3 + 68 \times 27.0 + \cdots + 74 \times 16.4 - 10 \times 65.1 \times 32.73)/9$$
$$= (2422.50 + 1836.00 + \cdots + 1213.60 - 10 \times 2130.723)/9$$
$$= (21211.40 - 21307.23)/9 = -95.83/9 = -10.6478$$

Interpretation

If one variable tends to increase as the other increases, such as systolic and diastolic blood pressure, the covariance is positive and large; large values of one are multiplied by large values of the other, which makes a very large sum. If one tends to decrease as the other increases, as with PSA and prostate density, the covariance is negative and large; large positive values of one multiply large negative values of the other. Conversely, if increases and decreases in one variable are unrelated to those of the other, the covariance tends to be small.

CORRELATION COEFFICIENT

The covariance could be very useful in indicating a shared behavior or independence between the two variables, but there is no standard for interpreting it. The covariance can be standardized by dividing by the product of standard deviations of the two variables. It is then called the *correlation coefficient,* designated ρ (Greek rho, rhymes with *snow*) for a correlation between two populations and r for a correlation between two samples. Thus, the formulas for calculating correlation coefficients ρ (population) or s (sample), respectively, are

$$\rho_{xy} = \frac{\sigma_{xy}}{\sigma_x \sigma_y} \quad \text{or} \quad r_{xy} = \frac{s_{xy}}{s_x s_y}. \tag{3.10}$$

Interpretation

This standardized covariance, the correlation coefficient, may be easily interpreted. If either variable is perfectly predictable from the other, the correlation coefficient is 1.00 when they both increase together and -1.00 when one increases as the other decreases. If the two variables are independent, that is, a value of one provides no information about the value of the other, the correlation coefficient is 0. A correlation coefficient of 0.10 is rather low, showing little predictable relationship, whereas 0.90 is rather high, showing that one increases rather predictably as the other increases.

Example

Continuing the age and volume example from before, we note that the standard deviation of age is $s_x = 6.9992$ and that of volume is $s_y = 15.9351$. The covariance was calculated as $s_{xy} = -10.6478$. Then

$$r_{xy} = -10.6478/(6.9992 \times 15.9351) = -0.0956.$$

This result would tell us that, for our 10 patients in Table DB1.1, volume tends to decrease as age increases, but in a very weak relationship. We should note that $n = 10$ is too small of a sample to provide us with much statistical confidence in our result.

Caution

We must remember that correlation methods measure relationship only along a straight line. If one variable increases when the other increases but not in a straight line, for example, as when weight is predicted by height (weight is a power of height) or when teenage growth depends on time (growth is a logarithm of time), the correlation may not be high despite good predictability. More sophisticated prediction methods are needed for curvilinear relationships. Elementary information on this topic appears in Chapter 8.

Furthermore, this linear correlation measures only the pattern of data behavior, not interchangeability of data. For example, temperature measured in the same patients using one thermometer in degrees Celsius and another in degrees Fahrenheit would have an almost perfect correlation of 1.0, but the readings could not be intermingled.

3.3. PICTORIAL SUMMARIES, ONE VARIABLE

COMMON TYPES

Graphs and charts allow us to visualize distributions and other properties of data. From a chart, we can often get a rough idea of a mean, a standard deviation, or a proportion. Although there are several types of charts, the most common are the bar chart (which was introduced in Chapter 2 and is shown again in Fig. 3.1), the histogram, the pie chart, the line chart, and, for two variables, the scattergram. Currently, the mean-and-standard-error chart and the box-and-whisker chart are also seen frequently in medical literature.

MAKING A BAR CHART

When forming a bar chart, the choice of intervals is important. It is, to some extent, an art. An unfortunate choice of intervals can change the apparent pattern of the distribution. Enough intervals should be used so that the pattern will be affected minimally by altering the beginning and ending positions. The beginning and ending positions should be chosen to be convenient in reading the bar chart and should relate to the meaning of the variable being charted. Recall the prostate volume data depicted in Fig. 2.1F. Figure 3.2A shows the prostate volume data allocated to 6 intervals rather than the 24 shown in Fig. 2.1F; useful information is obscured by the lack of detail. Figure 3.2B shows the data allocated to 48 intervals; the viewer is distracted by too much detail. Furthermore, if we had fewer data, say, only 50 or so values, the excess number of intervals would obscure the distribution pattern as badly as too few. The choice of number, width, and starting points of intervals arise from the user's judgment. They should be considered carefully before forming a bar chart.

HISTOGRAM

The histogram appears much like the bar chart but differs in that the number of observations lying in an interval is represented by the *area* of a rectangle (or bar) rather than its height. If all intervals are of equal width, the histogram is no different from the bar chart (except perhaps cosmetically). However, if one interval had been chosen that is twice the width of the others, the rectangle height over that interval must be half the height it would be in a bar chart.

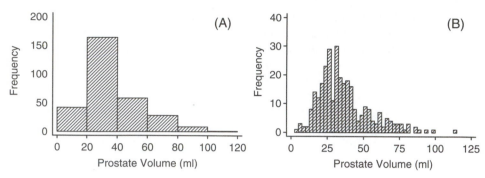

Figure 3.2 The effects on a bar chart displaying a frequency distribution of too few intervals (A) and of too many intervals (B).

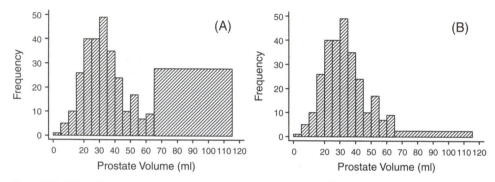

Figure 3.3 The effect, when intervals are unequal, of representing frequency in an interval by height (using a bar chart) (A) as opposed to representing that frequency by an area (using a histogram) (B).

This area-in-lieu-of-height presentation avoids a possible misinterpretation by the viewer. For example, if the prevalence of a disease in an epidemic is displayed monthly for several months but the last rectangle represents only 15 days, its height per infected patient would be doubled in a histogram to convey the true pattern of prevalence.

Example

Figure 3.3 illustrates the requirement for using areas in lieu of heights for unequal intervals. Suppose in recording prostate volumes the intervals had been organized by 5-ml increments until 65 ml, and a final interval of 65 to 115 was used. Then the bar chart of volumes (see Fig. 2.1F) would be amended so that the last bar would have height 28, as in Fig. 3.3A. If we adjust by converting the height to area, we obtain Fig. 3.3B, which gives a more accurate depiction.

Pie Chart

A pie chart represents proportions rather than amounts. Its main use is to visualize the *relative* prevalence of a phenomenon rather than its absolute prevalence. It also has the advantage of avoiding the illusion of sequence that sometimes is

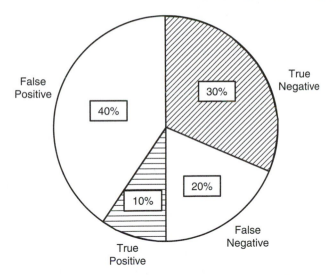

Figure 3.4 A pie chart representing proportions of biopsy results as predicted by the digital rectal examination.

implied by the order of bars in a bar chart, regardless of whether or not it was intended. To draw a pie chart, the user must allocate the 360° of a circle to the components in proportion to their prevalence. A prevalence of 20% is shown by $0.2 \times 360° = 72°$ of angle about the center of the circle.

Example

Let us look at the prediction of biopsy result by digital rectal examination (DRE) from Table DB1.1. There are 30% true-negative (DRE−, BIOP−), 20% false-negative (DRE−, BIOP+), 10% true-positive (DRE+, BIOP+), and 40% false-positive (DRE+, BIOP−) results. Figure 3.4 visually conveys these percentages of results in a pie chart. The first piece of pie includes 30% of $360° = 108°$.

LINE CHART

In a bar chart, if we connected the center of the bar tops by line segments and then erased the bars, we would have a form of line chart. The main use of a line chart is to convey information similar to a bar chart but for intervals that form a sequence of time or order of events from left to right. In Fig. 2.1F, we intend no progression of frequency in logical sequence as prostate volumes increase. In contrast, the frequencies of patients per age follow an interesting pattern as age progresses.

Example

Figure 3.5 depicts frequencies for our 301 patients for 5-year age intervals. We note that the frequencies increase successively to about the mid-60s, and then decrease successively thereafter. It is not difficult to conjecture the forces causing

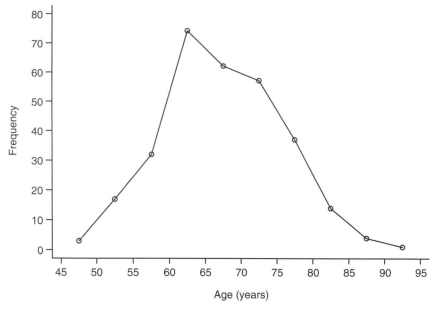

Figure 3.5 A line chart representing frequencies of men with prostate problems according to age. The points *(circles)* showing frequencies are positioned over the centers of the defined age intervals and are then connected by straight-line segments.

this data pattern in these patients who have presented for possible prostate problems. Prostate problems are rare in young men and increase in frequency with age. However, starting in the late 60s, men are dying of other causes in increasing numbers; therefore, the number of survivors available to present is decreasing with age.

Relation of a Line Chart to a Probability Distribution

Conceptually, it is important to observe that, as the sample size increases and the width of the intervals decreases, the line chart of a sample distribution approaches the picture of its probability distribution.

MEAN-AND-STANDARD-ERROR CHART

A diagram showing a set of means to be compared, augmented by an indication of the size of uncertainty associated with each mean, is appearing more and more frequently in medical articles.

Showing the Means

When a relationship among several groups is of interest, a lot of information is given by a plot in which groups are given positions on the horizontal axis and means are shown by vertical height above each position. Increasing time or increasing amount of a drug might appear on the horizontal axis. The clinical response to this sequence is shown by the mean. For example, postoperative pain rated by patients on a visual analog scale may be related to the amount of pain-relieving drug used. Four standard levels of drug dose, starting with zero, could be positioned on the horizontal scale, and mean pain rating could be shown vertically over each respective dose. In this case, we would expect the means to be decreasing with increasing dose. In other cases, they could be increasing or going up then down or even scattered in no perceptible pattern. We can tell a great deal about the process going on from the pattern.

Showing the Variability About the Means

We must ask whether, for example, the downward pattern of pain with increasing drug is meaningful, because we can make it look huge or minuscule by altering the vertical scale. A crucial part of the information is how different the means are relative to the variability in the data. The means may be decreasing, but with such small decrements relative to the data variability that it might have happened by chance and we do not accept the decreasing pattern as shown to be meaningful. A useful solution is to show the associated uncertainty as "whiskers" on the means, that is, as lines up and down from the mean indicating variability by their lengths. This variability depicted may be standard deviation, standard error, or some related measure.

Example

Figure 3.6 shows prostate volume between 0 and 50 ml for 291 patients in the 50- to 89-year age range separated into decades of age: 50s, 60s, 70s, and 80s. The means are shown by solid circles. The whiskers indicate about 2 standard errors above and below the mean, which includes 95% of the data on an idealized distribution (see Chapter 4 for a more detailed discussion of this topic). We can see by inspection that the mean volumes appear to increase somewhat by age but that there is so much overlap in the variability that we are not sure this increase is a dependably real phenomenon from decade to decade. However, we would take a small risk in being wrong by concluding that the increase from the youngest (50s) to the oldest (80s) is a real change.

Effect of Irregular Data

Does this chart tell the full story? If the data per group are distributed in a fairly symmetric and smooth bell-type curve, most of the relevant pattern may be discerned. However, if the data are distributed irregularly and/or asymmetrically, this chart actually covers up important relationships and may lead to false conclusions. That is because the assumption of regularity and symmetry are made for this chart, and as usual, relationships are distorted when assumptions are violated. Charts that are "data dependent" rather than "assumption dependent," such as the box-and-whisker charts discussed next, often will provide a better understanding of the data.

BOX-AND-WHISKER CHART

A practical way to explore and understand available data is to diagram them such that they display not only the typical aspects (e.g., distribution center and spread) but also atypical characteristics (e.g., asymmetry, data clumps, and

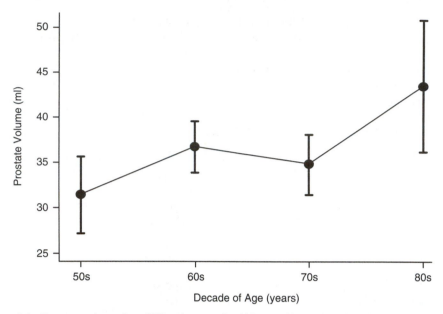

Figure 3.6 Prostate volume (in milliliters) means for 297 men allocated to their decades of age, with attached whiskers representing about 2 standard deviations (exactly 1.96) above and below the respective means.

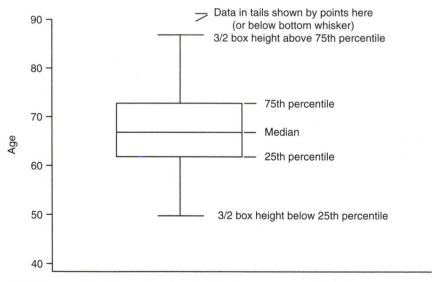

Figure 3.7 Representative box-and-whisker plot showing component definitions.

outlying values). The box-and-whisker chart does this rather well, although no technique displays all the vagaries of data. Figure 3.7 shows such a diagram, representing aspects of the distribution of age for the 301 urologic patients. As indicated in the diagram, the box includes the 25th to the 75th percentiles of data, with the median (the 50th percentile) as an intermediate line. If the median is in the center of the box, the middle half of the data is nearly symmetric with the median not far different from the mean. An off-center median indicates asymmetry. The whiskers extend the plot out to another indicator of spread, somewhat closer to the tails of the data distribution. This indicator might be a range not including outliers or might be a defined percentile enclosed. In Fig. 3.7, it is 3/2 of the box height, which would include about 90% of the data in a large normal sample. Whisker lengths that are similar and are about half the semibox length are further evidence of symmetry and a near-normal distribution. Unequal whisker lengths indicate asymmetry in the outer parts of the data distribution. Whisker lengths shorter or longer than 3/2 the box length indicate tendency toward a "flat-topped" or "peaky" distribution. A short whisker attached to a long box portion or vice versa shows evidence of "lumpy" data. Finally, the presence of data far out in the tails, as well as their distance out, is shown by dots above and below the whisker ends in Fig. 3.7. Other characteristics can be detected with a little experience in using the box-and-whisker chart.

Effect of No Assumptions

Charts like the box-and-whisker with no distributional assumptions, such as was used in the mean-and-standard-error chart, are completely dependent on the data available at that moment, and general distributional characters cannot be easily inferred; they are designed to describe the sample, not the population. This is at once their great strength and their great weakness. They must be used with care.

Example

Consider the sample of 291 urologic patients with PSA levels of 50 ng/ml or less included in the 50- to 89-year age range separated into decades of age: 50s, 60s, 70s, and 80s. Figure 3.8 shows a box-and-whisker chart of PSA levels by age decade. Immediately obvious are the many high values stretching out above the whiskers, with none symmetrically below. That, in addition to the longer whisker on the upper side, indicates a strong skewness. These data suggest that PSA has a narrower distribution in the 50s, becoming worse with age. However, by the time the 80s are reached, the worst cases appear to have died, so the distribution narrows again, except for a couple of extreme cases. It should be apparent that a mean-and-standard-error chart would not be appropriate for these data.

Showing Sample Size

One piece of information lacking from the chart in Fig. 3.8 is sample size. Might some of the differences be attributable to a small number of patients in the group? A variation on the box-and-whisker chart is to draw the width of the box proportional to sample size so that relative sample sizes (not actual numbers) can be compared. Figure 3.9 is a redrawing of Fig. 3.8 with box widths proportional to sample size. The smaller size in the 50s and especially in the 80s lends less credence to the results for these age decades.

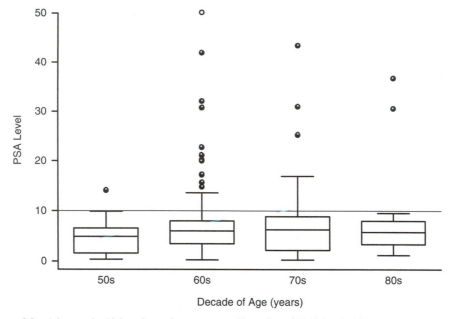

Figure 3.8 A box-and-whisker chart of prostate-specific antigen (PSA) level of 291 urologic patients in the 50- to 89-year age range separated into decades of age: 50s, 60s, 70s, and 80s. Readings above the line drawn at PSA level 10 show a high risk for prostate cancer.

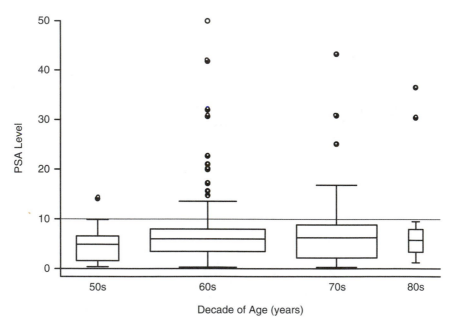

Figure 3.9 A redrawing of Fig. 3.8 with widths of boxes proportional to number of data in the respective age decade.

3.4. PICTORIAL SUMMARIES, TWO VARIABLES

DEPICTING THE RELATIONSHIP BETWEEN VARIABLES

Suppose we have two types of readings on each patient, perhaps heart rate and blood oxygen level. We may examine either one alone by single variable methods. What interests us here is to explore how the data act (i.e., how they are distributed) in the two dimensions simultaneously. Are these variables independent or correlated? If they are not independent, what is the nature of their relationship? Graphical representation will often display subtleties not apparent from summary statistics.

SCATTERPLOT

The simplest depiction is to plot the pair of readings for each patient on perpendicular axes. We can see whether points appear to be randomly scattered or clustered. If clustered, we can see the locations and shape of these clusters.

Example

We might ask whether there is a relationship between age and prostate volume for patients at risk for prostate cancer. (Let us omit volumes greater than 65 ml, because many of the larger glands are due to benign prostate hypertrophy, biasing the cancer risk group.) In Fig. 3.10, volumes are plotted on the vertical

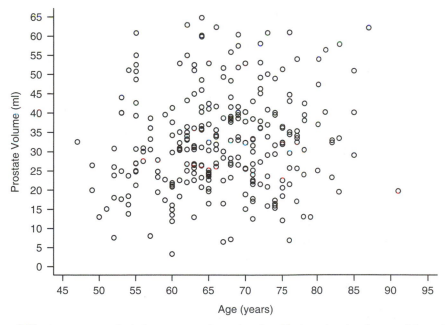

Figure 3.10 A scattergram depicting prostate volumes less than 65 ml as related to the age of the patient.

axis and age on the horizontal axis. A pattern of some sort might have suggested a relationship, for example, if prostate volumes tended to reduce with age. However, no obvious pattern appears. If you were given an age, would this plot allow you to suggest an average volume different from the overall average? No. We would say the two variables appear independent.

TWO-DIMENSIONAL FREQUENCY DISTRIBUTION

For one variable, we provided its frequency distribution by dividing the axis into intervals and showing the number of cases in each interval (bar chart, line chart, etc.). For two variables, we could divide the space into rectangles and count the cases in each rectangle. However, to show the number of cases by heights, we need a third dimension. Although a three-dimensional (3-D) solid model could be developed out of plaster, wood, or some other material, it is not very practical; we would like to show the 3-D image in two dimensions, that is, on a page. This can be done, and there exist computer software packages to assist.

Challenges in Three-Dimensional Imaging

The user is challenged to manage several characteristics at once, including scaling and aspect. The viewer should be able to read the variable or frequency values for any point in the three dimensions. It is especially difficult to show the height of a column, because most columns are not flush against the scale depiction. The viewer must extrapolate the height from a scale somewhat removed that depends on the aspect angles. In addition, the extrapolation may be distorted by perspective. (Distances appear smaller as they recede from the viewer in a 3-D model, so they may be drawn smaller in the 2-D image.) Another issue is the ability to see data farther back in the image that are obscured by data in front. It may be possible to swap ends on one or both axes or to change the aspect. When the data are "bumpy," it is usually impossible to render an image that shows all the subtleties.

Example of Low Correlation

Figure 3.11 shows a 3-D bar chart of the data from Fig. 3.10. It gives the viewer some perception of the joint frequency distribution of age and prostate volume. Age intervals are chosen as decades, and volume is measured in 10-ml intervals. What is the height of the tallest column? We cannot even see the bottom. A viewer could approximate the height by adroit use of dividers and straight edge. What we can see is that the volume distribution patterns (not the actual column heights) for the respective age decades are not that different. This implies a low correlation between age and volume, and indeed, the calculation of the correlation coefficient yields $r = 0.05$.

Example of High Correlation

Figure 3.12 shows a 3-D bar chart of PSA against prostate-specific antigen density (PSAD). The plot was limited to PSA levels less than 20 ng/ml, because the few larger PSA values would have compressed the graphical area of interest

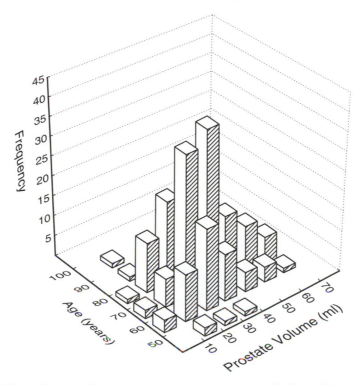

Figure 3.11 Three-dimensional bar chart of age by prostate volume showing little evidence of correlation.

too much for characteristics to be discernible. The direction of the PSA axis was reversed to avoid obscuring much of the result behind higher bars. Because PSA is a major part of the calculation of PSAD, we would expect a high correlation; the calculation of the correlation coefficient yields $r = 0.70$. It can be seen that data frequency is high when the two variables lie along the diagonal line from left to right, and it is sparse when one is large and the other is small, which is evidence of an association.

STATISTICAL GRAPHS IN EXPLORING DATA

The use of pictorial summaries have two main purposes. First, as descriptive statistics, they present a visual image of the data all at once, which not only helps describe the interrelationships among the data but also allows the viewer to retain this image. A visual image in memory often provides a more complete picture than does remembering a smattering of table entries, even when the table contains the same information. Second, pictorial summaries may suggest the forces giving rise to patterns of data. These perceived patterns are not sufficient evidence in themselves for scientific conclusions, but they suggest hypotheses to be posed that can then be tested with quantitative results. Graphs and charts are formidable tools in *data exploration.*

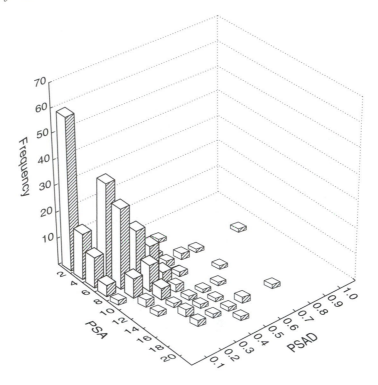

Figure 3.12 Three-dimensional bar chart of prostate-specific antigen level by prostate-specific antigen density showing evidence of considerable correlation.

3.5. GOOD GRAPHING PRACTICES

Although graphing practices is not a central focus of this book, a few comments may be in order. Books having such a focus are available, and the reader is encouraged to seek more complete advice from these sources.

The goal of graphing is to transmit information. When constructing graphs, we are well advised to concentrate on that goal. We should avoid whatever does not contribute to that goal, because it is likely to distract the viewer from the central message. Data should be presented in a way that relates to clinical understanding, including format and data intervals. The size and darkness of images and labels should be consistent and balanced, with the major components of the message slightly, not greatly, larger and/or darker. Be sure all components are labeled— axes, numbers on axes, responses from different groups, and so on; if the viewer is left wondering what an image is or means, the overall message is degraded. In his first grand book of several works, Edward Tufte[87] encouraged minimizing the amount of ink. Avoid displaying two dimensions in a 3-D graph, despite such a practice being default in certain software packages. Avoid "cutesy" axes relating to the topic at hand that contribute confusion but no information. For example, the author has seen airline economic data drawn with the axis in the shape of an airplane fuselage and tail, preventing the viewer from reading data off the graph. Graphing practices seen weekly in new magazines and national newspapers tend to attempt drama at the expense of accuracy and are seldom models to emulate. In another grand book, Darrell Huff[32] showed the use of two dimensions to distort a one-dimension comparison: the height of money bag images represented relative average worker income in two nations, but the viewer perceives the relative area of the bags, roughly the square of height.

CHAPTER EXERCISES

3.1. From DB7, find the (a) mean, (b) median, (c) variance, (d) standard deviation, (e) 1st quartile, (f) 3rd quartile, and (g) SEM of the bone density. Round all answers to one decimal place after calculation.

3.2. From DB10, find the (a) mean, (b) median, (c) variance, (d) standard deviation, (e) 1st quartile, (f) 3rd quartile, and (g) SEM of the distance covered in a triple hop on the operated leg.

3.3. From DB15, find the (a) mean, (b) median, (c) variance, (d) standard deviation, (e) 1st quartile, (f) 3rd quartile, and (g) SEM of temperatures at depth 1 for each treatment.

3.4. From DB12, find the mode age.

3.5. From DB10, find the two-sample SEM of distance covered using the operated leg sample and the nonoperated leg sample.

3.6. From DB10, find (a) the covariance and (b) the correlation coefficient of distance covered between the operated and nonoperated legs.

3.7. From DB10, find (a) the covariance and (b) the correlation coefficient of seconds to perform the hops between the operated and nonoperated legs.

3.8. From DB14, find the correlation coefficient of exhaled nitric oxide (eNO) between before exercise and 20 minutes after.

3.9. From DB7, construct a bar chart of patient ages.

3.10. From DB7, construct a histogram of bone density in the intervals 80 to <140, 140 to <160, 160 to <180, and 180 to <200.

3.11. From DB7, construct pie charts of patients' (a) sexes and (b) ages (grouped 17–19, 20–22, 23–25, >25).

3.12. From DB14, construct a bar chart of the 20-minute eNO change.

3.13. From DB14, construct a histogram of the 20-minute eNO change with all values greater than 5 in one interval.

3.14. From DB14, construct a pie chart of the four categories: male with exercise-induced bronchoconstriction (EIB), female with EIB, male without EIB, and female without EIB.

3.15. From DB11, construct a line chart of number rats surviving malaria by day number for the three treatments.

3.16. From DB14, construct a line chart for eNO means over time for groups with and without EIB.

3.17. From DB3, construct (a) a mean-and-standard-error chart (±1.96 SEM) and (b) a box-and-whisker chart for serum theophylline levels at baseline, 5 days, and 10 days.

3.18. From DB14, construct (a) a mean-and-standard-error chart (±1.96 SEM) and (b) a box-and-whisker chart of eNO for EIB patients across time.

3.19. From DB3, construct a scattergram of serum theophylline level at 10 days depending on level at baseline.

3.20. From DB14, construct a scattergram of 20-minute eNO change by age.

3.21. From DB14, construct a 3-D bar chart of 20-minute eNO change by age using 5-year intervals for age and 10-unit intervals for eNO change.

Chapter 4

Confidence Intervals and Probability

4.1. OVERVIEW

BASIS OF A CONFIDENCE INTERVAL

We have observed that hematocrit (Hct) values (measured in percent) for healthy patients are not all the same; they range over an interval. What is this interval? We know that extreme values occasionally arise in healthy patients. We cannot specify an interval that will always include only healthy patients and exclude only unhealthy patients. The best we can do is to find an interval that most frequently includes healthy patients and excludes unhealthy patients. This is an expression of relative frequency or likelihood. We might say that the interval should include 95% of the healthy population, which is to say that a randomly chosen healthy patient has a 0.95 probability of being within the interval. This leads to the term *confidence interval*, because we are 95% confident that a healthy patient will be within the interval.

ERROR RATE

We let the symbol α represent the probability that a healthy patient's Hct will be outside the healthy interval; in this case, $\alpha = 5\%$. When a patient's Hct does fall outside the healthy interval, we think, "It is likely, but not certain, that the patient has arisen from an unhealthy population. We must compare the Hct with other indicators to derive a complete picture."

ESTIMATING THE INTERVAL

In Section 2.7, it was noted that probabilities of occurrences correspond to areas under portions of probability distributions. Thus, if we know the distribution of Hcts, we can find the Hct values outside of which 5% of the area (2.5% in each tail) will occur. The mechanism for this will be seen in the following sections.

A GENERAL STATEMENT OF CONFIDENCE

The confidence interval need not be constrained to 95% probability, and it may apply to any distribution. A general statement for a confidence interval on an individual observation drawn randomly from a population, for any probability distribution and any desired interval, may be stated as follows (with indentations designed to help identify subconcepts):

> The probability
> > that a randomly drawn observation
> > > from a given probability distribution
> >
> > is contained in a specified interval
>
> is given by
> > the area of the distribution under the curve over that interval.

(4.1)

OTHER USES FOR PROBABILITIES

The use of probabilities is by no means limited to confidence intervals. Suppose we needed to know the chance of encountering a healthy patient with an Hct less than 0.30. If we knew the Hct probability distribution, we could calculate this chance as the area under that distribution to the left of the horizontal axis value of 0.30. Rarely must we calculate such probability values directly, because we may use computers and/or tables to find them. Sections in this chapter and in Chapter 13 describe the method for finding probabilities from the distributions commonly met in statistics, which were introduced in Section 2.8.

USE OF THE NORMAL DISTRIBUTION

The most frequently used distribution in biostatistics is the normal one, for both biological and mathematical reasons. In biology, a great many data sets naturally follow the normal distribution, at least approximately. In mathematics, because of the Central Limit Theorem introduced in Chapter 2, any mean of a data set follows the normal, at least approximately. Thus, our look at probability begins with the normal distribution.

4.2. THE NORMAL DISTRIBUTION

THE STANDARD NORMAL

Recall from Section 2.8 that the normal distribution is a perfect case of the famous bell curve. Although an infinite number of cases of the normal exist, we need deal with only one: the standard normal. Any normally distributed variable or sample of observations becomes a standard normal variable, symbolized z, by subtracting the mean from each value and dividing by the standard deviation. Thus, *z represents the number of standard deviations away from the mean*, positive for above the mean and negative for below the mean. Figure 4.1 shows a standard normal distribution with $z = 1.96$ (frequently seen in practice) and the

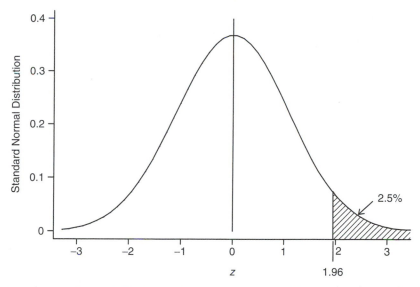

Figure 4.1 A standard normal distribution shown with a 2.5% α and its corresponding $z = 1.96$. The α shown is the area under the curve to the right of a given or calculated z. For a two-tailed computation, α is doubled to include the symmetric opposite tail. A two-tailed $\alpha = 5\%$ lies outside the ± 1.96 interval, leaving 95% of the area under the curve between the tails.

corresponding area $\alpha = 2.5\%$ under the curve to the right of that z. (Addition of a similar area in the left tail would provide $\alpha = 5\%$ in total.)

Table of the Standard Normal

Table I (see Tables of Probability Distributions at the back of the book) contains selected values of z with four areas that are often used: (1) the area under the curve in the positive tail for the given z, that is, one-tailed α; (2) the area under all except that tail, that is, $1 - \alpha$; (3) the areas combined for both positive and negative tails, that is, two-tailed α; and (4) the area under all except the two tails, that is, $1 - \alpha$. Table 4.1 shows a segment of Table I.

PROBABILITY OF CERTAIN RANGES OCCURRING

Prostate volume has been posed as a possible indicator of prostate cancer. Because the population we wish to examine is composed of patients with possible prostate cancer, we omit five patients with known benign prostate hypertrophy (BPH) as lying in a different population. The remaining prostate volumes (shown in Fig. 2.1F) are distributed not too far from normal. These remaining 296 volumes have a mean of 35.46 ml and a standard deviation of 16.35 ml. Let us round them to 35 and 16 ml to make the arithmetic easy in this illustration. Two patients present to you, one with a volume of 59 ml and one with a volume of 83 ml. How typical are they? They are 1.5 and 3 standard deviations above the mean, respectively. By looking in Table 4.1 under a z of 1.5, we find a one-tailed α of 0.067 and a two-tailed α of 0.134. Approximately 6.7% of the area under the curve lies to the right of the z-value. This tells us that 6.7% of healthy patients have prostate volumes at least this much greater than the mean, and similarly,

Table 4.1

Table 4.1

Segment of Normal Distribution of Table I[a]

z (no. standard deviations to right of mean)	One-tailed applications		Two-tailed applications	
	One-tailed α (area in right tail)	$1 - \alpha$ (area except right tail)	Two-tailed α (area in both tails)	$1 - \alpha$ (area except both tails)
0	.500	.500	1.000	.000
.50	.308	.692	.619	.381
1.00	.159	.841	.318	.682
1.281	*.100*	*.900*	*.200*	*.800*
1.50	.067	.933	.134	.866
1.645	*.050*	*.950*	*.100*	*.900*
1.960	*.025*	*.975*	*.050*	*.950*
2.00	.023	.977	.046	.934
2.326	*.010*	*.990*	*.020*	*.980*
2.50	.006	.994	.012	.088
3.00	.0013	.9987	.0026	.9974

[a]For selected distances (z) to the right of the mean, given are (1) one-tailed α, the area under the curve in the positive tail; (2) one-tailed $1 - \alpha$, the area under all except the tail; (3) two-tailed α, the areas combined for both positive and negative tails; and (4) two-tailed $1 - \alpha$, the area under all except the two tails. Entries for the most commonly used areas are in italics.

13.4% have volumes this deviant from (larger or smaller than) the mean. This indicates a large prostate, but not an obviously atypical one. In contrast, the second patient's z-value of 3 yields a one-tailed α of 0.0013; only about 1 in 1000 healthy subjects would have a prostate so large by chance alone. We are led to conclude that this patient is unlikely to have arisen from the normal population; he more likely arose from a population with atypically large prostates.

USING THE STANDARD NORMAL

Another way to look at the same information is to transform the data to the standard normal by the relationship $z = (x - \mu)/\sigma$. Because $\mu = 35$ ml and $\sigma = 16$ ml, our prostate with volume 59 becomes $z = (59 - 35)/16 = 1.5$, which refers to the same position in the table. Similarly, the volume 83 becomes $z = (83 - 35)/16 = 3$. A standard normal z-value represents the number of standard deviations from the mean.

CRITICAL VALUE

If we wanted to select out as abnormal all patients with prostate volumes in the 1% upper tail, that is, to identify the upper 1%, what would be our *critical*, or separating, value? By looking at Table I, we find that $\alpha = 0.01$ is paired with $z = 2.326$. This indicates that all patients with volumes more than about 2.3 standard deviations greater than the mean would be classified as abnormal. The actual critical volume will be $x = \mu + 2.3\sigma = 35 + 2.3 \times 16 = 71.8$ ml. A critical value is often informally called a cut point.

4.3. CONFIDENCE INTERVAL ON AN OBSERVATION FROM AN INDIVIDUAL PATIENT

EXAMPLE

Now we can look at a confidence interval on Hct with actual numbers. Suppose, for example, we should find that healthy Hct values arise from a $N(47,3.6^2)$ probability distribution. [Recall that $N(47,3.6^2) = N(47,12.96)$ symbolizes a normal distribution with mean $\mu = 47$ and standard deviation $\sigma = 3.6$.] Table 4.1 shows that when $1 - \alpha$ (the area except for both tails) is 95%, the area is enclosed by 1.96 standard deviations above and below the mean. This implies that the upper limit of the healthy interval lies at $\mu + 1.96\sigma = 47 + 1.96 \times 3.6 = 54$. Similarly, the lower limit lies at $\mu - 1.96\sigma = 47 - 1.96 \times 3.6 = 40$. Our healthy interval is 40% to 54%. If we have a healthy patient, we would bet 95 to 5 (or 19 to 1) that the Hct will be in the interval, which is to say we are 95% confident that the interval will include a healthy patient.

We can restate the form of Eq. (4.1) for the Hct case as follows:

The probability
that a randomly drawn Hct
from $N(47,3.6^2)$
is contained in the interval (40,54)
$= 95\%$.

A briefer and easier way to say the same thing would be

$$P[40 < \text{Hct} < 54] = 0.95.$$

CAUTION

An important caveat is implicit in the aforementioned method but is so often overlooked that we should note it explicitly. If the frequency distribution involved is other than a normal curve, the often seen "95% confidence contained in ± 2 standard deviations" statement does *not* apply; the limits must be found from the appropriate distribution for that case.

4.4. CONCEPT OF A CONFIDENCE INTERVAL ON A DESCRIPTIVE STATISTIC

THE MOST FREQUENT USE OF A CONFIDENCE INTERVAL IS ON A MEAN

In the preceding section, we used a distribution of patients to obtain a confidence interval, in relation to which we interpreted an observation from a single patient. This observation may be thought of as a sample of size 1 from a population with known μ and σ. Although this is often useful in clinical practice, in research, we are interested in the confidence interval on the estimate of a population statistic, most often a mean or, less often, a standard deviation.

Confidence Interval Defined

A random sample is a set of observations drawn from a population, in which the method of drawing is random. When we calculate a descriptive statistic from a random sample, we obtain an estimate of the equivalent population statistic. m and s are our best estimates of μ and σ, respectively. But what is the accuracy of these estimates? How much confidence do we have that the estimate is "on target"? We want an interval about an estimate that will tell us if the width includes the target. *A confidence interval is an interval about an estimate, based on its probability distribution, which expresses the confidence, or probability, that that interval contains the population statistic being estimated.*

General Form for a Confidence Interval on a Statistic

A general form may be given in the same pattern as Eq. (4.1):

The probability
  that a population statistic
    from a distribution of estimates of that statistic
  is contained in a specified interval
is given by
  the area of the distribution over that interval.
(4.2)

Common Form for a Confidence Interval on a Statistic

Any estimate of a statistic from a random sample follows a probability distribution. We have already noted that a sample mean follows a normal distribution and a sample variance follows a chi-square distribution. The estimate usually will be toward the center of the distribution, with only the rare cases lying in the tails. To express the confidence, we find the critical values that separate the tails from the main body of the distribution; we are confident that the interval between these critical values contains the population statistic to the extent of the proportion of the curve contained in the main body between the tail areas. The concept in Eq. (4.2) usually is written in the following format:

$$P[\text{lower critical value} < \text{population statistic} < \text{upper critical value}] = 1 - \alpha,$$
(4.3)

which implies that the interval defined by the critical values, excluding α proportion of the curve, will enclose the population statistic with probability $1 - \alpha$.

One Tail Is Sometimes Used

Although most confidence intervals are formed as intervals excluding both tails, an occasional case occurs in which we want to exclude only one tail. Chapter 14 further examines this option.

4.5. CONFIDENCE INTERVAL ON A MEAN, KNOWN STANDARD DEVIATION

CONFIDENCE INTERVAL EXAMPLE

In Section 4.2, we took the 296 non-BPH prostate volumes as a population with $\mu = 35$ ml and $\sigma = 16$ ml. We also know from Sections 2.8 and 2.9 that a sample mean m is distributed normal and the population standard deviation of the sample mean of m (standard error of the mean [SEM]), σ_m, is $\sigma/\sqrt{n}$.

Suppose we did not know μ but wanted to describe it as well as possible from the sample of 10 volumes in Table DB1.1. We would estimate μ by calculating m ($= 32.73$ ml), we would find $\sigma_m (= 16/\sqrt{10} = 5.06$ ml), and then we would use these calculations to put a confidence interval on μ.

To find a 95% confidence interval with 2.5% of unusual cases in each tail, we find the end points of the interval as $m \pm 1.96\sigma_m = 32.73 \pm 1.96 \times 5.06 = 22.81$ and 42.65 ml. In the format of Eq. (4.3),

$$P[22.81 < \mu < 42.65] = 0.95.$$

We are 95% confident that the population mean is included in the interval of 22.81 to 42.65 ml; indeed, the population mean of 36.47 ml is so included.

METHOD FOR 95% CONFIDENCE INTERVAL

If 95% confidence is wanted, we look in Table I (or Table 4.1) for 0.950 under "two-tailed $1 - \alpha$ (area except both tails)." To its left in the first column, that is, under "z (no. standard deviations to right of mean)," we find 1.960. Thus, our critical values, which include 95% of the curve, are the sample mean $m \pm 1.96 \times \sigma_m$, the standard deviation of m. (Recall that σ_m also is called the population SEM.) In symbols,

$$P[m - 1.96 \times \sigma_m < \mu < m + 1.96 \times \sigma_m] = 0.95. \tag{4.4}$$

METHOD FOR OTHER CONFIDENCE LEVELS

For any other level of confidence (e.g., 90% or 99%) for a sample mean with a known population standard deviation, we follow the same pattern, but looking in the 0.90 or 0.99 row in Table I (or Table 4.1). We may express this in general terms by denoting the confidence as $1 - \alpha$ so that $\alpha/2$ denotes the area in each tail. By substituting $1 - \alpha$ for 95% and $z_{1-\alpha/2}$ for 1.96 in Eq. (4.4), we obtain the more general confidence statement:

$$P[m - z_{1-\alpha/2}\sigma_m < \mu < m + z_{1-\alpha/2}\sigma_m] = 1 - \alpha. \tag{4.5}$$

ADDITIONAL EXAMPLE

An orthopedist is experimenting with the use of Nitronox as an anesthetic in the treatment of children's arm fractures.[26,27] He anticipates that it may provide an attractively short procedure. He treats $n = 50$ children and records the treatment time in minutes. He finds $m = 26.26$ and $\sigma = 7.13$. (The sample size is large enough to use the calculated standard deviation as σ.) He wants a 95% confidence interval on mean treatment time. He will require $\sigma_m = \sigma/\sqrt{n} = 7.13/\sqrt{50} = 1.008$. From Eq. (4.4),

$$P[m - 1.96 \times \sigma_m < \mu < m + 1.96 \times \sigma_m]$$
$$= P[26.26 - 1.96 \times 1.008 < \mu < 26.26 + 1.96 \times 1.008]$$
$$= P[24.28 < \mu < 28.24] = 0.95.$$

The orthopedist is 95% confident that the mean time to treat is between about 24 and 28 minutes.

4.6. THE t DISTRIBUTION

WHY WE NEED t

Often, we need to look at a confidence interval on a mean when we do not know the population standard deviation and must estimate it from a small sample. In this case, the normal distribution does not apply. We need the t distribution, introduced in Section 2.8. This is a distribution similar in appearance to the normal, but it gives a slightly wider confidence interval to compensate for lack of accuracy in the standard deviation. Recall that the smaller the sample, the wider the distribution, so the particular member of the family of t distributions to be used depends on the sample size or a variation of it, the degrees of freedom (df). Let us examine the t distribution in more detail.

THE NATURE OF t

The t distribution is a symmetric, bell-shaped distribution very much like the standard normal distribution, but with a larger standard deviation. As in the standard normal, t is given by the deviation of the observation from the mean divided by the standard deviation. As the t is used for smaller samples in which the mean and standard deviation are estimated, $t = (x - m)/s$. In methods of inference using t involving one sample, df equals the sample size (n) less one, or $df = n - 1$. In inference involving two samples, $df = n - 2$. When df is infinite, the t is the standard normal. As the number of df grows smaller, the t distribution grows wider. Because the t distribution is similar to the normal distribution except for an adjustment for df, methods of inference for small samples using small sample standard deviations follow logic identical to the normal case, except that we remember to look up the areas under the tail for the appropriate df.

THE t PICTURED

Figure 4.2 shows a t distribution for 9 df with 2.5% of the area under the curve's right tail shaded. Note that it is similar in appearance to the normal distribution depicted in Fig. 4.1, except that the 2.5% critical value lies 2.262 s (sample

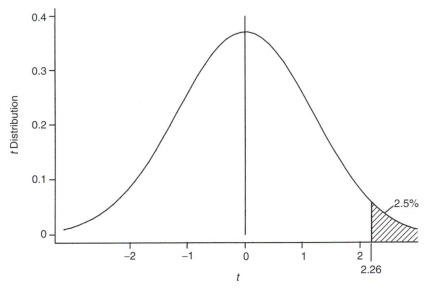

Figure 4.2 A *t* distribution shown with $\alpha = 2.5\%$ and its corresponding *t* for 9 *df*. The α shown is the area under the curve to the right of a given or calculated *t*. For a two-tailed computation, α is doubled to include the symmetric opposite tail. As pictured for 9 *df*, 2.5% of the area lies to the right of $t = 2.26$. For two-tailed use, the frequently used $\alpha = 5\%$ lies outside the interval $\pm t$, leaving 95% of the area under the curve between the tails.

standard deviations) to the right of the mean rather than 1.96 σ (population standard deviations).

TABLE OF *t* PROBABILITIES

Table II (see Tables of Probability Distributions) contains selected distances (*t*) away from the mean for the most commonly used one- and two-tailed α and $1 - \alpha$ areas under the curve for various *df*. These *t*-values correspond to italicized *z*-values in Table I. Table 4.2 shows a portion of Table II that may be used to follow the examples.

EXAMPLE OF CONFIDENCE INTERVAL
FOR AN INDIVIDUAL PATIENT USING *t*

Let us follow the prostate volume example as in the preceding section on the standard normal distribution, except that we shall use *m* and *s* from the small sample of Table DB1.1 rather than μ and σ. (Actually, a small skewed sample such as this would be treated better by the rank-order methods of Chapter 16, but the example serves to illustrate *t*.) For the 10 volumes, $m = 32.7$ ml and $s = 15.9$ ml. Because we are dealing with only one sample, $df = n - 1 = 9$. We therefore need look only at the $df = 9$ row in Table 4.2. The prostate volume 59 ml yields $t = (59 - 32.7)/15.9 = 1.65$, which places it 1.65 standard deviations above the mean. This lies between the 1.38, which falls under $\alpha = 0.10$ on the 9 *df* row, and 1.83, which falls under $\alpha = 0.05$. We can conclude that between

Table 4.2

Segment of *t* Distribution of Table II[a]

	.10	.05	.025	.01	.005	.001	.0005
One tailed α (right tail area)	.10	.05	.025	.01	.005	.001	.0005
One tailed $1 - \alpha$ (except right tail)	.90	.95	.975	.99	.995	.999	.9995
Two-tailed α (area both tails)	.20	.10	.05	.02	.01	.002	.001
Two-tailed $1 - \alpha$ (except both tails)	.80	.90	.95	.98	.99	.998	.999
$df = 5$	1.476	2.015	2.571	3.365	4.032	5.893	6.859
9	1.383	1.833	2.262	2.821	3.250	4.297	4.781
10	1.372	1.812	2.228	2.764	3.169	4.144	4.587
14	1.345	1.761	2.145	2.624	2.977	3.787	4.140
20	1.325	1.725	2.086	2.528	2.845	3.552	3.850
30	1.310	1.697	2.042	2.457	2.750	3.385	3.646
40	1.303	1.684	2.021	2.423	2.704	3.307	3.551
60	1.296	1.671	2.000	2.390	2.660	3.232	3.460
100	1.290	1.660	1.984	2.364	2.626	3.174	3.390
∞	1.282	1.645	1.960	2.326	2.576	3.090	3.291

[a]Selected distances (*t*) to the right of the mean are given for various degrees of freedom (*df*) and for (1) one-tailed α, area under the curve in the positive tail; (2) one-tailed $1 - \alpha$, area under all except the positive tail; (3) two-tailed α, areas combined for both positive and negative tails; and (4) two-tailed $1 - \alpha$, area under all except the two tails.

5% and 10% of patients will have volumes greater than this patient. Similarly, the 83-ml prostate yields $t = 3.16$, which is between $\alpha = 0.01$ and $\alpha = 0.005$; less than 1% of patients will have volumes this large. We do not tabulate *t* for as many possible values as we do the standard normal, because a full table would be required for every possible *df*. We can calculate on a computer $\alpha = 0.067$ for $t = 1.65$ and $\alpha = 0.006$ for $t = 3.16$ if we need them.

CRITICAL VALUE FOR *t*

The critical value above which 1% of prostate volumes occur is found by using the 9 *df* *t*-value in the one-tailed $\alpha = 0.01$ column; namely, $t = 2.82$. We calculate the volume at 2.82 standard deviations above the mean; that is, $m + 2.82s = 32.7 + 2.82 \times 15.9 = 77.54$. We expect that no more than 1% of volumes will exceed 77.5 ml.

4.7. CONFIDENCE INTERVAL ON A MEAN, ESTIMATED STANDARD DEVIATION

EXAMPLE

Suppose we wanted to use the 10 data from Table DB1.1 to put a 95% confidence interval on the population mean prostate volume. We have found that the 10 prostate volumes in Table DB1.1 have $m = 32.73$ ml and $s = 15.92$ ml.

We calculate $s_m = s/\sqrt{n} = 15.92/3.16 = 1.87$. To find the *t*-value for 95% confidence interval, we look under 0.95 for "two-tailed $1 - \alpha$ (except both tails)" in Table 4.2 for 9 *df* to find $t_{1-\alpha} = 2.262$. Because we had to use a sample SEM rather than that for the population, the confidence interval based on the *t* distribution extends more than 1.96 SEMs above and below the mean, as it would if it were based on the normal distribution. By substituting for *m* and s_m in the formula introduced in the next paragraph,

$$P[m - t_{1-\alpha/2}s_m < \mu < m + t_{1-\alpha/2}s_m]$$
$$= P[32.73 - 2.262 \times 1.87 < \mu < 32.73 + 2.262 \times 1.87]$$
$$= P[28.50 < \mu < 36.96] = 0.95.$$

We note from the first table in DB1 (showing means and standard deviations) that $\mu = 36.47$, which is included in the interval.

METHOD

More often than not in medical applications, we do not know σ and must estimate it by a small sample *s*. Finding a confidence interval on μ using *m* and *s* follows the same logic as using *m* and σ, except that we use a *t* table rather than a normal table. Because the *t* distribution has a greater spread than the normal, the confidence interval will be slightly wider. If we want 95% confidence, we look under 0.95 for "two-tailed $1 - \alpha$ (except both tails)" in Table II for the appropriate *df*. Replacing the σ and *z* symbols in Eq. (4.5) with the equivalent *s* and *t* symbols, we obtain

$$P[m - t_{1-\alpha/2}s_m < \mu < m + t_{1-\alpha/2}s_m] = 1 - \alpha. \tag{4.6}$$

(We know from Sections 2.8 and 4.6 that *m* follows a *t* distribution when we use *s* instead of σ, and we know from Section 2.9 that the sample standard deviation of *m*, that is, s_m, is $s/\sqrt{n}$. Recall that s_m is also called the sample SEM.) Chapter 14 provides a treatment of one-sided confidence statements.

ADDITIONAL EXAMPLE

A dermatologist is studying the efficacy of tretinoin in treating 15 $(=n)$ women's postpartum abdominal stretch marks.[77] Tretinoin was used on a randomly chosen side of the abdomen and a placebo on the other side. Neither the patient nor the investigator knew which side was medicated. The patient rated the improvement on each side on a 10-cm-long visual analog scale, and ratings were recorded as a reading between 0 and 10. The difference, treated-side rating minus untreated-side rating, indicating the excess of improvement with tretinoin versus the placebo, was calculated. $m = -0.33$ and $s = 2.46$. $s_m = s/\sqrt{n} = 2.46/\sqrt{15} = 0.64$. From Table II or Table 4.2, the 95% *t*-value for 14 *df* is 2.145. When these values are substituted in Eq. (4.6), the confidence interval evolves as

$$P[m - t_{1-\alpha/2}s_m < \mu < m + t_{1-\alpha/2}s_m]$$
$$= P[-0.33 - 2.145 \times 0.64 < \mu < -0.33 + 2.145 \times 0.64]$$
$$= P[-1.70 < \mu < 1.04] = 1 - \alpha.$$

The dermatologist is 95% confident that the mean stretch-mark improvement is between -1.7 and $+1.0$. The sample average is negative (untreated side better), the confidence interval includes 0, and the upper confidence bound of 1 is not very important clinically; evidence is insufficient to conclude a benefit from tretinoin in this particular medical application.

4.8. THE CHI-SQUARE DISTRIBUTION

WHY WE NEED CHI-SQUARE

Although the mean is the statistic on which we most frequently focus, we sometimes do focus on measures of variability. Suppose the amount of active ingredient in a medicinal capsule is crucial: less than $\mu - c$ mg fails to work, and more than $\mu + c$ mg damages the patient. We need a confidence interval on the standard deviation to be confident that the probable variability in content is not too large. Because the standard deviation is a square root, which is difficult to work with mathematically, confidence intervals on variability are found on the variance s^2 and afterward are converted to standard deviation units. As noted in Section 2.8, the sample variance s^2, drawn randomly from a normal population, when multiplied by the constant df/σ^2, follows a chi-square distribution, so we need chi-square to find confidence intervals on the variance. (df for simple confidence intervals will be $n - 1$.)

CHI-SQUARE PICTURED

Remember that the chi-square, being composed of squares, cannot be negative; however, it can grow to any large size. Thus, it is like a lopsided bell curve with the right tail stretched out (right skewed). Figure 4.3 shows the chi-square distribution for 9 df.

TABLES OF CHI-SQUARE

Table III (see Tables of Probability Distributions) provides the chi-square values that yield commonly used values of α, that is, the probability that a randomly drawn value from the distribution lies in the tail demarked by the tabulated chi-square value. Table 4.3 shows a segment of Table III, which may be used to follow the examples. Because the chi-square distribution is asymmetric, we cannot take an area in one tail and expect the other tail to be the same. Finding areas in a tail is much the same as in the *t*: the desired area in the tail specifies the column, the *df* specifies the row, and the *critical* chi-square value (the value demarking the tail area) lies at the row–column intersection. Table III provides

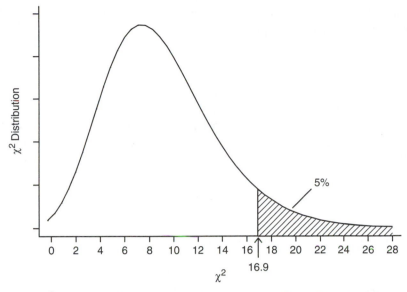

Figure 4.3   The χ^2 (chi-square) distribution for 9 *df* with a 5% α and its corresponding chi-square value of 16.9. The α probability is shown as the shaded area under the curve to the right of a critical chi-square, in this case, representing a 5% probability that a value drawn randomly from the distribution will exceed a critical chi-square of 16.9.

Table 4.3

Segment of Chi-square Distribution, Right Tail, of Table III[a]

α (area in right tail)	.10	.05	.025	.01	.005	.001	.0005
$1 - \alpha$ (except right tail)	.90	.95	.975	.99	.995	.999	.9995
df = 1	2.71	3.84	5.02	6.63	7.88	10.81	12.13
3	6.25	7.81	9.35	11.34	12.84	16.26	17.75
5	9.24	11.07	12.83	15.08	16.75	20.52	22.15
7	12.02	14.07	16.01	18.47	20.28	24.35	26.02
9	14.68	16.92	19.02	21.67	23.59	27.86	29.71
10	15.99	18.31	20.48	23.21	25.19	29.58	31.46
15	22.31	25.00	27.49	30.57	32.81	37.71	39.73
20	28.41	31.41	34.17	37.57	39.99	45.31	47.46
25	34.38	37.65	40.65	44.31	46.93	52.65	51.93
30	40.26	43.77	46.98	50.89	53.68	59.68	62.23
40	51.80	55.76	59.34	63.69	66.76	73.39	76.11
50	63.17	67.51	71.42	76.16	79.50	86.66	89.56
60	74.40	79.08	83.30	88.38	91.96	99.58	102.66
100	118.50	124.34	129.56	135.81	140.16	149.41	153.11

[a]Selected χ^2 values (distances above zero) are given for various degrees of freedom (*df*) and for (1) α, the right tail area under the curve, and (2) $1 - \alpha$, the area under all except the right tail.

chi-square values for the more commonly used right tail; Table IV (see Tables of Probability Distributions) provides values for the left tail.

EXAMPLE

We saw that $\sigma = 16.35$ ml for the population of prostate volumes (excluding known BPH). How large a standard deviation s would we have to observe in a sample size of 10 to have a 5% or less probability that it could have been so large by chance alone? From either Table 4.3 or Figure 4.3, the critical chi-square value (the value demarking the tail area) for a 5% tail for $n - 1 = 9$ df is $\chi^2 = 16.9$. That is,

$$\chi^2 = df \times s^2/\sigma^2 \text{ or } 16.9 = 9s^2/16.35^2 = 0.033667s^2.$$

We solve to find $s^2 = 501.98$ or $s = 22.40$. A sample standard deviation exceeding 22.4 would have less than 5% probability of occurring by chance alone.

4.9. CONFIDENCE INTERVAL ON A VARIANCE OR STANDARD DEVIATION

EXAMPLE

Using the 10 prostate volumes from Table DB1.1, what is a 95% confidence interval on the population standard deviation, σ? We have calculated $s = 15.92$ ml, which has $n - 1 = 9$ df. We know the probability distribution of s^2 but not of s; therefore, we shall find the interval on the population variance, σ^2, and take square roots: $s^2 = 15.92^2 = 253.4464$. Looking in Table III or Table 4.3 under right tail area $= 0.025$ for 9 df, we find 19.02. Similarly in Table IV, under left tail area $= 0.025$ for 9 df, we find 2.70. Substituting these values in the formula for confidence limits on a variance [see Eq. (4.7)], we find:

$$P[s^2 \times df/\chi_R^2 < \sigma^2 < s^2 \times df/\chi_L^2]$$

$$= P[253.45 \times 9/19.02 < \sigma^2 < 253.45 \times 9/2.70]$$

$$= P[119.93 < \sigma^2 < 844.83] = 0.95.$$

Taking square roots within the brackets, we obtain

$$P[10.95 < \sigma < 29.07] = 0.95.$$

We note from the first table of DB1 (showing means and standard deviations) that the population standard deviation (excluding known BPH to make the distribution more symmetric) is 16.35 ml, falling well within the confidence interval.

METHOD

We know (from Section 2.8) that sample variance s^2 drawn randomly from a normal population is distributed as chi-square (multiplied by the constant σ^2/df). We make a confidence-type statement on s^2 from this relationship, excluding

2.5% in each tail as usual, and use some algebra to put it in the form of a statement on σ^2, arriving at the expression:

$$P[s^2 \times df/\chi_R^2 < \sigma^2 < s^2 \times df/\chi_L^2] = 0.95, \qquad (4.7)$$

where χ_R^2 is the critical value for the right tail found from Table III and χ_L^2 is that for the left tail found from Table IV. We have to calculate separately a critical value for each tail of the chi-square distribution (using both Tables III and IV) because the distribution is not symmetric like the normal or t. For an exact chi-square, the sample must be taken from a normal distribution. However, if the population is at all close to normal, even roughly bell-shaped, the approximation is close enough to use this confidence method.

CONFIDENCE ON σ

To find the confidence on σ rather than σ^2, we just take square roots of the components within the brackets.

CONFIDENCE OTHER THAN 95%

We are not confined to 95% confidence, of course. For any general $1 - \alpha$ confidence, we find the two critical chi-square values that cut off $\alpha/2$ in each tail: χ_L^2 for the left and χ_R^2 for the right.

ADDITIONAL EXAMPLE

We are investigating the reliability of a certain brand of tympanic thermometer (temperature measured by a sensor inserted into the patient's ear).[63] The standard deviation will give us an indication of its precision. We want a 95% confidence interval on this precision, that is, a range outside of which the standard deviation would appear no more than 5 times in 100 random readings. Sixteen readings (°F) were taken on a healthy patient at 1-minute intervals. Data were 95.8, 97.4, 99.3, 97.1, ..., yielding $s = 1.23$. $s^2 = 1.23^2 = 1.51$. We use Eq. (4.7), that is, $P[s^2 \times df/\chi_R^2 < \sigma^2 < s^2 \times df/\chi_L^2] = 0.95$. The chi-square value from Table III for 97.5% area except right is $\chi_R^2 = 27.49$ for 15 df, and the value from Table IV for area except left is $\chi_L^2 = 6.26$. We find

$$P[s^2 \times df/\chi_R^2 < \sigma^2 < s^2 \times df/\chi_L^2]$$
$$= P[1.51 \times 15/27.49 < \sigma^2 < 1.51 \times 15/6.26]$$
$$= P[0.8239 < \sigma^2 < 3.6182] = 0.95.$$

Taking square roots within the bracket, we obtain

$$P[0.9077 < \sigma < 1.9022] = 0.95.$$

We are 95% sure that the standard deviation of this thermometer will occur within the 0.9 to 1.9°F range.

4.10. OTHER FREQUENTLY SEEN CONFIDENCE INTERVALS AND PROBABILITIES

WHAT PROBABILITIES ARE POSSIBLE?

Any datum that has a component of chance in its selection follows some probability distribution or other, although often we do not know what it is. The number and forms of probability distributions are unlimited. Fortunately, in the common medical applications of statistics, we meet only a few of these distributions. In this book, we need only six, which were introduced in Section 2.8. The normal, t, and χ^2 have been examined in this chapter, and the other three—F, binomial, and Poisson—are discussed in Chapter 13.

WHAT CONFIDENCE INTERVALS ARE POSSIBLE?

A confidence interval may be placed on any statistic for which the probability distribution is known. Fortunately, there are only a few statistics of frequent interest in medical statistics. Confidence intervals on an observation from an individual patient, on μ using σ, on μ using s, and on σ itself, have been examined in this chapter. Others that are commonly used include confidence intervals on proportions and on correlation coefficients, which are examined in Chapter 14.

CHAPTER EXERCISES

4.1. A probability distribution for variable x is expressed as N(2,9). How would you transform it to N(0,1), the standard normal distribution for variable z?

4.2. For the N(2,9) distribution, what are the values on the x-axis (horizontal) that enclose 95% of the area under the curve?

4.3. In DB12, the age distribution of patients undergoing carinal resections is approximately normal with mean = 47.8 years and standard deviation = 14.8 years. The sample is large enough to take the standard deviation as if it were the population σ. You have a 12-year-old patient whose tracheal carina requires resection. Does this patient fall within a 95% confidence limit of age for individual patients, or is this patient improbably young?

4.4. In DB13, let us take the standard deviation of the 104 laboratory International Normalized Ratio (INR) values as the population σ. Find m and σ. Find the 95% confidence interval on an individual patient. Does this confidence interval match the patient "in-range" interval?

4.5. The orthopedist in the Nitronox example in Section 4.5 also is interested in the patients' pain, which he has measured using the CHEOPS rating form. For his $n = 50$ patients, $m = 9.16$ and $\sigma = 2.04$. Find the (a) 95%, (b) 99%, and (c) 90% confidence intervals on mean pain rating.

4.6. For the laboratory INR values of Exercise 4.4, find the 95% confidence interval on the mean.

4.7. Among the indicators of patient condition after pyloromyotomy (correction of stenotic pylorus) in neonates is time (hours) to full feeding.[21] A surgeon wants to place a 95% confidence interval on mean time to full feeding. He has readings from $n = 20$ infants. Some of the data are 3.50, 4.52, 3.03, 14.53, ... He calculates $m = 6.56$ hours and $s = 4.57$ hours. What is the 95% confidence interval?

4.8. An emergency medicine physician samples the heart rate of $n = 8$ patients after a particular type of trauma. The mean is $m = 67.75$ beats/min, and the standard deviation is $s = 9.04$ beats/min. What is the 95% confidence interval on the mean?

4.9. In the kidney cooling data of DB15, the baseline (time 0) means are $m_1 = 34.40$ for saline infusion and $m_2 = 36.12$ for ice slush, each based on 12 readings. (Use two decimal places throughout for easier calculation.) The respective estimated standard deviations are $s_1 = 3.25$ and $s_2 = 1.17$. Find 95% confidence intervals on the means for each treatment. Are they similar?

4.10. The surgeon referenced in Exercise 4.3 is concerned about the variability of time to full feeding in neonates. Even if the mean is satisfactory, if the variability is too large, the outlying neonates on the longer side of the scale would be at risk. Recall that $n = 20$, $m = 6.56$ hours, and $s = 4.57$ hours. Establish 95% confidence intervals on (a) the variance and (b) the standard deviation.

4.11. The emergency department physician of Exercise 4.8 is concerned not only about the average heart rate of the trauma patients but also about the variability. Even if the mean is satisfactory, if the standard deviation is too large, outlying patients would be at risk. Recall that $n = 8$, $m = 67.75$ beats/min, and $s = 9.04$ beats/min. Establish 95% confidence intervals on (a) the variance and (b) the standard deviation.

4.12. Using the sample sizes and standard deviations given in Exercise 4.9, find 95% confidence intervals on the standard deviation of the baseline data for each treatment. Are they similar?

Chapter 5

Hypothesis Testing: Concept and Practice

5.1. HYPOTHESES IN INFERENCE

A CLINICAL HYPOTHESIS

Suppose we want to know if a new antibiotic reduces the time for a particular type of lesion to heal. We already know the mean time μ to heal in the population (i.e., the general public) without an antibiotic. We take a sample of lesion patients from the same population, randomizing the selection to ensure representativeness (see Section 1.6), and measure the time to heal when treated with the antibiotic. We compare our sample mean time to heal, say, m_a (a for antibiotic), with μ. Our *clinical hypothesis* (not statistical hypothesis) is that the antibiotic helps; that is, $m_a < \mu$. To be sure that our comparison gives us a dependable answer, we go through the formal sequence of steps that composes scientific inference.

DECISION THEORY INTRODUCED

In general terms, the decision theory portion of the scientific method uses a mathematically expressed strategy, termed a *decision function* (or sometimes decision rule), to make a decision. This function includes explicit, quantified gains and losses to reach a conclusion. Its goal is to *optimize* the outcome of the decision—that is, to jointly maximize gains and minimize losses. Gains might be factors such as faster healing, less pain, or greater patient satisfaction. Losses might be factors such as more side effects or greater costs—in time, effort, or inconvenience, as well as money. Any number of possible decision functions exist, depending on the strategy selected, that is, on the gains and losses chosen for inclusion and their relative weightings. Is financial cost of treatment to be included? Is pain level more or less important than eventual overall satisfaction? In a given situation, quite different "optimum" decisions could be reached, depending on the decision function chosen. Those who wish to apply outcomes derived from an investigator's use of decision theory should note that a personal or financial agenda may be involved in the choice of elements and weightings

75

used in the decision function. To safeguard against such agendas, a user should accept only decision functions with natures that have been clearly and explicitly documented.

OPERATIONS RESEARCH

In most applications, the decision problem is expressed as a strategy to select one of a number of options based on a criterion of minimum loss. The loss might include risk of making a wrong decision, loss to the patient (financial cost, as well as cost in pain; decreased quality of life; or even death), and/or financial or time cost to the investigator/institution. In the industrial, business, and military fields, applied decision theory most often has come under the heading of operations research (or operational analysis [British]). In medicine, some forms of applied decision theory using multiple sources of loss in the decision strategy appear under the heading of outcomes analysis, which is introduced in Chapter 8.

DECISION MAKING BY TESTING A STATISTICAL HYPOTHESIS

Chapter 1 notes that medicine's major use of statistical inference is in making conclusions about a population on the basis of a sample from that population. Most often, we form statistical hypotheses, usually in a form different from the clinical hypothesis, about the population. We use sample data to test these hypotheses. This procedure is a special case of two-option decision theory in which the decision strategy uses only one loss, that of the risk (probability) of making an erroneous decision. More specifically, the loss is measured as known, controlled probability of each of two possible errors: choosing the first hypothesis when the second is true and choosing the second when the first is true.

HOW WE GET FROM OBSERVED DATA TO A TEST

From the data, we calculate a value, called a *statistic*, that will answer the statistical question implied by the hypothesis. For example, to answer the time-to-heal question, we may compare the statistic m_a (mean time to heal using the antibiotic) with the parameter μ (mean time to heal for the untreated population). Such statistics are calculated from data that follow probability distributions; therefore, the statistics themselves will follow a probability distribution. Section 2.8 notes that a mean, the time-to-heal statistic here, is distributed (at least approximately) normal. Areas under the tails of such distributions provide probabilities, or risks, of error associated with "yes" or "no" answers to the question asked. Estimation of these error probabilities from the sample data constitutes the test.

THE NULL HYPOTHESIS

The question to be answered must be asked in the form of a hypothesis. This *statistical hypothesis*, which differs from the clinical hypothesis, must be stated carefully to relate to a statistic, especially because the statistic must have a known or derivable probability distribution. In most applications, we want to test for a difference, starting with the *null hypothesis*, symbolized H_0, that our sample is not different (hence, "null") from known information. (The hypothesis to be tested must be based on the known distribution, in this case, of the established healing time, because it cannot be based on an unknown antibiotic healing time. This is explained further later.) In testing for a *difference*, we hypothesize that the population mean time to heal with an antibiotic, μ_a (which we estimate by our sample mean, m_a), does not differ from the mean time to heal without the antibiotic; that is, H_0: $\mu_a = \mu$. In contrast to testing for a difference, we could test for *equivalence*. In this case, we would hypothesize a specified decrease in healing times versus those seen without the antibiotic; that is, H_0: $\mu_a = \mu -$ decrease. (The historical name and symbol are carried over even though the hypothesis is no longer truly null.) Most of this chapter focuses on the more familiar difference testing. Equivalence testing is addressed later.

THE ALTERNATE HYPOTHESIS

After forming the null hypothesis, we form an *alternate hypothesis*, symbolized H_1, stating the nature of the discrepancy from the null hypothesis if such discrepancy should appear. Hypotheses in words may be long and subject to misunderstanding. For clarity, they usually are expressed in symbols, where the symbols are defined carefully.

FORMS THE HYPOTHESES CAN TAKE

To answer the time-to-heal question, we hypothesized that the population mean, μ_a, from which our sample is drawn does not differ from μ. Our alternate hypothesis is formed logically as *not* H_0. In this case, we truly believe that the antibiotic *cannot lengthen* the healing, so the alternative to no difference is shortened healing. Thus, our statistical hypotheses are

$$H_0: \mu_a = \mu \text{ and } H_1: \mu_a < \mu. \tag{5.1}$$

The alternative here is known as a *one-sided* hypothesis. If we believed the antibiotic could either shorten *or* lengthen the healing, we would have used a *two-sided* hypothesis, H_l: $\mu_a \neq \mu$. More generally, when a decision is made about using or not using a medical treatment, the sidedness is chosen from the decision possibilities, not physical possibilities. If we alter clinical treatment only for significance in the positive tail and not in the negative tail, a one-tailed test is appropriate.

WHY THE NULL HYPOTHESIS IS NULL

It may seem a bit strange at first that our primary statistical hypothesis in testing for a difference says there is no difference, even when, according to our clinical hypothesis, we believe there is one and might even prefer to see one. The reason lies in the ability to calculate errors in decision making. When the hypothesis says that our sample is not different from known information, we have available a known probability distribution and therefore can calculate the area under the distribution associated with the erroneous decision: A difference is concluded when, in truth, there is no difference. This area under the probability curve provides us with the risk for a false-positive result. The alternate hypothesis, on the other hand, says just that our known distribution is not the correct distribution, not what the alternate distribution is. Without sufficient information regarding the distribution associated with the alternate hypothesis, we cannot calculate the area under the distribution associated with the erroneous decision: No difference exists when there is one, that is, the risk for a false-negative result.

A NUMERICAL EXAMPLE

As a numerical example, consider the sample of the 10 prostate volumes in Table DB1.1. Suppose we want to decide whether the population mean μ_v from which the sample was drawn is the same as the mean of the population of 291 remaining volumes. Because the mean of sample volumes may be either larger or smaller than the population mean, the alternative is a two-sided hypothesis. Our hypotheses are

$$H_0: \mu_v = \mu \text{ and } H_1: \mu_v \neq \mu. \tag{5.2}$$

m estimates the unknown μ_v, and if H_0 is true, m is distributed $N(\mu, \sigma_m^2)$. (μ is the mean and σ the standard deviation of the 291 volumes, and σ_m is the standard error of the mean [SEM] $\sigma/\sqrt{n}$ [see Section 2.9].) Standardization of m provides a statistic z, which we know to be distributed $N(0,1)$, that is,

$$z = \frac{m - \mu}{\sigma_m} = \frac{m - \mu}{\sigma/\sqrt{n}}. \tag{5.3}$$

We know that $m = 32.73$, $\mu = 36.60$, and $\sigma = 18.12$; $n = 10$ is the size of the sample we are testing. By substituting in Eq. (5.3), we find $z = -0.675$; that is, the sample mean is about two thirds of a (population) standard deviation below the population mean. The following small excerpt from Table I (see Tables of Probability Distributions) shows the two-tailed probabilities for the 0.60 and 0.70 standard deviations:

z (no. standard deviations to right of mean)	Two-tailed α (area in both tails)
0.60	0.548
0.70	0.484

The calculated z lying between these two values tells us that the probability of finding a randomly drawn normal observation more than 0.675σ away from the mean is a little more than 0.484, or about 0.5. We conclude that we have a 50% chance of being wrong if we decide that the sample mean did not arise from this distribution. There is not enough evidence to conclude a difference. This result may be visualized on a standard normal distribution as in Fig. 5.1.

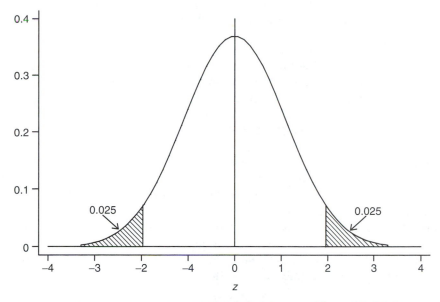

Figure 5.1 A standard normal distribution showing 2.5% tail areas adding to 5% risk for error from concluding that a random value did not arise from this distribution (risk for a false-positive result). The 0.025 area under the right tail represents the probability that a value drawn randomly from this distribution will be 1.96σ or farther above the mean.

THE CRITICAL VALUE

A "cut point" between whether a statistic is significant or nonsignificant is termed a *critical value.* In symbols, the critical values of z leading to a two-tailed α are $z_{\alpha/2}$ and $z_{1-\alpha/2}$. By symmetry, we need calculate only the right tail $z_{1-\alpha/2}$ and use its negative for the left. The decision about significance of the statistic z maybe stated as a formula. z is not significant if $-z_{1-\alpha/2} < z < z_{1-\alpha/2}$, or, using Eq. (5.3),

$$-z_{1-\alpha/2} < \frac{m - \mu}{\sigma_m} < z_{1-\alpha/2}. \tag{5.4}$$

THE MOST COMMON TYPES OF STATISTIC BEING TESTED AND THEIR ASSOCIATED PROBABILITY DISTRIBUTIONS

A previous paragraph demonstrated how a hypothesis having a general pattern of a sample mean divided by a known population SEM follows a standard normal distribution. In the same fashion, a hypothesis about a mean when the standard deviation must be estimated (the population standard deviation is unknown) has the same general pattern; namely, a sample mean divided by a *sample* SEM, so it follows a t distribution. A hypothesis about a standard deviation uses a variance, which follows a chi-square distribution. And finally, a hypothesis comparing two standard deviations uses a ratio of two variances, which follows an F distribution. Thus, a large number of statistical questions can be tested using just these four well-documented probability distributions.

CONFIDENCE INTERVALS ARE CLOSELY RELATED

Confidence intervals are closely related to hypothesis tests procedurally, in that null and alternate hypotheses could state that the interval does and does not, respectively, enclose the population mean. However, the use differs: A confidence interval is used to estimate, not to test.

ASSUMPTIONS IN HYPOTHESIS TESTING

Section 2.7 shows that various assumptions underlie hypothesis testing and that the result of a test is not correct when these assumptions are violated. Furthermore, the methods do not tell the user when a violation occurs or by how much the results are affected. Results appear regardless. It is up to the user to verify that the assumptions are satisfied. The assumptions required vary from test to test, and which apply are noted along with each test methodology. The most common assumptions are that errors are normal and independent from one another and, for multiple samples, that variances are equal.

THE MEANING OF "ERROR"

An "error" in an observation does not refer to an error in the sense of a mistake but rather to the deviation of the individual observation from the typical. The term *error*, and often the term *residual*, is used for historical reasons; if you find the term *error* referring to observations confuses your reading, read "deviation from typical" instead.

THE ASSUMPTION OF INDEPENDENCE OF ERRORS

When a set of observations is taken, we assume that knowledge of the error on one tells us nothing about the error on another; that is, *the errors are independent from one another*. Suppose we take temperature readings on a ward of patients with bacterial infections. We expect the readings to be in the 38° to 40°C range, averaging about 39°C. We assume that finding a 39.5°C reading for one patient (i.e., an error of 0.5°C greater than the average) tells us nothing about the deviation from average we will find for the patient in the next bed. How might such an assumption be violated? If the ward had been filled starting from the far end and working nearer as patients arrived, the patients at the far end might be improving and have lower temperatures. Knowledge of the temperature error of a particular patient might, indeed, give us a clue to the temperature error in the patient in the next bed. The assured independence of errors is a major reason for incorporating randomness in sampling. This assumption is made in almost all statistical tests.

THE ASSUMPTION OF NORMALITY OF ERRORS

Second, many, but not all, tests also assume that *these errors are drawn from a normal distribution.* A major exception is rank-order (or nonparametric) tests. Indeed, the avoidance of this assumption is one of the primary reasons to use rank-order tests.

THE ASSUMPTION OF EQUALITY OF STANDARD DEVIATIONS

The third most frequently made assumption occurs when the means of two or more samples are tested. It is assumed that *the standard deviations of the errors are the same.* This assumption is stated more often as requiring equality of variances rather than of standard deviations. The term *homoscedasticity* sometimes encountered in a research article is just a fancy word for equality of variances.

5.2. ERROR PROBABILITIES

POSING AN EXAMPLE

We have a mean of prostate-specific antigen (PSA) readings from a patient group, and we want to compare it with the mean of healthy patients to predict the presence or absence of prostate cancer. The null hypothesis, H_0, states that cancer is absent; that is, our new sample mean is not different from the healthy mean.

TYPE I (α) ERROR

The null hypothesis may be rejected when it should have been accepted; we conclude that our patients are not healthy when they are. Such an error is denoted a *Type I error*. Its probability of occurring by chance alone is denoted α. Formally, $\alpha = P[\text{rejecting } H_0 | H_0 \text{ true}]$ (read: "probability of rejecting H_0 given H_0 is true"). The conclusion of a difference when there is none is similar to the *false-positive* result of a clinical test. Properly, α is chosen before data are gathered so that the choice of critical value cannot be influenced by study results.

TYPE II (β) ERROR AND THE POWER OF A TEST

Alternatively, the null hypothesis may be accepted when it should have been rejected; we conclude that our patients are healthy when they are not. Such an error is denoted a *Type II error*. Its probability of occurring by chance alone is denoted β. Formally, $\beta = P[\text{accepting } H_0 | H_0 \text{ false}]$. The conclusion of no difference when there is one is similar to the *false-negative* result of a clinical test. $1 - \beta$ is the *power* of the test, which is referred to often in medical literature and is used especially when assessing the sample size required in a clinical study (introduced in Chapter 7). Power is the probability of detecting a difference that exists.

p-VALUE

After the data have been gathered, new information is available: the value of the decision statistic, for example, the (standardized) difference between control and experimental means. The error of rejecting the null hypothesis when it is true that can now be estimated using sample data is termed the *p-value*. If the *p*-value is smaller than α, we reject the null hypothesis; otherwise, we do not have enough

Table 5.1

Relationships Among Types of Error and Their Probabilities as Dependent on the Decision and the Truth

		Decision	
		H_0 true	H_0 false
Truth	H_0 true	Correct decision True negative (probability $1 - \alpha$)	Type I error False positive (probability α)
	H_0 false	Type II error False negative (probability β)	Correct decision True positive (probability $1 - \beta$)

evidence to reject it. The size of the p-value can give a clue about the relationship of the null hypothesis to the data. A p-value near 0 or near 1 leaves little doubt as to the conclusion to be drawn from the study, but a p-value slightly exceeding α may suggest that further study is warranted. A listing of the actual p-value in a study result often adds information beyond just indicating a value greater or less than α. The next section further addresses this issue.

RELATION AMONG TRUTH, DECISION, AND ERRORS

Type of error depends on the relationship between decision and truth, as is depicted in Table 5.1.

LOGICAL STEPS IN A STATISTICAL TEST

The logic of a statistical test of difference used historically is the following: (1) We choose an α, the risk for Type I error, that we are willing to accept. (2) Because we know the distribution of the statistic we are using, for example, z or t, we use a probability table to find its critical value. (3) We take our sample and calculate the value of the statistic arising from it. (4) If our statistic falls on one side of the critical value, we do not have the evidence to reject H_0; if on the other side, we do reject H_0.

THE CRITICAL VALUE AND THE ERRORS ILLUSTRATED

Illustrative null and alternate distributions and their α and β values are shown in Fig. 5.2.

CHOOSING α BUT NOT β IS A COMMON REALITY

We usually know or have reason to assume the nature of the null distribution, but we seldom know or have enough information to assume that of the alternate distribution. Thus, instead of minimizing both risks, as we would wish to do, we fix a small α and try to make β small by taking as large a sample size as we can.

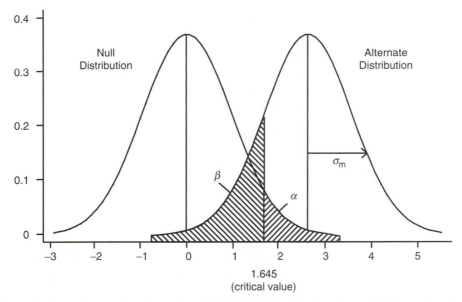

Figure 5.2 Depiction of null and alternate distributions along with the error probabilities arising from selecting a particular critical value. To the right of the critical value is the rejection region; to the left is the no-evidence-to-reject region.

In medical applications, choosing $\alpha = 5\%$ has become commonplace, but 5% is by no means required. Other risks, perhaps 1% or 10%, may be chosen as long as they are stated and justified.

ADDENDING A CONFIDENCE INTERVAL

Increasingly, readers of medical literature are asking for a confidence interval on the statistic being tested to "give assurance of adequate power." Simple algebra can change Eq. (5.4) into the confidence interval Eq. (4.5). Significance in Eq. (5.4) is equivalent to observing that the confidence interval does not cross zero. Thus, the confidence interval does not add to the decision-making process. It does not tell us what the power is. However, it often provides useful insights into and understanding of the relationships among the parameter estimates involved.

ACCEPTING H_0 VERSUS FAILING TO REJECT H_0

Previously, it was common in the medical literature to speak of accepting H_0 if the statistic fell on the "null side" of the critical value (see Fig. 5.2). Users often then were mystified to see that a larger sample led to contradiction of a conclusion already made. The issue is one of evidence. The conclusion is based on probabilities, which arise from the evidence given by the data. The more accurate statement about H_0 is the following: The data provide inadequate evidence to infer that H_0 is probably false. With additional evidence, H_0 may be properly rejected. The interpretation is that our databased evidence gives inadequate cause to say that it is untrue, and therefore we act in clinical decision making as if it were true. The reader encountering the "accept H_0" statement should recall this admonition.

TESTING FOR A DIFFERENCE VERSUS EQUIVALENCE

Testing for a difference, for example, between mean times to resolve infection caused by two competing antibiotics, has a long history in medical statistics. In recent years, the "other side of the coin" has been coming into practice, namely, testing for equivalence. If either antibiotic could be the better choice, the equivalence test is two-sided. However, in many cases, we have an established (gold) standard and ask if a new competitor is as good as the standard, giving rise to a one-sided equivalence test. A one-sided test is often termed a *nonsuperiority* or *noninferiority* test. For example, we may have an established but expensive brand-name drug we know to be efficacious. A much cheaper generic drug appears on the market. Is it equally efficacious? The concept is to pose a specific difference between the two effects, ideally the minimum difference that is clinically relevant, as the null hypothesis, H_0. If H_0 is rejected, we have evidence that the difference between the effects is either zero or at most clinically irrelevant, and we accept the proposition that the generic drug is equally effective. The methodology of equivalence testing is more fully addressed in Chapter 21; at this stage of learning, the student need only be aware of this option.

5.3. TWO POLICIES OF TESTING

A BIT OF HISTORY

During the early development of statistics, calculation was a major problem. It was done by pen in the Western cultures and by abacus (7-bead rows) or soroban (5-bead rows) in Eastern cultures. Later, hand-crank calculators and then electric ones were used. Probability tables were produced for selected values with great effort, the bulk being done in the 1930s by hundreds of women given work by the U.S. Works Project Administration during the Great Depression. It was not practical for a statistician to calculate a p-value for each test, so the philosophy became to make the decision of acceptance or rejection of the null hypothesis on the basis of whether the p-value was bigger or smaller than the chosen α (e.g., $p \geq 0.05$ vs. $p < 0.05$) without evaluating the p-value itself. The investigator (and reader of published studies) then had to not reject H_0 if p were not less than α (result not statistically significant) and reject it if p were less (result statistically significant).

CALCULATION BY COMPUTER PROVIDES A NEW OPTION

With the advent of computers, calculation of even very involved probabilities became fast and accurate. It is now possible to calculate the exact p-value, for example, $p = 0.12$ or $p = 0.02$. The user now has the option to make a decision and interpretation on the exact error risk arising from a test.

CONTRASTING TWO APPROACHES

The later philosophy has not necessarily become dominant, especially in medicine. The two philosophies have generated some dissension among statisticians. Advocates of the older approach hold that sample distributions only approximate the probability distributions and that exactly calculated p-values are not accurate anyway; the best we can do is select a "significant" or "not significant" choice. Advocates of the newer approach—and these must include the renowned Sir Ronald Fisher in the 1930s—hold that the accuracy limitation is outweighed by the advantages of knowing the p-value. The action we take about the test result may be based on whether a not-significant result suggests a most unlikely difference (perhaps $p = 0.80$) or is borderline and suggests further investigation (perhaps $p = 0.08$) and, similarly, whether a significant result is close to the decision of having happened by chance (perhaps $p = 0.04$) or leaves little doubt in the reader's mind (perhaps $p = 0.004$).

OTHER FACTORS MUST BE CONSIDERED

The preceding comments are not meant to imply that a decision based on a test result depends solely on a p-value, the *post hoc* estimate of α. The *post hoc* estimate of β, the risk of concluding a difference when there is none, is germane. And certainly the sample size and the clinical difference being tested must enter into the interpretation. Indeed, the clinical difference is often the most influential of values used in a test equation. The comments on the interpretation of p-values relative to one another do hold for adequate sample sizes and realistic clinical differences.

SELECT THE APPROACH THAT SEEMS MOST SENSIBLE TO YOU

Inasmuch as the controversy is not yet settled, users may select the philosophy they prefer. I tend toward the newer approach.

5.4. ORGANIZING DATA FOR INFERENCE

THE FIRST STEP: IDENTIFY THE TYPE OF DATA

After the raw data are obtained, they must be organized into a form amenable to analysis by the chosen statistical test. We often have data in patient charts, instrument printouts, or observation sheets and want to set them up so that we can conduct inferential logic. The first step is to identify the type of data to be tested. Section 1.4 notes that most data could be typed as categorical (we count the patients in a disease category, for example), rank-order (we rank the patients in numerical order), or continuous (we record the patient's position on a scale). Although there is no comprehensive rule for organizing data, an example of each type will help the user become familiar with the terminology and concepts of data types.

Table 5.2

**Format for Recording Joint Occurrences
of Two Binary Data Sets**

		DRE result		
		0	1	Totals
Biopsy result	0			
	1			
	Totals			

DRE, digital rectal examination.

CATEGORICAL DATA

In the data of Table DB1.1, consider the association of a digital rectal examination (DRE) result with a biopsy result. We ask the following question: "Does DRE result give some indication of biopsy result, or are the two events independent?" Both results are either positive or negative by nature. We can count the number of positive results for either, so they are nominal data. For statistical analysis, we almost always want our data quantified. Each of these results may be quantified as 0 (negative result) or 1 (positive result). Thus, our variables in this question are the numerical values (0 or 1) of DRE and the numerical values (0 or 1) of biopsy. The possible combinations of results are 0,0 (0 DRE and 0 biopsy); 0,1; 1,0; and 1,1. We could set up a 2×2 table (Table 5.2) in the format of Table 5.1.

ENTERING THE DATA

After setting up the format, we go to the data (from Table DB1.1) and count the number of times each event occurs, entering the counts into the table. It is possible to count only some of the entries and obtain the rest by subtraction, but it is much safer to count the entry for every cell and use subtraction to check the arithmetic. The few extra seconds are a negligible loss compared with the risk of publishing erroneous data or conclusions. Entering the data produces Table 5.3. Our data table is complete, and we are ready to perform a categorical test.

Table 5.3

Format of Table 5.2 with Table DB1.1 Data Entered

		DRE result		
		0	1	Totals
Biopsy result	0	3	4	7
	1	2	1	3
	Totals	5	5	10

DRE, digital rectal examination.

RANK DATA

Remaining with Table DB1.1, suppose we ask whether average prostate-specific antigen density (PSAD) is different for positive versus negative biopsy. PSAD is a continuous-type measure; that is, each value represents a point on a scale from 0 to a large number, which ordinarily would imply using a test of averages with continuous data. However, PSAD values come from a very right-skewed distribution. (The mean of the PSADs for the 301 patients is 0.27, which lies far from the center of the range 0–4.55.) The assumption of normality is violated; it is appropriate to use rank methods.

CONVERTING CONTINUOUS DATA TO RANK DATA

Sometimes ranks arise naturally, as in ranking patients' severity during triage. In other cases, such as the one we are addressing now, we must convert continuous data to rank data. To rank the PSAD values, we write them down and assign rank 1 to the smallest, rank 2 to the next smallest, and so on. We associate the biopsy results and sort out the ranks for biopsy 0 and biopsy 1. These entries appear here. Our data entry is complete, and we are ready to perform a rank test.

PSAD	PSAD ranks	Biopsy results	Ranks for biopsy = 0	Ranks for biopsy = 1
0.24	6	0	6	
0.15	3	0	3	
0.36	9	1		9
0.27	8	1		8
0.22	5	1		5
0.11	1	0	1	
0.25	7	0	7	
0.14	2	0	2	
0.17	4	0	4	
0.48	10	0	10	

CONTINUOUS MEASUREMENT DATA

Suppose we ask whether average PSA is different for positive versus negative biopsy results. For each patient, we would record the PSA level and a 0 or 1 for biopsy result, ending with data as in columns 5 and 8 of Table DB1.1. Prostate-specific antigen level is a continuous measurement; that is, each value represents a point on a scale from 0 to a large number. Section 4.2 notes that PSA is not too far from normal when patients with benign prostatic hypertrophy are excluded. Thus, a continuous data–type test of averages would be appropriate. We would need means and standard deviations. Their method of calculation was given in Chapter 3. One way to display our data in a form convenient to use is shown here. Our data entry and setup are complete, and we are ready to perform a means test.

PSA	Biopsy result	PSA for biopsy $= 0$	PSA for biopsy $= 1$
7.6	0	7.6	
4.1	0	4.1	
5.9	1		5.9
9.0	1		9.0
6.8	1		6.8
8.0	0	8.0	
7.7	0	7.7	
4.4	0	4.4	
6.1	0	6.1	
7.9	0	7.9	
		$m = 6.54, s = 1.69$	$m = 7.23, s = 1.59$

WHAT APPEARS TO BE A DATA TYPE MAY NOT ACT THAT WAY IN ANALYSIS

Suppose a list of cancer patients have their pathologic stage rated T1, T2, T3, or T4. Each stage is more severe than the previous one, so that the ratings are ordered data. It appears that your analysis will be a rank-order method. But wait! When you examine the data, you find that all the patients are either T2 or T3. You could classify them into low or high categories. You actually analyze numbers of patients (counts) by category, using a categorical method. This phenomenon may also occur with continuous data, where actual data readings may force the type into rank-order or categorical form, thereby changing the method of analysis. The investigator must remain aware of and sensitive to such a data event.

5.5. EVOLVING A WAY TO ANSWER YOUR DATA QUESTION

FUNDAMENTALLY, A STUDY IS ASKING A QUESTION

Clinical studies ask questions about variables that describe patient populations. A physician may ask about the PSA of a population of patients with prostate cancer. Most of these questions are about the characteristics of the probability distribution of those variables, primarily its mean, standard deviation, and shape. Of most interest in the prostate cancer population is the average PSA. Three stages in the evolution of scientific knowledge about the population in question are discussed in Section 1.2: description (the physician wants to know the average PSA of the cancerous population), explanation (the physician wants to know if and why it is different from that in the healthy population), and prediction (the physician wants to use PSA to predict which patients have cancer).

DESCRIPTION AND PREDICTION

When we know little about a distribution at question, the first step is to describe it. We take a representative sample from the population and use the statistical summarizing methods of Chapter 3 to describe the sample. These sample summaries estimate the characteristics of the population. We can even express our confidence in the accuracy of these sample summaries by the confidence methods of Chapter 4. The description step will not be pursued again in this chapter. Prediction combines the result of the inference with a cause-explanatory model to predict results for cases not currently in evidence. This is a more sophisticated stage in the evolution of knowledge that is discussed further in Chapter 8. This chapter addresses statistical testing, which leads to the inference from sample to population.

TESTING

We often want to decide (1) if our patient sample arose from an established population (Does a sample of patients who had an infection have the same average white blood count [WBC] after treatment with antibiotics as the healthy population?) or (2) whether the populations from which two samples arose are the same or different (Does a sample of patients treated for infection with antibiotics have the same average WBC count as a sample treated with a placebo?). In these examples, the variable being used to contrast the differences is mean WBC. Let us subscript the means with h for healthy patients, a for patients treated with antibiotics, and p for patients treated with a placebo. Recall μ represents a population mean and m represents a sample mean. Then, (1) contrasts m_a with μ_h, and (2) contrasts m_a with m_p.

STEPS IN SETTING UP A TEST

These contrasts are performed by statistical tests following the logic of inference (see Section 5.1) with measured risks of error (see Section 5.2). The step-by-step logic adhered to is as follows (assuming the simplest case of one question to be answered using one variable):

(1) Write down the question you will ask of your data. (Does treating my infected patients with a particular antibiotic make them healthy again?)

(2) Select the variable on which you can obtain data that you believe best to highlight the contrasts in the question. (I think WBC count is best to show the state of health.)

(3) Select the descriptor of the distribution of the variable that will furnish the contrast (e.g., mean, standard deviation). (Mean WBC will provide the most telling contrast.)

(4) Write down the null and alternate hypotheses indicated by the contrast. ($H_0: \mu_a = \mu_h$; $H_1: \mu_a \neq \mu_h$.)

(5) Write down a detailed, comprehensive sentence describing the population(s) measured by the variable involved in that question. (The healthy population is the set of people who have no current or chronic infections affecting their WBC count. The treated population is the set of people who have the

infection being treated and no other current or chronic infection affecting their WBC count.)

(6) Write down a detailed, comprehensive sentence describing the sample(s) from which your data on the variable will be drawn. (My sample is selected randomly from patients presenting to my clinic who pass my exclusion screen [e.g., other infections].)

(7) Ask yourself what biases might emerge from any distinctions between the makeup of the sample(s) and the population(s). Could this infection be worse for age, sex, cultural origin, and so on, of one patient than another? Are the samples representative of the populations with respect to the variable being recorded? (I have searched and found studies that show that mortality and recovery rates, and therefore probably WBC count, are the same for different sexes and cultural groups. One might suspect that elderly persons have begun to compromise their immune systems, so I will stratify my sample [see Section 1.6] to ensure that it reflects the age distribution at large.)

(8) Recycle steps 1 through 7 until you are satisfied that all steps are fully consistent with each other.

(9) In terms of the variable descriptors and hypotheses being used, choose the most appropriate statistical test (see Chapter 12 provides some detail on such choice) and select the α level you will accept.

(10) If your sample size is not preordained by availability and/or economics, satisfy yourself that you have an adequate sample size to answer the question (see also Chapter 7).

At this point, you are ready to obtain your data (which might take hours or years).

ADDITIONAL EXAMPLE

We should like to test (and lay to rest) the following assertion: "People who do not see a physician for a cold get well faster than people who do."

(1) Question being asked: Does seeing a physician for a cold retard the time to heal?

(2) Variable to use: length of time (days) for symptoms (nasal congestion, etc.) to disappear.

(3) Descriptor of variable that will provide contrast between patients who see a physician (group 1) and those who do not (group 2): mean number of days μ_1 (estimated by m_1) and μ_2 (estimated by m_2).

(4) Hypotheses: $H_0: \mu_1 = \mu_2$; $H_1: \mu_1 > \mu_2$.

(5) Populations: Populations are the sets of people in this nation who have cold symptoms but are otherwise healthy and see a physician for their condition (population 1) and those who do not (population 2).

(6) Samples: For sample 1, 50 patients will be randomly chosen from those who present at the walk-in clinic of a general hospital with cold symptoms but evince no other signs of illness. The data for sample 2 is more difficult to obtain. A random sample of five pharmacies in the area is taken, and each is monitored for customers who have signs of a cold. Of these customers, a random sample of 10 who state they are not seeing a physician is taken from each pharmacy,

with the customers agreeing to be included and to participate in follow-up by telephone.

(7) Biases and steps to prevent them: There are several possible sources of bias, as there are in most medical studies. However, the major questions of bias are the following: (a) Are patients and customers at and in the vicinity of our general hospital representative of those in general? (b) Are customers who buy cold medicines at pharmacies representative of cold sufferers who do not see physicians? We can answer question (a) by analyzing the demographics statistically after our study is complete, whereas question (b) requires a leap of faith.

(8) Recycle: These steps seem to be adequately consistent as they are.

(9) Statistical test and α: Two-sample t test with $\alpha = 0.05$.

(10) Sample size: Chapter 7 explains that we need not only (a) α ($= 0.05$), but also (b) the power ($1 - \beta$, which we take to be power $= 0.80$), (c) the difference between means that we believe to be clinically meaningful (which we choose as 2 days), and (d) σ_1 and σ_2. We estimate σ_1 to be 3 from a pilot survey of patients with cold symptoms who were followed up by telephone and assume σ_2 is the same. Using the methodology of Section 7.4, we can find that we need at least 36 in each sample; our planned 50 per group is a large enough sample.

Now we may begin to collect our data.

CHAPTER EXERCISES

5.1. In DB4, a hypothesis to be investigated is that protease inhibitors reduce pulmonary admissions. Is this a clinical or a statistical hypothesis? What would be a statement of the other type of hypothesis?

5.2. In DB14, we ask if the mean ages of patients with and without exercise-induced bronchoconstriction (EIB) are different. Write down the null and alternate hypotheses.

5.3. A clinical hypothesis arising from DB3 might be: The mean serum theophylline level is greater at the end of the antibiotic course than at baseline. (a) What probability distribution is associated with this hypothesis? (b) What assumptions about the data would be required to investigate this hypothesis? (c) State in words the Type I and Type II errors associated with this hypothesis. (d) How would the probability (risk) of these errors be designated? How would the power of the test be designated?

5.4. A clinical hypothesis arising from DB9 might be: The standard deviation of platelet counts is 60,000 which gives an approximate 95% confidence range of 240,000 [mean $\pm 2 \times 60{,}000$]); this hypothesis is equivalent to supposing that the variance is $60{,}000^2 = 3{,}600{,}000{,}000$. (a) What probability distribution is associated with this hypothesis? (b) What assumptions about the data would be required to investigate this hypothesis? (c) State in words the Type I and Type II errors associated with this hypothesis. (d) How would the probability (risk) of these errors be designated?

5.5. A clinical hypothesis arising from DB5 might be: The variances (or standard deviations) of plasma silicon level before and after implant are different. (a) What probability distribution is associated with this hypothesis?

(b) What assumptions about the data would be required to investigate this hypothesis? (c) State in words the Type I and Type II errors associated with this hypothesis. (d) How would the probability (risk) of these errors be designated?

5.6. In DB14, healthy subjects showed a mean decrease of 2.15 ppb exhaled nitric oxide (eNO) from before exercise to 20 minutes after exercise. A 5% significant test of this mean against a theoretical mean of 0 (paired *t* test) yielded $p = 0.085$. Make an argument for the two interpretations of this *p*-value discussed in Section 5.3.

5.7. Of what data type is: (a) The variable Respond versus Not Respond in DB6? (b) The variable Nausea Score in DB2? (c) The variable Platelet Count in DB9?

5.8. In DB10, rank the seconds to perform the triple hop for the operated leg, small to large.

5.9. From DB14, set up tables as in Section 5.4: (a) EIB frequency by sex; (b) for the six EIB patients, 5-minute eNO differences by sex as ranks; (c) for the six EIB patients, 5-minute eNO differences by sex as continuous measurements.

5.10. Using the 2×2 table in DB2, follow the first nine steps of Section 5.5 in setting up a test to learn if the drug reduces nausea score.

5.11. From DB13, we want to learn if our clinic results agree with those of the laboratory. Follow the first nine steps of Section 5.5 in setting up a test.

Chapter 6

Statistical Testing, Risks, and Odds in Medical Decisions

6.1. OVERVIEW

EACH DATA TYPE HAS A DIFFERENT FORM OF TEST

We have noted repeatedly that there are three types of data: categorical, rank-order, and continuous. Each requires its own form of statistical testing. This chapter introduces the unique character of each type and then explains and exemplifies the statistical test most frequently used for that type in medical studies.

CATEGORICAL FORM

To compare two variables using categorical data, we form two-way tables of counts, with one variable representing rows and the other representing columns. We test the proposition that knowledge of the counts in one variable's categories tells us something about the counts in the other variable's categories; that is, the two variables are not independent. Furthermore, if one variable represents a proposed event (diagnosis; acquiring a disease) and the other represents the known outcome of this event ("truth"), the table becomes a truth table, allowing risks and odds to be calculated, such as sensitivity and specificity, among others.

RANK-ORDER FORM

To compare two groups that are in rank-order form, we attach ranks to the data combined over the two groups and then add the rank values for each group separately, forming rank sums. If the group rankings are not much different, the rankings from the two groups will be interleaved and the rank sums will not be much different. If one group has most of its members preceding the other in rank, one rank sum will be large and the other small. Probabilities of rank sums have been tabulated, so the associated *p*-value can be found in the table and the decision about the group difference made.

CONTINUOUS FORM

Whether or not a difference between means exists most often is the focus in comparing two groups with data in continuous form. Our first inclination is to look at the difference between means. However, this difference depends on the scale. The offset distance of a broken femur appears larger if measured in centimeters than if measured in inches. The distance must be standardized into units of data variability. We divide the distance between means by a measure of variability and achieve a statistic (*z* if the population variability is known or closely estimated; *t* if it is estimated by small samples). The risk of concluding a difference when there is none (the *p*-value) is looked up in a table, and the decision as to the group difference is made.

6.2. CATEGORICAL DATA: BASICS

NOMINAL DATA: CATEGORIES AND COUNTS

In Table DB1.1, the variables DRE, TRU, and BIOP (digital rectal examination, transurethral ultrasound, and biopsy, respectively) can be quantified by associating their outcomes with 0 or 1. If we had recorded patient job occupation (e.g., construction worker, secretary, professional, cook), we could categorize "job" as A, B, C, D, ... or 0, 1, 2, 3, 4, ... Data of this sort that are placed into named categories (as opposed to being measured as a point on a scale or being ranked in order) usually are referred to as *categorical* or *nominal* data (see discussion of data types in Section 1.4). When categorical data are being used, the basic statistic, the *count*, is obtained by counting the number of events per category. In the Table DB1.1 data, there are five positive DREs and three positive BIOPs. The symbol for number of "successes" used in calculated formulas is *n* with the appropriate subscript. Unsubscripted *n* represents the total number summed over all categories.

PROPORTIONS AND PERCENTAGES

Another important statistic obtained from categorical data is the *proportion* of data in that category, which is the count in a category divided by the total number in the variable. The proportion of positive DREs is $5/10 = 0.50$ and that of positive BIOPs is $3/10 = 0.30$. Multiplying by 100 yields the percentage (denoted %), which in the case of BIOP is 30%. Percentage is useful in that most of the public is used to thinking in terms of percentages, but statistical methods have been developed for proportions. The probability distribution underlying binary (two-category) data is the binomial distribution (see Section 2.8). The symbol used for the sample proportion was *p*. To use the binomial distribution in a medical application, we need a theoretical or population proportion to specify the member of the family of binomials with which we are dealing. This theoretical proportion is symbolized π, in keeping with the convention of using Greek letters for population values and Roman letters for sample values.

CHOOSING COUNTS VERSUS PROPORTIONS

When should tables of counts be tested and when should proportions? When all cells of the table of counts can be filled in, use the methods for counts. Interestingly, the number of degrees of freedom (*df*) for tests of these tables is the minimum number of cells that must be filled to calculate the remainder of cell entries using the totals at the side and bottom. For example, for a 2×2 table with the sums at the side and bottom, only one cell needs to be filled and the others can be found by subtraction; it has 1 *df*. Tests of proportions are appropriate for cases in which certain table totals are missing but for which proportions can still be calculated or are given from other sources.

CATEGORIES VERSUS RANKS

Consider cases for which counts or proportions are obtained for each sample group. If the order in which these groups are written down can be swapped about without affecting the interpretation, for example, group 1: high hematocrit (Hct); group 2: high white blood cell (WBC) count; group 3: high platelet count, then analysis is constrained to categorical methods. If, however, sample groups fall into a natural order in which the logical implication would change by altering this order, for example, group 1: low Hct; group 2: normal Hct; and group 3: high Hct, rank methods are appropriate. Categorical groups may be formed by dividing up the scale on which continuous data occur. If we were to categorize age by decade (50–59, 60–69, and 70–79 years), we would have age groupings, which we could name 1, 2, and 3. These groups could be considered as categories and categorical methods used. However, they fall into a natural rank order, as group 1 clearly comes before group 2, and so on. Rank methods give better results than categorical methods. Although rank methods lose power when there are a large number of ties, they are still more powerful than categorical methods, which are not very powerful. *When ranking is a natural option, rank methods should be used.*

ORGANIZING DATA FOR CATEGORICAL METHODS

If the relationship between two variables is at question (e.g., are they independent or associated?), the number of observations falling simultaneously into the categories of two variables is counted. For example, the categories of the variables DRE and BIOP may be counted at the same time. This gives rise to the four possible categories, with a count for each: both positive, both negative, and two counts of one positive, one negative. These counts may be organized as shown in Table 6.1, with the counts from data Table DB1.1 filled in, where the DRE result is thought of as predicting the biopsy result. We can see that 3 is the number of negative DREs, contingent on also being a negative biopsy. Such tables are called *contingency tables*, because the count in each cell is the number

Table 6.1

Contingency Table of Simultaneous Biopsy and Digital Rectal Examination Results from Table DB1.1 Data

		DRE		
		1	0	Totals
BIOP	1	1	2	3
	0	4	3	7
	Totals	5	5	10

BIOP, biopsy; DRE, digital rectal examination.

in that category of that variable contingent on also lying in a particular category of the other variable.

SYMBOLS ARE NEEDED FOR CONTINGENCY TABLE COMPUTATIONS

Symbols to represent the numbers in tables such as Table 6.1 are needed for formulas, and they must be capable of being used in tables with more than four categories. For example, if we tried to predict BIOP by prostate-specific antigen (PSA) categories <4, 4–10, and >10, we would have a 2×3 table. To be general, r could denote the number of rows and c the number of columns; we then would have an $r \times c$ table. Because we do not know the values of r and c until we specify a particular case, we need to symbolize the cell numbers and the marginal (row and column) totals in a way that can apply to a table of any size.

SYMBOLS USED IN CONTINGENCY TABLES

The symbolism biostatisticians have almost been forced to use over the decades by practicality is n (for "number") with two subscripts: the first subscript indicates the row and the second, the column. Thus, the number of cases observed in row 1, column 2 is n_{12}. The row and column totals are denoted by placing a dot in place of the numbers we summed to get the total. Thus, the total of the first row would be $n_{1.}$, and $n_{.2}$ would be the total of the second column. The grand total can be denoted just n; sometimes $n_{..}$ is used. Table 6.2 repeats Table 6.1 with the various n symbols attached for illustration.

6.3. CATEGORICAL DATA: TESTS ON 2 × 2 TABLES

TESTING A CONTINGENCY TABLE WILL ANSWER THE QUESTION: ARE THE ROW CATEGORIES INDEPENDENT OF THE COLUMN CATEGORIES?

Less technically, does knowing the outcome of one of the variables give us any information about the associated outcome of the other variable? Another way to express the question would be to ask, are the data homogeneous?

Table 6.2

Table 6.2

Contingency Table of Simultaneous Biopsy and Digital Rectal Examination Results from Table DB1.1 Data[a]

		DRE		
		1	0	Totals
BIOP	1	$n_{11} = 1$	$n_{12} = 2$	$n_{1.} = 3$
	0	$n_{21} = 4$	$n_{22} = 3$	$n_{2.} = 7$
	Totals	$n_{.1} = 5$	$n_{.2} = 5$	$n = 10$

[a]n symbols are included to illustrate cell counts, marginal totals, and grand total.
BIOP, biopsy; DRE, digital rectal examination.

Table 6.3

Contingency Table of Simultaneous Biopsy and Digital Rectal Examination Results from 301 Patients[a]

		DRE		
		1	0	Totals
BIOP	1	$n_{11} = 68$	$n_{12} = 27$	$n_{1.} = 95$
	0	$n_{21} = 117$	$n_{22} = 89$	$n_{2.} = 206$
	Totals	$n_{.1} = 185$	$n_{.2} = 116$	$n = 301$

[a]n symbols are included to illustrate cell counts, marginal totals, and grand total.
BIOP, biopsy; DRE, digital rectal examination.

Is Prediction (Diagnosis) of Prostate Cancer by Digital Rectal Examination Better Than Chance?

Table 6.3 shows a contingency table in a format similar to that of Table 6.2 in which data for all 301 patients have been tabulated. Does knowing a patient's DRE help us in predicting whether or not he will have a positive biopsy, or would we do as well just to choose randomly? Let us conceive of a spinner in the center of a circle with circumference divided into 301 equal parts. We color a sector (slice of pie) of 95 parts red to match the 95 positive biopsies. To predict a patient's biopsy result randomly, we spin. If the spinner stops on red, we choose positive; otherwise, we choose negative. If the DRE adds no information, the ratio of positive biopsy prediction to total for a positive DRE should be about the same as the 95/301 ratio our spinner would give. We would say that the biopsy result is independent of DRE outcome.

The Chi-square Test of Contingency Is a Common Way to Test Independence

Different tests exist to answer the question of independence of the two variables. The time-honored test and the one seen most frequently in the medical literature is the *chi-square test of contingency*.

WE SHOULD UNDERSTAND WHERE THE METHOD COMES FROM

We can better understand the chi-square method if we survey the ideas underlying it. For the data of Table 6.3, the question may be expressed statistically as follows: Are the numbers of DRE positive and negative predictions, contingent on actual outcome, distributed the same as they would be without the outcome information, that is, as is the right-hand column (the right margin)?

EXPECTED VALUES

This last question gives rise to what is termed an *expected value*. If the occurrence of predicted positive cases is unrelated to biopsy outcome, the number of correct positive predictions we would *expect* to occur (call it e_{11} to match n_{11}) in ratio to all positive predictions ($e_{11}/n._1$) is the same as the number of either prediction for positive biopsy in ratio to total cases ($n_1./n$), or $e_{11}/n._1 = n_1./n$. By multiplying both sides by $n._1$, we find the *expected number* of correct positive predictions to be *its row sum multiplied by its column sum divided by the total sum*. In general, for row i and column j, the expected value of the ijth entry is given by

$$e_{ij} = \frac{n_i.n._j}{n}. \tag{6.1}$$

For example, the expected value for the top left position in Table 6.3 is $95 \times 185/301 = 58.4$, in contrast to the observed value of 68. The expected values e_{12}, e_{21}, and e_{22}, respectively, are 36.6, 126.6, and 79.4.

THE BASIS OF THE CHI-SQUARE TEST OF CONTINGENCY

The chi-square test of contingency is based on the differences between the observed values and those that would be expected if the variables were independent. If these differences are small, there is little dependence between the variables; large differences indicate a dependence. The actual chi-square statistic is the sum of these differences squared in ratio to the expected value. A small chi-square statistic arises if the observed values are close to the values we would expect if the two variables were unrelated. A large chi-square statistic arises if the observed values are rather different from those we would expect from unrelated variables. If the chi-square statistic is large enough that it is unlikely to have occurred by chance, we conclude that it is significant and that the rows variable is not totally independent of the columns variable. However, it does *not* follow that one can be well predicted by the other.

WE WOULD HOPE THAT DIGITAL RECTAL EXAMINATION IS NOT INDEPENDENT OF THE BIOPSY OUTCOME

This chi-square test should tell us if the DRE result is independent of the biopsy outcome. We would hope that it is not; the two should be closely related if we are to detect prostate cancer using the DRE. Let us detail the calculations in the test so we can answer this question.

CALCULATION FOR THE CHI-SQUARE TEST OF CONTINGENCY

The chi-square calculation is based on the difference between the observed cell count and the cell count that would be expected if the rows and columns were independent. The expected count for a cell is its row total times its column total divided by the grand total, as given in Eq. (6.1). After the four e_{ij} are determined, the chi-square is found using the simple pattern: sum of [(observed − expected)^2/expected]. The symbolic form of this pattern is given as

$$\chi^2 = \sum_i^2 \sum_j^2 \frac{(n_{ij} - e_{ij})^2}{e_{ij}}. \tag{6.2}$$

One $\sum$ symbol tells us to add across the two rows, and the other tells us to add over the two columns. The p-value can be found by calculation or from tables of chi-square probabilities.

YATES' CORRECTION

The reader should note that sometimes the chi-square statistic is calculated as the sum of [(|observed − expected| − 0.5)^2/expected], where the 0.5 term, called Yates' correction, is subtracted to adjust for the counts being restricted to integers. It was used previously to provide a more conservative result for contingency tables with small cell counts. Currently, Fisher's exact test provides a better solution to dealing with small cell counts and is preferred. For larger cell counts, Yates' correction alters the result negligibly and may be ignored. Thus, the chi-square form Eq. (6.2) is used in this book.

THE CHI-SQUARE TEST FOR THE DIGITAL RECTAL EXAMINATION VERSUS BIOPSY DATA

Let us test the null hypothesis that DRE result and biopsy result are independent; that is, knowledge of the DRE result tells us nothing about the biopsy result. The expected values, e_{ij}, were noted just following Eq. (6.1). The chi-square statistic is calculated $\chi^2 = [(68 - 58.4)^2/58.4] + \cdots + [(89 - 79.4)^2/79.4] = 5.998$. Table III (right tail of the chi-square distribution; see Tables of Probability Distributions) will tell us if this chi-square value is significant. A fragment of Table III is shown as Table 6.4. The statistic 5.98 is between 5.02 and 6.63, associated with $\alpha = 0.025$ and 0.01, respectively; therefore, the p-value is about 0.015, a statistically significant result. (Exact calculation on a computer tells us it is $p = 0.014$.) We reject the null hypothesis of independence and conclude that the DRE result does indeed give us some information about the presence of prostate cancer. Note the wording: It gives us some information; it does not tell us how good this information is.

Table 6.4

A Fragment of Table III, the Right Tail of the Chi-square Distribution[a]

α (area in right tail)	0.10	0.05	0.025	0.01
$df = 1$	2.71	3.84	5.02	6.63

[a] χ^2 values (distances to the right of 0 on a χ^2 curve) are tabulated for 1 *df* for four values of α. *df*, degrees of freedom.

ASSUMPTION REQUIRED FOR THE CHI-SQUARE TEST OF CONTINGENCY

The chi-square method should not be used for extremely small samples. It really is an approximation to more exact methods (see Chapter 15 for a more detailed discussion of methods), and the approximation is not dependable for small samples. In particular, it is valid *only* if the expected value (row sum × column sum ÷ total sum) of every cell is at least 1 and a minimum count of 5 appears in every cell.

THE EFFECT OF SAMPLE SIZE ON CONTINGENCY TESTS

A significant chi-square result indicates that the types of category are not independent, but it does not indicate how closely they are associated, because the significance is influenced by both sample size and association. To illustrate this, Table 6.5 has each cell count in Table 6.3 replaced by one-fifth of its value, rounded to maintain integers. We have approximately the same pattern of proportions, which gives about the same level of association (37% correct positive prediction for the full sample and 38% correct positive prediction for the smaller one). However, now the chi-square has shrunk from 5.98 to 1.70. From Table 6.4, we can see that $\chi^2 = 2.71$ is associated with $\alpha = 0.10$, so the *p*-value for $\chi^2 = 1.70$ will be greater than 0.10. (Calculating it directly yields $p = 0.192$.) Reduction of the sample size while keeping the same general pattern of relationship between the variables has caused the *p*-value to grow from a significant 0.014 to a nonsignificant 0.192. Contingency test results must be interpreted with care.

Table 6.5

Contingency Table of Simultaneous Biopsy and Digital Rectal Examination Results with Each Cell Showing One-Fifth the Count Arising from the 301 Patients[a]

		DRE		
		1	0	Totals
BIOP	1	$n_{11} = 14$	$n_{12} = 5$	$n_{1.} = 19$
	0	$n_{21} = 23$	$n_{22} = 18$	$n_{2.} = 41$
	Totals	$n_{.1} = 37$	$n_{.2} = 23$	$n = 60$

[a] New cell counts are rounded to integers. The pattern of proportions is approximately maintained, but the *p*-value loses its significance because of the smaller sample size.
BIOP, biopsy; DRE, digital rectal examination.

WHAT DO WE DO IF WE HAVE THE PERCENTAGES BUT NOT THE COUNTS?

Just note that tests of proportion exist, proportion being just percent with the decimal point moved two places. We can test a sample proportion against either a known population proportion or another sample proportion. For example, suppose we are concerned with the efficacy of an antibiotic in treating pediatric otitis media accompanied with fever. We know that, for example, 85% of patients get well within 7 days without treatment. We learn that of 100 such patients, 95% of those treated with the antibiotic got well within 7 days. Is this improvement statistically significant? Note that we cannot fill in a contingency table because we do not have the number of untreated patients or the total number of patients. However, we do have the theoretical proportion and therefore can answer the question using a test of proportions (see Chapter 15 for a more detailed discussion of this topic).

6.4. CATEGORICAL DATA: RISKS AND ODDS

TRUE AND FALSE POSITIVE AND NEGATIVE EVENTS

A special case of the contingency table is the case in which one variable represents a prediction that a condition will occur and the other represents the outcome, that is, "truth." Suppose we are predicting the occurrence of a disease among a sample of patients by means of a clinical test (or, alternatively, from the fact of exposure or nonexposure to a disease). If we also know the outcomes of the test (or occurrence of the disease), we can count the possible relations between predictions and outcomes and array them as n_{11} through n_{22} in a 2×2 *truth table*, as illustrated in Table 6.6. Four possible situations are named in the table: (1) *true positive*, the event of a predicted disease being present; (2) *false negative*, the event of predicting no disease when disease is present; (3) *false positive*, the event of a predicted disease being absent; and (4) *true negative*, the event of predicting no disease when disease is absent. These four concepts are used a great deal in several fields of clinical medicine and should be noted well. The *n* values are the counts of these events' occurrences and are considered extensively in the remainder of this section.

Table 6.6

Truth Table Showing Counts of the Prediction of Presence or Absence of a Malady as Related to the Truth of That Presence or Absence

		Prediction (exposure or test result)		
		Have disease	Do not have disease	
TRUTH	Have disease	n_{11} true positive	n_{12} false negative	$n_1.$(Yes)
	Do not have	n_{21} false positive	n_{22} true negative	$n_2.$(No)
		$n._1$ (Predict Yes)	$n._2$ (Predict No)	n (or $n..$)

Table 6.7

Table of Biopsy Results Predicted by Digital Rectal Examination and by Prostate-Specific Antigen Density > 0.14 as Related to the True Outcomes of the Biopsies for 301 Patients

		Predicted by DRE			Predicted by PSAD > 0.14		
		+	−		+	−	
	+	68 (n_{11})	27 (n_{12})	95	75 (n_{11})	20 (n_{12})	95
Truth (biopsy)	−	117 (n_{21})	89 (n_{22})	206	88 (n_{21})	118 (n_{22})	206
		185	116	301	163	138	301

DRE, digital rectal examination; PSAD, prostate-specific antigen density.

EXAMPLE DATA

In Table 6.7, the format of Table 6.6 is used to show DRE and PSAD values of the 301 prostate patients in predicting biopsy results. The DREs are symbolized + for predicted positive biopsy and − for predicted negative biopsy. By using a PSAD value of 0.14 as the critical value (decision criterion), PSAD >0.14 predicts a positive biopsy and PSAD ≤0.14 predicts a negative biopsy.

PROBABILITY AND ODDS DEFINED

The probability (defined in Section 2.2) of an event is estimated by the number of ways the event in question can occur in ratio to the number of ways any event can occur. If r of n patients have a disease, the chance that one of those patients chosen randomly has the disease is r/n. The *odds* that the randomly chosen patient has the disease is the number of ways the event can occur in ratio to the number of ways it does not occur: $r/(n - r)$. For example, the probability of a one-spot on the roll of a die is 1/6; the odds are 1-to-5.

SENSITIVITY, SPECIFICITY, ACCURACY, AND ODDS RATIO

Definitions of a number of concepts related to estimates of correct and erroneous predictions from the sample truth table are given in Table 6.8. Column 1 gives the names of the concepts, and column 2 gives the formulas for computing them from the sample truth table. Columns 3 and 4 give population definitions and the relationships these concepts are estimating, that is, what they would be if all data in the population were available. These are the values that relate to error probabilities in statistical inference, as discussed in Sections 5.1 and 5.2. As before, a vertical line (|) is read "given." Columns 5 and 6 show numerical estimates of the probabilities and odds in predicting biopsy results from DRE and PSAD values calculated from the counts displayed in Table 6.7. Interpretations of these outcomes appear later. (A number of other related concepts, e.g., positive and negative predictive values, appear in Chapter 15.)

FALSE-POSITIVE RATE AND FALSE-NEGATIVE RATE

The false-positive rate is the relative frequency of not cancer when cancer is predicted. This value is the *p*-value of the test, the posttest estimate of α. The DRE estimates this chance as 117/206 = 57%, and PSAD, as 88/206 = 43%.

Table 6.8

Concepts of False Positive, False Negative, Sensitivity, Specificity, Accuracy, and Odds Ratio Based on Sample Error Rates Arising from the Format of Table 6.6[a]

Names of estimates	Values found from a sample truth table	Population definitions of probabilities and odds of values being estimated		Examples of prediction	
	Computing formulas for samples	Definitions	Relationships	By DRE	By PSAD
False-positive sample rate (p-value)	$n_{21}/n_{2.}$	Probability of a false positive (α)	P(predict yes | no)	0.568	0.427
False-negative sample rate	$n_{12}/n_{1.}$	Probability of a false negative (β)	P(predict no | yes)	0.284	0.210
Sensitivity	$n_{11}/n_{1.}$	Probability of a true positive: $power\ (1-\beta)$	P(predict yes | yes)	0.716	0.790
Specificity	$n_{22}/n_{2.}$	Probability of a true negative $(1-\alpha)$	P(predict no | no)	0.433	0.573
Accuracy	$(n_{11}+n_{22})/n_{..}$	Overall probability of a correct decision	P(predict no | no or yes | yes)	0.522	0.641
Odds ratio	$\dfrac{n_{11}/n_{21}}{n_{12}/n_{22}}$ or $n_{11}n_{22}/n_{12}n_{21}$	Odds of a disease when predicted in ratio to odds of the disease when not predicted	$\dfrac{ratio(Yes/No \mid predicted\ yes)}{ratio(Yes/No \mid predicted\ no)}$	1.916	5.028

[a]The third and fourth columns show the population entities these values are estimating. The last two columns show biopsy outcome predictions of these values from the data of Table 6.7.

DRE, digital rectal examination; PSAD, prostate-specific antigen density.

The false-negative rate is the relative frequency of cancer when not cancer is predicted, estimated as $27/95 = 28\%$ by DRE and $20/95 = 21\%$ by PSAD. It is the posttest estimate of β.

SENSITIVITY

The sensitivity, $1 - $ (false-negative rate), is the sample estimate of the chance of detecting cancer when it is present. Sensitivity is the sample estimate of the frequently met *power* of the decision and is related to power in the same way that the *p*-value is related to α. The sensitivity of DRE and PSAD is $68/95 = 72\%$ and $75/95 = 79\%$, respectively.

SPECIFICITY

The specificity, $1 - $ (false-positive rate), is the posttest estimate of the chance of correctly classifying the patient as free of cancer, that is, ruling out cancer when it is absent. The specificity of DRE and PSAD is $89/206 = 43\%$ and $118/206 = 57\%$, respectively. From the results so far, we see that PSAD has both a higher sensitivity and a higher specificity than DRE.

ACCURACY

Sometimes the interest is in the rate of overall accuracy, that is, the total percentage correct, combining the true-positive and true-negative results. Accuracy is calculated as true-positive plus true-negative counts, divided by the total count. The accuracy of DRE is $(68 + 89)/301 = 52\%$, and the accuracy of PSAD is $(75 + 118)/301 = 64\%$. However, this definition of accuracy assumes the errors of a false positive and a false negative have the same weight, that is, the same importance in the decision process, which they often do not. For example, a false negative in a test of carcinoma implies a missed cancer, whereas a false positive implies an unnecessary biopsy; clearly the two errors are not equal in severity. Accuracy must be used judiciously and interpreted carefully.

ODDS RATIO

The odds ratio (OR) gives the odds of cancer when predicted in ratio to the odds of cancer when not predicted, indicating the usefulness of the prediction method. The PSAD OR of $(75 \times 118) \div (20 \times 88) = 5.03$ indicates that the PSAD odds of cancer when predicted is 5 times its odds when not predicted. The PSAD OR is more than 2.5 times the DRE OR of $(68 \times 89) \div (27 \times 117) = 1.92$, indicating that, for these data, the PSAD is a much better predictor of prostate cancer than DRE. The OR gives us some indication of the level of association between the two variables; tests of association exist but are not addressed in this section (if curious, see Chapter 15 for more details).

6.5. RANK DATA: BASICS

WHAT ARE RANKS?

Ranked data are data entries put in order according to some criterion: smallest to largest; worst to best; cheapest to most expensive. Such rank-order data may arise from ranking events directly, such as a surgeon ranking in order of difficulty the five types of surgery performed most often in his specialty. Alternatively, rank-order data may arise from putting already recorded continuous-type quantities in order, such as ordering the PSA of Table DB1.1 from smallest to largest. These latter ranks would appear as follows:

Patient number:	1	2	3	4	5	6	7	8	9	10
PSA:	7.6	4.1	5.9	9.0	6.8	8.0	7.7	4.4	6.1	7.9
PSA rank:	6	1	3	10	5	9	7	2	4	8

RANKING CATEGORIZED DATA THAT FALL INTO A NATURAL ORDER

Note that if PSA values were categorized into (1) PSA <4, (2) PSA 4–10, and (3) PSA >10, the data could still be ranked and analyzed using rank-order methods, although there would be so many ties that the analysis would be much less sensitive.

WHEN DO WE USE RANKS?

Continuous measurements contain more information than ranks, and ranks contain more information than counts. When the user can rank events but cannot measure them, it is obvious that rank-order statistical methods are to be used. The question of when and why to use ranks arises primarily with using rank-order methods for data on which continuous measurements are available.

WHY DO WE USE RANKS?

Statistical methods using continuous measurements on variables depend on the probability distributions of those variables. We assume certain properties of those probability distributions, such as that we are sampling from a normal distribution. When we have small sample sizes, our sample frequency distributions are insufficient to tell us if the assumptions are justified. Rank-order methods do not require as stringent assumptions about the underlying distributions. Furthermore, even when we have larger samples, we may have evidence that the assumptions are not satisfied. Recall that the distribution of prostate volumes (compare Fig. 2.1) is skewed to the right. Using methods that assume a normal distribution would violate this assumption, because the sample distribution is not the same shape as the assumed distribution, causing the probability calculations to be wrong. Rank-order methods, not subject to the skew, would not violate the assumption.

SUMMARY OF WHEN TO USE RANK-ORDER METHODS

Rank-order methods should be used (1) when the primary data consist of ranks, (2) when samples are too small to form conclusions about the underlying probability distributions, or (3) when data indicate that necessary assumptions about the distributions are violated.

TIES IN RANKED DATA

If ranks are tied, the usual procedure is to average the tied ranks. For example, a sample of heart rates in increasing order is 64, 67, 67, 71, 72, 76, 78, 89. We have the ranks 1 through 8 to assign. However, the second and third heart rates are the same. We average 2 and 3 to assign 2.5 to each. The ranks are 1, 2.5, 2.5, 4, 5, 6, 7, 8. Some statisticians prefer to assign the potential ranks to tied values randomly to avoid ties, but this technique introduces a bit of false information. Ties disturb the theory of rank methods somewhat, but they are still approximately correct.

6.6. RANK DATA: THE RANK-SUM TEST TO COMPARE TWO SAMPLES

WHAT IS BEING TESTED

Given two samples, the hypothesis being tested is whether or not the value for a randomly chosen member of the first sample is probably smaller than one of the second sample, a slightly technical concept. For practical purposes, the user may think of it informally as testing whether or not the two distributions have the same median.

STEPS IN CONDUCTING THE TEST

Steps for conducting the rank-sum test are given in the left-hand column that follows. In the parallel position in the right-hand column, a numerical example is given.

Example

Returning to Table DB1.1, we ask, are the PSA levels different for the $n_1 = 3$ positive biopsy results and the $n_2 = 7$ negative ones? (In this test, the symbol n_1 is always assigned the smaller of the two n's.) The PSA levels, their ranks (small to large), and the biopsy results are as follows:

PSA:	7.6	4.1	5.9	9.0	6.8	8.0	7.7	4.4	6.1	7.9
Ranks:	6	1	3	10	5	9	7	2	4	8
Biopsy:	0	0	1	1	1	0	0	0	0	0

Steps in method:

(1) Satisfy yourself that the sample has been drawn such that it represents the population and such that observations are independent from one another. This step is pure judgment based on the way the data have been collected. If these requirements are violated, statistics will not help.

(2) Specify α and hypotheses, which usually are null—distributions are not different—and alternate—distributions are different.

(3) Name the sample sizes n_1 and n_2; n_1 is the smaller.

(4) Combine the data, keeping track of the sample from which each datum arose.

(5) Rank the data.

(6) Add the ranks of the data from the smaller sample and name it T.

(7) If $n_2 \leq 8$, calculate $U = n_1 n_2 + n_1(n_1 + 1)/2 - T$. If $n_2 > 8$, calculate $\mu = n_1(n_1 + n_2 + 1)/2$, $\sigma^2 = n_1 n_2(n_1 + n_2 + 1)/12$, and $z = (T - \mu)/\sigma$.

(8) Look up the p-value from the appropriate table: Table IX for U or Table I for z.

(9) Reject the null hypothesis if $p < \alpha$; do not reject the null hypothesis if $p \geq \alpha$.

Steps in example:

(1) We ask if PSA is a risk factor for prostate cancer. Do our urologic patients with cancer have PSA levels different from those without cancer? We judge that the data are independent and that the sample is adequately representative.

(2) $\alpha = 0.05$. Hypotheses are as at left.

(3) $n_1 = 3$; $n_2 = 7$.

(4) First and third rows in preceding data display.

(5) Second row in preceding data display.

(6) The rank sum T for positive biopsies is $3 + 10 + 5 = 18$.

(7) $n_2 = 7$; find $U = 3(7) + 3(4)/2 - 18 = 9$.

(8) Table 6.9 gives the portion of Table IX including $n_2 = 7$, $n_1 = 3$, and U varies from 7 to 11. The p-value for $U = 9$ is 0.834.

(9) p is much greater than α. We have no significant evidence that the PSA distributions are different for positive and negative biopsy results.

Table 6.9

Portion of Table IX, Rank-Sum U Two-Tailed Probabilities

	n_2:			6					7				
	n_1:	1	2	3	4	5	6	1	2	3	4	5	6	7
	7			0.714	0.352	0.178	0.094			0.516	0.230	0.106	0.052	0.026
	8			0.904	0.476	0.246	0.132			0.666	0.316	0.148	0.074	0.038
U	9				0.610	0.330	0.180			0.834	0.412	0.202	0.102	0.054
	10				0.762	0.428	0.240			1.00	0.528	0.268	0.138	0.072
	11				0.914	0.536	0.310				0.648	0.344	0.180	0.098

For two samples of size n_1 and $n_2 (n_2 > n_1)$ and the value of U, the entry gives the p-value.

WHY U IS CALCULATED FROM T

The Mann–Whitney U is tabulated in this book rather than the rank-sum T, because U requires a much smaller table size.

A ONE-TAILED TEST

If we have some sound nonstatistical reason why the result of the test must lie in only one tail, such as a physiologic limit preventing the other tail (a knee does not bend forward), we can halve the tabulated value to obtain a one-tailed *p*-value.

THE NAME OF THIS TEST

This test may be referred to in the literature as the rank-sum test, Mann–Whitney U test, Wilcoxon rank-sum test, or Wilcoxon–Mann–Whitney test. Mann and Whitney published what was thought to be one test, and Wilcoxon published another. Eventually, they were seen to be only different forms of the same test.

6.7. CONTINUOUS DATA: BASICS OF MEANS

THE MEAN IS STUDIED MORE THAN OTHER CHARACTERISTICS

The mean is the most important characteristic of a population, but certainly not the only one. The standard deviation and various aspects of the shape are often crucial for reasons discussed later.

THE MOST FREQUENT QUESTION: ARE TWO MEANS THE SAME?

The basic question asked is either if a sample mean is the same as a population mean, or if two samples have the same mean. The question is answered by testing the null hypothesis that the means are equal and then accepting or rejecting this hypothesis. This concept is discussed in Sections 5.1 and 5.2.

THE ROBUSTNESS CONCEPT

Tests of means were developed under the assumption that the sample was drawn from a normal distribution. Whereas usually not truly normal, a distribution that is roughly normal in shape is adequate for a valid test. That is because the test is moderately *robust*. Robustness is an important concept. A robust test is one that is affected little by deviations from the underlying assumptions. If a small-to-moderate sample is too far from normal in shape, the calculation of error probabilities, based on the assumed distribution, will lead to erroneous decisions; use rank-order methods. If a large sample is too far from normal, a statistician may be able to find a suitable data transformation to reshape the distribution, such as a logarithm, square, square root, or other transformation. In particular, tests of means are only moderately robust and are especially sensitive to outliers, whereas tests of variance are much more robust.

OTHER ASSUMPTIONS: INDEPENDENTLY SAMPLED DATA AND EQUAL VARIABILITY

A normal shape to the frequency distributions is not the only assumption made in tests of means. Together with most other tests of hypotheses, means tests assume that the data being used are independent of each other and that the standard deviations are the same. Independence implies that knowledge of the value of one datum will provide no clue as to the value of the next to be drawn. An example of violating this assumption would be pooling repeated blood pressure measurements on several patients with hypertension. Knowledge of a pressure for one patient would give some information about what pressure would be expected from the next reading if it arose from the same patient, but it would give no information if the patient were different; thus, the readings are not all independent. An example of different standard deviations might arise upon comparing WBC counts from a sample of healthy patients with those from a sample of patients with bacterial infections. The infected sample might be much more variable than the healthy sample. Techniques exist to adjust for unequal standard deviations, but not much can be done to salvage data with dependencies. Independent data and equal standard deviations are assumed in the remainder of this chapter.

THE ALTERNATE HYPOTHESIS MUST SPECIFY A ONE- OR TWO-TAILED TEST

The null hypothesis states that the mean of the population from which the sample is drawn is not different from a theorized mean or from the population mean of another sample. We must also choose the alternate hypothesis, which will select a two-tailed or a one-tailed test. We should decide this before seeing the data so that our choice will not be influenced by the outcome. We often *expect* the result to lie toward one tail, but expectation is not enough. If we are sure the other tail is impossible, such as for physical or physiologic reasons, we unquestionably use a one-tailed test. Surgery to sever adhesions and return motion to a joint frozen by long casting will allow only a positive increase in angle of motion; a negative angle physically is not possible. A one-tailed test is appropriate. There are cases in which an outcome in either tail is possible, but a one-tailed test is appropriate. When making a decision about a medical treatment, that is, whether or not we will alter treatment depending on the outcome of the test, the possibility requirement applies to the alteration in treatment, not the physical outcome. If we will alter treatment only for significance in the positive tail and it will in no way be altered for significance in the negative tail, a one-tailed test is appropriate.

6.8. CONTINUOUS DATA: NORMAL (z) AND t TESTS TO COMPARE TWO SAMPLE MEANS

THE NORMAL TEST AND THE t TEST ARE TWO FORMS OF THE TWO-SAMPLE MEANS TEST

This section concentrates on the type of means test most frequently seen in the medical literature: the test for a difference between two means. We will look at two subclasses: the case of known population standard deviations, or samples large enough that the sample standard deviations are not practically different from known, and the case of small-sample estimated standard deviations. The means test uses a standard normal distribution (z distribution) in the first case and a t distribution in the second, for reasons discussed in Section 2.8. (Review: The means are assumed normal. Standardizing places a standard deviation in the denominator. If the standard deviation is known, it behaves as a constant and the normal distribution remains. If the standard deviation is estimated from the sample, it follows a probability distribution and the ratio of the numerator's normal distribution to this distribution turns out to be t.)

THE STEPS TO FOLLOW: EXAMPLE OF z TEST

The steps to follow for either the z or the t test are rather straightforward. In the following list, these steps are given in the left column with an example in the right column.

Steps in method:

(1) Satisfy yourself that the sample has been drawn such that it represents the population and such that observations are independent from one another. This step is pure judgment based on the way the data have been collected. If these requirements are violated, statistics will not be helpful.

(2) Make quick frequency plots of the two samples' basic data to check the normality and equal standard deviation assumptions. If one or both are violated, use rank-order methods.

(3) Choose the z test or t test as appropriate. You have n data split between two samples of size n_1 and n_2. If σ (the population standard deviation under the null hypothesis) is known or if n is large (>50 or 100), use z. If σ is unknown and n is small, use t.

(4) Specify null and alternate hypotheses. The null hypothesis usually will be H$_0$: $\mu_1 = \mu_2$. Select the alternate

Steps in example, normal (z) test:

(1) We ask if age is a risk factor for prostate cancer. Are our urologic patients with cancer older than those without cancer? We judge that the data are independent and the sample is adequately representative.

(2) Figure 6.1 shows a plot of the two frequency distributions. They appear adequately normal in shape with equivalent standard deviations.

(3) Of $n = 301$, $n_1 = 206$ biopsy results were negative and $n_2 = 95$ were positive. The samples are large enough to use the z test. From the data, we calculate $m_1 = 66.59$ years, $s_1 = 8.21$ years, $m_2 = 67.14$ years, and $s_2 = 7.88$ years.

(4) H$_0$: $\mu_1 = \mu_2$. Because it is possible for patients with positive biopsy results to be either older or younger (although

as $H_1: \mu_1 \neq \mu_2$ for a two-sided test, or $H_1: \mu_1 < \mu_2$ or $H_1: \mu_1 > \mu_2$ for a one-sided test.

(5) Choose an appropriate α and look up the associated critical value from Table I (z) or Table II (t) with $n_1 + n_2 - 2df$. For a two-sided test, use the "two-tailed α" heading. For a one-sided test, use "one-tailed α."

(6) Calculate as appropriate the z or t statistic. The test is in the form of a standardized difference between means, that is, the difference of $m_1 - m_2$ divided by the standard error σ_d or s_d. Calculate z by

$$z = (m_1 - m_2)/\sigma_d, \qquad (6.3)$$

where

$$\sigma_d = \sigma \sqrt{\frac{1}{n_1} + \frac{1}{n_2}}, \qquad (6.4)$$

or t by

$$t = (m_1 - m_2)/s_d, \qquad (6.5)$$

where

$$s_d = \sqrt{\left(\frac{1}{n_1} + \frac{1}{n_2}\right)\left[\frac{(n_1-1)s_1^2 + (n_2-1)s_2^2}{n_1 + n_2 - 2}\right]}. \qquad (6.6)$$

(7) Reject or do not reject the null hypothesis depending on where the statistic lies relative to the critical value.

(8) If a statistical software package is available, the actual p-value may be calculated to facilitate further interpretation of the decision (as in Section 5.2).

we might not expect younger), we choose a two-sided alternate, $H_1: \mu_1 \neq \mu_2$.

(5) We choose $\alpha = 0.05$. Table 6.10 shows a fragment of Table I, the normal distribution. Under "Two-tailed applications," for $\alpha = 0.05$, we find the critical $z = 1.96$.

(6) We want to use the z form. From the data set statistics in the first table of DB1, the standard deviation for the 301 ages is 8.10; we take it as σ. Then from Eq. (6.5), $\sigma_d = 8.10 \times \sqrt{\frac{1}{206} + \frac{1}{95}} = 1.00$, and from Eq. (6.4), $z = (m_1 - m_2)/\sigma_d = (66.59 - 67.14)/1 = -0.55$.

(7) The statistic $z = -0.55$ is well within the null hypothesis acceptance bounds of ± 1.96. We do not reject the null hypothesis.

(8) $p = 0.582$. We would be more likely wrong than not to conclude a difference. We feel justified in dismissing this question and not revisiting it with larger samples.

COULD THE t TEST HAVE BEEN USED INSTEAD OF THE z TEST?

What would have been the effect of using the t test instead of the normal test of means in examining the age difference between patients with positive and negative biopsy results? Is the t test appropriate? It is. The normal test assumes that the variances are obtained from the entire population, which often is an infinite number, so the t really is an approximation to the normal.

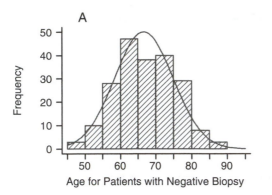

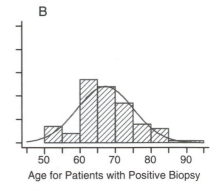

Figure 6.1 A plot of age distributions by biopsy result for 301 patients: (A) negative biopsy result (B) positive biopsy result. Normal curves fitted to the data are superposed.

Table 6.10

Fragment of Table I, Normal Distribution

z (no. standard deviations to right of mean)	One-tailed applications		Two-tailed applications	
	One-tailed α (area in right tail)	1 − α (area except right tail)	Two-tailed α (area in both tails)	1 − α (area except both tails)
1.90	0.029	0.971	0.054	0.946
1.960	*0.025*	*0.975*	*0.050*	*0.950*
2.00	0.023	0.977	0.046	0.934

Table 6.11

Fragment of Table II, *t* Distribution

One tailed α (right tail area)	0.10	0.05	0.025	0.01	0.005	0.001	0.0005
Two-tailed α (area both tails)	0.20	0.10	0.05	0.02	0.01	0.002	0.001
$df = 40$	1.303	1.684	2.021	2.423	2.704	3.307	3.551
60	1.296	1.671	2.000	2.390	2.660	3.232	3.460
100	1.290	1.660	1.984	2.364	2.626	3.174	3.390
∞	1.282	1.645	1.960	2.326	2.576	3.090	3.291

df, degrees of freedom.

By comparing Tables I and II, we can see that the critical *t*-values for ∞ *df* are the same as the normal values. Let us compare the critical values for a two-tailed α of 5%.

The normal critical value is 1.96. With samples of 206 negative biopsy results and 95 positive results, $df = (206 − 1) + (95 − 1) = 299$. We want to use Table II, *t* distribution, a fragment of which is shown as Table 6.11. Probabilities for 299 *df* will lie between the rows for 100 *df* and an unmeasurably large *df* (symbolized by infinity, ∞). Under the column for two-tailed α = 0.05, the *t* critical value will lie between 1.98 and 1.96, or about 1.97, quite similar to the 1.96 critical value for the normal. The following section illustrates the steps in performing the two-sample *t* test for these data.

FOLLOWING THE STEPS FOR THE *t* TEST

The first four steps and the choice of α are identical to the z test for these data. Finding the critical value in the fifth step was addressed in the previous paragraph; our critical value is 1.97.

Step six is actual calculation. The formula in Eq. (6.7) for the standard error of the mean in the case of small samples gives

$$s_d = \sqrt{\left(\frac{1}{n_1} + \frac{1}{n_2}\right)\left[\frac{(n_1 - 1)s_1^2 + (n_2 - 1)s_2^2}{n_1 + n_2 - 2}\right]}$$

$$= \sqrt{\left(\frac{1}{206} + \frac{1}{95}\right)\left[\frac{205 \times 8.21^2 + 94 \times 7.88^2}{206 + 95 - 2}\right]}$$

$$= 1.01.$$

t is then the simple calculation of Eq. (6.6):

$$t = (m_1 - m_2)/s_d = (66.59 - 67.14)/1.01 = -0.54.$$

For the z test, the z statistic was -0.55, which is well within the H_0 acceptance region of ± 1.96. Similarly, to follow step seven for the t test, the t statistic falls well within the H_0 acceptance region of ± 1.97. We note that, for samples of more than 100, the difference between the methods is negligible. Indeed, on calculating the exact p-value, we find it to be 0.582, which is identical to that for the z test.

6.9. OTHER TESTS OF HYPOTHESES

TESTS OF HYPOTHESES EXIST FOR MANY OTHER RESEARCH QUESTIONS

In this chapter, an understanding of the concepts of testing is at issue, not use of the right method for the reader's specific problem. A wide variety of tests exists for various types of data (counts, ranks, measurements), various sample groupings (paired vs. unpaired data; groups of one, two, and more than two), and various distributional characteristics at question (proportions, order precedence, means, variability, shape). The reader who completes Part I of this book and seeks to find the appropriate test for a particular question will find some guidance in Table 10.1 and will be taken through a number of specific methods in Chapters 15 through 20.

CHAPTER EXERCISES

6.1. In DB4, test the contingency table using chi-square to decide whether or not access to protease inhibitors has reduced the rate of pulmonary admissions.

6.2. In DB6, test the contingency table using chi-square to decide whether or not titanium-containing tattoo ink is harder to remove than other inks.

6.3. In DB2, test the 2×2 contingency table using chi-square to decide whether or not nausea score is reduced by the drug.

6.4. In DB13, combine below-range with above-range result categories (last two columns) to form an out-of-range category. Form an in-range versus out-of-range 2×2 contingency table and use χ^2 to decide if clinic and laboratory result categories were discrepant.

6.5. A general surgeon noted a number of errors in his hospital's diagnosis of appendicitis. Of the next 200 patients with abdominal pain, 104 turned out to have had appendicitis and 96 turned out not to have had appendicitis. Of those with appendicitis, 82 were correctly diagnosed; of those without appendicitis, 73 were correctly diagnosed. (a) Form a truth table. Calculate and state the interpretation of (b) the false-positive rate, (c) the false-negative rate, (d) sensitivity, (e) specificity, (f) accuracy, and (g) the OR for this hospital's clinical diagnosis of appendicitis.

6.6. Using the results of Exercise 6.4, term a patient in range as negative and out of range as positive. The laboratory results are taken as "truth," so the properly formatted result forms a truth table. For the clinic's test results, calculate and state the interpretation of (a) the false-positive rate, (b) the false-negative rate, (c) sensitivity, (d) specificity, (e) accuracy, and (f) the OR.

6.7. In DB3, calculate the difference: serum theophylline level at baseline minus at 5 days; assign ranks to this new variable and perform the rank-sum test to learn if there is evidence that the reduction in level caused by the drug differs between men and women.

6.8. In DB7, rank bone density and perform the rank-sum test to learn if there is evidence that the bone density is different between men and women.

6.9. Using DB14, plot the frequency distributions of the 5-minute exhaled nitric oxide (eNO) differences for male and female subjects. Is this distribution close enough to normal in shape to use a *t* test? Rank the differences and use the rank-sum test to learn if there is a response difference between sexes. (Because $n_2 > 8$, use the formulas in step 7 of the example in Section 6.6.)

6.10. In DB12, the sample is large enough to take the sample standard deviation as if it were a population standard deviation σ. For convenience, carry the calculations to only two decimal places. Perform a normal (z) test to learn if there is evidence that (a) mean age and (b) extent of carinal resection is different for the patients who survived versus those who died.

6.11. In DB3, use the difference between theophylline levels at baseline minus at 5 days and perform a *t* test to learn if there is evidence that the mean reduction in level caused by the drug differs between men and women. Compare the result with the result of Exercise 6.5.

6.12. In DB7, perform a *t* test to learn if there is evidence that the mean bone density is different between men and women. Compare the result with the result of Exercise 6.6.

Chapter 7

Sample Size Required for a Study

7.1. OVERVIEW

THE BIGGER THE SAMPLE, THE STRONGER THE STATISTICAL CONCLUSIONS

How large a sample do we need? Speaking statistically, *the larger the better*. Larger samples provide better estimates, more confidence, and smaller test errors. The best way to choose a sample size is to take *all the data* our time, money, support facilities (hospital, animal laboratories), and ethics of patient use will permit. Then why all the attention to methods for estimating the minimum sample size required? The primary purpose is to *verify that we will have enough* data to make the study worthwhile. If we can manage only 50 subjects and the sample size requirement methods show we need 200, we should not undertake the study. Another purpose arises in cases in which cost (funds and/or resources) or subject ethics mandates a minimum sample size; that is, to ensure that we do not sample many more than required to answer the question the study poses.

CONCEPT

Estimation of the minimum sample size required for a decision is not a single unique method, but the concepts underlying most methods are similar. Figure 7.1 shows distributions for null and alternate hypotheses in a test of means for samples of size n (see Fig. 7.1A) and $4n$ (see Fig. 7.1B). The size of the difference between treatment outcomes that will answer the clinical question being posed is often termed *clinical significance* or, better, *clinical relevance*. For such a clinically important difference between means, the sizes of error probabilities (α and β) are illustrated by the shaded areas in Fig. 7.1. When the sample is, say, quadrupled, the standard error of the mean is halved, so the curves become more slender, overlap less, and consequently yield smaller error probabilities. The method of estimating a minimum sample size is to specify the

115

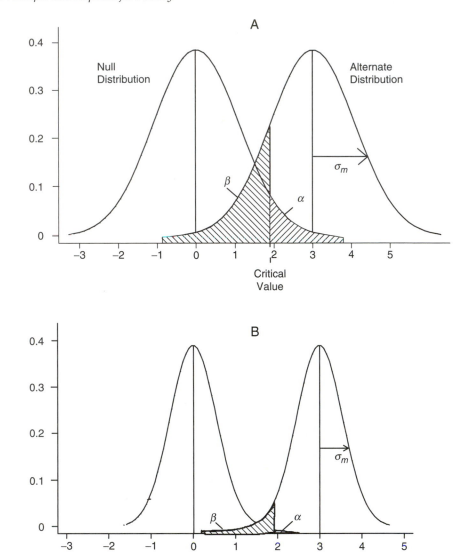

Figure 7.1 Distributions for null and alternate hypotheses in a test of means (A) with sizes of associated error probabilities indicated by *shaded areas* (see Section 5.2). If the sample size is quadrupled, $\sigma_m = \frac{\sigma}{\sqrt{n}}$ shrinks to half, yielding an equivalent diagram (B) with more slender curves. Note how much the errors α and β have shrunk. The method of estimating a minimum sample size to yield specified error sizes for detecting a given distance between means is to start with those error sizes and that distance and "back-solve" the relationships to find the associated n.

difference between means and the required chances of error and then to find the *n* that satisfies these specifications.

THE TERM *POWER ANALYSIS*

A note on terminology: $1 - \beta$ is called "power." As β is the chance of falsely accepting the null hypothesis, power is the chance of correctly accepting the alternate hypothesis. This approach to estimating the minimum sample size required is often termed *power analysis* because it depends on α (the chance of erroneously accepting the alternate hypothesis) and β ($1 -$ power).

VALUE OF VERY SMALL SAMPLES

Much is said about small samples in medicine. Although derision directed toward generalizing from "samples of size one" (i.e., anecdotal data) is just, nevertheless, the first datum encountered in a situation of complete ignorance provides the greatest amount of information from one datum that an investigator will encounter. Consider the appearance of a new disease for which there is no knowledge whatsoever about the incubation period. The first patient for whom we know the time of exposure and the time of appearance of the disease syndrome provides a great deal of useful information. This first incubation period datum gives us an order of magnitude, improves our study planning, and suggests many hypotheses about disease patterns such as infectious sources.

EFFECT OF INCREASING THE SAMPLE SIZE

Let us see what happens in estimating the minimum sample size as we gradually increase our sample size from 1 to its final n. Intuitively, we observe that the informative value per datum decreases as the number of data increase. Consider how much more is learned from data 1–5 than from data 101–105. There are, in fact, theoretical reasons we will not pursue to claim that the amount of information increases as the square root of the sample size (i.e., $\sqrt{n}$). For example, to double the information about an event, we must quadruple the data.

CONVERGENCE

At some point, enough data are accumulated that the sample distribution differs only negligibly from the population distribution, and we can treat the results as if they came from the population; the sample is said to *converge* to the population.

CHOOSING TEST SIDEDNESS

Notably, sample size estimation may be based on a one- or two-sided alternate hypothesis, just as is the test for which the sample size is being estimated. In the methods of this chapter, the more commonly used two-sided form of the normal tail area ($z_{1-\alpha/2}$) is given. For one-sided cases, just replace $z_{1-\alpha/2}$ by $z_{1-\alpha}$ wherever it appears. For example, replace the two-tailed 5% α's $z = 1.96$ by the one-tailed 5% α's $z = 1.645$. The effect on the patient should be considered in selecting sidedness. When a two-sided test is appropriate, a one-sided test doubles the error rate assigned to the chosen tail, in which case too large a number of healthy subjects will be treated as ill and too small a number of ill patients will not be treated. Choice of a two-sided test when a one-sided one is appropriate creates the opposite errors.

CHOOSING TEST PARAMETERS

Although $\alpha = 5\%$ and power = 80% ($\beta = 20\%$) have been the most commonly selected error sizes in the medical literature, a 20% β is larger than is appropriate in most cases. Furthermore, a β/α ratio of 4/1 may affect the patient. When the false positive is worse for the patient than the false negative, as in a case of testing a drug used for a non–life-threatening disease that has severe negative side effects, the common choices of $\alpha = 5\%$ and $\beta = 20\%$ are not unreasonable. In contrast, in testing treatments for cancer, failing to treat the cancer, which has rate β (false negatives), is more serious for the patient than unnecessary cancer tests, which has rate α (false positives), and the ratio β/α should be decreased.

CLINICAL RELEVANCE AND PATIENT CARE

The clinical relevance, often denoted d or δ, usually is the statistical parameter that most influences the sample size. It may also affect patient care. Because a larger difference will require a smaller n, the temptation exists to maximize the difference to allow for a small enrollment with subsequent early closure. This clearly is statistical tampering, because the choice is made on statistical rather than on clinical grounds and begs an ethical question: If the proposed new therapy really is *so much* better than the current one, how may the researcher in good faith not offer the patient the superior course? The difference should be chosen with patient care as part of the consideration.

7.2. IS THE ESTIMATE OF MINIMUM REQUIRED SAMPLE SIZE ADEQUATE?

A "SAFETY FACTOR" IS ADVISABLE

Whatever method we use to estimate the sample size required for a study, this sample size should ideally be estimated on the *results* of that study. Thus, we cannot know the sample size until we do the study, and we do not want to do the study until we know the sample size. The usual solution to this Catch-22 is to use inputs to the sample size equation drawn from other sources, such as a pilot study or parameter estimates quoted in the literature. Because these are not the actual data from our study, there is a *wrong data* source of possible error in our estimate. Furthermore, sample size is estimated on the basis of data that are subject to randomness, so there is also a *randomness* source of possible error in the estimate. Because of these errors from different data or chance, the α and β errors on which we based our sample size estimate may be larger than we anticipated. Because the estimated sample size represents the *minimum allowable*, whatever the method used, *we should add a "safety factor" to the estimated required sample size* to account for these two sources of possible error. The size of this safety factor is an educated guess. The more uneasy we are about how well the data used in the sample size estimate will agree with the ultimate study data, the larger the safety factor we should allow. The resulting estimate of sample size is a mixture of guess and statistics, a number that may be taken only as a rough indicator.

7.3. SAMPLE SIZE IN MEANS TESTING

WHAT INPUTS ARE NEEDED

Section 6.8 introduces a test for the difference between the means of samples from two normal distributions. If we are planning a study that will require such a test, we want to estimate how many data are needed. To follow the development of a method of such estimation, let us look at the simpler case of testing a sample mean against a known population mean. Values used in minimum sample size estimation are the error risks (α and β), the standard deviation of the population data (σ), and the difference wanted to detect between the two means being tested. This difference, often denoted d or δ, is the clinical relevance—the difference clinically important to detect. The minimum sample size depends more on this difference than on the other inputs. The sample size grows large quickly as this difference grows small.

THE LOGIC BEHIND THE METHOD

An emergency medicine physician knows from large-sample historical evidence that mean heart rate (HR) from a particular healthy population is $\mu = 72$ beats/min, with standard deviation $\sigma = 9.1$ beats/min. The physician wants to know whether mean HR from a population of patients who have just been exposed to a particular type of toxin is the same or greater. She plans to collect data from the next sample of patients who present with that toxic exposure and to test the sample mean against the population mean. She needs to estimate the minimum sample size she will need. She designates the unknown population mean μ_s (subscript s for "sample's distribution"). The hypotheses tested will be

$$H_0 : \mu_s = \mu \text{ versus } H_1 : \mu_s > \mu.$$

From the data that will be forthcoming, the value of μ_s will be estimated by the sample mean m. Thus, to decide whether $\mu_s = \mu$, we test m against μ. Figure 7.2 shows the two distributions involved. Because they are standardized, the horizontal axis quantities are given by mean HR differences divided by the standard deviation. The left distribution is the null, with its mean μ standardized to zero, indicated by a vertical line. The right distribution is a possible alternate, with its mean μ_s, estimated by m, indicated by a vertical line at about 3. σ is the standard deviation from the population data, so σ_m, the standard error of the mean (the standard deviation of distribution shown in Fig. 7.2) is $\sigma/\sqrt{n}$, where n is the sample size we seek. We use a form similar to a test for a significant difference between means, $m - \mu$, except the error risks are inputs and the n is output. For this discussion, we use $\alpha = 5\%$ and $\beta = 20\%$ (power $= 1 - \beta = 80\%$). The critical value (here, $\mu + 1.645\,\sigma_m$) is the position separating the two types of error, shown in Fig. 7.2 as the number of standard errors above μ that yields a 5% α (area under the tail of the null distribution; 1.645 is the value from the normal

119

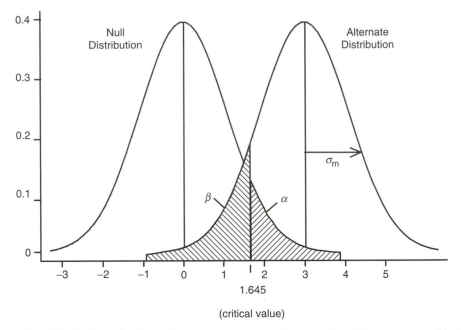

Figure 7.2 Distributions of null and alternate hypotheses showing Types I and II errors separated by the critical value, chosen as 1.645 (standard normal value) to provide $\alpha = 5\%$. We specify the difference in standard errors between the means (here about 3) that we want the test to detect. The error probabilities are dictated by the overlap of distributions, which, in turn, is dictated by the standard error $\sigma_m = \sigma/\sqrt{n}$. As we increase n, σ_m and, therefore, the areas under the overlap giving the error sizes shrink until the specified error probabilities are reached. The resulting n is the minimum sample size required to detect the specified difference in means.

table, Table I [see Tables of Probability Distributions], for $\alpha = 5\%$). Similarly, β is the area under the tail of the alternate distribution and is specified by the number of standard errors below m that the critical value lies, or $m - 0.84\,\sigma_m$ (0.84 is the value from Table I for 20% in the tail area). Because the axis value of these two expressions both equal the critical value, we set them equal to each other; that is, $\mu + 1.645\,\sigma/\sqrt{n} = m - 0.84\,\sigma/\sqrt{n}$. Solving the equation for n yields:

$$n = \frac{(z_{1-\alpha} + z_{1-\beta})^2 \sigma^2}{d^2} = \frac{(1.645 + 0.84)^2 \sigma^2}{(m - \mu)^2}. \tag{7.1}$$

Other formulas for minimum required sample size follow equivalent logic. In particular, the formula for the case of two means, which are introduced in the next section, uses a formula recognizably similar to Eq. (7.1).

WHAT IF σ IS UNKNOWN?

Note that the methods given here primarily use normal distribution theory and the population or large-sample σ rather than a small-sample s. If s is all we have, we just use it in place of σ. The reason for this is twofold. First, if n is unknown, we have no degree of freedom (df) to use in finding the appropriate t-values to use in the calculation. Second, the nicety of using t would be lost in the grossness of the approximation, because the process depends on pilot data or results from other studies and not on the data to be used in the actual analysis.

7.4. MINIMUM SAMPLE SIZE ESTIMATION FOR A TEST OF TWO MEANS

EXAMPLE FROM ORTHOPEDICS

Two types of artificial knee are to be compared for range of motion (measured in degrees). Theoretically, either could give a greater range, so a two-sided alternate hypothesis is appropriate. The hypotheses to be tested become H_0: $\mu_1 = \mu_2$ versus H_1: $\mu_1 \neq \mu_2$. A journal article on the first type of knee gave $m_1 = 112°$ with $s_1 = 13°$, and an article on the second type gave $m_2 = 118°$ with $s_2 = 11°$. If we want to perform a prospective, randomized clinical trial to decide whether a 6° difference is statistically significant, what is the minimum number of patients receiving each knee we must record?

THE METHOD IN GENERAL

We ask whether μ_1 (estimated by m_1) is different from μ_2 (estimated by m_2). Choose the smallest distance d (clinically relevant difference) between m_1 and m_2, that is, $d = m_1 - m_2$, to be detected with statistical significance. From the medical literature or pilot data, find σ_1^2 and σ_2^2, estimated by s_1^2 and s_2^2, if necessary. Choose the risk required of an erroneous rejection of H_0 (α) and of a correct rejection of H_0 (power, or $1 - \beta$). Look up the z-values in Table I, $z_{1-\alpha/2}$ and $z_{1-\beta}$, for these two risks. Substitute the z-values and d, together with the standard deviations in Eq. (7.2), to find n_1 ($= n_2$), the minimum sample size required in *each* sample. [Note how similar Eq. (7.2) is to Eq. (7.1).] For a one-sided test, substitute $z_{1-\alpha}$ for $z_{1-\alpha/2}$.

$$n_1 = n_2 = \frac{\left(z_{1-\alpha/2} + z_{1-\beta}\right)^2 \left(\sigma_1^2 + \sigma_2^2\right)}{d^2} \tag{7.2}$$

THE ORTHOPEDIC EXAMPLE USING THE METHOD

We chose $d = m_1 - m_2 = -6°$ and found the values $\sigma_1^2 = 169$ and $\sigma_2^2 = 121$ from the literature. We choose α as 5% and $1 - \beta$ (the power) as 80%. From Table I (with interpolation), $z_{1-\alpha/2} = 1.95$ and $z_{1-\beta} = 0.84$. Substitution in Eq. (7.2) yields

$$n_1 = n_2 = \frac{\left(z_{1-\alpha/2} + z_{1-\beta}\right)^2 \left(\sigma_1^2 + \sigma_2^2\right)}{d^2} = \frac{2.8^2 \times 290}{6^2} = 63.16.$$

The required minimum sample size is 64. For the reasons given in Section 7.2, inclusion of a few more patients would be advisable.

7.5. OTHER SITUATIONS IN WHICH MINIMUM SAMPLE SIZE ESTIMATION IS USED

SAMPLE SIZE METHODS EXIST FOR CONFIDENCE LIMITS AND STATISTICAL TESTS

Two basic uses for minimum sample size estimation are in confidence limits and statistical tests. In confidence limits, we estimate a parameter, for example, m estimating μ, and we want some idea of how large a sample we need to find 95% (or other) confidence limits on the parameter. In testing, we pose a clinically relevant effect, for example, difference between two means, and we want some idea of how large a sample we need to detect this difference with statistical significance.

SAMPLE SIZE METHODS HAVE BEEN DEVELOPED FOR SOME BUT NOT ALL REQUIREMENTS

Methods for minimum sample size estimation have been developed for some of the most frequently used statistical techniques, but not for all techniques. In the reference part of this book, specifically Chapter 22, minimum sample size methods may be found for the following statistical techniques:

Categorical data
Confidence interval on a proportion
Test of a sample proportion against a theoretical proportion
Test of two sample proportions (equivalent to using a 2×2 contingency table)
Continuous data
Confidence interval on a mean
Confidence interval on a correlation coefficient
Test on one mean (normal distribution)
Test on two means (normal distributions)
Test on means from poorly behaved distributions
Test on means in the presence of clinical experience but no objective prior data

Minimum sample size estimation methods for other statistical techniques exist; some have been developed with considerable thoroughness and rigor, whereas others have not. To list all of them would be misleading, because new approaches appear in the statistical literature from time to time. Some of the more common statistical techniques for which sample size methods exist are tests on $r \times c$ contingency tables, tests on variances (standard deviations), one-way analysis of variance (a test on three or more means), and tests on regression models.

CHAPTER EXERCISES

7.1. Should minimum sample size estimation be based on a one-sided or a two-sided test, assuming the underlying distributions are normal, of (a) a difference in nausea scores between ondansetron hydrochloride and placebo in DB2 and of (b) the difference between assay types in DB8?

7.2. Assess the clinical effects of false-positive and false-negative outcomes and specify whether β should be the usual 4α in estimating minimum sample size for a test of (a) the effect of ondansetron hydrochloride on nausea after gallbladder surgery in DB2 and (b) a change in plasma silicon level after implant in DB5.

7.3. DB7 contains very small samples of bone density measures for men and women. On the basis of the standard deviations of those samples, estimate the minimum sample size required to detect a difference of 10 units of bone density between the means for men and women with $\alpha = 0.05$ and power = 0.80 ($\beta = 0.20$).

7.4. In DB12, the extent (in centimeters) of the carinal resection appears to affect survival. Using the standard deviations given for surviving (died = 0) and dying (died = 1) patients, estimate the minimum sample size required for a test between means of resection extent of 0.37 cm for those two groups with $\alpha = 0.05$ and power = 0.80 ($\beta = 0.20$).

7.5. Suppose we could take a set of International Normalized Ratio (INR) readings from the clinic and another set from the laboratory, as in DB13, but not paired. We want to test for a two-sided difference between means, where a difference of 0.25 INR is clinically relevant. We take $\alpha = 0.05$ and power = 0.90. $\sigma_{clinic} = 0.54$ INR and $\sigma_{lab} = 0.63$ INR. How many readings are needed?

7.6. Suppose we did not have the exercise-induced bronchoconstriction and exhaled nitric oxide data as in DB14 but found a 5-minute difference standard deviation of 7 parts per billion (ppb) for both healthy subjects and asthmatics. We want to test for a difference in means between these two groups using $\alpha = 0.05$ and power = 0.80, where we take the clinically relevant difference to be 5 ppb. Assuming equal numbers of healthy subjects and asthmatics, what minimum sample size is required?

7.7. In DB3, the mean for the 16 patients' differences between baseline and 5-day serum theophylline level is 0.993 mg/dl, with standard deviation 3.485. For such a difference to be significant with $\alpha = 0.05$ and power = 0.80, $n = 50$ pairs would be required [using a formula slightly different from Eq. (7.2) that can be found in Chapter 22]. Generating 3 random samples of 50 on a computer from a normal distribution with the same mean and standard deviation, we find t tests yielding p-values of 0.040, 0.042, and 0.072. Comment on the adequacy of the estimated minimum sample size.

Chapter 8

Statistical Prediction

8.1. WHAT IS A "MODEL"?

PATTERNS IN DIAGNOSIS, CAUSE, AND OUTCOMES

In medicine, we need to recognize the *pattern* of a malady (diagnosis), we need to understand its *pattern* of development (etiology), and we need to be able to predict its *pattern* of response to treatment (outcome). These patterns, which underlie every stage of medical practice, historically have been only implicit. They were developed intuitively through experience and not recognized formally even by the practitioner in earlier days. They represented the *art* of medicine. As medicine evolves from art more and more toward science, these patterns are beginning to be recognized and formalized. When a pattern is formalized into word-based logic, it is often termed a *paradigm*. When a pattern is formalized into quantitative logic, it is often termed a *model*. (These terms are used differently by different people. The reader must ascertain how a writer is using them, and the writer should define clearly the terms being used.)

Definition of Model

In the context of this book, a *model* is a *quantitative representation of a relationship or process.* A patient's white blood cell (WBC) count at a fixed interval of time after infection depends on the dose level of infecting bacteria. The WBC count and the density of infecting organisms are both quantities. The relationship can be represented by a rising curve. We might conjecture, as a starting point, that it looks similar to Fig. 8.1. At this point, we are only conjecturing about the pattern or shape of the relationship, so values are not yet assigned to the axes. This would be a *model,* not of real data but as a conceptual preliminary step. The next section refines our model and makes it more realistic with the acquisition of data.

A MODEL IS A LIMITED TOOL

A model expresses some important aspect between two (or more) variables. It does not purport to express all the subtleties. The model may indicate that WBC tends to increase with dose level of a certain bacterium. It may not take

125

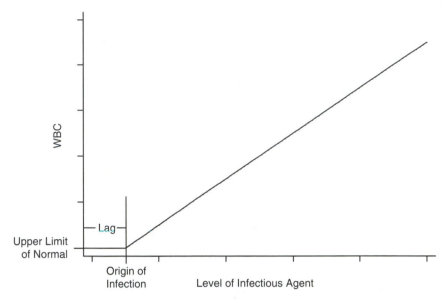

Figure 8.1 A symbolic relationship between white blood cell count and an infectious agent.

into account the time since initial exposure or the state of the patient's immune system. The model is a tool, a window through which we can see the aspect of the relationship we are questioning.

WHAT WE NEED TO KNOW TO FORM THE WHITE BLOOD CELL MODEL

Suppose we have data on infection dosages and the related WBC counts at various points in time. Some questions we might ask are the following:

1. How big an initial dose is required to cause a WBC response?
2. Does WBC count increase as a straight line with dose of infection?
3. What is the gradient of this line; that is, how steep is it?

The following section explains that the dose at which the WBC count exceeds normal is the dose required for a response, that a straight line fit might be appropriate if the data do not curve as they increase (there are tests for this fit), and that the gradient is given by the amount of increase in WBC count per unit increase in infection level. Later sections in this chapter review curved models and the use of more than one indicator at a time.

8.2. STRAIGHT-LINE MODELS

A LINE IS DETERMINED BY TWO DATA

A mathematical straight line on a graph with x- (horizontal) and y- (vertical) axes is defined by two pieces of information, for example, a point and a slope. If we think of a straight-line segment as a piece of straight wire, we can anchor it at a point about which it can rotate. Then we can specify its slope leading out of that point, and the wire is fixed. The point is the location of a known *x, y* pair. The slope is given by the amount of rise the line makes in the y direction for each unit of movement in the x direction. We can see this fixing occurring in Fig. 8.1. The point occurs where the dose of infection begins to affect the upper limit of the normal WBC count range. If we then specify the rate of increase of the WBC count as caused by the dose of infection, we have the slope, or gradient. The line will be determined. Similarly, two points will specify a line.

FITTING A LINE

Note that this line is a mathematical certainty and is not subject to statistical variation. When real data are available, we must use these data to *estimate* rather than to determine the two pieces of information that specify the line. Estimation of a point and a slope from data subject to variability is termed *fitting* the line to the data.

CHOOSING THE FORM FOR THE WHITE BLOOD CELL EXAMPLE

To exemplify this process of data and estimation, let us look at some leuko-cyte counts as related to counts of a specific bacterium resulting from a culture. Figure 8.2 shows 30 data of this sort. (Both *x* and *y* are counts $\times 10^9$.) A straight line fit is a reasonable model. Most statistical straight-line fits use one of two forms: intercept and slope, or mean and slope. (Simple algebra can change one form to another.) The intercept is the point, say, β_0, at which the line crosses the

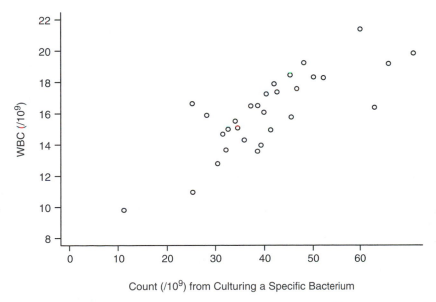

Figure 8.2   White blood cell count data as dependent on culture count of initial infectious agent at a fixed period after exposure.

y-axis, that is, the (x, y) point $(0, \beta_0)$. If we denote the slope by β_1, the intercept-and-slope form would be $y = \beta_0 + \beta_1 x$. In this example, a zero value of culture count is out of the realm of reality, because all of the data come from infected patients. We would do better to choose the defining point as the mean (m_x, m_y), because it lies amid the data. In this case, the mean-and-slope form would be

$$y - m_y = \beta_1(x - m_x). \tag{8.1}$$

We understand that m_x is the sample mean of the culture count and m_y is the sample mean WBC count. We are left with the manner of estimating the slope, which is considered in the next two sections.

8.3. WHAT IS "REGRESSION" (AND ITS RELATION TO CORRELATION)?

PREDICTION

A frequent use of a straight-line fit is to predict a dependent variable on the y-axis by an independent variable on the x-axis. In the leukocyte example, WBC count can be predicted by the infection culture count. Of course, because the fit is estimated from variable statistical data rather than exact mathematical data, the prediction also will be an estimate, subject to probability and susceptible to confidence interval statements.

HOW TO CALCULATE A FIT

It seems intuitively obvious that a fit should use all the data and satisfy the most important mathematical criteria of a good fit. The simplest and most commonly used fitting technique of this sort is named *least squares.* The name comes from minimizing the squared vertical distances from the data points to the proposed line. A primitive mechanical way of thinking about it would be to imagine the points as tacks, each holding a rubber band. A thin steel bar representing the line segment is threaded through all the rubber bands. The tension on each rubber band is the square of the distance stretched. The bar is shifted about until it reaches the position at which the total tension (sum of tensions of each rubber band) is a minimum. In the form of Eq. (8.1), the point in the point–slope fit is given by the sample means of x and y, and the slope can be shown mathematically to be estimated by the sample estimates of the x,y covariance divided by the x variance, or s_{xy}/s_x^2. Let us denote by b values the estimates of the β values, to conform with the convention of Greek letters for population (or theoretical) values and Roman letters for sample values. Then, $b_1 = s_{xy}/s_x^2$, which yields

$$y - m_y = b_1(x - m_x) = \frac{s_{xy}}{s_x^2}(x - m_x). \tag{8.2}$$

For the data plotted in Fig. 8.2, $m_x = 40.96$, $m_y = 16.09$, $s_x^2 = 163.07$ ($s_x = 12.77$), and $s_{xy} = 25.86$. Substituting these quantities in Eq. (8.2), we obtain Eq. (8.3)

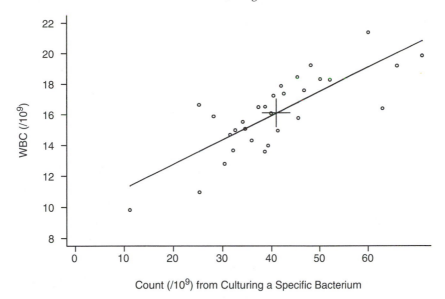

Figure 8.3 White blood cell count data as dependent on culture count with least squares (regression) fit superposed. Crosshairs show the point composed of means, about (41,16). Note that the line rises about 1.6 units for each horizontal advance of 10 units.

as the fit:

$$y - 16.09 = 0.1586(x - 40.96) \qquad (8.3)$$

Figure 8.3 shows the data with the fit superposed.

THE TERM *REGRESSION*

The term *regression* has a historical origin that is unnecessary to remember. It may well be thought of as just a name for fitting a model to data by least squares methods. Interestingly, however, the name arose from a genetic observation. Evidence that certain characteristics of outstanding people were genetic anomalies rather than evolutionary changes was given by fitting the characteristics of the children of outstanding people to those of their parents and grandparents. Because the children's characteristics could be better predicted by their grandparents' characteristics than their parents', the children were said to have "regressed" to the more normal grandparent state.

CORRELATION AS RELATED TO REGRESSION

Correlation was introduced in Section 3.2. Clearly, both correlation and the straight-line regression slope express information about the relationship between x and y. However, they clearly are not exactly the same because the sample correlation r, by Eq. (3.10), is

$$r = \frac{s_{xy}}{s_x s_y} = \frac{\text{cov}(x, y)}{sd(x)sd(y)} \qquad (8.4)$$

129

and the sample slope b_1, from Eq. (8.2), is

$$b_1 = \frac{s_{xy}}{s_x^2} = \frac{\text{cov}(x, y)}{sd(x)sd(x)}. \tag{8.5}$$

Simple algebra will verify that

$$r = b\frac{s_x}{s_y}. \tag{8.6}$$

What does this difference imply? In correlation, the fit arises from simultane-ously minimizing the distances from each point perpendicular to the proposed line. In regression, the fit arises from simultaneously minimizing the vertical (y) distances from each point to the proposed line. Thus, regression makes the assumption that *the x measurements are made without random error.* Another difference is that regression may be generalized to curved models, whereas cor-relation is restricted to straight lines only. A difference in interpretation is that correlation primarily is used to express how closely two variables agree (the width of an envelope enclosing all the points), whereas regression can indicate this (the amount of slope in the fit) and also provide a prediction of the most likely y-value, and a confidence interval on it, for a given x-value. Both can be subjected to hypothesis tests, but tests on regression are more incisive, can be generalized to curved models, and can often be related to other statistical methods (such as analysis of variance).

ASSUMPTIONS

As in most of statistics, certain assumptions underlie the development of the method. If these assumptions are violated, we can no longer be sure of our conclusions arising from the results. In the case of least squares regression fits, we assume not only that x-values are measured without error but also that, for each x, the distribution of data vertically about the regression line is approxi-mately normal and that the variability of data vertically about the regression line is the same from one end of the line segment to the other.

8.4. ASSESSING AND PREDICTING RELATIONSHIPS BY REGRESSION

DOES HOSPITAL STAY FOR MENTAL PATIENTS RELATE TO IQ?

At a mental hospital, a psychologist's clinical experience suggests to him that brighter patients seem to stay in the hospital longer. If true, such a finding might result from brighter patients being harder to treat. The psychologist collects intelligence quotient (IQ) measurements (x) and days in the hospital (y) for the next 50 patients that he treats.[22] Using Eqs. (3.1), (3.3), and (3.6) to calculate means, variances, and the covariance, he finds $m_x = 100$, $m_y = 10.5$, $s_x = 10$, $s_y = 11.2$, and $s_{xy} = 39.43$, from which Eq. (8.4) can be used to find

$$r_{xy} = 39.43/(10 \times 11.2) = 0.35.$$

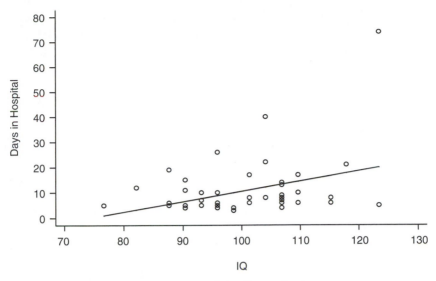

Figure 8.4 Regression of days in the hospital on IQ for 50 mental patients.

The correlation coefficient of 0.35 is not decisive, but it is large enough to believe that some relationship exists. (Tests of significance on correlation coefficients are reviewed in Chapter 24.) The psychologist wants to regress days in the hospital on IQ, test his ability to predict, and assess how much of the causal forces that decide days in the hospital may be attributed to IQ.

REGRESSION AND PREDICTION

Using Eq. (8.2), the psychologist finds that the regression line follows the equation

$$y - 10.5 = 0.3943(x - 100). \tag{8.7}$$

The data from the 50 patients with the regression line superposed are shown in Fig. 8.4. What would be the best prediction of days in the hospital for a patient with an IQ of 90? With an IQ of 110? Substituting 90 for x in Eq. (8.7), he finds about 6.6 days to be the expected length of hospital stay for a patient with an IQ of 90. Similarly, about 14.4 days is the best prediction of length of hospital stay for a patient with an IQ of 110.

ASSESSMENT OF THE REGRESSION MODEL

How much of the possible causal influence on length of hospital stay does IQ represent? A good indication of this is given by a statistic named the *coefficient of determination,* designated R^2, which is nothing more than the square of the correlation coefficient when only one predictor is used. In this case, $r = 0.35$, $R^2 = 0.12$, implying that IQ represents only about 12% of the causal influence. Our psychologist would be inclined to conclude that IQ is a real but rather minor influence on length of hospital stay. Would this conclusion be justified? We have not yet considered the satisfaction of the assumptions on which this regression analysis was based. Let us consider the three assumptions, assisted by the graph in Fig. 8.4. (Actual tests of assumptions can be made and are considered further in Chapter 24.)

Is the assumption that the x-values are measured without error satisfied? The patient's true IQ is a fixed but unknown value. The measured IQ is an effort to estimate the true IQ, but its accuracy certainly is recognized as less than perfect. In addition to accuracy missing the mark, measured IQ varies from time to time in a patient and varies among types of IQ tests. Therefore, no, the assumption of exact x-values is not well satisfied. Is the assumption of normal shape about the line for each x satisfied? There are too few data to be sure. From the appearance in Fig. 8.4, a case might be made toward the center, but data in the tails appear to be rather skewed. Satisfaction of this assumption is questionable. Finally, is the assumption of equal variability for each x-value satisfied? A glance at Fig. 8.4 shows that the data spread out more and more as we move to the right along the regression line. This assumption seems violated. So where does violation of the assumptions underlying the method leave the investigator? It is possible to adjust for some of the violations by using more sophisticated statistics, but the sample size per IQ value is still too small. The psychologist cannot trust his flawed results. About all he can report is that some informal pilot results exist, suggesting a small influence by IQ on the length of hospital stay, on the basis of which he will carry out a more carefully designed study with a much larger sample and seek the assistance of a biostatistician for both planning and analysis.

8.5. OTHER QUESTIONS THAT CAN BE ANSWERED BY REGRESSION

CURVED LINE MODELS

Just as a model can be a segment of a straight line, a segment of a curved line can be used as well if the data are better fit by it or, preferably, if known physiologic or medical relationships follow a curve. The curve may be a segment of a parabola (number of cells of a fetus or cancer may increase as the square of time units), a logarithmic form [number of bacteria surviving an antibiotic may decrease as constant $- \log(x)$], or even a cyclic pattern (aspergillosis is shown in Chapter 23 to follow a sine wave over the seasons).

MULTIPLE VARIABLES SIMULTANEOUSLY

Furthermore, multiple variables may be used as predictors. Number of days a patient must remain in the hospital may be predicted by body temperature, bacterial count, presence or absence of emesis, presence or absence of neurologic signs, and so on. No single one may be useful as a predictor, but several in consort may provide sufficient accuracy to assist in hospital bed planning.

PREDICTING ONE OF TWO STATES

The regression described earlier assumes the *y* variable is a continuous-type variable. Suppose we want to predict outcome as one of two states—for example, survival or death, cure or no cure, disease type A or other. There is a technique termed *logistic regression* in which a transformation can "spread" the two states onto an interval on the y-axis, after which ordinary regression methods may be used on the transformed data. As the transformation is logarithmic in nature, the technique is a *log*istic (log-is-tic) regression. It has come to be mispronounced logistic (lo-jis-tik), although it has nothing to do with the logistics of materiel (see Section 24.10 for a definition of this transformation).

IMPLICATION OF THIS SECTION FOR THE READER

The cases curvilinear regression, multiple regression, and logistic regression introduced in this section are pursued further in Chapter 24. At this stage, the reader should remember their names and what sorts of problems they treat so as to recognize the names or the classes of analysis when meeting them.

8.6. CLINICAL DECISIONS AND OUTCOMES ANALYSIS

MEASURES OF EFFECTIVENESS AND OUTCOMES ANALYSIS

Clinical decisions are usually made on the basis of a *measure of effectiveness* (MOE), or often more than one. There is no unique MOE. The effectiveness of a treatment may be measured in terms of the probability of cure, the probability of cure relative to other treatments, the risk of debilitating side effects, the cost of treatment relative to other treatments, and so on. What might be a best decision when applying one MOE may not be with another. When quantified MOEs first began to be used, measures easy to quantify were used, for example, the reduction in number of days for a patient to achieve a certain level of improvement. However, these simple measures usually omitted other important considerations, such as those just mentioned. Investigators began to be dissatisfied with settling for measurably reduced edema following a knee injury when the really important issue was whether the patient could walk again. Measures of effectiveness of this sort represent the eventual "outcome" of the treatment rather than an interim indicator of progress. Their use therefore has come to be termed *outcomes analysis*. Certainly, outcomes often are more difficult to quantify and take much longer to obtain, but without doubt they are better indicators of the patient's general health and satisfaction.

A COMBINATION MEASURE OF EFFECTIVENESS

How should a surgeon's effectiveness be measured? Some possible MOEs are time in surgery, time to heal, and patient survival rate, among others. Most would agree that some combination of measures would be best; that is, wisely chosen indicators are combined in a balanced way to form a resultant MOE.

However, there is no formula to build MOEs. Like so many aspects of medicine, it is a matter of thoughtful judgment. We must start with the question: *What are we really trying to accomplish?* Is our goal a perfect surgery? Patient survival rate? Minimum pain? Minimum cost? We might think of adding several measures together, but we must remember that they are mostly in different units of measurement and represent different levels of relative importance. We must weight each component to adjust for units and importance. A serious question is how levels of relative importance are rated. Indeed, there is controversy about weighing the cost of treatment against the chance of it being effective. The best an investigator can do is use good judgment and allow the readers to accept or reject it for themselves.

EXAMPLE POSED

A dramatic illustration is the issue of quality of life for a terminally ill patient. Consider an elderly male stroke patient. He experienced a cerebral embolism followed by cerebral hemorrhage. He is severely hemiplegic with aphasia. Consciousness and dysphagia are sporadic. Currently, his airway is intubated and he is fed by intravenous infusion. Brain edema contraindicates corticosteroids and anticoagulant drugs, but he is receiving a diuretic. Furthermore, he is uninsured, with costs being paid by his adult children, who have marginal incomes. Let us attempt to build an MOE for him.

EXAMPLE FORMED

An overly simple illustration might be to maximize days of survival in which the patient would rather live than die: n_g (number of good days) minus n_b (number of bad days), or MOE $= n_g - n_b$. If the patient has more good than bad days, we keep him alive. If he stops having good days, our MOE grows smaller each day and we would maximize the remaining value of the MOE by immediately ceasing life support. But there are other relevant indicators. Our MOE might contain p, the probability of (partial) recovery. These are medical judgments suggested by the percentages reported in epidemiologic studies tempered by the characteristics of this case. Earlier, we used MOE $= n_g - n_b$. Including p, it becomes MOE $= pn_g - (1-p)n_b$, where n_g is the total number of good days if he recovers plus those if he does not recover, and n_b is the total number of bad days if he recovers plus those if he does not recover. Suppose we judge p to be 0.30. We judge his life expectancy, given recovery, to be a little more than 4 years or about 1500 days and, given no recovery, to be 30 days. We judge that if he recovers, he will have 60% good days. If he does not recover, he will have 100% bad days. Then, $n_g = 60\%(1500) + 0\%(30) = 900$ and $n_b = 40\%(1500) + 100\%(30) = 630$. MOE $= pn_g - (1-p)n_b = 0.3(900) - 0.7(630) = 270 - 441 = -171$. On average, our patient will live a life of poor quality in which he would rather be dead. We have not yet factored in cost. We do not want to destroy his family's financial future. (The effects of this decision may ripple through two or three generations, for example, costing his grandchildren a college education.) Our MOE must add together quality of life levels and money. To do this, we would have to judge the relative importance of a good day in terms of money units (e.g., $1000) of cost to the family, and this would stretch our judgment rather far. In this case, we don't have to, because we have already shown that life-sustaining medical intervention will generate primarily misery for the patient.

DIFFERENT MEASURES OF EFFECTIVENESS

This MOE could certainly be built in a different way, which is quite acceptable if it serves the same purpose, just as houses built in different ways may acceptably keep the inhabitant warm, dry, and safe. Every practitioner would find it helpful to try building some of his MOEs. It makes us think through the issues involved and rank them in importance. It gives us a basis for decision in those tortured cases in which we are just not sure. Perhaps most important, it reduces our number of erroneous decisions.

CHAPTER EXERCISES

8.1. It is believed that respiration rate in infants decreases with age. To examine that belief, respiration rate was recorded for 232 infants from 0 to 24 months.[50] The scatter diagram of respiration rate against age is shown (see accompanying figure). Form a conceptual model.

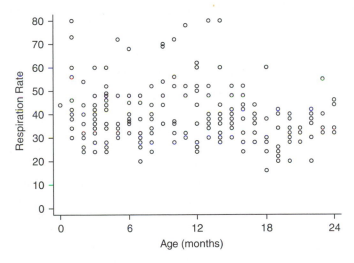

8.2. Summary statistics for the data in Exercise 8.1 (where subscripts a and r represent age and respiration rate, respectively) are $m_a = 10.9$; $m_r = 39.0$; $s_a = 6.7$; $s_r = 11.8$; and $s_{ar} = -15.42$. What is the correlation coefficient between age and respiration rate? How is this correlation coefficient interpreted?

8.3. What is the equation for the straight-line regression of r on a for Exercise 8.1? Lay a thin paper over the scattergram of respiration rate by age and sketch in this line.

8.4. What is the coefficient of determination? How is this coefficient of determination interpreted?

8.5. In DB14, we investigate if the change in exhaled nitric oxide (eNO) caused by exercise is related to the initial (preexercise) eNO, or if they are independent. The scatter diagram is shown (see accompanying figure). Form a conceptual model.

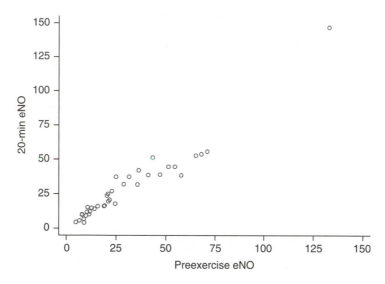

8.6. Summary statistics for the data in Exercise 8.5 (where subscripts 0 and 20 represent preexercise eNO [0 minutes] and eNO at 20 minutes, respectively) are approximately $m_0 = 29.3$; $m_{20} = 28.1$; $s_0 = 25.6$; $s_{20} = 25.3$; and $s_{0,20} = 621.5$. Using these values, what is the correlation coefficient between age and respiration rate? How is this correlation coefficient interpreted?

8.7. What is the equation for the straight-line regression prediction of eNO_{20} by eNO_0 for Exercise 8.5? Lay a thin paper over the scattergram of respiration rate by age and sketch in this line.

8.8. What is the coefficient of determination for the data of Exercise 8.5? How is this coefficient of determination interpreted?

8.9. Two types of medication exist to treat a certain medical condition. Medicine A costs x dollars for a course of treatment and has always been successful. Medicine B costs y dollars but is unsuccessful in 1 of 10 cases. When results with medicine B are unsuccessful, medicine A must be used. (There is no contagion or deterioration in medical condition because of the delay.) The question is which medicine should be used, inconvenience to the patient aside. (a) What is the MOE? (b) Set up a relationship between the two medicines such that when the relationship holds true, medicine B should be used. (c) If $x = \$100$ and $y = \$75$, what is the gain per patient by using medicine B?

Chapter 9

Epidemiology

9.1. THE NATURE OF EPIDEMIOLOGY

DEFINITIONS

An *epidemic* is the occurrence in a community or region of cases of an illness, specific health-related behavior, or other health-related events clearly in excess of normal expectancy.[40] *Epidemiology* is the study of the occurrence of illness in populations. To the epidemiologist comparing the rate of occurrence with expectancy, 2 cases of plague in a city of 6 million people may constitute an epidemic, whereas 10,000 cases of influenza may not. Epidemiology started with the analysis of mortality recordings, from which surprising insights into the patterns of public disease and health events immediately demonstrated its importance. It long concentrated on mortality from infectious diseases (e.g., smallpox), then evolved to include any disease (e.g., cardiovascular disease, cancer) and morbidity of diseases, and finally broadened to include any condition threatening public health (e.g., smoking, mercury in a water or food supply).

EPIDEMIOLOGY COMPARED WITH OTHER BRANCHES OF MEDICINE

Epidemiology, like all of medicine, seeks to understand and control illness. The main conceptual difference between epidemiology and other branches of medicine is its primary focus on health in the group rather than in the individual, from which arises the concept of *public health*. The main methodologic difference between epidemiology and other branches of medicine is its attention to rates and distributions. Basic to epidemiology is thinking probabilistically rather than in mutually exclusive categories. The epidemiologist is concerned not with a patient who is sick or well but rather with the proportion ill from a catchment of patients and the probability that the proportion will grow. To have a baseline for estimating these proportions and probabilities, the epidemiologist continually is concerned with *denominator data*.

STAGES OF SCIENTIFIC KNOWLEDGE IN EPIDEMIOLOGY

The accrual of knowledge in epidemiology follows the same stages as in other fields of medicine, first met in the beginning words of this book: description, explanation, and prediction. Let us consider these stages in sequence. In *description,* the etiology of a public health disease or condition is documented. This stage generates basic epidemiologic data and generates the scientific hypotheses to be tested. In *explanation,* the epidemiologic data are assessed to explain the outbreak of public disease. The scientific hypotheses are tested. The next stage is *prediction,* which integrates the test results and provides a model of the course of an epidemic, which can be used in prevention or control of the outbreak and even in assessment of the efficacy of potential treatments or measures of prevention and control.

EPIDEMIOLOGY IS AN ECLECTIC SCIENCE

Epidemiology must interact with many fields of study to isolate causal factors: human and veterinary clinical medicine in their various specialties; biological, biochemical, and genetics investigations; social, economic, and demographic factors; geographic and historic patterns; and the personal habits of individuals. The epidemiologist must be a general health scientist.

9.2. SOME KEY STAGES IN THE HISTORY OF EPIDEMIOLOGY

The respective beginnings of the descriptive, explanatory, and predictive stages in epidemiology may be followed over three centuries. Quantified description in epidemiology began in Great Britain in 1662, when John Graunt published a tabular analysis of mortality records. In 1747, James Lind, also in Great Britain, pioneered isolating causes by experimentation, finding the nutritional rather than infectious basis of scurvy. Although epidemiology was to wait another century for formal predictive modeling, the theory of contagion allowed the predictive stage of epidemiologic science to begin on an informal basis in the 1800s. An example is the 1847 work of Ignaz Semmelweis in Austria (see also Section 15.6, which recounts Semmelweis' origin of antiseptic surgery), who traced the source of the highly fatal disease puerperal fever by epidemiologic methods and showed how to control it.

9.3. CONCEPT OF DISEASE TRANSMISSION

SPECTRUM OF DISEASE

The sequence of events from exposure to a disease to resolution, often death, is termed the *spectrum of disease.* It arises from the interaction among three main factors: the host, the agent, and the environment. The *host* includes primary hosts in which a parasite develops, for example, intermediary hosts (the host is a component in the disease cycle) and carrier hosts (host only transports the disease in space or time). The host as a member of the epidemiologic catchment is characterized by the susceptibility to acquiring the disease, which includes behavior leading to exposure, immunization or resistance, and the like. The *agent* is not necessarily an infectious agent. It may simply consist of a deficiency of necessary nutrient (vitamin C in the case of scurvy), the excess of a deleterious substance (tobacco smoke), or a metabolic condition (metabolic acidosis). More dramatically, it may consist of microbes (the bacterium *Yersinia pestis* in the case of plague) or toxins (mercury poisoning, or *yusho;* the name arose from the classic episode in Minamata, Japan). The *environment* as a component in the spectrum is characterized not only by the physical and biological properties (in malaria: still water to breed mosquitos, access by the mosquitos to previously infected hosts, exposure of the catchment member to the infected mosquito) that affect disease transmission but also by the social properties (as in sexually transmitted diseases).

MODES OF TRANSMISSION

The classic book by Lilienfeld[44] classifies two modes of transmission: common-vehicle epidemics, in which the vehicle is water, food, air, and so forth, and serial transfer epidemics, in which the vehicle is a host-to-host transfer, such as infectious transmission by touching, coughing, and so on.

HERD IMMUNITY

Rates of infectious disease transmission are not constant. The pattern of an epidemic (depending considerably on incubation period) is usually an initial surge of infected patients, continuing until a substantial number of the epidemiologic catchment is immune. At this point, the rate slows, the return cycle to the infectious agent is interrupted, and the epidemic wanes. This high proportion of immunity within a demographic catchment has been termed *herd immunity*. When the herd immunity reaches a critical level, the rate of change of new infection changes from positive to negative and the epidemic starts to decline. The population's susceptibility to epidemic may be reduced by increasing the herd immunity, for example, by vaccination. It should be noted that herd immunity is by no means the only mechanism to alter the rate of new infection. Many occurrences may alter it. For example, it was noted as early as the 1700s that a yellow fever epidemic often was terminated by a change in the weather.[45]

9.4. DESCRIPTIVE MEASURES

INCIDENCE AND PREVALENCE

Incidence and prevalence have been used informally, although not incorrectly, in previous chapters, assuming the student is familiar with the everyday connotations of the terms. In epidemiology, the terms are used with technical precision, as indeed they should be in all fields of medicine. The incidence rate of a disease is the rate at which new cases of the disease occur in the epidemiologic population. The prevalence rate of the disease is the proportion of the epidemiologic population with that disease at a point in time. Thus, during an influenza epidemic in a certain city in November, the prevalence indicates how much of the population is sick and the incidence indicates how rapidly the epidemic is increasing. Incidence and prevalence rates are usually given per 1000 members of the population, unless otherwise indicated.

MORTALITY RATE

Similar to incidence rate is mortality rate, the rate at which the population is dying rather than becoming ill. For a disease resulting in certain death, the mortality rate at the end of the duration period of the disease would be the same as the incidence rate at the beginning of the period.

FORMULAS

Incidence rate must be expressed in relation to an interval in time, for example, as "2000 new cases of illness per month." This is because the number of new cases will be zero at a point in time, that is, when the interval goes to zero. Prevalence rate, on the other hand, can be measured at a point in time, although sometimes a short interval must be used to allow for the time to obtain the prevalence data. Let us denote the number of individuals in the epidemiologic population as n, the number of new cases in a specified interval as n_{new}, and the number of cases present at any one point in time by $n_{present}$. Then, incidence rate I is given by

$$I = 1000 \times \frac{n_{new}}{n}, \qquad (9.1)$$

and the prevalence rate P is

$$P = 1000 \times \frac{n_{present}}{n}. \qquad (9.2)$$

If we denote by n_{dying} the number of patients dying during the specified interval, the mortality rate M becomes

$$M = 1000 \times \frac{n_{dying}}{n}. \qquad (9.3)$$

In the remainder of this chapter, the terms *incidence* and *prevalence* are used when incidence rate and prevalence rate are actually intended, because this is the general custom in medical articles, at least outside strict technical usage in epidemiologic articles.

CERVICAL CANCER IN THE ACORNHOEK REGION OF THE TRANSVAAL[81]

The population served is given as 56,000. From diagrams of sex ratios and assumptions about the child/adult ratio, we believe the adult female population served to be 13,000. (A further subtraction should be made for adult women who never present to medical services, but this unknown number will be ignored for this illustration.) The study covered the 9 years from 1957 to 1966. Cervical cancer was seen in 53 cases, or about 6 per year. Substituting in Eq. (9.1) yields $I = 1000 \times 6/13,000 = 0.462$. The yearly incidence of cancer of the cervix is about 0.5. This type of cancer in women rarely goes into remission. To find prevalence, let us assume, quite arbitrarily, that a woman survives, on average, 2 years after diagnosis. Then there will be about 12 women with cancer of the cervix surviving at any one time, and Eq. (9.2) yields $P = 1000 \times 12/13,000 = 0.923$, or about 0.9. After the first 2 years, M, from Eq. (9.3), is the same as I.

THE ODDS RATIO

The odds ratio (OR), which was introduced in Section 6.4, also is used in epidemiology. Let us examine an epidemiologic study of the interrelation between occurrences of cervical cancer and schistosomiasis.[67] (Schistosomiasis is a parasitic infection common in parts of the Third World that is contracted during immersion in river or stream water.) Table 9.1 reproduces Table 1 from a study by Riffenburgh and colleagues.[67] Some clinicians in Africa have suggested a protective effect against cancer of the cervix by schistosomiasis. Do the figures support this? Let us calculate the cervical cancer OR for schistosomiasis. Odds of cervical cancer given schistosomiasis, $101/165 = 0.6121$, in ratio to odds of cervical cancer given not schistosomiasis, $5212/2604 = 2.0015$, provide OR $= 0.3058$. These figures indicate that a woman is more than 3 times as likely ($1/0.3058 = 3.27$) to have cervical cancer if she does not have schistosomiasis! The figures appear to support the claim of a protective effect. However, the epidemiologic OR has been derived from biased sampling. The rate of schistosomiasis *without* cervical cancer is markedly *over*reported. The often overwhelmed, undertrained African clinician who finds the presence of *Schistosoma haematobium* often ceases to look further; the patient's complaints have been explained adequately, and the cervical cancer is not detected. Thus, we could expect that many of the patients reported in the lower left cell of the table should be moved to the cell above it. Furthermore, the rate of cervical cancer *with* schistosomiasis is markedly *under*reported for an entirely different reason: Schistosomiasis tends to be acquired at an earlier age (mean $= 30$ years), and these patients are less likely to live to the age at which cervical cancer tends to

Table 9.1

Frequencies of Occurrence of *Schistosoma haematobium* Cervicitis, Cervical Cancer, Both, and Neither Pooled from Two Similar Studies in Africa

Observed frequencies	With schistosomiasis	Without schistosomiasis	Totals
With cervical cancer	101	5212	5313
Without cervical cancer	165	2604	2769
Totals	266	7816	8082

Reproduced from Riffenburgh RH, Olson PE, Johnstone PA. Association of schistosomiasis with cervical cancer: detecting bias in clinical studies. *East Afr Med J* 1997; 74:14–16, with permission.

appear (mean = 45 years). If these two reports were corrected, the OR would be considerably changed. It should be clear to the student that high-quality data are crucial prerequisites to meaningful epidemiologic results.

9.5. TYPES OF EPIDEMIOLOGIC STUDIES

BASIC VARIABLES

All epidemiologic studies are framed in terms of *exposures* and *outcomes.* In simplest terms, the exposure is the putative cause under study, and the outcome is the disease or other event that may result from the exposure.

EXPERIMENTAL STUDIES AND INTERVENTION TRIALS

The investigator assigns exposure (or nonexposure) according to a plan. The concept of an experimental *control,* introduced in Section 1.6 and further explored in Section 10.2, is used for nonexposed subjects in experimental studies. In *clinical trials,* patients already with disease are subjects; an example would be James Lind's study of scurvy. In *field trials,* patients without disease are subjects; an example would be a vaccine trial in a population. *Community intervention trials* are field trials that are conducted on a community-wide basis.

NONEXPERIMENTAL OR OBSERVATIONAL STUDIES

In nonexperimental or observational studies, the investigator has no influence over who is exposed. Two types of such studies are cohort studies and case-control studies, both of which are examined in the context of clinical research in Section 10.3. Each study design has inherent strengths and weaknesses.

Cohort Studies

Cohort studies are also termed *follow-up* or *incidence* studies. In these studies, outcomes among two or more groups initially free of that outcome are compared. Subjects may be selected randomly or according to exposure. Indeed, if outcome comparison is the purpose, the study by definition is a cohort study. Cohort studies may be prospective (if exposed and unexposed subjects are enrolled before outcome is apparent) or retrospective (if subjects are assembled according to exposure after the outcome is known). Cohort studies are particularly well suited to evaluating a variety of outcomes from a single exposure (e.g., smoking). Population-based rates or proportions, or relative risk (RR; see discussion in Section 15.4), may be computed.

Case–Control Studies

Case–control studies are also termed case–referent studies or, loosely and confusingly, sometimes retrospective studies. In case–control studies, the cases and noncases of the outcome in question are compared for their antecedent exposures. If exposure comparison is the purpose, the study by definition is a case–control study. The investigator generally has no influence over these antecedent exposures. The appropriate measure of risk in these studies is the exposure OR (see also Sections 6.4 and 15.4).

Prevalence or Cross-Sectional Studies

Prevalence or cross-sectional studies are investigations of an entire population enrolled, regardless of exposure and outcome, with exposure and outcome ascertained at the same time. Effectively, these are "snapshots" of a population, where analysis may be performed as a cohort study or a case–control study.

INFERRING CAUSATION

Identifying causal relationships in observational studies can be difficult. If neither chance nor bias is determined to be a likely explanation of a study's findings, a valid statistical association may be said to exist between an exposure and an outcome. Statistical association between two variables does not establish a cause-and-effect relationship. The next step of inferring a cause follows a set of logical criteria by which associations could be judged for possible causality, which was first described by Sir Bradford Hill in 1965.

Evidence Supporting Causality

Seven criteria currently in widespread use facilitate logical analysis and interpretation of epidemiologic data.

(1) *Size of effect.* The difference between outcomes given exposure and those not given exposure is termed *effect.* Large effects are more likely to be causal than small effects. Effect size is estimated by the RR. (Relative risk is the probability of having a disease when it is predicted in ratio to the probability of having the disease when the prediction is not having it; see Section 15.4 for further details.) As a reference, a RR >2.0 in a well-designed study may be added to the accumulating evidence of causation.

(2) *Strength of association.* Strength of association is based on the p-value, the estimate of the probability of rejecting the null hypothesis. A weak association is more easily dismissed as resulting from random or systematic error. By convention, $p < 0.05$ is accepted as evidence of association.

(3) *Consistency of association.* A particular effect should be reproducible in different settings and populations.

(4) *Specificity of association.* Specificity indicates how exclusively a particular effect can be predicted by the occurrence of potential cause. Specificity is complete where one manifestation follows from only one cause.

(5) *Temporality.* Putative cause must precede putative effect.

(6) *Biological gradient.* There should be evidence of a cause-to-outcome process, which frequently is expressed as a dose–response effect, the term being carried over from clinical usage.

(7) *Biological plausibility.* There should be a reasonable biological model to explain the apparent association.

Further information on the evidence list can be found in elsewhere in the literature.[25,47,69]

9.6. AN INFORMAL APPROACH TO PUBLIC HEALTH PROBLEMS

GATHERING AND ASSESSING EPIDEMIOLOGIC INFORMATION

The essence of epidemiologic methodology can be expressed informally rather simply and briefly, encapsulating it in the classic journalistic guideline: answer the questions what, who, where, when, and how. Let us examine these steps through the example of the discovery of a smallpox vaccine. Suppose you are working with Edward Jenner in the 1790s encountering smallpox.

What

The first step is to characterize a *case definition*. We must describe the disease with all the signs, symptoms, and other properties that characterize it in minute detail. We do not know which variable or combination of variables will relate to others. In this example, smallpox is characterized by chills, high fever, backache, headache, and sometimes convulsions, vomiting, delirium, and rapid heart rate; after a few days, these symptoms retreat and papules erupt, becoming pustules that leave deep pockmarks. Complications include blindness, pneumonia, and kidney damage. The mortality rate approaches 250 per 1000 people. (We have no means to assess its viral origin; we do not even know viruses exist.)

Where

We investigate the incidence by geographic area. The disease never appears in the absence of an infected patient. Investigating in greater detail where the disease occurs, that is, at the household level, we find that no one contracts the disease without having been in proximity to an infected patient. It appears that the infection passes only from person to person. However, it seems to pass with *any* type of contact: airborne vaporized fluids, touch, or second-level touch (touch of things touched by an infected person).

When

We record the times an infected patient had contact with a previously infected patient. The period from contact to the origin of symptoms is identified as between 1 and 2 weeks, implying a 10 ± 3-day incubation period. Infectious contact may be from the very beginning to the very end of symptoms.

Who

Only humans are affected. We gather demographic data, estimating the incidence of smallpox among all sorts of groups. We compare incidence by sex, age groupings, ethnic origin, socioeconomic level, and occupation. We gather biological, physical, and genetic data and estimate incidences among groupings within these variables. We gather personal data, estimating incidences by cleanliness habits, diet, food and drink types, food and water sources, and methods for disposal of bodily and food waste products. We find little difference among groupings. We are beginning to think that we must experiment with controlling food, water, waste, and so on, when a secondary occupational analysis turns up a surprising result: Dairymaids do not contract smallpox!

How

We now further isolate the events and characteristics. We examine what is different about dairymaids that leads to immunity. We find that they differ little in any respect from the usual smallpox victim, except in socioeconomic level and disease history. Socioeconomic level has already been ruled out. In the disease history, we find that most dairymaids have undergone a course of cowpox, a mild form of pox that has some similarities. Exposure to cowpox conveys immunity! We are well on the way to a control for this heinous disease.

ISOLATING THE CAUSE

The key to epidemiologic detective work is isolating the cause, or at least a variable associated with the cause, which can start a fruitful chain of reasoning. If we understand the biology of the disease, we can pose hypotheses to test experimentally. However, we seldom understand the biology without clues leading to theory. To isolate a causal clue, sampling must be undertaken and done with great care and in great detail. A strategy should be planned, trading off ease, cost, and speed of sampling the potential variables with the likelihood that each is a causal clue. Sometimes factors can be isolated physically, sometimes by data analysis, comparing counts by groupings, and sometimes only statistically, inferring relationships probabilistically from odds ratios. At times, variables may be introduced experimentally that provide contrasts not available by sampling. Representativeness in sampling (see initial discussion in Section 1.6) is crucial. The effect of a sampling bias is illustrated dramatically in the schistosomiasis and cervical cancer example in Section 9.4.

9.7. ANALYSIS OF SURVIVAL AND CAUSAL FACTORS

LIFE TABLES LIST SURVIVAL THROUGH TIME

The life table is historic, having been used for survival analysis during the 1700s by Daniel Bernoulli in Switzerland. A life table gives the proportion of a demographic group surviving to the end of each time interval. If no patients are lost to follow-up, the proportion simply is surviving number divided by initial number. When a patient is lost to follow-up before dying, we face a dilemma: We do not know if or when that patient died, but we have useful information on a period during which we know the patient lived and we do not want to jettison that information. Consequently, we include the patient in the base number to calculate survival up to the point at which the patient was lost, but afterward we drop that person from the base number in calculating survival. Patients removed from the database without knowledge of their survival status are termed *censored*.

SURVIVAL CAN BE USED TO SEE PATTERNS OF EVENTS OTHER THAN LIFE AND DEATH THROUGH TIME

In this treatment, survival is discussed as representing a patient's remaining alive. However, replacement of "time to death" by "time to onset of illness" provides a window on morbidity rather than mortality. Replacement of "time to death" by "time to fail" provides a window on the reliability of medical instruments.

DATA AND CALCULATIONS REQUIRED FOR A LIFE TABLE

Basic data for a life table on a number n of patients are (1) time intervals; (2) *begin*, the number at the beginning of each time interval; (3) *died*, the number dying in each time interval; and (4) *lost*, the number lost to follow-up in each time interval. Each time interval may be of any length, perhaps 2 weeks or 6 months. The rest of the table arises from calculations.

Calculations are more exact when the precise time of death or loss to follow-up occurs; but, to be practical, we often know only at the end point of an interval of time that the death or loss occurred during the interval. Epidemiologists often make sophisticated adjustments for the unknown point of time in the interval, but this book takes the simplest form in which death or loss is taken as occurring at the end of the interval. Survival may be calculated by formulas. Perhaps the easiest is as follows. When loss to follow-up occurs, we will enter a second line for that time interval to list it. The remaining calculations are as follows: (5) We fill in a column for *end*, the number at the end of each time interval, which is *begin* − *died* − *lost*. Either *died* or *lost* (or both) will be 0. (6) The proportion surviving, S, is the proportion surviving up to the current period multiplied by the proportion surviving that period, that is, S for last period × (*end* for this period ÷ *end* for last period).

LIFE TABLE FOR INFANT MALARIA

Table 9.2 provides basic and calculated data for a life table on the malarial morbidity of 155 infants in the Cameroon[41] (born of mothers without malarial infection of their placentas). Note that survival in this example is not thought of as remaining alive but as remaining disease free. S for the first period (>0–13 weeks) is $1.00 \times (141/155) = 0.9097$. In the second period (>13–26 weeks), 23 died and 3 were lost to follow-up, who were separated on two lines, with died first. S for that period is $0.9097 \times (118/141) = 0.7613$. At the end of that period, we subtracted the 3 *lost* subjects, leaving 115. Because we assumed that they remained alive to the end of the period, they did not reduce the survival rate, but they are removed for calculating survival rate in the next period. For the third period (>26–39 weeks), *end* is divided by the value of *end* just above it, which has had the 3 *lost* subjects removed. S for the third period is $0.7613 \times (98/115) = 0.6488$. Calculations are continued in this fashion. At the end of the first year, 58% of the infants remain free of malaria. This also may be interpreted as an estimated 0.58 probability that an infant randomly chosen at the outset will remain disease free longer than 1 year.

GRAPHING SURVIVAL INFORMATION

The graphical display of survival data was developed by E. L. Kaplan and P. Meier in 1958. One mode of display is to graph the survival data from the life table against the time intervals. A survival datum stays the same for the period of an interval, so the graph is a horizontal line over that interval. At the end of the interval, a vertical drop shows the reduction in survival to the next interval. Thus, the survival curve has a stepped pattern. The survival curve for Table 9.2 is shown in Fig. 9.1. Note that the number of subjects lost to follow-up (censored)

Table 9.2

Life Table on Malaria *(Plasmodium falciparum)* Morbidity of 155 Infants Born in Cameroon from 1993–1995

Interval (weeks)	Begin	Died	Lost	End	S (survival rate)
0 (outset)	155	0	0	155	1.0000
>0–13	155	14	0	141	0.9097
>13–26	141	23	0	118	0.7613
	118	0	3	115	
>26–39	115	17	0	98	0.6488
>39–52	98	11	0	87	0.5760
	87	0	4	83	
>52–65	83	14	0	69	0.4788
>65–78	69	24	0	45	0.3123
	45	0	2	43	
>78–91	43	17	0	26	0.1888
>91–104	26	4	0	22	0.1598

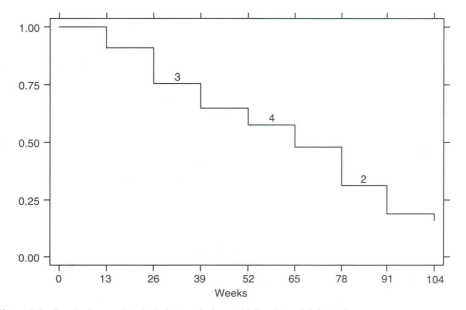

Figure 9.1 Survival curve for the infant malaria morbidity data of Table 9.2.

is shown as a small integer over the line for the period in which they were lost, distinguishing between those lost and those dying.

KAPLAN–MEIER CURVES

Kaplan and Meier developed this form of survival graph in connection with a nonparametric estimate of the survival function at each death or censoring time. For small samples, this Kaplan–Meier product limit method provides a more accurate representation of the survival pattern than does the life table, and a statistical software package with Kaplan–Meier capability should be used when actual death and censoring times are available. If computer software is used, it will provide a Kaplan–Meier survival graph. For large samples in which the number of losses per time interval is large and the width of the interval is small, the graph of the life table data is approximately the same. Survival graphs based on life tables are much simpler to understand, so they will be used here and in Chapter 25.

CONFIDENCE INTERVALS

A method exists to find a confidence interval for a Kaplan–Meier survival curve. It appears as two similar stepped patterns enclosing the survival curve. The calculation of confidence intervals on survival curves is addressed further in Chapter 25.

ADDITIONAL CONCEPTS

Two additional statistical methods useful in epidemiology should be noted: the log-rank test and correlation through time.

The Log-Rank Test Compares Two Survival Curves

If we also had examined the data of infant morbidity for cases in which the placentas had been infected with malaria, we might ask if the two curves are in probability the same or different. If they are different, we would conclude that placental infection in the mother affects the morbidity of the child. The log-rank test, a form of chi-square (χ^2), is calculated much the same way as the χ^2 statistic in Eq. (6.2). The difference between survival number expected if the two curves are the same and survival number observed is squared, divided by expected, and added. A critical value for the resulting χ^2 statistic is found from Table III (see Tables of Probability Distributions). The log-rank test is addressed further in Chapter 25.

Serial Correlation

Recall from Chapter 3 that a correlation coefficient is the adjusted covariance between two matching data sets. (Covariance is the sum of cross-products of the deviation of observations from their means.) If the matching data sets are observations not taken at a point in time but taken sequentially through time, the correlation between them is termed *serial correlation*. A serial correlation in which the second of the matching sets is a repeat of the first, but starting at a different point in time, is designated *autocorrelation;* if the second is a different variable, the serial correlation is a *cross-correlation.*

Cross-correlation

The correlation of infant malarial morbidity between malaria-free mothers and infected mothers through the 104 weeks is a cross-correlation. It tells us how closely related the two variables are through time. The correlation is based on the difference between them, not on their individual behavior through time. Thus, if they both increase and decrease together, the correlation is high even though the pattern may not be simple. Serial correlation can also be calculated with one of the sets lagged behind the other. For example, the appearance of symptoms of a disease having a 2-week incubation period can be correlated with exposure to the disease, where exposure observations are paired with the symptom observations that occurred 2 weeks later. By varying the lag, we may be able to find the incubation period. Alternatively, knowing the incubation period, finding lagged cross-correlations with several potential exposure candidates may help identify how the exposure occurred.

Autocorrelation

Observations through time may be correlated with themselves. A set of observations through time is taken as the first set, and the same set is taken as the second, except lagged. If the autocorrelation coefficient retreats from 1.00 as the lag increases and then returns to nearly 1.00, we know that we have a periodically recurring disease. If that lag is 12 months, the disease is seasonal. A plot of autocorrelation coefficients on the vertical axis with different lags on the horizontal axis is termed a *correlogram.* Periodicities in a time series can be seen easily in a correlogram as time values at which the autocorrelation coefficient reapproaches 1.00.

CHAPTER EXERCISES

9.1. How is epidemiology similar to clinical diagnosis? How is it different?

9.2. From Section 9.2, extract and list the major steps in the evolution of epidemiologic methodology.

9.3. Name two diseases in which there can be no herd immunity.

9.4. In a Canadian study including the effect of caffeine consumption on fecundability,[10] 2355 women drank caffeine-containing drinks and 89 did not. Of those who consumed caffeine, 236 conceived within 6 months of beginning pregnancy attempts. Of those who did not consume caffeine, 13 became pregnant. Calculate the 6-month incidence of pregnancy for the two groups. Note carefully that the population being considered is women attempting pregnancy, not the general population. Among this group, prevalence might be thought of as the rate at which pregnancy occurs at all, whereas the population attempting it is equivalent to the population at risk in morbidity studies. A total of 1277 couples attempting pregnancy were followed for an extensive period, and 575 were successful. Of these, 304 of the women consumed caffeine. Calculate the prevalence of caffeine consumption among women who became pregnant. Furthermore, calculate the OR (fecundity ratio) of a caffeine-consuming woman becoming pregnant.

9.5. A 2003 article[15] addresses the occurrence of acute gastroenteritis caused by Norwalk-like viruses among crew members on the U.S. Navy vessels *Peleliu* (2800 crew members) and *Constellation* (4500 crew members) visiting ports in Southeast Asia. During a period of rapid increase in the disease, 28 cases appeared over 5 days on *Peleliu* and 31 cases appeared over 14 days on *Constellation*. Calculate the incidence during the initial increase for the two ships. During an ensuing stable period (before a reduction arising from herd immunity), an average of 16.56 cases were seen on *Peleliu* and 29.37 cases on *Constellation*. Calculate the prevalence during this stable period. The total number of cases for the outbreak was 162 on *Peleliu* and 425 on *Constellation*. Calculate the OR for *Constellation* as compared with *Peleliu*.

9.6. Suppose you are working with James Lind in 1747, attempting to isolate the cause of scurvy. *What:* After a long period at sea, a number of crew members are suffering from scurvy, characterized by weakness, fatigue, listlessness, bleeding gums, loose teeth, and subcutaneous bleeding. *Where:* You note that scurvy most often occurs aboard ships, never in a fertile countryside. Although the hold is stuffy, sailors get considerable fresh air working topside. Food and human waste are emptied over the side daily, not remaining to decay. Water and food are stored and are the same for all, except that the officers eat better. Bathing is available frequently, but in seawater, not freshwater. *When:* Scurvy never occurs at the beginning of a voyage, but always after a lengthy time without landfall. Once begun, scurvy only worsens aboard ship and improves only on return to shore. *Who:* Anyone aboard ship is susceptible, but the incidence rate for officers is much lower. Crew members are in close contact over extended periods. Some contract it, but others do not. Therefore, you conclude that it is not contagious. (You have no knowledge of other limits to contagion such as immunization.) *How:* How do you further your investigation?

9.7. Table 9.3 gives the basic life table data[42] for the survival of 370 women with diabetes mellitus. Complete the table. What is the estimate of probability that a woman with diabetes survives more than 10 years?

9.8. Sketch a rough survival graph from the life table survival results in Exercise 9.7.

9.9. Data[35,66] on the survival of women with untreated breast cancer give rise to a life table with format and essential data as in Table 9.4. Complete the table. What is the probability that a woman with untreated breast cancer will survive 5 years?

9.10. Sketch a rough survival graph from the life table survival results of Exercise 9.9.

Table 9.3

Survival Data of 370 Women in Rochester, Minnesota, Having Adult-Onset Diabetes Mellitus Who Were Older Than 45 Years at Onset During 1980–1990

Interval (years)	Begin	Died	Lost	End	S (survival rate)
0 (outset)	370	0	0	370	1.0000
>0–2		33	0		
>2–4		20	0		
		0	3		
>4–6		18	0		
>6–8		22	0		
		0	2		
>8–10		20	0		

Table 9.4

Survival Data of 1824 Women with Untreated Breast Cancer

Interval (years)	Begin	Died	Lost	End	S (survival rate)
0 (outset)	1824	0	0	1824	1.000
1		545	0		
2		816	0		
3		274	0		
4		88	0		
5		44	0		
6		24	0		
		0	2		
7		5	0		

9.11. Give one example each of a medical phenomenon that could be detected using cross-correlation without lag, cross-correlation with lag, and (lagged) autocorrelation.

9.12. Give another example each of a medical phenomenon that could be detected using cross-correlation without lag, cross-correlation with lag, and (lagged) autocorrelation.

Chapter 10

Reading Medical Articles

10.1. ASSESSING MEDICAL INFORMATION FROM AN ARTICLE

Two Primary Goals

The two primary goals in reading the medical literature are keeping up with new developments and searching for specific information to answer a clinical question. The mechanisms to satisfy these goals are not very different. Keeping up is accruing general information on a specialty or subspecialty, whereas searching is accruing information about a particular question; the distinction is merely one of focus.

Ways to Improve Efficiency in Reading Medical Articles

There are several ways to improve efficiency in reading medical articles.

(1) Allow enough time to *think* about the article. A fast scan will not ferret out crucial subtleties. It is what a charlatan author would wish the reader to do, but it disappoints the true scientist author.

(2) From the title and beginning lines of the abstract, identify the central question about the subject matter being asked by the author. Read with this in mind, searching for the answer; this will focus and motivate your reading.

(3) Ask yourself, if I were to do a study to answer the author's question, how would I do it? Comparing your plan with the author's will improve your experiment planning if the author's plan is better than yours, or it will show up weaknesses in the article if it is worse than yours.

(4) A step process in answering a study question was posed in Section 5.5. Verify that the author has taken these steps.

(5) Read the article repeatedly. Each time, new subtleties will be discovered and new understanding reached. Many times we read an article that appears solid on a first perusal only to discover feet of clay by the third reading.

(6) When seeming flaws are discovered, ask yourself: Could I do it better? Many times we read an article that appears to be flawed on a first perusal, only to find on study and reflection that it is done the best way possible under difficult conditions.

The reader may rightly protest that there is not sufficient time for these steps for each of the myriad articles appearing periodically in a given field. A fast perusal of articles of minor importance is unavoidable. My advice is to weigh the importance of articles appearing in a journal and select those, if any, for solid study that may augment basic knowledge in the field or change the way medicine is practiced. There will not be many.

10.2. KEEP IN MIND HOW A STUDY IS CONSTRUCTED

SCIENTIFIC DESIGN OF MEDICAL STUDIES

A medical study design may be involved, approaching the arcane. Breaking it into the components used in constructing a study will simplify it. The basic steps are:

- Specify, clearly and unequivocally, a question to be answered about an explicitly defined population.
- Identify a measurable variable capable of answering the question.
- Obtain observations on this variable from a sample that represents the population.
- Analyze the data with methods that provide an answer to the question.
- Generalize this answer to the population, limiting the generalization by the measured probability of being correct.

CONTROL GROUPS AND PLACEBOS

A frequent mechanism to pinpoint the effect of a treatment and to reduce bias is to provide a *control* group having all the characteristics of the experimental group except the treatment under study. For example, in an animal experiment on the removal of a generated tumor, the control animals would be surgically opened and closed without removing the tumor so that the surgery itself will not influence the effect of removing the tumor. In the case of a drug efficacy study, a control group may be provided by introducing a *placebo*, a capsule identical to that being given the experimental group except lacking the study drug.

VARIABLES

A *variable* is just a term for an observation or reading giving information on the study question to be answered. Blood pressure (BP) is a variable giving information on hypertension. Blood uric acid level is a variable giving information on gout. The term *variable* may also refer to the symbol denoting this observation or reading. In study design, it is essential to differentiate between independent and dependent variables. An *independent* variable is a variable that, for the purposes of the study question to be answered, occurs independently of the effects being studied. A *dependent* variable is a variable that depends on, or more exactly is influenced by, the independent variable. In a study on gout, suppose we ask if blood uric acid (level) is a factor in causing pain. We record blood uric acid level as a measurable variable that occurs in the patient. Then we record pain as reported by the patient. We believe blood uric acid level is predictive of pain. In this relationship, the blood uric acid is the independent variable and pain is the dependent variable.

EVIDENCE AND PROOF

The results of a single study are seldom conclusive. We rarely see "proof" in science. As evidence accrues from similar investigations, confidence increases in the correctness of the answer. The news media like to say, "The jury is still out." In a more accurate rendition of that analogy, the jurors come in and lodge their judgment one at a time—with no set number of jurors.

10.3. STUDY TYPES

Different types of study imply different forms of design and analysis. To evaluate an article, we need to know what sort of study was conducted.

REGISTRY

A *registry* is an accumulation of data from an uncontrolled sample. It is not considered to be a "study." It may start with data from past files or with newly gathered data. It is useful in planning a formal study to get a rough idea of the nature of the data: typical values to be encountered, the most effective variables to measure, the problems in sampling that may be encountered, and the sample sizes required. It does not, however, provide definitive answers, because it is subject to many forms of bias. The fact of needing information about the nature of the data and about sampling problems implies the inability to ensure freedom from unrepresentative sampling and unwanted influences on the question being posed.

CASE–CONTROL STUDY

A *case–control study* is a study in which an experimental group of patients is chosen for being characterized by some outcome factor, such as having acquired a disease, and a control group lacking this factor is matched patient for patient. Control is exerted over the selection of cases but not over the acquisition of data within these cases. Sampling bias is reduced by choosing sample cases using factors independent of the variables influencing the effects under study. It still lacks evidence that chance alone selects the patients and therefore lacks assurance that the sample properly represents the population. There still is no control over how the data were acquired and how carefully they were recorded. Often, but not always, a case–control study is based on prior records and therefore is sometimes loosely termed a *retrospective study*. Case–control studies are useful in situations in which the outcomes being studied either have a very small incidence, which would require a vast sample, or are very long developing, which would require a prohibitively long time to gain a study result.

155

COHORT STUDY

A *cohort study* starts by choosing groups that have already been assigned to study categories, such as diseases or treatments, and follows these groups forward in time to assess the outcomes. To try to ensure that the groups arose from the same population and differ only in the study category, their characteristics, both medical and demographic, must be recorded and compared. This type of study is risky, because only the judgment of what characteristics are included guards against the influence of spurious causal factors. Cohort studies are useful in situations in which the proportion in one of the study categories (not in an outcome as in the case–control study) is small, which would require a prohibitively large sample size.

CASE–CONTROL CONTRASTED WITH COHORT STUDIES

The key determinant is the sequence in which the risk factor (or characteristic) and the disease (or condition) occur. In a cohort study, experimental subjects are selected for the risk factor and are examined (followed) for the disease; in a case–control study, experimental subjects are selected for the disease and are examined for the risk factor.

RANDOMIZED CONTROLLED TRIAL

The soundest type of study is the *randomized controlled trial* (RCT), often called a *clinical trial*. An RCT is a true experiment in which patients are assigned randomly to study category, such as clinical treatment, and are then followed forward in time (making it a *prospective study*) and the outcome is assessed. (A fine distinction is that, in occasional situations, the data can have been previously recorded and it is the selection of the existing record that is prospective rather than the selection of the not-yet-measured patient.) An RCT is *randomized*, meaning that the sample members are allocated to treatment groups by chance alone so that the choice reduces the risk of possibly biasing factors. In a randomized study, the probability of influence by unanticipated biases diminishes as the sample size grows larger. An RCT should be *masked* or *blinded* when practical, meaning that the humans involved in the study do not know the allocation of the sample members, so they cannot influence measurements. Thus, the investigator cannot judge (even subconsciously) a greater improvement in a patient receiving the treatment the investigator prefers. Often, both the investigator and the patient are able to influence measurements, in which case both might be masked; such a study is termed *double-masked* or *double-blinded study*.

PAIRED AND CROSSOVER DESIGNS

Some studies permit a design in which the patients serve as their own controls, as in a "before-and-after" study or a comparison of two treatments in which the patient receives both in sequence. For example, to test the efficacy of drugs A and B to reduce intraocular pressure, each patient may be given one drug for a period of time and then (after a "washout" period) the other. A *crossover* design is a type of paired design in which half the patients are given drug A followed by B and the other half is given drug B followed by A; this prevents any effect of the first treatment carrying over into the second and contaminating the contrast between A and B. Of course, the allocation of patients to the A-first versus B-first groups must be random.

10.4. SAMPLING BIAS

Bias, or lack of representativeness, has been referred to repeatedly in this chapter, for good reason. The crucial step in most design aspects lies in the phrase "a sample that represents the population." Sampling bias can arise in many ways. Clear thinking about this step avoids most of the problems.

Although true randomness is a sampling goal, too often it is not achievable. In the spirit of "some information is better than none," many studies are carried out on convenience samples that include biases of one sort or another. These studies cannot be considered conclusive and must be interpreted in the spirit in which they were sampled.

SOURCES OF BIAS

Let us consider biases that arise from some common design characteristics. The terms used here appear in the most common usage, although nuances occur and different investigators sometimes use the terms slightly differently.

The sources of bias in a study are myriad and no list of possible biases can be complete. Some of the more common sampling biases to be alert to are given in the following list. (Some other biases that are unique to integrative literature studies, or meta-analyses, are addressed in Section 10.6.)

(a) *Bias resulting from method of selection.* Included would be, for example, patients referred from primary health care sources or advertisements for patients (biased by patient awareness or interest), patients who gravitate to care facilities that have certain reputations, and assignment to clinical procedures according to therapy risks.

(b) *Bias resulting from membership in certain groups.* Included would be, for example, patients in a certain geographic region, in certain cultural groups, in certain economic groups, in certain job category groups, and in certain age groups.

(c) *Bias resulting from missing data.* Included would be patients whose data are missing because of, for example, dropping out of the study because they got well or not responding to a survey because they were too ill, too busy, or illiterate.

(d) *State-of-health bias* (Berkson's bias). Included would be patients selected from a biased pool, that is, people with atypical health.

(e) *Prevalence/incidence bias* (Neyman's bias). Included would be patients selected from a short subperiod for having a disease showing an irregular pattern of occurrence.

(f) *Comorbidity bias.* Included would be patients selected for study that have concurrent diseases affecting their health.

(g) *Reporting bias.* Some socially unacceptable diseases are underreported.

In the end, only experience and clear thought, subjected when possible to the judgment of colleagues, can provide adequate freedom from bias.

10.5. STATISTICAL ASPECTS WHERE ARTICLES MAY FALL SHORT

Apart from bias, there are several statistical areas in which journal articles may fall short. Some of those giving the most frequent problems are addressed here.

CONFUSING STATISTICAL VERSUS CLINICAL SIGNIFICANCE

Statistical significance implies that an event is unlikely to have occurred by chance; clinical significance implies that the event is useful in health care. These are different and must be distinguished. A new type of thermometer may measure body temperature so accurately and precisely that a difference of 1/100 degree is detectable and statistically significant, but it is certainly not clinically important. In contrast, a new treatment that increases recovery rate from 60% to 70% may be very significant clinically but is associated with a level of variability that prevents statistical significance from appearing. When "significance" is used, its meaning should be designated explicitly if not totally obvious by context. Indeed, we might better use the designation clinically important or clinically relevant.

VIOLATING ASSUMPTIONS ON WHICH STATISTICAL METHODS ARE BASED

The making of assumptions, more often implicit than explicit, is discussed in Section 2.7. If data are in the correct format, a numerical solution to an equation will always emerge, leading to an apparent statistical answer. The issue is whether or not the answer can be believed. If the assumptions are violated, the answer is spurious, but there is no label on it to say so. It is important for the reader to note whether or not the author has verified the assumptions.

GENERALIZING FROM POORLY BEHAVED DATA

How would we interpret the mean human height, which occurs as a bimodal distribution? If we made decisions based on the average of such a bimodal distribution, we would judge a typical man to be abnormally tall and a typical woman to be abnormally short. Authors who use descriptors (e.g., m, s) from a distribution of sample values of one shape to generalize to a theoretical distribution of another shape mislead the reader. The most frequently seen error is using the mean and standard deviation of an asymmetric (skewed) distribution to generate confidence intervals that assume a symmetric (normal) distribution. We see a bar chart of means with little standard error whiskers extending above, implying that they would extend below symmetrically; but do they? For example, the preoperative plasma silicone level in DB5 is skewed to the right; it has a mean of about 0.23, with standard deviation 0.10. For clinical use, we want to know above what level the upper quarter of the patients is. From the data, the 75th percentile is 0.27, but generalization from N(0.23,0.01)[*] claims the 75th percentile to be about 0.30. The risk-of-danger level starts higher using the normal assumption than is shown by the data. The author should verify for the reader the shape of the sample distributions that were used for generalization.

FAILURE TO DEFINE DATA FORMATS, SYMBOLS, OR STATISTICAL TERMS

An author labels values in a table as means but adds a plus/minus ($\pm$) symbol after the values, for example, 5.7 ± 1.2. Are the 1.2 units s, standard error of the mean, 1.96σ, $t_{1-\alpha/2} \times s$, or something else? Beware of the author who does not define formats and symbols. A further problem is the use of statistical terms. Articles sometimes fail to define terms clearly, for example, referring to "data samples compared for significance using the general linear model procedure." The general linear model is quite general and includes a host of specific tests. Examination of the data and results leads this statistician to conclude that the authors in this case compared pairs of group means using the F test, which is the square of the t test for two means. The authors could have said, "Pairs of means were compared using the t test." However, this may be a wrong conclusion. Are we at fault for failing to understand the jargon used? No. It is incumbent on authors to make their methodologies clear to the reader. Beware of authors who "snow" rather than inform.

USING MULTIPLE RELATED TESTS THAT CUMULATE THE p-VALUE

The means of a treatment group are compared with those of a placebo group for systolic BP, diastolic BP, heart rate (HR), and white blood cell count using four t tests, each with a 5% chance of indicating there is a difference when there is not (the p-value). Because the treatment may affect all of these variables, the risk for a false difference accumulates to approximately 20%. (Four such tests yield a risk $= 1 - (1 - p)^4 = 0.185$.) If we performed 20 such tests, we would be almost sure that at least 1 positive result is spurious. The solution to this problem is to use a multivariate test that tests the means of several variables simultaneously, such as Hotelling's T^2 or Mahalanobis' D^2 (methods, unfortunately, outside the realm of this book). Can multiple t or other tests ever be used? Of course. When independent variables do not influence the same dependent variable being tested, they may be tested separately.

---

[*]Recall the symbolism: normal distribution with mean 0.23 and variance 0.01 (standard deviation 0.10).

CHOOSING INAPPROPRIATE STATISTICAL TESTS

Too many investigators choose the wrong method to test a hypothesis, usually through ignorance, occasionally by machination. This alters the risk associated with a conclusion (e.g., the *p*-value) and may even change the conclusion of the investigation. Although some involved studies require the judgment of an experienced professional statistician for design and analysis, most studies may be assessed for appropriate choice of test using Table 10.1 (or using an expanded version of this table, which appears on the inside back cover of this book). Some of these methods appear in Part I of this book, and the remaining methods are discussed in Part II. The use of this table is explained in the table's legend.

10.6. EVOLVING TERMS: META-ANALYSIS, MULTIVARIABLE ANALYSIS, AND OTHERS

A number of terms and methods new to mainstream medicine are beginning to be seen in journal articles, one of the more frequent being *meta-analysis*. Let us define some of these terms; then the concept of meta-analysis will be examined in more detail.

MULTIPLE, MULTIVARIATE, AND MULTIVARIABLE

Multi- connotes several or many, but technical uses vary. *Multiple* usually just implies "more than two." Sometimes it connotes multivariate, apparently when the adjective "multivariate" coupled with the noun it modifies generates too many syllables, as in "multiple regression." *Multivariate* implies *one dependent* variable depending on *more than one independent* variable. Methods satisfying that definition are collectively called *multivariate analysis;* unfortunately, much of multivariate analysis is beyond what is attempted in this book. The clinician will not encounter it daily but will find it useful to have seen the definition before encountering it. Consider DB12 on carinal resection, for which we would like to identify risk factors for patient death or survival, the dependent variable. To decide if extent of resection (independent variable) will separate the surviving group from the dying group (dependent variable), we would test mean extent for the two groups using a *t* test. However, we could test the surviving group's mean extent, age, prior surgery, and intubation against the dying group's means for those variables using a generalization of the *t* test, the T^2 test of Harold Hotelling. The concept is much the same, except we are using several independent variables simultaneously. Multiple regression is another form of multivariate analysis. In simple regression, a dependent variable (e.g., HR) is predicted from an independent variable (perhaps minutes of stair climbing). In multiple regression, a dependent variable (HR) is predicted from several independent variables simultaneously (minutes of stair climbing, patient age, number of days per week the patient exercises). One step further is the case of *more than one dependent variable*, which is termed *multivariable analysis* and is quite far afield from the level of this book. If, in DB3, we wish to predict the dependent variables 5- and 10-day

Table 10.1

A First-Step Guide to Choosing Statistical Tests That Appear in Part I

Type of data: Questions about:		Counts (Nominal) Proportions		Ranks Position in distribution	Continuous measurements (including discrete)		
					Averages		Spread
		p not near 0 or 1	p near 0 or 1	Not required	Normal curve	Far from normal curve	
Assumed distributions:		Binomial					
Small sample	Single or paired sample	Binomial table	Poisson table	Signed-rank test	Normal test if using σ; t if using s	Go to rank methods	χ^2
	Two samples	Form a 2 × 2 contingency table and use Fisher's exact or χ^2 test		Rank-sum test	Normal test if using σ_1, σ_2; t if s_1, s_2	Go to rank methods	F
Large sample	Single or paired sample	Normal approx	Poisson approx	Normal approx to signed-rank	Normal test if σ known or if large n; t if use estimate s		χ^2
	Two samples	Form a 2 × 2 contingency table and use Fisher's exact or χ^2 test		Normal approx to rank-sum	Normal test if σ_1, σ_2 known or if large n; t if use estimates s_1, s_2		F

Select column by specifying the type of data, the question being asked of these data, and the distribution being assumed. Select row by specifying sample size and number of samples. The row–column intersection provides the most apparent test fitting these conditions. This selection does not satisfy all requirements and must be considered as only tentative. An expanded table is reviewed more fully in Chapter 12 (see Table 12.1) and also appears on the inside back cover of this book. χ^2, chi-square.

serum theophylline levels simultaneously from the independent variables baseline serum level and age, we would use a form of multivariable analysis. The several terms discussed in this paragraph are arcane enough to be used without proper care in some articles; therefore, the reader must be the careful one.

THE CONCEPT OF META-ANALYSIS

Meta-analysis is a pooling of data from various journal articles to enlarge the sample size, thus reducing the sizes of the Types I and II errors. It treats the different articles as replications of a single study, wherein lies its hazard. The crucial questions are which articles are acceptable to pool and how to pool them.

STEPS TO CONDUCT A META-ANALYSIS

A meta-analysis should be developed as follows:

(1) Define the inclusion/exclusion criteria for admitting articles.
(2) Search exhaustively and locate all articles addressing the issue.
(3) Assess the articles against the criteria.
(4) Quantify the admitted variables on common scales.
(5) Aggregate the admitted databases. The reader may see one of several aggregation methods referred to, including nonparametric or omnibus combination, vote counting, weighted pooling, analysis of variance combining (hopefully, after homogeneity tests), linear or regression models, correlation coefficient combining, or clustering. These aggregation methods are not discussed here. When an author of an article has referred to one of these methods, it may be enough to note that the author is aware of the requirement for methodical aggregation and has attempted it. The meta-analysis that gives no indication of how the databases were combined should be viewed with some reserve.

BIASES THAT MAY INFILTRATE AN INTEGRATIVE LITERATURE REVIEW

Several biases may infiltrate an integrative literature review.

(1) *Lack of original data.* Are original data given? A bias may occur from omitting data that do not support the author's agenda.

(2) *Lack of scientific rigor.* Various biases can creep in when one article has been done less rigorously than another. Among other biases are varying enthusiasm in interpreting results, varying quality of randomization, and downplaying or omitting outliers.

(3) *Inadequate reporting policy.* Authors tend to avoid submitting negative findings and journals tend to reject them when they are submitted, leading to overestimation of the success of an approach.

CRITERIA FOR AN ACCEPTABLE META-ANALYSIS

Because a meta-analysis depends a great deal on the investigator's judgment, clear criteria are crucial. A minimum list of criteria follows, and the reader should verify, so far as it is possible, that each criterion has been met.

(1) The study objectives were clearly identified.
(2) Inclusion criteria of articles in general and specific data to be accepted were established before selection.
(3) An active effort was made to find and include all relevant articles.
(4) An assessment of publication bias was made.
(5) Specific data used were identified.
(6) Assessment of article comparability (e.g., controls, circumstances) was made.
(7) The meta-analysis was reported in enough detail to allow replication.

Even after a careful meta-analysis, limitations may remain, for example, innate subjectivity, aggregation of data of uneven quality, and the forcing of results into a mold for which they were not intended.

The book *Statistical Methods for Meta-Analysis*, by Hedges and Olkin,[24] provides an excellent reference for further information regarding meta-analysis. Section V of *Medical Uses of Statistics*, by Bailar and Mosteller,[2] is a quite readable additional reference.

10.7. SELECTION OF STATISTICAL TESTS TO USE IN A STUDY

We read a journal article and see that the author used the frequently seen *t* test. Was this test appropriate? To answer this question, we need to examine the type of data, the question to be asked of the data, the number of arms or groups in the study, the size of the sample, and other characteristics. Although there is no easy "formula" to choosing a test, a cross-categorization table may be helpful. Table 10.1 provides a first-step guide to choosing the appropriate test for the most common situations, which are found in Part I of this book. It is, of course, far from exhaustive. Directions for using the table are given in the legend. An expanded table appears inside the book's back cover (see also Table 12.1), together with more complete directions on how to use the table. In these latter tables, references to sections in the book where the methods may be found are described and exemplified. The more complete table will also aid in choosing analysis methods when planning the reader's own studies.

CHAPTER EXERCISES

10.1. Choose a medical article from your field. Evaluate it using the guidelines given in Section 10.1.
10.2. From the discussion in Section 10.3, what sort of study gave rise to (a) DB2? (b) DB14? What are the independent and dependent variables in each?

10.3. Is the study represented by DB6 an RCT? Why or why not?

10.4. Might sex and/or age differences in the independent variables have biased the outcomes for (a) DB2? (b) DB14? What can be done to rule out such bias?

10.5. From Table 10.1, what test for age bias for DB14 data would you choose? For sex bias?

10.6. Find a medical article in which meta-analysis is used and evaluate it using the guidelines of Section 10.6.

Answers to Chapter Exercises, Part I

CHAPTER 1

1.1. Many such questions are possible. Some examples follow. (a) What is the average PSA level for the 301 patients? The sample frequency distribution? What is the correlation coefficient between age and PSA? (b) Is the average PSA level different for patients with positive and with negative biopsy results? Is the correlation coefficient between age and PSA statistically significant, or could it have happened by chance alone? (c) Can PSA predict (within limits) biopsy outcome and therefore serve as a risk factor? Can age predict (within limits) a patient's PSA and therefore serve as a risk factor?

1.2. Many such questions are possible. Some examples follow. (a) What is the mean difference in eNO over 20 minutes for patients with and without EIB? The median? What is the correlation between eNO difference and relative humidity? (b) Is the mean difference in eNO over 20 minutes different for patients with and without EIB? Is the correlation between eNO difference and relative humidity significant in probability? (c) Can eNO before exercise predict eNO change over 20 minutes? Can relative humidity predict eNO change over 20 minutes?

1.3. Phase II.

1.4. $m = 0, f = 1$ (or vice versa).

1.5. Recording to the nearest 10,000 would be adequate; to the nearest 25,000 would not be accurate. The clinical decision for a range of 156,000 to 164,000 recorded as 160,000 would not be affected, but the clinical decision for a range of 135,000 to 185,000 recorded as 160,000 might be.

1.6. (a) Sex. (b) Age, bone density, UCSF norm. (c) Yes.

1.7. (a) Below-, in-, and above-range values. (b) INR readings. (c) Yes, -0.70 is the smallest (rank 1) and 0.62 is the largest (rank 104).

1.8. DB2: nausea scores.

1.9. He would like to generalize to humankind as a whole, or at least to humankind in the United States. He is not justified in doing so based on the information given, because there is no indication of how the sample was drawn. (For example, it might have been drawn from deck sailors who are constantly in the sun and wind and have developed different skin properties, which would form a sampling bias.)

CHAPTER 2

2.1. (a) 1–<1.5, 1.5–<2, 2–<2.5, ..., 5–<5.5, 5.5–6.

(b) Tally can be verified by comparing frequencies with plot in (d).

(c) Median = 2.5.

(d)

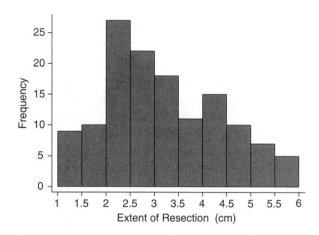

(e) Distribution is skewed to the right and does not grow negligible in the tails but is not dramatically abnormal.

(f) Mean appears to be about 3. Mode is 2.25. Mode < median < mean, as expected in a right-skewed distribution.

2.2. (a) 6–10, 11–15, 16–20, ..., 76–80.

(b) Tally can be verified by comparing frequencies with plot in (d).

(c) Yes, median of accumulating data converges to final median of 51.

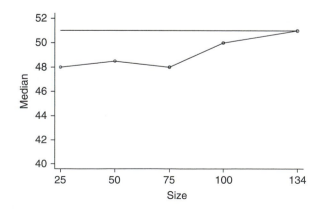

(d)

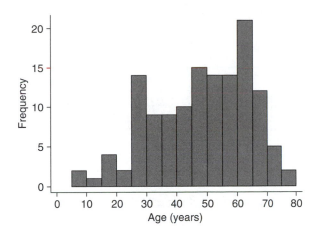

(e) Distribution is a little skewed to the left, a little too "bulky" in the center but is not dramatically abnormal.

(f) The mean appears to lie in the 46–50 interval. The mode is 62.5. The mean < median < mode, as expected in a left-skewed distribution.

2.3 (a) 1–<1.25, 1.25–<1.50,..., 4–4.25.

(b) Tally can be verified by comparing frequencies with plot in (d).

(c) Median = 2.38.

(d)

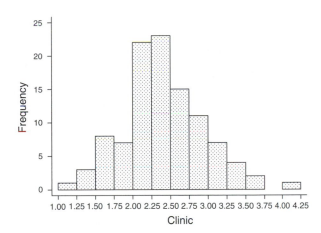

(e) Mean of 2.40 is almost the same as median. Although this does not imply symmetry in all respects, as can be seen from inspection of the distribution in (d), it does imply that the distribution is not skewed.

2.4. (a) 1–<1.25, 1.25–<1.50,..., 3.50–3.75.

(b) Tally can be verified by comparing frequencies with plot in (d).

(c) Median = 2.15.

(d)

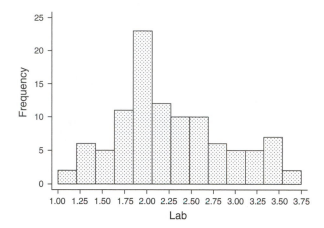

(e) Mean of 2.28 is somewhat to right of median, implying that the data extend farther to the right of the median than to the left, indicating a slight skew.

2.5. (a) Variance $= 0.0467$. (b) Standard deviation $= 0.2160$.

2.6. (a) Variance $= 0.009167$. (b) Standard deviation $= 0.0957$.

2.7. The sample estimate we would use is the mean (averaged over patients) of changes in serum theophylline levels from baseline to 5 days. We assume this mean is drawn from a normal distribution. We choose the probability of a false positive (concluding there is a real change when there is not), usually (but not always) 0.05. We use the t probability table (for this small sample size of 16) to establish an interval about the theoretical mean of zero (no difference), within which there is probably no difference. We reject the hypothesis of no difference if the calculated t falls outside this interval or conclude no difference was shown if t falls within the interval.

2.8. The sample estimate we would use is the mean (averaged over patients) of changes in eNO from before exercise to 20 minutes after. We assume this mean is drawn from a normal distribution. We choose the probability of a false positive (concluding there is a real change when there is not), usually (but not always) 0.05. We use the t probability table (for this small sample size of 39) to establish an interval about the theoretical mean of zero (no change), within which there is probably no difference. We reject the hypothesis of no change if the calculated t falls outside this interval or conclude no change was shown if t falls within the interval.

2.9. Chi-square, because chi-square is based on squared data. (Thus, variances, e.g., are distributed as chi-square.)

2.10. For plot, see 2.3(d). The distribution is more similar to a normal.

2.11. $df = n - 1 = 29$. Subtract 0, the theoretical mean difference if preoperative and postoperative arise from identical distributions; divide by the standard deviation of the sample mean (see discussion of standard error of the mean in Section 2.9). The cut point is called the *critical value*.

2.12. $df = n - 1 = 103$.

2.13. In order appearing in database, ranks are: 20, 19, 17, 6, 15, 18, 13, 14, 2, 16, 12, 11, 8, 4, 7, 10, 3, 5, 1, 9.

2.14. In order appearing in database, ranks are: 3, 15, 5, 12, 4, 16, 8, 14, 10, 2, 11, 17, 18, 13, 6, 9, 7, 1.

2.15. Proportion is not close to zero: binomial.

2.16. Proportion is very close to zero: Poisson.

2.17. SEM $= 0.2160/\sqrt{4} = 0.1080$.

2.18. SEM $= 0.2868/\sqrt{8} = 0.1014$.

2.19. Covariance.

CHAPTER 3

3.1. (a) 154.2, (b) 154.1, (c) 576.4, (d) 24.0, (e) 140.1, (f) 164.9, (g) 5.7.

3.2. (a) 452.75, (b) 462.5, (c) 8416.2140, (d) 91.7399, (e) 372.5, (f) 532.0, (g) 32.4350.

3.3.

Treatment	(a)	(b)	(c)	(d)	(e)	(f)	(g)
1	31.0	31.8	16.8	4.1	28.6	33.2	0.84
2	15.1	9.2	165.2	12.9	5.4	24.8	2.62

3.4. 62.

3.5. 39.04.

3.6. (a) 5539.75, (b) 0.9823.

3.7. (a) 0.20007, (b) 0.8398.

3.8. 0.963.

3.9.

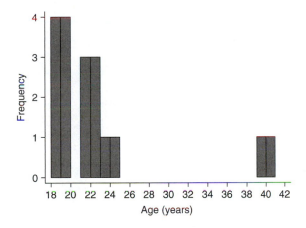

3.10.

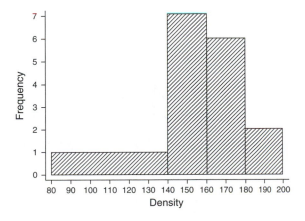

3.11. (a)

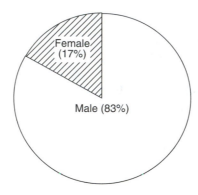

(b)

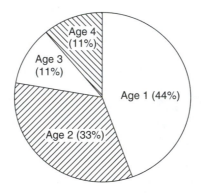

3.12.

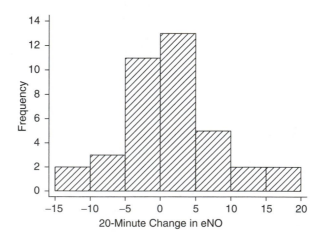

3.13.

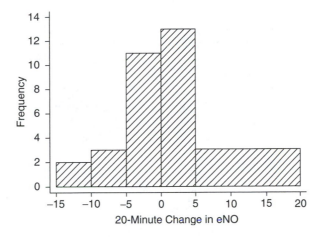

3.14.

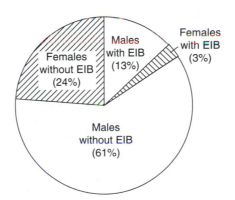

3.15.

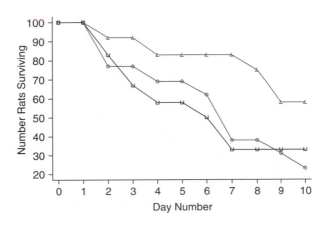

3.16.

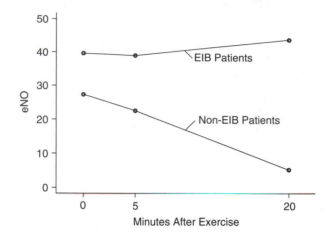

3.17. (a)

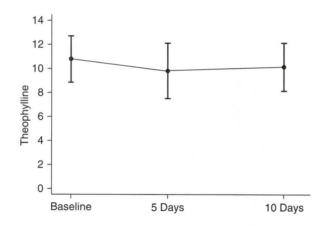

(b)

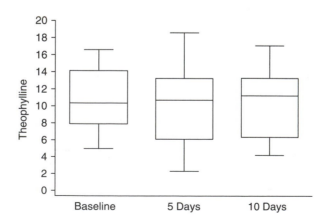

3.18. (a)

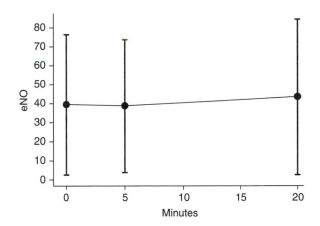

(b)

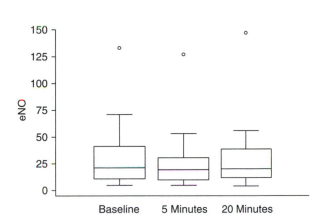

Comment on 3.18: Note how the one patient with outlying values distorts the means and magnifies the standard deviations so that 3.18(a) gives little of the information about the data pattern seen in 3.18(b) and renders the result misleading.

3.19.

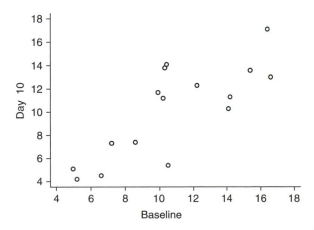

3.20.

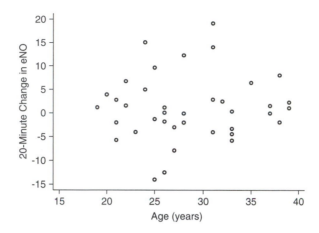

3.21.

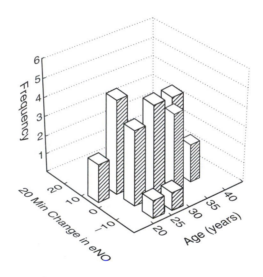

CHAPTER 4

4.1. Mean is 2, standard deviation is 3. $z = (x - 2)/3$.

4.2. $2 \pm 1.96 \times 3$, or $-3.88, 7.88$.

4.3. The 95% confidence limits on individuals are 16.8 to 78.8. Your patient's age is 2.27 standard deviations below the mean; only 1.2% of patients would be younger (using Table I). Your patient is improbably young.

4.4. The confidence interval on an individual patient would be $1.96 \times \sigma$ or 1.23 units below and above the mean of 2.28 units. Thus, we would be 95% confident that an individual patient would lie in the interval 1.05–3.51 INR units. This databased confidence interval is much larger than the in-range interval.

4.5. Using Table I, (a) (8.59, 9.73); (b) (8.42, 9.90); (c) (8.69, 9.63).

4.6. Confidence limits for mean INR, based on 104 patients, are $2.28 \pm 1.96 \times 0.63/\sqrt{104}$, leading to $P[2.16 < \mu < 2.40] = 0.95$. The confidence interval on the mean is much smaller than the individual patient in-range interval, as would be expected; but, interestingly, it lies close to the lower in-range limit rather than lying in the center.

4.7. Using Table II with 19 *df*, (4.42, 8.70).

4.8. Using Table II with 7 *df*, (60.19, 75.31).

4.9. Confidence limits for treatment 1 are $34.40 \pm 2.201 \times 3.25/\sqrt{12}$, leading to $P[32.34 < \mu < 36.46] = 0.95$, and for treatment 2 are $36.12 \pm 2.201 \times 1.17/\sqrt{12}$, leading to $P[35.38 < \mu < 36.86] = 0.95$. The confidence interval on the mean for treatment 2 is notably smaller.

4.10. Using Table III in calculating the lower bound and Table IV in calculating the upper bound, with 19 *df*, (a) (12.08, 44.54); (b) (3.48, 6.67).

4.11. Using Table III in calculating the lower bound and Table IV in calculating the upper bound, with 7 *df*, (a) (35.73, 338.49); (b) (5.98, 18.40).

4.12. For treatment 1, $P[10.56 \times 11/21.92 < \sigma^2 < 10.56 \times 11/3.82] = p[5.30 < \sigma^2 < 30.41] = 0.95$, leading to $P[2.30 < \sigma < 5.51] = 0.95$. For treatment 2, $P[1.37 \times 11/21.92 < \sigma^2 < 1.37 \times 11/3.82] = P[0.69 < \sigma^2 < 3.95] = 0.95$, leading to $P[0.83 < \sigma < 1.99] = 0.95$. The confidence interval on the standard deviation for treatment 2 is notably smaller.

CHAPTER 5

5.1. It is a clinical hypothesis, stating what the investigator suspects is happening. A statistical hypothesis would be the following: Protease inhibitors do not change the rate of pulmonary admissions. By stating no difference, the theoretical probability distribution can be used in the test. (If there is a difference, the amount of difference is unknown, and thus the associated distribution is unknown.)

5.2. $H_0: \mu_w = \mu_{w/o}$; $H_1: \mu_w \neq \mu_{w/o}$.

5.3. (a) The t distribution is associated with small data sample hypotheses about means. (b) Assumptions include: The data are independent one from another. The data samples are drawn from normal populations. The standard deviations at baseline and at 5 days are equal. (c) The Type I error would be concluding that the baseline mean and the 5-day mean are different when, in fact, they are not. The Type II error would be concluding the two means are the same when, in fact, they are different. (d) The risk for a Type I error is designated α. The risk for a Type II error is designated β. The power of the test would be designated $1 - \beta$.

5.4. (a) The χ^2 distribution is associated with a hypothesis about the variance. (b) Assumptions include: The data are independent one from another. The data sample is drawn from a normal population. (c) The Type I error would be concluding that the platelet standard deviation is different from 60,000 when, in fact, it is 60,000. The Type II error would be concluding the platelet standard deviation is 60,000 when, in fact, it is not. (d) The risk for a Type I error is designated α. The risk for a Type II error is designated β.

5.5. (a) The F distribution is associated with a hypothesis about the ratio of two variances. (b) Assumptions include: The data are independent one from another. The data samples are drawn from normal populations. (c) The Type I error would be concluding the variance (or standard deviation) of serum silicon before the implant removal is different from the variance (or standard deviation) after when, in fact, they are the same. The Type II error would be concluding the before and after variances (or standard deviations) are the same when, in fact, they are different. (d) The risk for a Type I error is designated α. The risk for a Type II error is designated β.

5.6. The "above versus below" view would interpret this decrease as not significant, end of story. Exercise has no effect on the eNO of healthy subjects. The "level of p" view would say that, although the 2.15-ppb decrease has an 8% chance of being false, it also has a 92% chance of being correct. The power of the test should be evaluated. Perhaps the effect of exercise on eNO in healthy subjects should be investigated further.

5.7. (a) A categorical variable. (b) A rating that might be any type, depending on circumstance. In this case, treating it as a ranked variable is recommended, because it avoids the weaker methods of categorical variables and a five-choice is rather small for use as a continuous variable. (c) A continuous variable.

5.8. 1, 7, 8, 3, 5, 2, 4, 6.

5.9. (a) Exercise-induced bronchoconstriction (EIB) frequency by sex:

	Sex		
	1 (male)	2 (female)	Totals
0	23	9	32
1	5	1	6
Totals	28	10	38

(b) 5-minute eNO differences by sex as ranks:

5-minute difference	5-minute ranks	Sex	Ranks for males (1)	Ranks for females (2)
6.0	5	1	5	
−1.0	4	1	4	
7.1	6	1	6	
−3.9	1	2		1
−2.0	2.5	1	2.5	
−2.0	2.5	1	2.5	

(c) 5-minute eNO differences by sex as continuous measurements:

5-minute difference	Sex	Difference for males (1)	Difference for females (2)
6.0	1	6.0	
−1.0	1	−1.0	
7.1	1	7.1	
−3.9	2		−3.9
−2.0	1	−2.0	
−2.0	1	−2.0	
		$m = 1.62, s = 4.54$	$m = -3.9$, no s

5.10. (1) Does the drug reduce nausea score following gallbladder removal? (2 and 3) Drug/No Drug against Nausea/No Nausea. (4) H_0: nausea score is independent of drug use; H_1: nausea score is influenced by drug use. (5) The population of people having laparoscopic gallbladder removals who are treated for nausea with Zofran. The population of people having laparoscopic gallbladder removals who are treated for nausea with a placebo. (6) My samples of treated and untreated patients are randomly selected from patients who present for laparoscopic gallbladder removal. (7) A search of the literature did not indicate any proclivity to nausea by particular subpopulations. (8) These steps seem to be consistent. (9) A chi-square test of the contingency table is appropriate; let us use $\alpha = 0.05$. (10) (Methodology for step 10 is not given in Part I; Chapter 22 provides information for the student who wishes to pursue this.)

5.11. (1) Are our clinic's INR readings different from those of the laboratory? (2) Difference between clinic and laboratory readings. (3) Mean of difference. (4) H_0: mean difference $= 0$; H_1: mean difference $\neq 0$. (5) Population: All patients, past and future, subject to the current INR evaluation methods in our Coumadin Clinic. (6) Sample: the 104 consecutive patients taken in this collection. (7) Biases: The readings in this time period might not be representative. We can examine records to search for any cause of nonrepresentativeness. Also, we could test a small sample from a different time and test it for equivalence. (8) Recycle: These steps seem consistent. (9) Statistical test and α: paired t test of mean difference against zero with $\alpha = 0.05$.

CHAPTER 6

6.1. Critical value is 3.84. $\chi^2 = 5.40$, which is greater than 3.84. We decide that protease inhibitors are effective. (Actual $p = 0.020$.)

6.2. Critical value is 3.84. $\chi^2 = 8.32$, which is greater than 3.84. We decide that titanium-containing ink is harder to remove. (Actual $p = 0.004$.)

6.3. Critical value is 3.84. $\chi^2 = 9.32$, which is greater than 3.84. We decide that nausea score is reduced by the drug. (Actual $p = 0.002$.)

6.4. The 2×2 table is:

		Clinic		
		Out of range	In range	
Truth (Laboratory)	Out of range	30	27	57
	In range	3	44	47
		33	71	104

From Eq. (6.2), $\chi^2 = 25.44$, which is greater than 3.84, the 5% critical value of χ^2. Clinic results were discrepant from laboratory results.

6.5. (a)

		Diagnosis		
		Appendicitis	Not appendicitis	
Truth	Appendicitis	82	22	104
	Not appendicitis	23	73	96
		105	95	200

(b) 0.24; 24% of cases were diagnosed as appendicitis when, in fact, they were not.

(c) 0.21; 21% of cases were diagnosed as not appendicitis when, in fact, they were.

(d) 0.79; 79% of appendicitis cases were diagnosed properly.

(e) 0.76; 76% of cases without appendicitis were diagnosed properly.

(f) 0.78; 78% of all cases were diagnosed properly.

(g) 11.8; when appendicitis is diagnosed, odds that the diagnosis is correct are nearly 12 to 1.

6.6. (a) 0.06; 6% of clinical results were out of range when laboratory results were in range.

(b) 0.47; 47% of clinic results were in range when laboratory results were out of range.

(c) 0.53; 53% of out-of-range clinic results agreed with laboratory results.

(d) 0.94; 94% of in-range clinic results agreed with laboratory results.

(e) 0.71; 71% of clinic results overall agreed with laboratory results.

(f) 16.30; when a clinic result is in range, the odds of agreement with the laboratory result are more than 16 to 1.

6.7. $n_1 = 6$, $n_2 = 10(>8)$, $T = 72$, $\mu = 51$, $\sigma = 9.22$, $z = 2.278$, $z > 1.96$, the critical value for two-tailed $\alpha = 0.05$. (Actual $p = 0.023$.) Serum theophylline levels drop significantly more in women than in men.

6.8. $n_1 = 3$, $n_2 = 15(>8)$, $T = 23$, $\mu = 28.5$, $\sigma = 8.44$, $z = -0.652$, $z < -1.96$, the critical value for two-tailed $\alpha = 0.05$. (Actual $p = 0.514$.) There is inadequate evidence to state that there is a difference in bone density between men and women.

6.9. A frequency plot:

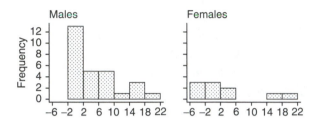

The frequency distributions are far from normal. Rank methods are appropriate. Ranking the data and abstracting the ranks for female subjects (the smaller group; $n_1 = 10$) leads to the following display:

5-minute differences	−5.4	−3.9	−3.9	0.1	0.3	1.0	2.0	4.0	17.5	21.1
Ranks	1	2.5	2.5	13	14.5	17.5	20	25	35	38

The sum of ranks is $T = 169$. $\mu = n_1(n_1 + n_2 + 1)/2 = 195$. $\sigma^2 = n_1 n_2(n_1 + n_2 + 1)/12 = 910$. $z = (T - \mu)/\sigma = -0.862$. We seek the area under the normal curve to the left of this z, that is, about 7/8 of a standard deviation below the mean, because of symmetry the same as the area above +0.862. From Table I, linear interpolation gives this value about 0.62 of the way from 0.212 to 0.184, or about 0.195. Doubling this value because the female 5-minute difference could have been either more or less than that for male subjects, we find the p-value for the test is about 0.390. (An exact calculation shows it to be 0.389.) We would have about a 4/10 chance of being wrong if we said there was a difference related to sex, so we conclude that there is no evidence of a sex difference.

6.10. (a) (Subscript s represents survivors, subscript d represents those who died.) $n_s = 117$, $n_d = 17$, $m_s = 48.05$, $m_d = 46.41$, $\sigma = 15.78$, SEM $= 4.10$, $z = 0.4$. $z < 1.96$, the normal's critical value for two-tailed $\alpha = 0.05$. (Actual $p = 0.345$.) The age difference between patients who survived and those who died is not significant in probability.

(b) $n_s = 117$, $n_d = 17$, $m_s = 2.82$, $m_d = 3.96$, $\sigma = 1.24$, SEM $= 0.32$, $z = -3.56$. $z < -1.96$, the normal's critical value for two-tailed $\alpha = 0.05$. (Actual $p < 0.001$.) The extent of resection is significantly greater in patients who died.

6.11. $n_w = 6$, $n_m = 10$, $m_w = 3.335$, $m_m = -0.412$, $s_w = 4.268$, $s_m = 2.067$, $s_d = 1.571$, $t_{14df} = 2.385$, $t > 2.145$, the critical value for two-tailed $\alpha = 0.05$. (Actual $p = 0.032$.) Serum theophylline levels drop significantly more in women than in men; indeed, levels tend to increase in men. The conclusion of the test agrees with the conclusion in Exercise 6.5.

6.12. $n_w = 3$, $n_m = 15$, $m_w = 148.97$, $m_m = 155.27$, $s_w = 15.79$, $s_m = 25.64$, $s_d = 15.57$, $t_{16df} = -0.405$, $t > -2.120$, the critical value for two-tailed $\alpha = 0.05$. (Actual $p = 0.691$.) There is inadequate evidence to state that there is a difference in bone density between men and women. The conclusion of the test agrees with the conclusion in Exercise 6.6.

CHAPTER 7

7.1. (a) One-sided. Ondansetron hydrochloride could decrease nausea score but does not seem to increase it. Also, the clinical decision not to use the drug is the same whether nausea remains the same or is increased.

(b) Two-sided. There is no reason to believe that the spectrophotometric readings will be greater in one assay than the other.

7.2. (a) A false-positive result implies that the patient takes the drug when it will not help nausea. A false-negative result implies the patient experiences nausea when it could have been reduced. The assessment depends on the side effects and the cost of the drug. If these should be minimal, a false negative is worse than a false positive, so α should be larger than β. If these are important clinically, the commonly used 5% α and 20% β trade-off should be rethought.

(b) A false-positive result, that is, implant appears to increase plasma silicone when it does not, may lead to unnecessary removal surgery. A false-negative result, that is, implant appears to be of no risk when in fact it is, implies allowing possible side effects. Both errors have undesirable clinical implications, but a β larger than α is reasonable.

7.3. $n_1 = n_2 = 71$.

7.4. $n_1 = n_2 = 141$.

7.5. $n_1 = n_2 = (1.96 + 1.28)^2(0.54^2 + 0.63^2)/0.25^2 = 115.6$. A minimum of 116 readings in each group for a total of 232 readings are required.

7.6. $n_1 = n_2 = 31$.

7.7. The minimum sample size estimates provide a sample size that will just barely reach the 5% α *on average*. It is not surprising that repeated sampling yields p-values on both sides of $\alpha = 0.05$. A slightly larger sample size than the minimum requirement would have increased the confidence that a significant p would occur.

CHAPTER 8

8.1. Respiration rate is the dependent variable. Age is the independent variable. The curve goes downward from left to right. No clear pattern is discernible; it would be sensible to start with a simple straight-line model, as in Eq. (8.1).

8.2. $\text{Corr}(r, a) = -0.195$. The form is a downward slope, as conjectured in Exercise 8.1. Some correlation is apparent, but it is not strong.

8.3. $r - 39.0 = -0.34(a - 10.9)$.

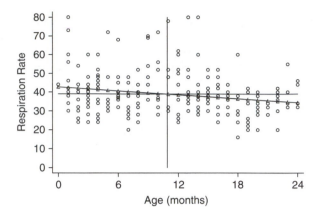

8.4. $R^2 = 0.038$. Age exerts only about 4% of the causal influence on respiration rate. (For the student's information, a t test on the slope of the regression line, which is addressed in Chapter 24, yields a p-value of 0.003, which is strongly significant. We can state that the influence of age on respiration rate is real, but it is small in comparison with other influences.)

8.5. Preexercise eNO increases with 20-minute eNO. It appears that a straight line will express the relationship. (Absent the outlying point in the top right, we might have tested for a curved relationship. When that point is included, we cannot suggest any pattern other than linear.)

8.6. The correlation coefficient $r = 0.96$. This is quite a high correlation coefficient. As baseline eNO increases, 20-minute eNO increases proportionally.

8.7. Using the mean-and-slope form, $eNO_{20} = 28.1 + 0.95(eNO_0 - 29.3)$.

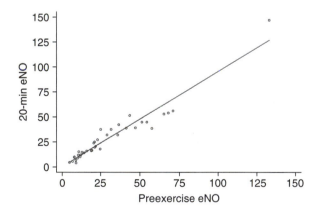

8.8. The coefficient of determination $R^2 = (0.96)^2 = 0.92$. It may be concluded that baseline eNO is quite a good predictor of 20-minute eNO.

8.9. (a) The measure of effectiveness (MOE) is the cost, because no other relevant factors are cited. (b) Using medicine A costs x. Using medicine B costs $y + 0.1x$. Therefore, use y if $y + 0.1x < x$, or if $y < 0.9x$. (c) \$15.

CHAPTER 9

9.1. In both epidemiology and clinical diagnosis, a cause of illness is being sought by trying to isolate a specific cause from among many potential causes. They are different in that clinical diagnosis seeks to treat an individual, whereas epidemiology seeks to treat a population.

9.2. The steps chosen are somewhat arbitrary. The student's attention to epidemiologic history evoked by searching and selecting steps has served the learning purpose.

9.3. Cancer caused by tobacco smoke or occupational exposure, vitamin or mineral deficiency, and airway obstruction caused by allergic reaction are three examples of many possible such diseases.

9.4. $I_{caff} = 1000 \times$ newly pregnant caffeine consumers/total caffeine consumers $= 1000 \times 236/2355 = 100$ per thousand. $I_{nocaff} = 1000 \times$ newly pregnant not caffeine consumers/total not caffeine consumers $= 1000 \times 13/89 = 146$ per thousand. $P_{caff} = 1000 \times$ pregnant women who consumed caffeine/all pregnant women $= 1000 \times 304/575 = 529$ per thousand. OR $=$ number pregnant who consume caffeine/number pregnant who do not consume caffeine $= 304/271 = 1.12$.

9.5. $I_{Pel} = 1000 \times 28/2800$ divided by 5 days $= 2.0$ per day per thousand. $I_{Con} = 1000 \times 31/4500$ divided by 14 days $= 0.5$ per day per thousand. $P_{Pel} = 1000 \times 16.56/2800 = 5.9$ cases per thousand. $P_{Con} = 1000 \times 29.37/4500 = 6.5$ cases per thousand. Odds of gastroenteritis on *Peleliu* were $162/2638 = 0.0614$ and on *Constellation* were $425/4075 = 0.1043$. OR of *Constellation* versus *Peleliu* were $0.1043/0.0614 = 1.6987$. A randomly chosen sailor was about 1.7 times as likely to acquire gastroenteritis on *Peleliu* as on *Constellation*.

9.6. Because all the sailors are exposed to the same air, food, water, and personal habits, and because no outside influences of any type occur, it appears that the cause is not something present, but something absent. You suspect diet. You choose a sample of patients (Lind chose 12), dividing them into experimental groups (Lind chose 2 per group for 6 treatments). You assign them six different diets, each including a component not regularly eaten aboard ship. It would be sensible to choose foods eaten regularly ashore but not at sea. (Lind did some of this, but some of his treatments were bizarre, such as drinking seawater.) Fortunately, one of your choices is fresh citrus fruits, which quickly cures that group, whereas no other group improves. You have found a cure for scurvy, despite not knowing that vitamins exist or that their deficiency can cause disease.

9.7.

Interval (years)	Begin	Died	Lost	End	S (survival rate)
0 (outset)	370	0	0	370	1.0000
>0–2	370	33	0	337	0.9108
>2–4	337	20	0	317	0.8567
	317	0	3	314	
>4–6	314	18	0	296	0.8076
>6–8	296	22	0	274	0.7476
	274	0	2	272	
>8–10	272	20	0	252	0.6926

The estimated probability that a woman with diabetes survives more than 10 years is 0.69.

9.8.

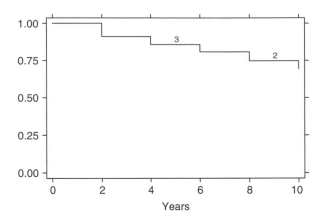

9.9.

Interval (years)	Begin	Died	Lost	End	S (survival rate)
0 (outset)	1824	0	0	1824	1.000
1	1824	545	0	1279	0.701
2	1279	816	0	463	0.254
3	463	274	0	189	0.104
4	189	88	0	101	0.055
5	101	44	0	57	0.031
6	57	24	0	33	0.018
	33	0	2	31	
7	31	5	0	26	0.014

The probability that a woman with untreated breast cancer will survive 5 years is about 3%.

9.10.

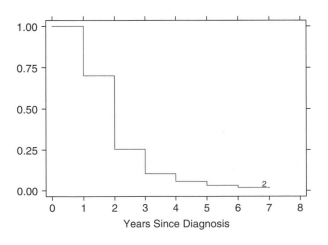

9.11. Many examples are possible. One each is given here for illustration. *Cross-correlation without lag:* Cases of yellow fever appearing in a certain tropical nation, suspected to arise from a mutated virus, seem to exhibit longer-lasting thrombocytopenia. From records, proportions of patients with below-normal platelet counts from before and after the appearance of the new strain are correlated over the course of the disease. The low correlation verifies the mutated behavior. *Cross-correlation with lag:* A group of people who recently traveled in a tropical nation contract a new type of viral disease. The dates of visit and of symptom onset are correlated with various lags. The lag that provides the largest cross-correlation coefficient estimates the average incubation period. *Autocorrelation:* Incidence of aspergillosis is recorded for several years (see Section 22.4 for data on this topic). The autocorrelation coefficient is calculated for a lag of 1 year and is seen to be high. It is clear that aspergillosis is a seasonal disease.

9.12. Many examples are possible. One each is given here for illustration. *Cross-correlation without lag:* A measure of pollen level in air correlates in time with prevalence of allergic reaction. *Cross-correlation with lag:* Days in hospital correlates with incidence of a certain nosocomial disease with an incubation period equal to the lag. *Autocorrelation:* Malarial symptoms in a patient correlate with themselves when the lag equals the reproduction cycle time of the parasite.

CHAPTER 10

10.1. There is no unique answer for this exercise.

10.2. (a) Randomized controlled trial. Drug versus placebo is the independent variable. Nausea scores are the dependent variables. (b) Cohort study. EIB is the independent variable. eNO differences are the dependent variables.

10.3. No. Among the reasons why not are the following: Rather than being a prospective study, the sample is a "convenience sample," that is, drawn from patients presenting already tattooed and who have decided on removal; there is no control group for comparison; and the investigator is not masked from the decision about "response" to the treatment.

10.4. (a) Yes. Nothing can be done; age and sex were not recorded. (b) Yes. Age and sex can be tested for bias.

10.5. Age: Continuous measurements, different averages for EIB versus non-EIB groups, assume normal, small sample, two groups, standard deviations estimated (i.e., s, not σ). Choose two-sample t test. Sex: Counts, small sample, two groups. Choose χ^2 test or Fisher's exact test of a contingency table.

10.6. This is no unique answer for this exercise.

Part II

A Reference Guide

Chapter 11

Using the Reference Guide

11.1. HOW TO USE THIS GUIDE

DESIGNED FOR THE RESEARCH CLINICIAN

Part II, A Research Guide, caters to the clinician reading an article or conducting a study. This part forms a "how to" guide to the more common statistical methods in medicine. It does not take the user through derivations or seldom used details but rather sticks to what the clinical investigator needs to do the basics of the job; if it is not in this book, see a biostatistician.

TWO WAYS TO ACCESS GUIDANCE

The reference part is designed to be accessed in two ways. If, while reading a medical article, the user sees a statistical method, the method and examples of its use may be found through the subject index. If the user has a research plan or actual data and needs the appropriate analytic method, a detailed table of contents is provided. A guide to appropriate statistical tests may be found by using a table of data characteristics paired with methods, which is located on the inside back cover and at the end of Chapter 12 (see Table 12.1). Whereas Part I is designed to be studied through in order from beginning to end, the reference part allows the user to "hop about"; that is, the user can go directly to the relevant stand-alone method with no requirement to read through lead-up material. A great effort has been made to avoid deluges of detail and to make the presentation user-friendly.

PART I PROVIDES THE BASICS THAT THE PART II USER NEEDS

The material found in Part I provides the foundation for the use of Part II as a reference. A user who has learned that material through another medium, such as personal study or an earlier course, need not study Part I, although it might be useful to scan it to refresh the memory or fill in any gaps in knowledge. A user who is not familiar with the material in Part I should study it to a reasonable level of understanding. In the same manner that the concepts of addition and multiplication are required to get far in arithmetic, the concepts of a random variable and its probability distribution are required to do much with statistics.

THE FORMAT IS TAILORED TO THE MEDICAL USER

Based on the observation that most statistics users page through a text to find an example first and then read the methodology, each new section starts with an example. After the example is posed, the method is given step by step, the example is completed, and then an additional example in a different medical specialty is given. Finally, so that the user can verify an understanding of the method, an exercise is posed. The exercise is worked out at the end of the chapter. The user, once familiar with the methods of a chapter, may review the specific chapters in Chapter Summaries. These summaries provide quick references so that the user does not have to ferret out formulas and related details from the text.

11.2. BASIC CONCEPTS NEEDED TO USE THIS GUIDE

A SUMMARY REVIEW

This section reviews definitions of the more basic concepts that were introduced in Part I. A reader finding a concept less than fully clear should read the equivalent section in Part I. The following subsections are given approximately in the order of Chapters 1 to 8 to facilitate finding these concepts. Alternatively, the reader may use the subject index.

TERMS AND SYMBOLS

Statistics: Development of probabilistic knowledge using observed quantities (data).
Types of data (statistical methods differ for each type of data):
   Continuous data: Positions on a scale, for example, heart rate or blood pressure.
   Rank-order data: Numbers of data put in order, for example, smallest to largest.
   Categorical data: Counts of members of categories, for example, number of male and female subjects.
Ratings: Quantified judgments about a patient or event on the basis of that event alone, not related to other events. (Ranks are quantified relative to other patients or events and all should be different, except for ties.)
Symbols: A sort of shorthand to allow relationships to be seen more easily.
Types of symbols:
   Name symbol: Defines a thing or concept, for example, x for a measurement or μ for a mean.
   Operator symbol: Defines an act or process, for example, $+$ or $\sum$ for addition.

Relationship symbol: Defines the relation between (sets of) symbols, for example, < or =.

Indicator symbol: A symbol, usually a subscript, that can represent any member of a family, for example, x_i representing the ith member of the x family.

Population: The entire set of subjects about which we want information.

Sample: A portion of a population available to estimate characteristics of the population.

Greek versus Roman symbols: Mostly, but not always, Greek is used for population values and Roman is used for sample values.

Representativeness: A sample should have the same characteristics as the population.

Bias: A violation of representativeness. (see Section 10.4 for types of biases)

Randomness: Selection of a sample member on the basis of chance alone.

Advantage of a large sample size: A sample grows increasingly representative of a population as its size approaches that of the population.

PROBABILITY AND FREQUENCY DISTRIBUTIONS

Frequency distribution: The way sample data are distributed along the scale of a variable divided into intervals, shown as number of data categorized into each interval.

Relative frequency (of an interval): The proportion of the total number of data categorized into that interval.

Probability (of an event): The proportion out of a total number of occurrences that a particular event occurs. Also, the proportion of area under a probability distribution curve associated with the occurrence of the event.

Probability distribution: The way theoretical data are distributed (usually smoothly) along the scale of a variable. Also, the frequency distribution of an entire population.

Parameter: A value of a variable that acts like a constant for the duration of the data set of interest, but varies from one data set to another, that is, the mean.

Quantiles: Separation points in a distribution divided so that the same number of entries (or area under the curve) falls into each quantile, for example, 4 for quartiles, 100 for percentiles.

Averages: Measures of the typical.

Mean: The sum of sample values divided by the number.

Median: The value such that half the data are less than and half are greater than that value.

Mode: The most frequently observed value or interval in a large data set.

Measures of spread:

Range: Highest value minus lowest value.

Variance: Average of squared deviations from the mean.

Standard deviation: Square root of variance, taken to put the measure of spread in the same dimensionality as the variable being measured.

Interquartile range: 3rd quartile minus 1st quartile, including the central half of the sample distribution. Sometimes, half of this value is used: the semi-interquartile range.

Measures of shape:

Skewness: Measure of asymmetry. One tail is stretched out.

Other: Rarely used is kurtosis, a measure of flatness or peaks of the distribution.

Standard distribution: Distribution of values with the mean subtracted from each and then divided by the standard deviation, yielding a distribution with 0 for its mean and 1 for its standard deviation.

Confidence interval: An interval (bounded by lower and upper bounds) on a parameter estimate that has a specified probability of enclosing the population (or theoretical) parameter value. The 95% confidence most commonly used implies a 5% risk for error.

Assumptions: All statistical methods require some assumptions to be valid, some requiring few, some, many. The assumptions should be verified before the method is used. The most common assumption is that all observations are drawn from the same probability distribution and are independent from one another. The assumption that this distribution is normal is a more stringent assumption required in some cases.

Robustness: How impervious a method is to violations of its assumptions.

Degrees of freedom (df): A difficult, multifaceted concept. Start by thinking of *df* as the number of observations in the sample minus the number of methodologic constraints.

The six probability distributions most frequently used in statistical methods:

Normal (also: Gaussian): A bell-shaped curve with certain specific properties.

Standard normal: A normal curve standardized to have mean 0 and standard deviation 1, sometimes symbolized $N(0,1)$.

t: Another bell-shaped curve looking much like the normal but a little more spread out.

Chi-square (symbolized χ^2): An asymmetric curve, starting at zero and extending indefinitely to the right, forming a right skew. The distribution of a sum of squares.

F: An asymmetric curve appearing somewhat like χ^2. The distribution of a ratio of two sums of squares.

Binomial: A unimodal curve arising from a set of binary events, for example, surviving or not. The curve is nearly symmetric if the binary events have probabilities close to 0.5, but becomes increasing skewed as the probabilities tend to 0 and 1.

Poisson: An approximation to the binomial when the binary probabilities are very close to 0 and 1.

Central Limit Theorem: The distribution of a mean of sample values is approximately normal, whatever the distribution of the values used to calculate the mean, and grows closer to normal as the sample size increases.

Critical value in a test: In testing a hypothesis, the cut point on the variable's axis where the hypothesis is rejected on one side but not on the other side.

Standard error of the mean (SEM): The standard deviation of a mean. A sample SEM grows small as the sample size grows large.

Joint distribution of two variables:
  Covariance: A measure of the joint variability of two variables.
  Correlation coefficient: The covariance standardized by dividing the two standard deviations, yielding a measure between −1 and 1. At 1 (or −1), the relation is perfect: the value of one variable gives the exact value of the other. At 0, the variables are unrelated.

DESCRIPTIVE STATISTICS

Numerical summaries: See previous paragraph for measures of average, spread, and shape. Methods of calculation appear in Section 3.1.

Pictorial summaries: Better understood as pictures than words; see Sections 3.3 and 3.4 for bar chart, histogram, pie chart, line chart, mean-and-standard-error chart, box-and-whisker chart, and scatter plot.

CONFIDENCE INTERVALS

The definition for confidence interval has been given earlier in this chapter. Chapter 4 describes different forms of confidence intervals and explains how to find the bounds of confidence from probability tables.

HYPOTHESIS TESTING

Clinical hypothesis: A statement of the relationship between variables in terms of what is expected to occur from a clinical viewpoint.

Statistical hypothesis: A statement of the relationship between variables in terms of what is expected to occur in statistical terminology.

Null hypothesis: A statement in statistical symbols that the sample is not different from known information. This allows probability calculations from the known information.

Alternate hypothesis: A statement in statistical symbols that the sample is different from known information. Probability calculations are difficult or impossible because the amount of difference is unknown.

Types of error:
  Type I (α) error: The error of rejecting a null hypothesis when it is true. (Similar to the false positive of a clinical test.) α is the probability of a Type I error.
  Type II (β) error: The error of failing to reject a null hypothesis when it is false. (Similar to the false negative of a clinical test.) β is the probability of a Type II error.

Power: $1 - \beta$. The probability of rejecting a null hypothesis when it is false.

TESTING, RISKS, AND ODDS

Test of contingency: In a 2 × 2 table of prediction (e.g., clinical diagnosis of appendicitis or not appendicitis) versus outcome (removed appendix was infected or not infected), a test of association between prediction and outcome. A χ^2 test is the most common. The numbers of rows and/or columns may be more than 2.

Sensitivity: From a 2 × 2 contingency table, the probability of a correct prediction of occurrence, for example, detecting a disease when it is present.

Specificity: From a 2 × 2 contingency table, the probability of a correct prediction of absence of the occurrence, for example, ruling out a disease when it is absent.

Odds of an occurrence: The frequency of occurrence in ratio to the frequency of nonoccurrence.

Odds ratio (OR): The odds of occurrence in ratio to the odds of non-occurrence.

Rank tests: Tests derived for comparing ranked data. When violation of assumptions invalidate the more stringent tests, the data may be ranked and rank tests used.

Rank-sum test: A rank test that compares two sample distributions.

Means test: A test that compares two sample means or a sample mean to a theoretical mean.

Normal means test: A means test for large samples.

t test: A means test for small samples.

SAMPLE SIZE

Minimum required sample size: The smallest sample size that will allow a successful test of hypothesis.

Power analysis: A name for the calculation of the minimum required sample size, so named because power is one of the values used in the calculation. Others are α, β, a clinically relevant test difference, and often the standard deviation of the raw data.

STATISTICAL PREDICTION

Model: A quantitative representation of a relationship or pattern of occurrence. A constant might represent hourly average heart rate of a healthy patient. An increasing straight line might represent that patient's minute-to-minute heart rate during the beginning of aerobic exercise.

Straight-line model (linear model): Two data define a straight line, most often the mean x, y point and the slope of the line.

Regression: In its simplest form, a method yielding the straight line that best fits a scatter of points in two dimensions.

Predictor: If knowing a value of x (the independent variable) provides some information about the value of y (the dependent variable), x is said to be a predictor of y. Regression provides a formula for this prediction.

Coefficient of determination (R^2): A measure of quality of regression prediction.

Multiple regression: The method of predicting a dependent variable by several independent variables simultaneously.

Logistic regression: The regression method for predicting a binary dependent variable, for example, disease free or recurrence. The name arises from the logarithmic transformation used in the calculation.

Measure of effectiveness (MOE): An indicator, usually a combination of simple measures, of how a treatment or protocol affects patients. Outcomes analysis is one form of using MOEs.

Chapter 12

Planning Medical Studies

12.1. THE SCIENCE UNDERLYING CLINICAL DECISION MAKING

THE SCIENTIFIC METHOD

Science is a collection of fact and theory resting on information obtained by using a particular method, which is therefore called the scientific method. This method is a way of obtaining information constrained by a set of criteria. The method is required to be objective; the characteristics should be made explicit and mean the same to every user of the information. The method should be unbiased, free of personal or corporate agendas; the purpose is to *investigate* the truth and correctness of states and relationships, not to prove them. The true scientific approach allows no preference for outcome. The method should involve the control of variables; ideally, it should eliminate as far as practicable all sources of influence but one so that the existence of and extent of influence of that one source is undeniable. The method should be repeatable; other investigators should be able to repeat the experiment and obtain the same results. The method should allow the accumulation of results; only by accumulation does the information evolve from postulate to theory to fact. The scientific method is the goal of good study design.

JARGON IN SCIENCE

Jargon may be defined as technical terminology or as pretentious language. The public generally thinks of it as the latter. To the public, *carcinoma* is jargon for *cancer*, but to the professional, technical connotation is required for scientific accuracy. We need to differentiate between jargon for pomposity and jargon for accuracy, using it only for the latter and not unnecessarily. The reason for this mention is that the same process occurs in statistics. Some statistical terms are used loosely and often erroneously by the public, who miss the technical implications. Examples are *randomness*, *probability*, and *significance*. Users of statistics should be aware of the technical accuracy of statistical terms and use them correctly.

195

EVIDENCE

The accumulating information resulting from medical studies is evidence. Some types of study yield more credible evidence than others. Anecdotal evidence, often dismissed by users seeking scientific information, is the least credible, yet is still evidence. The information that patients with a particular disease often improve more quickly than usual when taking a certain herb usually does give actual rates of improvement and does not indicate the rate of failure of the treatment. However, it may suggest that a credible study be done. It may serve as a candle in a dark room. The quality of the study improves as we pass through registries, case–control studies, and cohort studies, to the current "gold standard" of credibility: the randomized controlled prospective clinical trial (RCT) (see Sections 10.2 and 10.3 for more information on types of studies). It is incumbent on the user of evidence to evaluate the credibility of the cumulative evidence: number of accumulated studies, types of studies, quality of control over influencing factors, sample sizes, and peer reviews. Evidence may be thought of as the blocks that are combined to build the scientific edifice of theory and fact. The more solid blocks should form the cornerstones and some blocks might well be rejected.

EVIDENCE-BASED MEDICINE

Evidence-based medicine (EBM) melds the art and science of medicine. Evidence-based medicine is just the ideal paradigm of health care practice, with the added requirement that updated credible evidence associated with treatment be sought, found, assessed, and incorporated into practice. It is much the way we all think we practice, but it ensures consideration of the evidence components. It could be looked at somewhat like an airliner cockpit check; even though we usually mentally tick off all the items, formal guides verify that we have not overlooked something.

One rendition of the EBM sequence might be the following: (1) We acquire the evidence: the patient's medical history, the clinical picture, test results, and relevant published studies. (2) We update, assess, and evaluate the evidence, eliminating evidence that is not credible, weighting that remaining evidence according to its credibility, and prioritizing that remaining according to its relevance to the case at hand. (3) We integrate the evidence of different types and from difference sources. (4) We add nonmedical aspects, for example, cost considerations, the likelihood of patient cooperation, and the likelihood of patient follow-up. (5) Finally, we embed the integrated totality of evidence into a decision model.

12.2. THE OBJECTIVE OF STATISTICS

PRIMARY OBJECTIVE

The primary objective of statistics is to make a decision about a population based on a sample from that population. Recalling that the term *population* refers to all members of a defined group, the term *sample* to a subset of the population, and the symbol α to the chance of being wrong if we decide a treatment difference exists, we may restate the most common objective of statistics as follows: Based on a sample, we conclude that a treatment difference exists in the population if the risk of being wrong (a false-positive difference) is less than α.

For example, suppose that of 50 urgent care patients with dyspepsia who are given no treatment, 30 are better within an hour, and of 50 given a "GI cocktail" (antacid with viscous lidocaine), 36 are better within an hour. The treatment is effective in this sample, in that the conditions of 20% more treated than untreated patients showed improvement for these 100 patients. The question for statistics to answer is: Is it likely to work for all patients, or was the result for this sample "luck of the draw"?

WHAT STATISTICS WILL NOT DO FOR US

Statistics will not make uncertainty disappear. Statistics will not give answers without thought and effort. Statistics will not provide a credible conclusion from poor data; that is, to use an old maxim, it won't make a silk purse out of a sow's ear. It is worth keeping in mind that putting numbers into a formula *will* yield an answer, and the process *will not* inform the user whether the answer is credible. The onus is on the user to apply credible data to obtain a credible answer.

WHAT STATISTICS WILL DO FOR US

There is no remedy for uncertainty, but statistics allows you to *measure and control uncertainty*. This benefit is *one of the most crucial and critically important bases for scientific investigation.*

In addition, statistics can assist us to do the following:

- Clarify our exact question
- Identify the variable and the measure of that variable that will answer that question
- Verify that the planned sample size is adequate
- Test our sample for bias
- Answer the question asked, while limiting the risk for error in our decision

Other benefits of statistics include:

- Allowing us to follow strands of evidence obscured by myriad causes
- Allowing us to mine unforeseen knowledge from a mountain of data
- Providing the credibility for the evidence required in EBM
- Reducing the frequency of embarrassing mistakes in medical research

12.3. CONCEPTS IN STUDY DESIGN

EXPERIMENTAL DESIGN CAN REDUCE BIAS

The crucial step that gives rise to most of the design aspects is encompassed in the phrase "a sample that represents the population." Sampling bias can arise in many ways. Clear thinking about this step avoids many of the problems. Experimental design characteristics can diminish biases.

CONTROL GROUPS, PLACEBOS, AND VARIABLES

The concepts of control groups, placebos, and variables and the distinction between independent and dependent variables are discussed in Section 10.2 and should be reviewed.

COMMON TYPES OF STUDY DESIGNS

Common designs—namely, registry, case–control, cohort, RCT, and paired designs—and how they reduce bias are introduced in Section 10.3 and should be reviewed.

12.4. SAMPLING SCHEMES

PURPOSE OF SAMPLING SCHEMES

The major reason for different sampling procedures is to avoid bias. Common sources of bias are discussed in Section 10.4 and should be reviewed. Methods of sampling are myriad but most relate to unusual designs, such as complicated mixtures of variables or designs with missing data. The following paragraphs sketch four of the most basic methods, used with rather ordinary designs.

Simple Random Sampling

If a sample is drawn from the entire population so that any member of the population is as likely to be drawn as any other, that is, drawn at random, the sampling scheme is termed *simple random sampling* (see earlier).

Systematic Sampling

Sometimes we are not confident that the members sampled will be drawn with truly equal chance ("equilikely"). We need to sophisticate our sampling scheme to reduce the risk for bias. In some cases, we may draw a sample of size n by dividing the population into k equal portions and drawing n/k members from each division. For example, suppose we want 50 measurements of the heart's electrical conductivity amplitude over a 10-second period, where recordings are available each millisecond (msec). We could divide the 10,000 msec into 50 equal segments of 200 msec each and sample one member equilikely from each segment. Another example might be a comparison of treatments on pig skin healing. The physiologic properties of the skin vary by location on the pig's flank. An equal number of samples of each treatment is taken from each location, but the assignments are randomized within this constraint. These schemes are named *systematic sampling*.

Caution

The term *systematic sampling* is sometimes used to refer to sampling from a systematic criterion, such as all patients whose name starts with G, or sampling at equal intervals, such as every third patient. In the latter case, the position of the patient chosen in each portion is fixed rather than random. For example, the third, sixth, and so on patients would be chosen rather than one equilikely from the first triplet, another equilikely from the second triplet, and so on. Such sampling schemes are renowned for bias and should be avoided.

Stratified Sampling

In cardiac sampling, suppose a cusp (sharp peak) exists, say, 100 msec in duration, occurring with each of 12 heartbeats, and it is essential to obtain samples from the cusp area. The investigator could divide the region into the 1200 msec of cusp and 8800 msec of noncusp, drawing 12% of the members from the first portion and 88% from the second. Division of the population into not-necessarily-equal subpopulations and sampling proportionally and equilikely from each is termed *stratified sampling*. As another example, consider a sports medicine sample size of 50 from Olympic contenders for which sex, an influential variable, is split 80% male to 20% female athletes. For our sample, we would select randomly 40 male and 10 female members.

Cluster Sampling

A compromise with sampling costs, sometimes useful in epidemiology, is *cluster sampling*. In this case, larger components of the population are chosen equilikely (e.g., a family, a hospital ward) and then every member of each component is sampled. A larger sample to offset the reduced accuracy usually is required.

12.5. HOW TO RANDOMIZE A SAMPLE

RANDOM VERSUS HAPHAZARD ASSIGNMENT

It has been noted that randomization is one way to reduce bias. In contrast, haphazard assignment may introduce bias. Haphazard assignment is selection by some occurrence unrelated to the experimental variables, for example, by the order of presentation of patients, or even by whim. Haphazard assignment might possibly be random, but there is no way to guarantee it. Haphazard assignment should be avoided.

RANDOM NUMBER GENERATORS

Randomization is accomplished by using a random number generator to assign patients to groups. Any mechanism that produces a number chosen solely by chance is a random number generator.

Mechanical Generators

Historically, gambling mechanisms from Roman "bones" (early dice) to roulette wheels gave numbers not far from random but with small biases. Through the centuries, more accurate manufacturing procedures have yielded mechanisms with smaller biases. When gambling dens were unable to eliminate bias in their machines, they developed machines with adjustable biases so that the bias could be "reset" if a player used statistical methods to assess and take advantage of the bias. The flip of a coin weighted equally on the two sides has been shown to be almost but not quite "50-50." The side showing after a flip yields relative frequencies closer to 51-49 in favor of the side showing at the start. The bias in dice is so small as to be negligible for experimental design purposes. It is possible to obtain dice devised for this purpose. I have a set of three icosahedron dice guaranteed by the Japan Standards Association to produce unbiased numbers between 0 and 999. Spinners might be used, but they should be tested for bias beforehand.

Electronic Generators

Many modern computer software packages include random number generator programs. The theory is to produce an irrational number (decimal places continuing infinitely without repeating patterns), often using the natural number e to some power (denoted *seed*), and choosing a number far past the decimal point. Another number in the sequence is used as a new seed for the next draw, and this algorithm is repeated. Tests of randomness have been made and randomness verified.

Tables of Random Numbers

Many statistical table reference books and statistical methods textbooks contain tables of random numbers. Close your eyes and touch the table to select a starting position. Reading from there in any direction (horizontally, vertically, diagonally) will yield random numbers. You may use digits or pairs, triplets, and so on, of digits. Tests over many years have verified the randomness of these tables. A personal observation, not verified mathematically, is that during my half century of practice I have found fewer flawed samples using these tables than using other methods (especially computer algorithms). Flawed random assignments are discussed later in Randomly Generated Sets That Are Flawed.

ASSIGNING PATIENTS TO GROUPS

Suppose you want to assign randomly 20 patients to two groups, A and B. You do not need to randomize with the arrival of each patient. First, randomly assign group A or B to a sequence of 20 blank spaces. Then, as you encounter patients, the first patient is assigned to group A or B according to the first entry in the list, after which you cross off that entry; the next patient is assigned to the next entry, and so on.

How do you assign to group A versus B when a normal die has six sides and tables or computer algorithms yield digits from 0 to 9? You can select the sequence of assignment to the two groups by even–odd. The first assignment is to group A if the die shows a 1-spot, 3-spot, or 5-spot or if the table digit is 1, 3, 5, 7, or 9. The assignment is to group B for even outcomes. Alternatively, you could take the lower number half (digits 0–4) for group A and the upper number half (digits 5–9) for B. Suppose you have three groups: A, B, and C. Assign a patient to A if the digits are 0, 1, or 2, to B if they are 3, 4, or 5, and to C if they are 6, 7, or 8. Ignore 9; it is random independent, so its occurrence is independent of any other occurrence. Follow the same sort of assignment logic for any other randomization scheme you may need.

RANDOMLY GENERATED SETS THAT ARE FLAWED

Suppose we use a random number generator to randomly assign 20 patients to 2 groups. We aim for 10 in each group. We might obtain 10 and 10, or 9 and 11, or even 8 and 12. We say "luck of the draw" and continue. It is possible, although unlikely, to obtain 4 and 16, or even 0 and 20. The probability of the former is 0.00462055, a little less than 1 in 200 trials; the probability of the latter is $1/2^{20} = 0.0000009537$, a little less than 1 in 1 million trials. Suppose we do encounter such an event; it is rare but one that obviously will produce a flawed experiment. What do we do? Clearly, we must generate a new assignment. But what keeps the unscrupulous investigator from generating assignment after assignment until obtaining one that yields a desired result? We must document our procedures in our reporting of experimental methods. If the initial assignment is clearly flawed, any reputable investigator will agree that we took the right action. The requirement to report prevents us from abusing the system. Another solution is to have your local biostatistician produce a random, unflawed assignment for you.

Exercise 12.1 Use a random number generator to assign 20 patients to groups A and B 10 different times. What is the frequency distribution of number of patients in group A? Did you get any flawed sample assignments?

12.6. HOW TO PLAN AND CONDUCT A STUDY

STEPS THAT WILL AID IN PLANNING A STUDY

Planning a study is involved, sometimes seeming to approach the arcane, but it need not be daunting if well organized. Total time and effort will be reduced to a minimum by spending time in organization at the beginning. An unplanned effort leads to stomach-churning uncertainty, false starts, acquisition of useless data, unrecoverable relevant data, and a sequence of text drafts destined for the wastebasket. Where does one start?

(1) *Start with objectives.* (Do not start by writing the abstract.) Specify, clearly and unequivocally, a question to be answered about an explicitly defined population.

(2) *Develop the background and relevance.* Become familiar with related efforts made by others. Be clear about why this study will contribute to medical knowledge.

(3) *Plan your materials.* From where will you obtain your equipment? Will your equipment access mesh with your patient availability? Dry run your procedures to eliminate unforeseen problems.

(4) *Plan your methods and data.* Identify at least one measurable variable capable of answering your question. Define the specific data that will satisfy your objectives and verify that your methods will provide these data. Develop a raw data entry sheet and a spreadsheet to transfer the raw data to that will facilitate analysis by computer software.

(5) *Define the subject population* and verify that your sampling procedures will sample representatively.

(6) *Ensure that your sample size will satisfy your objectives.*

(7) *Anticipate what statistical analysis will yield results* that will satisfy your objectives. Dry run your analysis with fabricated data that will be similar to your eventual real data. Verify that the analysis results answer your research question.

(8) *Plan the bridge from results to conclusions.* In the eventual article, this is usually termed the *discussion*, which also explains unusual occurrences in the scientific process.

(9) *Anticipate the form* in which your conclusions will be expressed (but, of course, not what will be concluded). Verify that your answer can be generalized to the population, limiting the generalization by the measured probabilities of error.

(10) *Now you can draft an abstract.* The abstract should summarize all the foregoing in half a page to a page. After drafting this terse summary, review steps (1) through (9) and revise as required.

12.7. MECHANISMS TO IMPROVE YOUR STUDY PLAN

TRICKS OF THE TRADE

There exist devices that might be thought of as "tricks of the trade." The preceding section gives steps to draft a study. However, reviewers of study drafts typically see a majority of studies not yet thoroughly planned. Three devices to improve study plans, used by many investigators but seldom if ever written down, will render plans more "solid." Using them early in the writing game often prevents a great deal of grief.

(1) Work Backward Through the Logical Process

After verification that the questions to be asked of the study are written clearly and unequivocally, go to step (9) of the list in Section 12.6 and work backward. (a) What conclusions are needed to answer these questions? (A conclusion is construed as answering a question, such as "Is the treatment efficacious?" rather than providing the specific conclusion the investigator desires.) (b) What data results will I need, and how many data will I need to reach these conclusions? (c) What statistical methods will I need to obtain these results? (d) What is the nature and format of the data I need to apply these statistical methods? (e) What is the design and conduct of the study I need to obtain these data? (f) Finally, what is the ambiance in the literature that leads to the need for this study in general and this design in particular? When the investigator has answered these questions satisfactorily, the study plan will flow neatly and logically from the beginning.

(2) Analyze Dummy Data

If you had your data at this stage, you could analyze them to see if you had chosen the appropriate data and the right recording format needed for that analysis. However, although you do not have the data per se, you have a good idea what they will look like. You have seen numbers of that sort in the literature, in pilot studies, or in your clinical experience. Use a little imagination and make up representative numbers of the sort you will encounter in your study. Then subject them to your planned analysis. You do not need more than a few; you can test out your planned analysis with 20 patients rather than 200. You do not need to have the data in the relative magnitudes you would like to see; a correlation coefficient even very different from the one that will appear in your later study will still tell you whether or not the data form can be used to calculate a legitimate correlation coefficient. This is not lost time, because not only will you learn how to perform any analyses with which you are not intimately familiar, but when you obtain your actual data, your analysis will be much faster and more efficient. This step is worth its time spent to avoid that sinking feeling experienced when you realize that your study will not answer the question because the hematocrits from 200 patients were recorded as low, normal, or high rather than as percentages.

(3) Play the Role of Devil's Advocate

A device that is useful at the planning stage, but perhaps more so when the finished study is drafted, is to "put on the hat" of a reviewer and criticize your own work. This is not an easy challenge. It requires a complete mental reset followed by self-disciplined focus and rigid adherence to that mind-set. Indeed, it requires the investigator to use a bit of acting talent. Many recognized actors achieve success by momentarily believing they are the characters they are playing: in this case, the character is a demanding, "I've-seen-it-all," somewhat cynical reviewer. A number of little mechanisms can help. Note everything that can be construed as negative, however trivial. Sneer periodically. Mutter "Good Grief! How stupid!" at each new paragraph. The object is to try really hard to discover something subject to criticism. When you have finished, and only then, go back and consider which of the criticisms are valid and rewrite to preempt a reviewer's criticism. Remember that you would rather be criticized by a friend than by an enemy, and, *if* you carry off this acting job properly, you are being your own best friend. The most difficult problem to recognize and repair in a study draft is lack of clarity. As is often true of computer manual writers, if you know enough to explain, you know too much to explain clearly. The author once had a colleague who advised, "Say it like you would explain to your mother." Find a patient person who knows nothing about the subject and explain the study, paragraph by paragraph. This technique often uncovers arcane or confusing passages and suggests wordings that can clarify such passages.

12.8. HOW TO MANAGE DATA

MAKE A PLAN

As with other aspects of experimental design, a plan reduces work, time to accomplish, and errors. Assess what you have for raw data and what you want to finish with, and plan how to get from the former to the latter.

RAW DATA

Raw data may be in patients' charts or even scattered among charts, laboratory reports, and survey sheets. Data are often in words that must be coded into numbers. Data may be in different units. Numbers may be recorded to different accuracies. In short, raw data can be messy. The investigator must account for the discrepancies and coordinate the data to a common format. The raw data must be assessed for quality, and data lacking credibility must be eliminated. The first step is to assess the data in its several forms and plan a recording sheet that will amalgamate and coordinate the data. The data may be assessed and quality assured while being entered into the recording form so that a finished dependable product emerges ready for transfer to a format amenable to analysis.

DATA IN FORMAT FOR ANALYSIS

The goal is a quantified spreadsheet, most often with cases (e.g., patients) down the columns and variables (e.g., age, laboratory test result, survival) across rows. The entries for all variables to be used in the analysis should be numeric. For example, use 1 for male and 2 for female, not the letters M and F. For most analyses, all data should be lodged in a single spreadsheet, at least to start. If control data, for example, are placed in one spreadsheet and experimental data in another, the investigator will have to merge them and reverify correctness to perform a statistical test of difference between the groups. Different methods of statistical analysis may require different formats of data for the analysis to run. Unfortunately, the required format is seldom obvious, requiring the user to go to some effort to ferret it out. The format just noted with cases down the left side and variables across the top is the safest with which to begin. When two groups, A and B, of patients are being compared, place all patients down the side and form an additional column to designate the group, for example, 1 for group A and 2 for group B.

DATA QUALITY

It is essential to verify the correctness of data at entry, at each change of data format (which, of course, implies making the fewest changes possible to avoid unnecessary work and errors), and again before beginning analysis. Verifying data is tedious and boring but essential. A busy, sometimes tired investigator might read a raw datum as 29 but type 92 into the spreadsheet. Failing to discover and correct such little errors could weaken or even reverse the conclusion of the study, in some ways a worse outcome than losing the data, time, and effort altogether. All too often an investigator finds a datum to be erroneous near the end of a project and must redo the entire analysis, interpretation, and reporting text.

SOFTWARE FOR DATA MANAGEMENT

A number of data management software packages exist, but Microsoft Excel is by far the most commonly used in medical research. In addition, all statistical analysis software packages have their own spreadsheet capabilities. Data management packages allow ease of data manipulation, such as repositioning data subsets or filling in columns of repetitive numbers, but are limited in their statistical analysis capability. The better statistical packages have less capability and are more awkward in data manipulation but allow a nearly complete range of analysis. Investigators having involved data but simple analysis often stick with Excel for the entire project. Those having simple data but involved analyses often go directly to a statistical package. Most prepare their data with data management software and transfer to statistical software for analysis.

12.9. SETTING UP A TEST WITHIN A STUDY

TESTING AS A STAGE OF SCIENTIFIC KNOWLEDGE

Because biological events tend to include imperfectly controlled variables, we can describe, explain, and predict events only probabilistically. For example, infection with a particular agent will likely but not certainly cause a particular symptom. Section 1.2 describes these three stages. We first seek to *describe* events. For this, we use descriptive statistics, many of which are presented in Chapter 3, and we sometimes use statistical tests to decide which groupings should be presented descriptively. Much of the mechanics of tests, although pursuing different goals, are used in confidence intervals as part of description. When we are acquainted with the nature of the biological events, we seek to *explain* the events, inferring causal factors for these events. Tests to include or exclude causal factors compose much of the explanation. When we have identified the primary causes, we seek to *predict* these events from specific values and combinations of the causes. We use statistical tests to identify the significant predictors. Thus, tests tend to be used in all stages. The largest part of clinical decision making is based on tests. This section discusses setting up and conducting a test; the next section provides a start in the selection of an appropriate test to use in a variety of circumstances.

Steps in Setting Up a Test

Section 5.5 provides a sequence of steps in test development that a reader should look for when judging articles in the literature. A similar sequence, somewhat modified, is useful when investigators set up their own tests.

(1) Specify clearly, completely, and unequivocally the question you are asking of your data.

(2) Identify, specify in detail, and plan how to measure the variable(s) to answer that question.

(3) Review your definitions of population and sample, and verify the appropriateness of generalization.

(4) Review the sampling scheme to obtain your data.

(5) Specify exactly the null and alternate hypotheses.

(6) Select risks for Type I (α) and Type II (β) errors. (Recall, power $= 1 - \beta$.)

(7) Choose the form of the test (see Section 12.10).

(8) Verify that your sample size is adequate to achieve the proposed power.

(9) *At this point, obtain your data.*

(10) Identify and test possible biases.

(11) Perform the calculation of your test statistic and form your conclusion.

12.10. CHOOSING THE RIGHT TEST

A First-Step Guide to Selecting the Right Test

Table 12.1 provides a guide to selecting a statistical test for a given set of conditions. Table 10.1 is a much simplified version of Table 12.1. Table 10.1 is designed to assist the reader of medical journals in assessing the appropriateness of statistical test usage in the articles being read. Table 12.1 is expanded to include more situations that the investigator may encounter in designing experiments. (This table also appears on the inside back cover of this book.) Because there are more considerations than can be displayed in a single table, this guide must be taken only as a first suggestion to be further assessed for appropriateness. The tests themselves are introduced in Chapter 6 and are explored more thoroughly in Chapters 15 through 20.

Information Required to Choose a Test

To select a test appropriate to your question and your data, at a minimum, the following information is required. The method of using this information is explained in the legend to Table 12.1.

(1) What is the class of data that you have (categorical, rank-order, continuous)?

Table 12.1

A First-Step Guide to Choosing Statistical Tests Found in Part II[a]

Type of data: → Questions about:		Categorical (nominal data) — Proportions: p not near 0 or 1	Proportions: p near 0 or 1	Counted quantities	Ranks — Position in distribution: Not required	Continuous measurements — Averages: Normal curve	Averages: Far from normal curve	Spread: Chi-square	Distribution normality: Any	Distribution equality
Assumed distributions		Binomial or multinomial							Any	
Small sample	Single or paired sample	Binomial table **15.6**	Poisson table **15.7**	Match pairs (McNemar) **15.8**	Signed-rank test **16.2**	Normal test if σ; t if s **17.2**	Go to rank methods	Chi-square **19.2**	Shapiro-Wilk or K-S test **20.2**	
	Two samples	Form a 2×2 contingency table and use FET or χ^2 **15.2**			Rank-sum test **16.3**	Normal test if σ; t if s **17.3**		F test **19.3**		Two-sample K-S test **20.3**
	Three or more samples	Form an $r \times c$ contingency table and use FET or χ^2 **15.3**			Kruskal-Wallis/Friedman paired **16.4/5**	One-way analysis of variance, multiple comparisons **17.4**		Bartlett's test **19.4**		
Large sample	Single or paired sample	Normal approx **15.6**	Poisson approx **15.7**	Match pairs (McNemar) **15.8**	Signed-rank norm approx **16.6**	Normal test if σ known or if large n; t if s **17.2**		Chi-square **19.2**	K-S test or test of fit **20.2**	
	Two samples	Form a 2×2 contingency table and use FET or χ^2 **15.2**			Rank-sum normal approx **16.7**	Normal test if σ_1, σ_2 or if large n; t if s_1, s_2 **17.3**		F test **19.3**		Two-sample K-S Test **20.3**
	Three or more samples	Form an $r \times c$ contingency table and use FET or χ^2 **15.3**			Kruskal-Wallis/Friedman paired **16.4/16.5**	One-way analysis of variance with multiple comparisons **17.4**		Bartlett's test **19.4**		

[a]The column is selected by specifying the type of data you have, the question you are asking of these data, and the distribution you are willing to assume. The row is selected by specifying the sample size and the number of samples. The row-column intersection provides the most apparent test fitting these conditions. This selection does not satisfy all requirements, and therefore must be taken as only tentative. Numbers that appear in italic boldface indicate the specific sections in the text that discuss the topics presented.

(2) What sort of statistical property are you asking about (rates, position in distribution, averages, spread, distributional characteristics)?

(3) What is the population distribution from which data arise (binomial, multinomial, normal, irregular)?

(4) Is the sample small or large? (The definition varies with the test being used and the irregularity of the data. Rule of thumb: Up to 30 is small, more than 100 is large, and 30 to 100 depends on test.)

(5) Do you have one, two, or three or more arms (groups to test) in your design?

(6) How big are the risks for error that you will tolerate?

AN EXAMPLE FROM PROSTATE CANCER

An example might help the reader to use Table 12.1. Is the prostate-specific antigen (PSA) level for 15 cancer patients different after hormone therapy than before? We want to contrast PSA level before versus after therapy. Type of Data: PSA level is a continuous measurement; go to the Continuous Measurements heading in the right portion of the table. Question About: averages; go to the Averages heading. Assumed Distributions: We do a quick tally and find that the differences, although a little skewed, are not remarkably different from normal in shape; go to the Normal Curve subheading. Now we know what column to use; we go to the left margin to find the appropriate row. Our sample of 15 is small; go to the Small Sample heading. We have a before and after reading on each patient, so the data arise in pairs; go to the Single or Paired Sample heading, which designates the row. (*Single:* One reading per patient. *Paired:* If the data appear in pairs, e.g., before and after treatment or siblings for whom one is given a drug and the other is given a placebo, we are concerned with the difference between pair members, not the actual readings of the individuals. The differences provide one reading per pair; thus, they act as a single sample.) The intersection of the column and row yield the entry—"Normal if σ; t if s"—and directs us to Section 17.2. We proceed to Section 17.2 and conduct our test as directed.

12.11. STATISTICAL ETHICS IN MEDICAL STUDIES

ETHICS IN THE CONDUCT OF MEDICAL STUDIES IS A BROAD TOPIC

Ethics in medical research covers *inter alia* the contrast between research and practice; informed consent issues; the organization, responsibility, and conduct of institutional review boards and animal care committees; adherence to the World Medical Association's Helsinki Declaration and its revisions; issues in the trade-off between care for populations (epidemic containment, immunization, placebo groups in clinical trials) and care for individual patients; and the interaction between the practices of statistics and medicine. The latter issue includes *inter alia* integrity of statistical methods chosen, documentation and availability of statistical methods used, qualifications of a data analyst, patient data privacy, rules for halting study when efficacy is shown prematurely, random patient allocation to treatment groups in the presence of unequal uncertainty about treatment preference, and the issue that poorly done statistics can lead to poorer patient care. The resolution of many of these statistical issues is obvious or has been well treated. The latter issue is the principal focus of this section.

PATIENT PROTECTION REQUIREMENTS

Although patient protection in medical research is not directly a statistical issue, the statistical designer in medical research must remain aware of its implications. It often affects statistical design, for example, forbidding certain control arms, constraining sample sizes in some studies, and requiring early termination when shown to benefit patients. For example, it does not permit denying treatment to cancer patients even though knowing the progress of untreated cancer would provide a baseline against which to compare cancer treatments. Different nations have different standards, but most developed nations have similar standards. In the United States as of 2004, the basic documents specifying standards are the following: the Nuremberg Code, 1949 (developed as a result of the Nuremberg war crimes trials at the end of World War II); the Belmont Report, 1979 (on protection of human subjects); United States Code 10 USC 980 Humans as Experimental Subjects, 1999; Food and Drug Administration (FDA) 21 CFR (Code of Federal Regulations) 312 Investigational New Drugs; FDA 21 CFR 812 Investigational Device Exemptions; Department of Defense 32 CFR 219 Protection of Human Subjects; Department of Health and Human Services 45 CFR 46 Protection of Human Subjects, 1999; and finally, the 2003 implementation of 45 CFR 160/164 Health Insurance Portability & Accountability Act of 1996 (HIPAA) that establishes patient protected health information (PHI). New standards documents appear sporadically; therefore, the ethical investigator must stay updated and informed regarding standards.

PATIENT IDENTIFIERS IN DATA SHEETS

The Health Insurance Portability & Accountability Act of 1996 rules require safeguards to patient identity. In setting up a spreadsheet to record data for research, plan to identify patients with a code, not with names or personal numbers that an unscrupulous person could use to obtain medical information about the patient. If there are no further data to be obtained, all patient identifiers should be shredded or deleted electronically. If future access to patient identity may be needed, keep a master code separate from the data in a locked file or safe.

STATISTICAL CONTROL PARAMETERS AND SAMPLE SIZES ARE AT ISSUE

If the statistical design and analysis of a study leads to erroneous conclusions, patients' health care resulting from that study will be degraded. The choice of the appropriate statistical method is a topic throughout this book; the smaller topic that this section addresses is that of the choice of and interaction among error risks (α and β), test sidedness, and sample size.

EXAMPLES OF ETHICAL CONSIDERATIONS IN SPECIFYING SAMPLE SIZE TEST PARAMETERS

This section pursues two examples of ethical considerations. (1) In a trial investigating the effect of a new muscarinic agent on return of salivary function after head and neck irradiation, how many patients are required for the trial when randomized to control versus pilocarpine? (2) In an investigation of a recursive-partitioning model in early-stage breast cancer to determine the need for axillary sampling, based on historical control data, how many patients are required for statistical validity?

RELATIONSHIPS AMONG THE STATISTICAL PARAMETERS

Required sample size n may be estimated by methods of Chapters 7 and 22. An estimate of the standard deviation of the variable being used is obtained, and α and β are selected. The remaining choice is δ, the size of the difference between treatments that will answer the clinical question being posed (often based on the investigator's clinical experience). Medical journal reviewers generally seem to expect choices of $\alpha = 0.05$ and $\beta = 0.20$ (power = 0.80), although power = 0.90 is being seen increasingly. The error risk associated with a probability distribution's tail will be α or $\alpha/2$, depending on whether a one- or two-sided test is being used. In most cases, a computer operation using these parameters as inputs will provide n.

IMPLICATIONS OF α AND β

α is the risk of inferring a difference between experimental groups when there is, in fact, no such difference, and β is the risk of inferring no difference between groups when there is, in fact, such a difference. Choosing $\alpha = 0.05$ and $\beta = 0.20$ (power = 0.80) implies setting the rate of false negatives at 4 times the rate of false positives.

EFFECT ON PATIENTS FROM THE XEROSTOMIA STUDY

A false-positive outcome from the study implies inferring the new treatment to be better when it is not; using it unnecessarily exposes patients to the risk for possible serious negative side effects. A false-negative outcome implies inferring the new treatment to be no better when, in fact, it is, thereby failing to palliate xerostomia. The false-positive outcome is worse for the patient than is the false-negative outcome. The α/β ratio choice of 0.05/0.20 (or 1/4) is justifiable.

EFFECT ON PATIENTS FROM THE BREAST CANCER STUDY

In the other trial, however, the false-positive outcome implies axillary metastases that are not there and unnecessarily doing an axillary sampling. The false-negative outcome implies the absence of cancer when it is present and subsequently failure to offer appropriate therapy. In this case, *the false-positive outcome represents* less *loss to the patient than does a false-negative outcome.*

To set the probability of missing a cancer at 4 times the probability of an unnecessary sampling may not serve patients well. The investigator should take a higher α and lower β. Would this require an untenable increase in sample size required? If the error rates were reversed to $\alpha : \beta = 0.20 : 0.05$ (four unnecessary samplings expected per cancer missed), the sample size would increase no more than about 10%. However, if α should be mandated at 0.05, as is usual in medical research (based more on tradition than good reason), the 4/1 ratio would require β to be 0.0125, and the required sample size would increase by about 125%.

The Choice of α and β Should Involve Clinical Implications to the Patient

The choice of α and β, often a hard choice, follows as a matter of judgment. Thus, this decision is partly clinical, partly statistical. Furthermore, although the sample size selected will not directly affect patients not participating in the study, the selection of α and β must be carried through into the statistical testing after acquiring the data, and test results *may* affect patients in greater generality.

Effect of Test Sidedness on the Patient

After head-and-neck irradiation, can the muscarinic agent inhibit *or* enhance the return of salivary function, or can it *solely* enhance it? The choice of a one- versus two-sided test should be made before gathering data. The statistical reason is that an investigator must be a stern self-disciplinarian to choose a two-sided test once the data show on which side of a hypothesized mean the sample mean lies. However, there is an ethical reason as well. When a two-sided test is appropriate, a one-sided test doubles the error rate assigned to the chosen tail, which gives rise to two results. The first result is a statistical result of benefit to the investigator: A smaller δ is required to obtain significance in a given sample. The second result is a clinical outcome to the detriment of the patient: Too large a number of healthy patients will be treated as ill, and too small a number of ill patients will not be treated. Choosing a two-sided test when a one-sided test is appropriate creates the same classes of mistake but with opposite results.

Choosing Sidedness

Often the choice of sidedness is obvious: If we are subjecting the patient to treatments A and B and have no idea which one will be better, two-sidedness is appropriate. If, however, we expect the result associated with one side to be more likely (especially if we *prefer* that result to be chosen), sidedness should be thoughtfully selected. In this case, sidedness should be chosen when the study is initially conceived and should be chosen in answer to the following questions: Can a result on the nonexpected side *possibly* occur physically? If not, select a one-sided test. If so, could a result on the nonexpected (perhaps, nonpreferred) side *affect the patient?* If not, select a one-sided test; if so, select a two-sided test. Note that the choice here, although affecting α and β, is made on other grounds, which is why α and β selection and sidedness selection are discussed separately.

211

SELECTION OF THE CLINICAL DIFFERENCE δ

Another potential problem arises when δ is selected. When a study is being developed, a larger δ will require a smaller *n*. The bigger a difference between two group averages is, the easier it is to detect it. Therefore, for protocols that are expected to accrue slowly, or for which a low *n* is expected, the temptation exists to maximize δ to allow for a small enrollment with subsequent early closure. This is clearly statistical tampering, but it obscures an ethical question: If the proposed new therapy is really *so much* better than the current one, how may the researcher in good faith not offer the patient the superior course? This manipulation of δ poses an ethical dilemma, the answer to which is that we be honest.

EFFECT OF THE CLINICAL DIFFERENCE δ ON THE PATIENT

Even despite an honest selection of δ, the issue persists of its influence on patients affected by the study results. The smaller the actual δ, the less sensitive the analysis will be in finding a true treatment effect, even with a larger *n*; the larger the actual δ, the more sensitive the analysis will be. In the crush for funding and in a forest of competing alternative treatments, it is easier to try many small trials than one enormous one. Are we then reducing our likelihood of improving the lot of all patients to reduce the risk to the small group we include in our trial? We must assure ourselves in trial design that such a trade-off is based solely on the potential patient benefit involved, without influence of personal benefit to the investigator.

CONCLUSION

A medical study holds the potential, whether explicitly or implicitly, of a deleterious outcome for all the patients treated in the future as a result of the information emerging from that study. Ethical soundness in statistical design currently is rarely discussed in the context of study planning. In clinical studies, statistical ethical soundness must be sought side by side with clinical ethical soundness.

Exercise 12.2. For the question of DB5, should a one- or two-sided test be used?

ANSWERS TO EXERCISES

12.1. There is no unique answer for this exercise.

12.2. One-sided. Although the postoperative minus preoperative difference could be either positive or negative, adding silicone to the body is believed to be unable to decrease the plasma silicon level; thus, any decrease would be an anomalous chance outcome. Furthermore, it would not be harmful to the patient. The test should be to detect an increase.

Chapter 13

Finding Probabilities of Error

13.1. INTRODUCTION

THE RISKS FOR ERROR IN A CLINICAL DECISION

Section 2.7 indicates that probabilities of occurrences correspond to areas under portions of probability distributions. The probability of inferring an occurrence when it is absent, often the probability of a false-positive result on a medical test, usually is denoted α. The probability of inferring the absence of an occurrence when it is present, often the probability of a false-negative result, usually is denoted β. Rarely is it necessary to calculate these probabilities directly, because computers or tables can be used to find them.

RELATIONSHIP OF THIS CHAPTER TO CHAPTERS 2 AND 4

The distributions commonly used in statistics were introduced in Section 2.8. Chapter 4 explains how to use tables to find areas under the curves of three of these distributions: normal, t, and chi-square distributions. This section reviews the use of these three tables and explains how to find areas under the probability curve for the other three common distributions of statistics: F, binomial, and Poisson distributions.

13.2. THE NORMAL DISTRIBUTION

REVIEW OF SECTION 4.2: THE NORMAL DISTRIBUTION

When dealing with variables that follow normal distributions, we want to use the standard normal, symbolized z, which is obtained by subtracting the mean from each value and dividing by the standard deviation σ. *z represents the number of standard deviations away from the mean.* Figure 13.1 (reproduced from Fig. 4.1) shows a standard normal distribution with $z = 1.96$ (frequently seen in practice) and the corresponding area under the curve to the right of

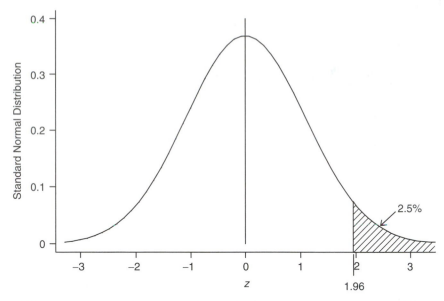

Figure 13.1 A standard normal distribution shown with a 2.5% α and its corresponding z. The α shown is the area under the curve to the right of a given or calculated z. For a two-tailed computation, α is doubled to include the symmetric opposite tail. As shown, 2.5% of the area lies to the right of $z = 1.96$. For a two-tailed use, the frequently used $\alpha = 5\%$ lies outside the ±1.96 interval, leaving 95% of the area under the curve between the tails.

that z. Table I (see Tables of Probability Distributions) contains values of z with four areas: α and $1 - \alpha$ for one tail of the distribution, and α and $1 - \alpha$ where α is split between the two tails. The most commonly used areas and their corresponding z-values are italicized in the table.

What Proportion of Carinal Resection Patients Are Younger Than 30 Years?

As further illustration of how to find normal probabilities, consider the age of patients undergoing resection of the tracheal carina (DB12). Of 134 patients, mean age is about 48 years, with a standard deviation (taken as σ for this sample size) of about 16 years. (Values are rounded to make the calculation easier in this formula.) What percentage of patients are younger than 30 years? Age 30 is 18 years less than (i.e., to the left of) the mean, or $18/16 = 1.125$ standard deviations below the mean. Our question becomes what proportion of the curve is more than 1.125σ below the mean? Because the distribution is symmetric, the result will be the same as the proportion more than 1.125σ above the mean. By using this property, we need have only one tail of the distribution tabulated. We look up $z = 1.125$ in Table I or Table 13.1, which is a segment of Table I. z lies one fourth of the way from 1.10 to 1.20, which yields a probability approximately one fourth of the way from 0.136 to 0.115, or about $0.136 - 0.005 = 0.131$. Thus, the chance of a patient younger than 30 years presenting for carinal resection is about 13 of 100.

Exercise 13.1. For a certain population of young healthy adults, diastolic blood pressure (DBP) follows a normal distribution with $\mu = 120$ and $\sigma = 5$ mm Hg.

Table 13.1

Table 13.1

Segment of Normal Distribution[a]

z (no. standard deviations to right of mean)	One-tailed applications		Two-tailed applications	
	One-tailed α (area in right tail)	$1 - \alpha$ (area except right tail)	Two-tailed α (area in both tails)	$1 - \alpha$ (area except both tails)
1.10	0.136	0.864	0.272	0.728
1.20	0.115	0.885	0.230	0.770

[a]For selected distances (z) to the right of the mean, given are (1) one-tailed α, the area under the curve in the positive tail; (2) one-tailed $1 - \alpha$, the area under all except the tail; (3) two-tailed α, the areas combined for both positive and negative tails; and (4) two-tailed $1 - \alpha$, the area under all except the two tails.

Table I in Tables of Probability Distributions provides an expanded version of this table.

You have a patient with DBP = 126 mm Hg. What percentage of the population has higher DBP measurements?

13.3. THE *t* DISTRIBUTION

REVIEW OF SECTION 4.6: THE *t* DISTRIBUTION

The *t* distribution looks and works much like the normal distribution. It is used where one would choose to use a normal distribution but where the standard deviation is estimated by s rather than being the known σ. Even the calculations are similar. For example, $(x - m)/s$ provides a standard t-value. The difference is that use of the estimated rather than the known standard deviation yields a less confident and therefore more spread out distribution, with the spread depending on the degrees of freedom (df), a variation of the sample size. Each df yields a member (a curve) of the t family. When using a t distribution arising from one sample, $df = n - 1$; for a t arising from two samples, $df = n - 2$. Figure 13.2 shows a t distribution for 7 df, with 2.5% of the area under the curve's right tail shaded. We note that it is similar in appearance to the normal distribution depicted in Fig. 13.1, except that the 2.5% critical value lies $2.36s$ (sample standard deviations) to the right of the mean rather than 1.96σ (population standard deviations).

What Proportion of Hamstring-Quadriceps Surgery Patients (DB10) Can Perform the Triple Hop in Less Than 1.95 Seconds?

By calculating from the database, we find $m = 2.70$ and $s = 0.53$. With $n = 8$ from one sample, $df = 7$. The postulated 1.95 seconds is x; therefore, $t = (1.95 - 2.70)/0.53 = -1.415$. Because the t distribution is symmetric, only the right tail has been tabulated; the area to the left of -1.415 is the same as the area to the right of 1.415. Table 13.2 provides a segment of the t table, Table II (see Tables of Probability Distributions). By looking in the row for 7 df, we find 1.415 in the first column under the heading for one-tailed $\alpha = 0.10$, implying

215

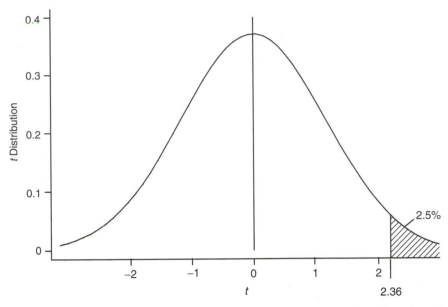

Figure 13.2 A *t* distribution with 7 *df* shown with $\alpha = 2.5\%$ and its corresponding *t*. The α shown is the area under the curve to the right of a given or calculated *t*. For a two-tailed computation, α is doubled to include the symmetric opposite tail. As shown for 7 *df*, 2.5% of the area lies to the right of $t = 2.36$.

Table 13.2

A Segment of the *t* Distribution[a]

	0.10	0.05	0.025	0.01	0.005	0.001
One-tailed α	0.10	0.05	0.025	0.01	0.005	0.001
One-tailed $1 - \alpha$	0.90	0.95	0.975	0.99	0.995	0.999
Two-tailed α	0.20	0.10	0.05	0.02	0.01	0.002
Two-tailed $1 - \alpha$	0.80	0.90	0.95	0.98	0.99	0.998
$df = 6$	1.440	1.943	2.447	3.143	3.707	5.208
7	1.415	1.895	2.365	2.998	3.499	4.785
8	1.397	1.860	2.306	2.896	3.355	4.501

[a]Selected distances (*t*) to the right of the mean are given for various degrees of freedom (*df*) and for (1) one-tailed α, area under the curve in the positive tail; (2) one-tailed $1 - \alpha$, area under all except the tail; (3) two-tailed α, areas combined for both tails; and (4) two-tailed $1 - \alpha$, area under all except the two tails.

that 10% of the area under the *t* curve falls to the right of 1.415. By using the symmetry, we conclude that 10% of the patients will have hop times less than 1.95 seconds.

Exercise 13.2. In DB10, the number of centimeters covered in the hop test using the operated leg has $m = 452.8$ and $s = 91.7$. What interval will include 95% of such patients (i.e., 2.5% in each tail)?

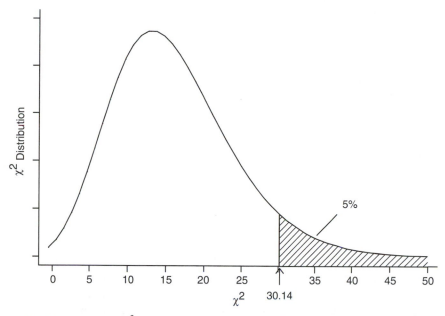

Figure 13.3 The chi-square (χ^2) distribution for 19 *df* with a 5% α and its corresponding χ^2 value of 30.14. The α probability is shown as the shaded area under the curve to the right of a critical χ^2, in this case representing a 5% probability that a value drawn randomly from the distribution will exceed a critical χ^2 of 30.14.

13.4. THE CHI-SQUARE DISTRIBUTION

REVIEW OF SECTION 4.4: THE CHI-SQUARE DISTRIBUTION

The chi-square (χ^2) statistic is composed of a sum of squares. It occurs in a test of a variance, the averaged sum of squares of deviations from a mean, and in a test of contingency, the weighted sum of squares of observed deviations from expected. Statistical applications involving χ^2, being composed of squares, must always have a positive value and can grow to any large size. It is rather like a bell curve strongly skewed to the right. Figure 13.3 shows the member of the χ^2 family of distributions having 19 *df*.

Is the Variability in Platelet Count Among a Sample of Wound Patients Treated with Growth Factor Different from That for Normal Patients?

In DB9, platelet count for $n = 20$ wound patients treated with growth factor has $m = 295,000$ and $s = 164,900$. Suppose we found extensive data in the medical literature that give the platelet count standard deviation as $\sigma = 60,000$ for normal patients. Is s so much larger than σ that it is improbable to have happened by chance? It turns out that the ratio of s^2 to σ^2, when multiplied by *df*, is a χ^2 statistic with $n - 1$ *df*, or

$$\chi^2 = \frac{df \times s^2}{\sigma^2}$$

217

Table 13.3

A Segment of the Chi-square Distribution, Right Tail[a]

α	0.10	0.05	0.025	0.01	0.005	0.001
$1 - \alpha$	0.90	0.95	0.975	0.99	0.995	0.999
$df = 17$	24.77	27.59	30.19	33.41	35.72	40.78
18	25.99	28.87	31.53	34.80	37.16	42.32
19	27.20	30.14	32.85	36.19	38.58	43.81
20	28.41	31.41	34.17	37.57	39.99	45.31

[a]Selected chi-square (χ^2) values (distances > 0), for various degrees of freedom (df), are given for (1) α, the area under the curve in the right tail; and (2) $1 - \alpha$, the area under all except the right tail.

(see Chapter 19 for further discussion of this topic). Thus, $\chi^2 = 19 \times 164{,}900^2/60{,}000^2 = 143.5$. Table 13.3 shows a segment of the right-tailed χ^2 table (see Table III in Tables of Probability Distributions). For the row $df = 19$, we can see that at $\chi^2 = 30.14$, 5% of the area under the curve lies to the right; at $\chi^2 = 36.19$, 1% of the area under the curve lies to the right; and at $\chi^2 = 43.81$, 0.1% of the area under the curve lies to the right. Because 143.5 is much larger than 43.81, the area under the curve to the right is much smaller than 0.1%. The chance that a standard deviation as large as 164,900 arose from a sample of patients with normal platelet counts is much less than 1 in 1000; we conclude that the variability in the sample of patients who have had growth factor is larger than in normal patients.

AN EFFECT OF ASYMMETRY IN THE CHI-SQUARE DISTRIBUTION

Table III (see Tables of Probability Distributions) provides the χ^2 values that yield commonly used values of α, that is, the probability that a randomly drawn value from the distribution lies in the right tail demarked by the tabulated χ^2 value. Because the χ^2 distribution is asymmetric, we cannot take an area in one tail and expect the area in the other tail to be the same. For areas under the left tail, another χ^2 table is needed. Table III provides χ^2 values for the more commonly used right tail, and Table IV (see Tables of Probability Distributions) provides values for the left tail. The mechanism of finding areas in a tail from the table is much the same as for the t: The desired area in the tail specifies the column, the df specifies the row, and the *critical* χ^2 value (the value demarking the tail area) lies at the row–column intersection.

Exercise 13.3. In DB7, the standard deviation of bone density of 18 patients with femoral neck fractures is $s = 24.01$. Take the standard deviation for the normal population as $\sigma = 16.12$. We want to know whether or not the variability for patients with a fracture is unusually larger than that for normals and therefore probably different. If we sampled repeatedly from the normal population, how often would we find an s this large or larger? (*Hint:* By calculating χ^2 as before, what percentage of the area under the curve is greater than that value of χ^2?)

13.5. THE F DISTRIBUTION

THE CONCEPT OF F

Section 2.8 notes that a ratio of two variances drawn from the same population has a distribution named F. Conceptually, F may be thought of as the size of the top variance relative to the bottom variance. An F of 2.5 indicates that the top variance is two and a half times the bottom variance. If the top is too much larger than the bottom, we believe it is unlikely that the two samples come from populations with the same variance. The probability distribution of F defines "too much."

F AND STANDARD DEVIATIONS

Variances, the squares of the respective standard deviations, are used in the computations because they are easier to deal with mathematically. However, the concepts and conclusions are the same for both variances and standard deviations. We may use variances in computations quite legitimately but make conclusions about the respective standard deviations.

F AND *df*

Each variance (multiplied by a constant) is distributed χ^2, and the resulting F distribution itself looks much like χ^2. Because both top and bottom variances have *df*, any particular F distribution is specified by a *pair* of *df*. The top *df* is always stated first. The member of the family of F distributions specified for the pair 2,6 *df* is shown in Fig. 13.4, with the upper 5% tail area shaded. We note that 5.14 is the critical F for 2,6 *df*; that is, 5% of the F distribution lies to the right of 5.14. Table 13.4, a segment of the F table (Table V in Tables of Probability Distributions), shows from where this critical value comes.

USING THE F TABLE

Table V in Tables of Probability Distributions (see also Table 13.4) provides the F-values that yield 5% values of α, that is, a 5% probability that a randomly drawn value from the distribution exceeds the tabulated F-value. Because the F distribution is asymmetric, we cannot take an F-value designating an area in one tail and expect the area in the other tail to be the same. However, the left tail is not used for the methods applied in this book. Finding areas in the right tail is much the same as in χ^2, except for having a pair of *df* rather than a single *df*. The *df* of the top variance specifies the column, the *df* of the bottom variance specifies the row, and the distance greater than zero (F) lies at the row–column intersection. Because *df* for both top and bottom must be used in tabulation, an F table can be given for only one α. Several pages of tables could be given, but because $\alpha = 5\%$ is commonly used in almost all medical research, only the 5% table is given here. Calculation of F for other α's may be found in any good statistical computer package or in books of tables.

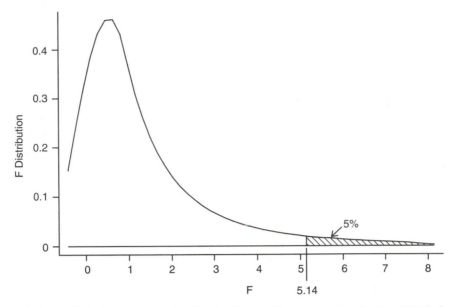

Figure 13.4   The F distribution for 2 and 6 *df* with a 5% α and its corresponding F-value of 5.14 is shown. The α probability is shown as the shaded area under the curve to the right of a given or calculated F, in this case representing a 5% probability that a value drawn randomly from the distribution will exceed an F of 5.14.

Table 13.4

A Segment of the F Table[a]

		Numerator *df*				
		1	2	3	4	5
Denominator *df*	3	10.13	9.55	9.28	9.12	9.01
	4	7.71	6.94	6.59	6.39	6.26
	5	6.61	5.79	5.41	5.19	5.05
	6	5.99	5.14	4.76	4.53	4.39

[a]Selected distances (F) are given for various degrees of freedom (*df*) for $\alpha = 5\%$, the area under the curve in the right tail. Numerator *df* appears in column headings, denominator *df* appears in row headings, and the F-value appears at the intersection of row and column in the table body.

Is Variability of Prostate Volumes Different for Patients with Positive and Negative Biopsy Results?

Let us calculate the variances for the prostate volumes found in Table DB1.1 for the groups of patients with positive and negative biopsy results. We want to know if the variability of volume for the seven patients with negative biopsy results is significantly larger than that for the three patients with positive biopsy results. Although variances are the statistics actually tested, think of the test as one of standard deviations. The sample sizes of 3 and 7 give the *df* pair 2,6, leading to the 5.14 entry in the F table. Because of the 5.14 critical value, the test requires the variance for negative biopsy volumes to be more than 5 times the size of the variance for positive biopsy volumes to show evidence that it is larger with no more than a 5% chance of being wrong. On calculating the variances, we find $s_-^2 = 326.89$ (standard deviation = 18.08) and $s_+^2 = 83.52$ (standard deviation = 9.14). The calculated F = 326.89/83.52 = 3.91. Because this is less than the critical F-value of 5.14, we would be wrong more than 5% of the time if we concluded that the disparity in variances (or standard deviations) occurred by causes other than chance alone.

F AND SAMPLE SIZE

The effect of sample size can be seen dramatically. If we had obtained the same variances from double the sample sizes, that is, 6 and 14 patients, we would have had 5,13 as the *df* pair, leading to a critical F (from Table V) of 3.03. Our 3.91 F ratio would have well exceeded this value, and we would reach the reverse conclusion: We would have enough evidence to say there is less than a 5% chance that the disparity in variances occurred by chance alone and a causal factor likely is present.

Exercise 13.4. In DB8, means of four glycosaminoglycan levels recorded for each of two types of assay were about 0.39 (type 1) and 0.44 (type 2), which are not very different. However, the standard deviations were 0.2421 (type 1) and 0.1094 (type 2). If the type 2 assay averages the same but is less variable, we would choose it as the preferable type of assay. The statistic F is the ratio of the two variances. Is there less than a 5% chance of finding an F this large or larger?

13.6. THE BINOMIAL DISTRIBUTION

BINOMIAL EVENTS DEFINED

Often we encounter situations in which only two outcomes are possible: ill or well, success or failure of a treatment, a microorganism that does or does not cause a disease. Let us denote π to be the probability of occurrence of one of the two outcomes (e.g., the first) on any random trial. (This symbol has nothing to do with the symbol for 3.1416... used in the geometry of circles.) If we have n opportunities for the outcome to occur (e.g., n patients), the binomial distribution will tell us how many occurrences of the outcome we would expect by chance alone.

BINOMIAL TABLE

Table VI (see Tables of Probability Distributions) gives the probability of n_0 occurrences of an event under scrutiny out of n trials, given an occurrence rate of π. Table 13.5 shows the segment of Table VI needed for the example.

Table 13.5

Values of the Cumulative Binomial Distribution, Depending on π (Theoretical Proportion of Occurrences in a Random Trial), n (Sample Size), and n_o (Number of Occurrences Observed)[a]

						π					
n	n_o	0.05	0.10	0.15	0.20	0.25	0.30	0.35	0.40	0.45	0.50
6	1	0.265	0.469	0.629	0.738	0.822	0.882	0.925	0.953	0.972	0.984
	2	0.033	0.114	0.224	0.345	0.466	0.580	0.681	0.676	0.836	0.891
	3	0.002	0.016	0.047	0.099	0.169	0.256	0.353	0.456	0.559	0.656
	4	0.000	0.001	0.006	0.017	0.038	0.071	0.117	0.179	0.255	0.344
	5		0.000	0.000	0.002	0.005	0.011	0.022	0.041	0.069	0.109
	6				0.000	0.000	0.001	0.002	0.004	0.008	0.016

[a]Given π, n, and n_o, the corresponding entry in the table body represents the probability that n_o or more occurrences (or alternatively that n_o/n proportion observed occurrences) would have been observed by chance alone.
Table VI in Tables of Probability Distributions provides an expanded version of this table.

Has the Success of Laser Trabeculoplasty Improved?

As an example, the long-term success of laser trabeculoplasty as therapy for open-angle glaucoma was examined in a Norwegian study.[13] At the end of 2 years, the failure rate was 1/3. Suppose you perform trabeculoplasty on six patients. At the end of 2 years, you find only one failure. What is the probability that your improved failure rate of 1/6 is due to more than chance, given the Norwegian rate of 1/3? Table 13.5 (or Table VI) gives the probability of n_o or more occurrences of the event under scrutiny out of n trials, given an occurrence rate of π. The table gives only a "$\geq$" value; to find the probability that n_o is 1 or 0, we must find $1 -$ (the probability that $n_o \geq 2$). The column is indicated by $\pi = 0.333$, lying between 0.30 and 0.35. The row is indicated by $n = 6$ and $n_o = 2$. The value for our π is bracketed by 0.580 and 0.681, about 0.65. Finally, we take $1 -$ the tabulated value, or 0.35, for our result. A more exact value, calculated with the aid of a computer, is 0.351, which is almost the same. With a 35% probability that our result could have occurred by chance alone, our small sample gives us inadequate evidence of an improved success rate.

Effect of Sample Size

Figure 13.5 shows a binomial distribution for $\pi = 1/3$ and $n = 12$, double the sample size of the example. (The values for 11 and 12 occurrences are not 0 but are too small to appear on the graph.) We can see by combining the first two bars that the probability of $n_o \leq 1$ is near to 5.5%, much closer to significance. A group of 60 patients with 10 failures rather than 6 patients with 1 failure would have given a probability of 0.003, which is strong evidence of improvement.

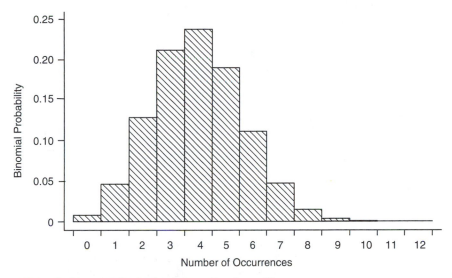

Figure 13.5 The binomial distribution for $n = 12$ and $\pi = 1/3$.

THE BINOMIAL FOR LARGER SAMPLES

The binomial distribution is difficult to calculate when n is not quite small. For medium to large n, we need an easier way to find the required probability. When π, the occurrence rate, is toward the center of the 0–1 interval, the normal distribution adequately approximates the binomial. The mean of the binomial sample is the observed occurrence rate $p = n_o/n$, number of occurrences observed $\div$ sample size, and the variance is $\pi(1 - \pi)/n$. Thus, the mean difference divided by the standard deviation follows approximately the standard normal distribution, as

$$z = \frac{p - \pi}{\sqrt{\dfrac{\pi(1 - \pi)}{n}}},$$

and probability questions may be addressed using Table I (see Tables of Probability Distributions). (The approximation can be slightly improved with an adjustment factor; see Section 15.6 for more detail.) The similarity to a normal shape can be seen in Fig. 13.5. If π is close to 0 or 1, the binomial becomes too skewed to approximate with the normal, but it can be approximated adequately by the Poisson distribution.

Exercise 13.5. In a relatively large study on subfascial endoscopic perforating vein surgery (SEPS),[55] 10%($= \pi$) of ulcers had not healed by 6 months after surgery. In a particular clinic, $n = 6$ SEPS were performed and $n_o = 2$ patients had unhealed ulcers after 6 months. What is the probability that, if the clinic shared the theoretical $\pi = 0.10$, 2 out of 6 patients would have had unhealed ulcers?

Table 13.6

Values of the Cumulative Poisson Distribution, Depending on $\lambda = n\pi$ (Sample Size $\times$ Theoretical Proportion of Occurrences in Random Trials) and n_o (Number of Occurrences Observed)[a]

n_o	0.1	0.2	0.3	0.4	0.5	0.6	0.7	0.8	0.9	1.0	1.1	1.2
							$\lambda(= n\pi)$					
1	0.095	0.181	0.259	0.330	0.394	0.451	0.503	0.551	0.593	0.632	0.667	0.699
2	0.005	0.018	0.037	0.062	0.090	0.122	0.159	0.191	0.228	0.264	0.301	0.337
3	0.000	0.001	0.004	0.008	0.014	0.023	0.034	0.047	0.063	0.080	0.100	0.121
4		0.000	0.000	0.001	0.002	0.003	0.006	0.009	0.014	0.019	0.026	0.034
5				0.000	0.000	0.000	0.001	0.001	0.002	0.004	0.005	0.008
6							0.000	0.000	0.000	0.001	0.001	0.002
7										0.000	0.000	0.000

[a]Given λ and n_o, the corresponding entry in the table body represents the probability that n_o or more occurrences (or alternatively that n_o/n proportion observed occurrences) would have been observed by chance alone.
Table VII in Tables of Probability Distributions provides an expanded version of this table.

13.7. THE POISSON DISTRIBUTION

POISSON EVENTS DESCRIBED

The Poisson distribution arises from situations in which there is a large number of opportunities for the event under scrutiny to occur but a small chance that it will occur on any one trial. The number of cases of bubonic plague would follow Poisson: A large number of patients can be found with chills, fever, tender enlarged lymph nodes, and restless confusion, but the chance of the syndrome being plague is extremely small for any randomly chosen patient. This distribution is named for Siméon Denis Poisson, who published the theory in 1837. The classic use of Poisson was in predicting the number of deaths of Prussian army officers from horse kicks from 1875 to 1894; there was a large number of kicks, but the chance of death from a randomly chosen kick was small.

POISSON TABLE

Table VII (see Tables of Probability Distributions) provides the α to test the hypothesis that n_o or more cases would occur by chance alone, given the occurrence rate $\lambda(= n\pi)$. Table 13.6 shows the segment of Table VII needed for the example.

Does Aluminum City Have a Higher Rate of Alzheimer's Disease Than Seattle?

As an example, a survey[72] of medical records of 23,000 people older than 60 years from a Seattle HMO revealed 200 participants with indications of Alzheimer's disease, a rate of 0.0087. A clinician in Aluminum City has 100

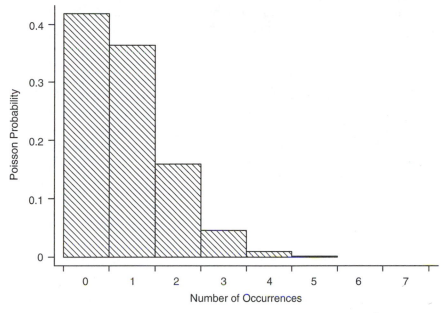

Figure 13.6 The Poisson distribution for an occurrence rate (λ) of 0.87.

patients older than 60 years and reports 4 with Alzheimer's disease. What would be the probability that this high rate is due to chance alone, if Aluminum City's rate were the same as that of Seattle? Table 13.6, or Table VII, provides the α to test the hypothesis that n_o or more cases would occur by chance alone, given the occurrence rate λ. The column in the table must be chosen using the value $\lambda = n\pi = 100 \times 0.0087 = 0.87$, which would lie between columns 0.8 and 0.9. The row is indicated by $n_o = 4$. The Poisson probability is between 0.009 and 0.014, about 0.012. The actual probability from a computer calculation is the same. This result indicates that the chance of being wrong in concluding a difference would be about 1 in 100, which provides adequate evidence to conclude that Aluminum City has a higher rate of Alzheimer's disease than Seattle.

THE APPEARANCE OF THE POISSON DISTRIBUTION

Figure 13.6 shows the member of the Poisson family of distributions for $\lambda = 0.87$, the occurrence rate in the example. We might note that theoretically the Poisson distribution has no greatest possible number of occurrences as does the binomial distribution.

THE POISSON FOR LARGER SAMPLES

Like the binomial, the Poisson distribution is difficult to calculate when n is more than a few. For medium to large n, when π, the occurrence rate, is close to 0 or 1, the normal distribution adequately approximates the Poisson. Again like the binomial, the mean of the Poisson sample is the observed occurrence rate $p = n_o/n$, the number of occurrences observed $\div$ sample size, but the variance is π/n. Thus, the mean difference divided by the standard deviation follows approximately the standard normal distribution, as

$$z = \frac{p - \pi}{\sqrt{\pi/n}},$$

and probability questions again may be addressed using Table I. Section 15.7 provides a more detailed discussion of this topic.

Exercise 13.6. The probability of acquired immunodeficiency syndrome (AIDS) developing in a person during the first year of infection with human immunodeficiency virus (HIV) is about 0.01. An investigator reviewed the records of 70 patients with HIV. During the year following their reported date of first possible exposure to another HIV-infected person, AIDS had developed in five. Does the investigator believe these patients' reporting?

ANSWERS TO EXERCISES

13.1. $z = (126 - 120)/5 = 1.20$. Your patient is 1.2σ above the mean. From Table 13.1, the area in the right tail is 0.115. Therefore, 11.5% of the population have greater DBP.

13.2. For 7 df, a two-tailed $\alpha = 0.05$ is obtained by $m \pm 2.365 \times s$, or the interval from 235.8 to 669.7.

13.3. $\chi^2 = 17 \times 576.48/259.85 = 37.71$. From the row for 17 df in Table 13.3 we see that a χ^2 value of 37.71 lies between 35.72, associated with an area under the curve in the right tail of 0.5%, and 40.78, associated with an area of 0.1%. By interpolation, which is approximate, the area associated with 37.71 will be about 0.4 of the way from 0.5% to 0.1%, or about 0.3%. If we sampled standard deviations repeatedly from a normal patient population, we would find a standard deviation of 24.01 or larger only about 3 times out of 1000.

13.4. For samples so small, numerator and denominator dfs both 3, the F ratio as shown in Table 13.4 would have to be larger than 9.28. For these types of assay, F = 0.0586/0.0120 = 4.88. As 4.88 < 9.28, we do not have evidence to conclude a difference in variability.

13.5. In Table 13.5, which displays the $n = 6$ block of Table VI, the probability 0.114 lies at the intersection of the column $\pi = 0.10$ and the row $n_o = 2$. The probability that there would be two or more patients with unhealed ulcers is more than 11%.

13.6. $\pi = 0.01$ and $n = 70$; therefore, the expected number of AIDS cases is $\lambda = 0.7$. $n_o = 5$ cases occurred. In Table 13.6, the intersection of the column $\lambda = 0.7$ and $n_o = 5$ shows the probability to be 0.001. There is only 1 chance in 1000 that the investigator received true reports.

Chapter 14

Confidence Intervals

14.1. OVERVIEW

REVIEW OF SECTION 4.1

We noted that hematocrit (Hct) values (measured in percent) for healthy patients are not all the same but rather range over an interval. What is this interval? Because we see an occasional very high or low value in a healthy patient, we want an interval that we are confident will include most of the healthy population; we name this a *confidence interval*. But what is the width of this interval? We might state that it should include 95% of the population. We denote by the probability that a healthy patient's Hct will be outside the healthy interval, usually $\alpha/2$ on each side of the interval. When a patient's Hct does occur outside the healthy interval, we think, "It is likely that the patient has arisen from an unhealthy population, but not certain. We must compare the Hct with other indicators to derive a total picture."

STATISTICS AND DISTRIBUTIONS IN CONFIDENCE INTERVALS

The statistic on which a confidence interval is seen most frequently is the population mean, estimated by the sample mean. Because the sample mean is distributed normal, the normal distribution is associated most often with the confidence interval. However, confidence intervals are not restricted to the mean. We can put confidence intervals on any statistic: an individual (Is my patient part of a known healthy population?), a standard deviation (Is the effect of a new drug less variable and therefore more dependable than an older one?), or others. The distribution from which the interval values are chosen is specified by the statistic on which the interval is desired: normal for a mean, chi-square for a variance (with the square root taken of the results to apply to standard deviations), binomial for a rate or proportion, and so on.

CONFIDENCE INTERVALS AND SYMMETRY

The reader of medical articles most often sees confidence intervals on means. A mean follows the normal distribution, which is symmetric, so a confidence interval on a mean is symmetric about that mean. It is often expressed as the mean plus or minus ($\pm$) the confidence distance above or below the mean. Sometimes a graph shows the height of the mean by a bar with a whisker above the bar showing the upper confidence bound, implying the lower bound to be an equal distance below the bar height. Notably, such expressions are *not acceptable for statistics arising from asymmetric distributions*. Confidence distances above and below the statistic are not equal for standard deviations (chi-square distribution), rates with mean other than 0.5 (binomial distribution), individual readings from skewed distributions, and some others. For these cases, *separate upper and lower limits* must be given, not plus or minus a single value.

LEVEL OF CONFIDENCE IN INTERVALS

The most frequently chosen interval is 95%, which appears to be becoming traditional in medicine. We may be guided, but should not be fettered, by tradition. Note that the greater the variability in the statistic's estimate, the wider the confidence interval will be. In situations encountering great variability, perhaps in certain psychiatric measures, the clinical need for a tighter interval might dictate a 90% interval. In situations characterized by small variability, perhaps in the control of electronic instruments, a 99% interval might be appropriate.

A CONFIDENCE INTERVAL CAN APPLY TO AN INDIVIDUAL PATIENT

We find the extent of resection of the tracheal carina in DB12 to be approximately normal with $m = 2.96$ and $s = 1.24$. The sample size is large enough to treat s as σ. In the normal table (Table I in Tables of Probability Distributions), we find that $1 - \alpha = 0.95$ is associated with $z = 1.96$, or the confidence interval on an observation extends $\pm 1.96\sigma$ on either side of m. Thus, the end points of the interval would be $2.96 - 1.96 \times 1.24 = 0.53$ and $2.96 + 1.96 \times 1.24 = 5.39$. This tells us that a new patient is 95% sure to have a resection between about 0.5 and 5.5 cm. In this particular instance, the confidence interval on an individual patient is not helpful clinically, because the surgically reasonably range is about 1–6 cm anyway. In other cases, such as choosing a healthy range for a laboratory test result, confidence intervals on individuals can be quite useful.

A CONFIDENCE INTERVAL ON A STATISTIC

In the previous paragraph, we used a distribution of patients to obtain a confidence interval in relation to which we interpreted an observation from a single patient. This observation may be thought of as a sample size of 1 from a population with known mean and standard deviation. Although this is often useful in clinical practice, in research we often are more interested in the confidence interval on the estimate of a population statistic, such as a mean or a standard deviation. In Chapter 4, Eq. (4.2) expresses a confidence interval separated into its primary and subordinate components as

> The probability
>> that a population statistic
>>> from a distribution of estimates of that statistic
>> is contained in a specified interval
> is given by
>> the area of the distribution within that interval. (14.1)

Denoting probability as P, the area under the probability curve outside the interval as α, and the interval end points as *critical values,* a somewhat simpler expression is obtained:

$$P[\text{lower critical value} < \text{population statistic} < \text{upper critical value}] = 1 - \alpha.$$
(14.2)

Critical values are those values that separate the tails from the main body of the distribution. All that is needed to implement Eq. (14.2) for any population statistic is the method to find the critical values, which is slightly different for each type of statistic. The methods to calculate the critical values are the focus for the remainder of this chapter.

ONE- VERSUS TWO-TAILED CONFIDENCE INTERVALS

Whereas most confidence intervals are formed as intervals excluding both tails, an occasional case occurs in which we want to exclude only one tail. The next section provides an example of excluding only one tail.

14.2. CONFIDENCE INTERVAL ON A MEAN, KNOWN STANDARD DEVIATION

EXAMPLES POSED

The extent of resection of the tracheal carina (DB12) has an approximately normal distribution with mean $m = 2.96$ and standard deviation $s = 1.24$. The sample size is large enough to treat s as σ. What would be a 95% confidence interval on the mean?

As a variation on this question, suppose we are concerned only with larger resections, which pose a threat because of their size; the smaller resections do not. Thus, we seek a critical value cutting off α proportion in only the upper tail, which could be thought of as a one-tailed confidence interval. Suppose we still want a 95% confidence.

METHOD

Equation (14.2) applied to means would be expressed as $P[$lower critical value $< \mu <$ upper critical value$] = 1 - \alpha$. The critical values on a population mean μ would be given by the sample mean m plus or minus a number of standard errors of the mean (SEMs) as indicated by areas under the normal curve. The SEM is given by dividing the standard deviation of the individual observations by the square root of the number of observations, or SEM $= \sigma/\sqrt{n}$. The number of SEMs is $z_{1-\alpha/2}$, the distance on the normal curve outside of which each tail $\alpha/2$ of the area lies. By substituting these elements in Eq. (14.2), we find

$$P[m - z_{1-\alpha/2}\sigma_m < \mu < m + z_{1-\alpha/2}\sigma_m] = 1 - \alpha. \qquad (14.3)$$

95% Confidence Interval

Table 14.1 is a segment of Table I that shows the normal areas most frequently used. If we want 95% confidence, we look in Table 14.1 for $1 - \alpha = 0.950$ under "Two-tailed applications." To its left in the first column (i.e., under z) we find 1.960. Thus, our critical values are the sample mean m plus and minus $1.96\sigma_m$. In symbols,

$$P[m - 1.96 \times \sigma_m < \mu < m + 1.96 \times \sigma_m] = 0.95. \qquad (14.4)$$

Table 14.1

The Most Used Entries from Table I, Normal Distribution[a]

z	One-tailed applications		Two-tailed applications	
	One-tailed α	$1 - \alpha$	Two-tailed α	$1 - \alpha$
1.645	0.050	0.950	0.100	0.900
1.960	0.025	0.975	0.050	0.950
2.326	0.010	0.990	0.020	0.980
2.576	0.005	0.995	0.010	0.990

[a]For distances (z) to the right of the mean, given are (1) one-tailed α, area under the curve in the positive tail; (2) one-tailed $1 - \alpha$, area under all except the tail; (3) two-tailed α, areas combined for both positive and negative tails; and (4) two-tailed $1 - \alpha$, the area under all except the two tails.
See Tables of Probability Distributions to review Table I.

One-Tailed Confidence Interval

If we are concerned with the mean occurring on one side only, say, the larger side, we would seek a critical value cutting off all α proportion in only the upper tail. Rewriting Eq. (14.3), we get

$$P[\mu < m + z_{1-\alpha}\sigma_m] = 1 - \alpha. \qquad (14.5)$$

For a $1 - \alpha = 95\%$ confidence interval, we look for 0.95 under the "One-tailed applications" column in Table 14.1 and find that $z_{1-\alpha}$ is 1.645. Substituting into Eq. (14.5), we obtain

$$P[\mu < m + 1.645\sigma_m] = 0.95. \qquad (14.6)$$

EXAMPLES COMPLETED

We want a 95% confidence interval on the mean extent of carinal resection. From Table 14.1 (or Table I) under "Two-tailed applications," we find that $1 - \alpha = 0.95$ is associated with $z = 1.96$, or the confidence interval on an observation extends $\pm 1.96\sigma_m$ on either side of m. $\sigma_m = \sigma/\sqrt{n} = 1.24/\sqrt{134} = 0.107$. Thus, the end points of the interval would be $2.96 - 1.96 \times 0.107 = 2.75$ and $2.96 + 1.96 \times 0.107 = 3.17$. By using the form in Eq. (14.4), the confidence interval would be expressed as

$$P[2.75 < \mu < 3.17] = 0.95$$

and voiced as "The probability that the interval 2.75 to 3.17 contains μ is 0.95."

To find the one-tailed confidence interval, we look under "One-tailed applications" in Table 14.1 (or Table I). For $1 - \alpha = 0.95$, we find $z_1 - \alpha$ (number of standard deviations to the right of the mean) is 1.645, and we substitute these and the values for m and σ_m into Eq. (14.6) to get

$$P[\mu < 2.96 + 1.645 \times 0.107] = P[\mu < 3.136] = 0.95.$$

We are 95% confident that the mean resection will not exceed 3.14 cm.

ADDITIONAL EXAMPLE

In the initial example of this section, we found a 95% confidence interval on the mean extent of resection of the tracheal carina (DB12). Let us find one on patient age for the same sample of patients: $m = 47.84$, and we take $\sigma = 15.78$, so $\sigma_m = 15.78/\sqrt{134} = 1.36$. By substituting in Eq. (14.4), we find

$$P[47.84 - 1.96 \times 1.36 < \mu < 47.84 + 1.96 \times 1.36] = P[45.17 < \mu < 50.51] = 0.95.$$

We are 95% sure that the interval 45.2–50.5 years will enclose the population mean age.

Exercise 14.1. Find the 90% and 99% confidence intervals on mean age of patients with carinal resection.

14.3. CONFIDENCE INTERVAL ON A MEAN, ESTIMATED STANDARD DEVIATION

EXAMPLE POSED

In DB5, plasma silicon level on $n = 30$ patients was measured before and again after silicone implantation. What is the 95% confidence level on the mean difference before (pre) versus after surgery (post)?

METHOD

Section 4.7 describes the method applied here. Let us briefly review it. The concept of a confidence interval on the mean using s is much the same as that using σ (see Section 14.2), the difference being that the t distribution (see Table II in Tables of Probability Distributions) rather than the normal is used to calculate the critical values. The form, seen previously as Eq. (4.6), is quite similar to Eq. (14.3):

$$P[m - t_{1-\alpha/2}s_m < \mu < m + t_{1-\alpha/2}s_m] = 1 - \alpha. \qquad (14.7)$$

Because the t distribution has a greater spread than the normal, the confidence interval will be slightly wider. If we want 95% confidence, we find the desired t-value in Table II at the intersection of the 0.95 column for "Two-tailed $1 - \alpha$ (except both tails)" and the row for the appropriate degree of freedom (df).

EXAMPLE COMPLETED

We want the 95% confidence level on the mean difference before versus after surgery for 30 patients. The mean difference is $m = 0.0073$ with standard deviation $s = 0.1222$. The SEM $= 0.1222/\sqrt{29} = 0.0227$. A segment of the t table (see Table II) appears as Table 14.2. The intersection of the column for two-tailed $1 - \alpha = 0.95$ with the row for $n - 1 = 29$ df gives $t_{1-\alpha/2} = 2.045$, which tells us that the critical values will be the sample mean plus or minus a little more than 2 SEMs. After substitution in Eq. (14.7), the confidence interval becomes

$$P[0.0073 - 2.045 \times 0.0227 < \mu < 0.0073 + 2.045 \times 0.0227]$$
$$= P[-0.039 < \mu < 0.054] = 0.95.$$

ADDITIONAL EXAMPLE

In DB3, serum theophylline levels were measured in patients with emphysema before they were given azithromycin, at 5 days, and at 10 days. What are 95% confidence intervals on mean serum theophylline levels at these three points in

Table 14.2

A Segment of Table II, t Table^a

| Two-tailed α | 0.20 | 0.10 | 0.05 | 0.02 | 0.01 | 0.002 | 0.001 |
Two-tailed $1 - \alpha$	0.80	0.90	0.95	0.98	0.99	0.998	0.999
$df = 15$	1.341	1.753	2.131	2.602	2.947	3.733	4.073
29	1.311	1.699	2.045	2.462	2.756	3.396	3.659

^{a}Selected distances (t) to the right of the mean are given for various degrees of freedom (df); for two-tailed α, areas combined for both positive and negative tails; and two-tailed $1 - \alpha$, area under all except the two tails.
See Tables of Probability Distributions to review Table II.

time? $n = 16$; $m_0 = 10.80$; $s_0 = 3.77$; $m_5 = 9.81$; $s_5 = 4.61$; $m_{10} = 10.14$; and $s_{10} = 3.98$. SEMs are 0.94, 1.15, and 1.00, respectively. From Table 14.2, the intersection of the column for two-tailed $1 - \alpha = 0.95$ with the row for 15 df gives $t_{1-\alpha/2} = 2.131$. After substitution in Eq. (14.7), these values yield the following confidence intervals:

$$P_o[10.80 - 2.131 \times 0.94 < \mu < 10.80 + 2.131 \times 0.94] = P_o[8.80 < \mu < 12.80]$$
$$= 0.95$$

$$P_5[9.81 - 2.131 \times 1.15 < \mu < 9.81 + 2.131 \times 1.15] = P_5[7.36 < \mu < 12.26]$$
$$= 0.95$$

$$P_{10}[10.14 - 2.131 \times 1.00 < \mu < 10.14 + 2.131 \times 1.00] = P_{10}[8.01 < \mu < 12.27]$$
$$= 0.95.$$

Exercise 14.2. Among the indicators of patient condition following pyloromyotomy (correction of stenotic pylorus) in neonates is time (hours) to full feeding.[21] A surgeon wants to place a 95% confidence interval on mean time to full feeding. The surgeon has readings from $n = 20$ infants. Some of the data are 3.50, 4.52, 3.03, 14.53, ... The surgeon calculates $m = 6.56$ and $s = 4.57$. What is the confidence interval?

14.4. CONFIDENCE INTERVAL ON A VARIANCE OR STANDARD DEVIATION

EXAMPLE POSED

The variability in platelet-derived growth factor (PDGF) has been estimated (DB9) as $s = 19,039.7$. How far from this may a sample standard deviation deviate before we suspect the influence of a new factor? We want a 95% confidence interval on the population σ.

METHOD

We know (from Sections 2.8 and 4.9) that s^2 calculated from a random sample from a normal population is distributed as $\chi^2 \times \sigma^2/df$. A confidence-type statement on σ^2 obtained from this relationship, excluding $1 - \alpha$ in each tail, is given by

$$P[s^2 \times df/\chi_R^2 < \sigma^2 < s^2 \times df/\chi_L^2] = 1 - \alpha, \qquad (14.8)$$

where χ_R^2 is the critical value for the right tail found from Table III (see Tables or Probability Distributions), and χ_L^2 is the critical value for the left tail found from Table IV (see Tables or Probability Distributions). The critical values are found separately because the chi-square distribution is asymmetric. To find confidence on σ, take the square roots of the components within brackets.

Example Completed

We asked for a 95% confidence interval on the standard deviation of PDGF. We know the probability distribution of s^2 but not that of s, so we shall find the interval on the population variance, σ^2, and take square roots. $s^2 = 19{,}039.7^2 = 362{,}510{,}180$. The distribution of s^2 requires df and critical values of χ^2. $df = n - 1 = 19$. Table 14.3 gives segments of the chi-square tables, Tables III (right tail) and IV (left tail). By looking in Table 14.3 under right tail area $= 0.025$ for 19 df, we find 32.85. Similarly, under left tail area $= 0.025$ for 19 df, we find 8.91. By substituting these values in Eq. (14.8), we find

$$P[s^2 df/\chi_R^2 < \sigma^2 < s^2 \times df/\chi_L^2]$$
$$= P[362{,}510{,}180 \times 19/32.85 < \sigma^2 < 362{,}510{,}180 \times 19/8.91]$$
$$= P[209{,}671{,}030 < \sigma^2 < 773{,}029{,}560] = 0.95.$$

Taking square roots within the brackets, we obtain

$$P[14{,}480.0 < \sigma < 27{,}803.4] = 0.95.$$

Additional Example

Is plasma silicon level (DB5) less variable after implantation? Whereas this question can be tested by a formal hypothesis test as in Chapter 19, we can make an informal comparison by contrasting the confidence intervals on σ_{pre} and σ_{post}. What are these confidence intervals? From Eq. (14.8), we need s^2, df, and the critical chi-squares: $s_{pre} = 0.098565$, $s_{pre}^2 = 0.009715$, $s_{post} = 0.076748$, $s_{post}^2 = 0.005890$, and $df = 29$. From Table 14.3, the intersections of columns $\alpha = 0.025$ and row $df = 29$ yields $\chi_R^2 = 45.72$ and $\chi_L^2 = 16.05$. Substitution in

Table 14.3
Segments of the Chi-square Tables[a]

	Right tail			Left tail		
α	0.05	0.025	0.01	0.01	0.025	0.05
$1 - \alpha$	0.95	0.975	0.99	0.99	0.975	0.95
$df = 18$	28.87	31.53	34.80	7.01	8.23	9.39
19	30.14	32.85	36.19	7.63	8.91	10.12
20	31.41	34.17	37.57	8.26	9.59	10.85
28	41.34	44.46	48.28	13.57	15.31	16.93
29	42.56	45.72	49.59	14.25	16.05	17.71
30	43.77	46.98	50.89	14.95	16.79	18.49

[a]Selected χ^2 values (distances above zero) are given for various degrees of freedom (df); α, the area under the curve in the indicated tail; and $1 - \alpha$, the area under all except the indicated tail. See Tables of Probability Distributions to review the chi-square tables: Tables III (associated with the right tail) and IV (associated with the left tail).

Eq. (14.8) yields

$$P[0.006162 < \sigma_{pre}^2 < 0.017554] = 0.95$$

and

$$P[0.003736 < \sigma_{post}^2 < 0.010642] = 0.95.$$

By taking square roots within the brackets, we find confidence intervals on the σ's to be

$$P[0.078 < \sigma_{pre} < 0.132] = 0.95$$

and

$$P[0.061 < \sigma_{post} < 0.103] = 0.95.$$

The confidence intervals are not very different from a clinical perspective.

Exercise 14.3. An infectious disease specialist samples the white blood cell count of $n = 19$ patients who contracted the same nosocomial infection in an orthopedic ward. She finds that the mean count is not unusual but suspects the standard deviation may be larger than normal: $s = 6000$. What is her 95% confidence interval on the standard deviation σ for the population of infected patients?

14.5. CONFIDENCE INTERVAL ON A PROPORTION

PROPORTIONS FALL INTO TWO TYPES

Proportions fall in the interval 0–1. However, methods must be divided into two types: cases in which the proportion lies toward the center of the interval, away from the extremes 0 and 1, which we may designate *central proportion,* and cases in which the proportion lies very close to 0 or 1, which we may designate *extreme proportion.* The reason is that they lead to different distributions (see later in Methods section).

EXAMPLE POSED: CENTRAL PROPORTION

Of our 301 urology biopsies from DB1, 95 had positive results, yielding a sample proportion $p = 0.316$. What is a reasonable range for positive rate? We want a confidence interval on the theoretical proportion π.

EXAMPLE POSED: EXTREME PROPORTION

Children with high lead levels are found in a certain hospital's catchment.[50] Of 2500 children sampled, 30 with high lead levels are found, yielding $p = 0.012$. How far may a rate deviate from this p before the hospital administration suspects an atypical situation? We want a confidence interval on the theoretical proportion π.

METHOD

The confidence interval we seek here is on the theoretical but unknown population proportion π, which we estimate by the sample proportion p.

Central Proportion

In this case, π is not close to 0 or 1 but is nearer to 0.5. It has been shown that p is distributed binomial, approximated by the normal with sample mean $\mu = p$ and standard deviation $\sigma = \sqrt{p(1-p)/n}$. The mean p and standard deviation σ are substituted in the 95% confidence interval pattern Eq. (14.4) to obtain

$$P[p - 1.96 \times \sigma - 1/2n < \pi < p + 1.96 \times \sigma + 1/2n] = 0.95, \qquad (14.9)$$

where the $1/2n$ components are continuity corrections to improve the approximation. If confidence other than 95% is desired, the form of Eq. (14.3) may be used in place of Eq. (14.4). In this case, 1.96 is replaced by the appropriate multiplier chosen from Table I as z (first column) corresponding to the desired two-tailed $1 - \alpha$ (last column).

Extreme Proportion

In this case, π is very near to 0 or 1. p is distributed Poisson, approximated by the normal with standard deviation estimated as the smaller of $\sigma = \sqrt{p/n}$ or $\sqrt{(1-p)/n}$. p and σ are substituted in the confidence interval pattern of Eq. (14.4) to obtain

$$P[p - 1.96 \times \sigma < \pi < p + 1.96 \times \sigma] = 0.95. \qquad (14.10)$$

Use Eq. (14.3) in place of Eq. (14.4) for confidence levels other than 95%.

EXAMPLE COMPLETED: CENTRAL PROPORTION

We know that a large-sample proportion not near 0 or 1 is distributed approximately normal with mean p, in this example 0.316. The standard deviation σ is given by

$$\sigma = \sqrt{\frac{p(1-p)}{n}} = \sqrt{\frac{0.316 \times 0.684}{301}} = 0.0268.$$

By using Eq. (14.9), we bracket π with 95% confidence as

$$P[p - 1.96 \times \sigma - 1/2n < \pi < p + 1.96 \times \sigma + 1/2n]$$
$$= P[0.316 - 1.96 \times 0.0268 - 0.00166 < \pi < 0.316 + 1.96 \times 0.0268 + 0.00166]$$
$$= P[0.262 < \pi < 0.370] = 0.95.$$

We are 95% confident that the proportion of positive prostate biopsies in the population of patients presenting with urologic problems lies between 26% and 37%.

Example Completed: Extreme Proportion

Because π is evidently close to zero, we use the normal approximation to the Poisson distribution: $\sigma = \sqrt{p/n} = \sqrt{0.012/2500} = 0.00219$. The hospital administration wants to be sure that it does not have too many high-lead children in its catchment and therefore chooses a 99% confidence interval. From Table 14.1, the 0.99 two-tailed $1 - \alpha$ yields a corresponding z of 2.576. Replacing the 1.96 in Eq. (14.10) with 2.576, we obtain

$$P[p - 2.576 \times \sigma < \pi < p + 2.576 \times \sigma]$$
$$= P[0.012 - 2.576 \times 0.00219 < \pi < 0.012 + 2.576 \times 0.00219]$$
$$= P[0.0064 < \pi < 0.0176] = 0.99.$$

By focusing on the right tail, the hospital administration may be 99.5% confident that no more than 1.8% of children in its catchment have high lead levels.

Additional Example

In a study[43] anesthetizing patients undergoing oral surgery by a combination of propofol and alfentanil, 89.1% of 110 patients rated the anesthetic as highly satisfactory or excellent. What are 95% confidence limits on π, the proportion of the population satisfied with the anesthetic? Because π, is not near 0 or 1, the normal approximation to the binomial is appropriate. $p = 0.891$ and $\sigma = \sqrt{0.891 \times 0.109/110} = 0.030$ are substituted in Eq. (14.9):

$$P[0.891 - 1.96 \times 0.030 - 1/220 < \pi < 0.891 + 1.96 \times 0.030 + 1/220]$$
$$= P[0.828 < \pi < 0.954] = 0.95.$$

We are 95% confident that at least 83% of patients will be quite satisfied with the combination of propofol and alfentanil and at least 5% will not.

Exercise 14.4.   Based on a sample of 226, a 5-year recurrence rate of basal cell carcinoma when lesions measured more than 20 mm was 0.261.[76] Put a 95% confidence interval on the population π for this proportion.

14.6. CONFIDENCE INTERVAL ON A CORRELATION COEFFICIENT

Example Posed

An orthopedist is studying hardware removal in broken ankle repair.[28] One of the variables is the maximum angle (degrees) of plantar flexion. The orthopedist wants to know whether age is related to plantar flexion. The sample correlation coefficient for the 19 patients is found to be 0.1945. What is the confidence interval on that coefficient?

METHOD

Confidence intervals on the population correlation coefficient ρ, estimated by $r = (s_{xy}/s_x s_y)$ for small sample sizes are usually too wide to be of much help. If we have a larger sample size, we can transform the correlation coefficient to have approximately the normal distribution. We recall that *ln* denotes "natural logarithm" and e denotes the "natural number," 2.71828. ... The old base-ten logarithms were used to facilitate certain difficult arithmetic in the days before computers and seldom are used anymore; ln and e are found on every computer and most handheld calculating machines. The normally transformed correlation coefficient has sample mean

$$m = \frac{1}{2} \ln \frac{1+r}{1-r}, \tag{14.11}$$

and the standard deviation is estimated as

$$\sigma = \frac{1}{\sqrt{n-3}}. \tag{14.12}$$

Solving the transformation Eq. (14.11) for r yields a ratio of exponential expressions, which, together with Eq. (14.12), can be entered into the pattern of Eq. (14.3). This inequality can then be solved mathematically to obtain a $1 - \alpha$ confidence interval as

$$P\left[\frac{1+r-(1-r)e^{\frac{2z_{1-\alpha/2}}{\sqrt{n-3}}}}{1+r+(1-r)e^{\frac{2z_{1-\alpha/2}}{\sqrt{n-3}}}} < \rho < \frac{1+r-(1-r)e^{\frac{2z_{1-\alpha/2}}{\sqrt{n-3}}}}{1+r+(1-r)e^{\frac{2z_{1-\alpha/2}}{\sqrt{n-3}}}}\right] = 1 - \alpha. \tag{14.13}$$

Equation (14.13) holds for nonnegative correlation coefficients. If r is negative, symmetry properties allow a simple solution. Find the confidence limits as if the r were positive (i.e., use $|r|$), then change the signs on the resulting limits and exchange their positions. This mechanism is illustrated in the additional example.

EXAMPLE COMPLETED

The correlation coefficient between maximum angle of plantar flexion and age for 19 patients was 0.1945. Equation 14.13 is not so intimidating when broken into pieces: $1 + r = 1.1945$, $1 - r = 0.8055$, $n - 3 = 16$, and $\sqrt{16} = 4$. For 95% confidence, $z_{1-\alpha/2} = 1.96$. The exponent of e will be $\pm 2 \times 1.96/4 = \pm 0.98$.

By substituting these components into Eq. (14.13), the orthopedist finds

$$P\left[\frac{1+r-(1-r)e^{\frac{2z_{1-\alpha/2}}{\sqrt{n-3}}}}{1+r+(1-r)e^{\frac{2z_{1-\alpha/2}}{\sqrt{n-3}}}} < \rho < \frac{1+r-(1-r)e^{\frac{2z_{1-\alpha/2}}{\sqrt{n-3}}}}{1+r+(1-r)e^{\frac{2z_{1-\alpha/2}}{\sqrt{n-3}}}}\right]$$

$$= P\left[\frac{1.1945-0.8055\times e^{0.98}}{1.1945+0.8055\times e^{0.98}} < \rho < \frac{1.1945-0.8055\times e^{-0.98}}{1.1945+0.8055\times e^{-0.98}}\right]$$

$$= P[-0.2849 < \rho < 0.5961] = 0.95.$$

Because the population correlation coefficient may be anywhere from about -0.28 to $+0.60$ (the confidence interval crosses zero), the orthopedist concludes that age has not been shown to be an influence.

ADDITIONAL EXAMPLE

In a study on increasing red blood cell mass before surgery, the correlation between Hct and serum erythropoietin levels was of interest. An inverse correlation was anticipated for patients with normal renal function. Levels were measured in $n = 126$ patients, and the correlation coefficient was calculated as $r = -0.59$.[18] Let us find a 95% confidence interval on the population correlation coefficient ρ. Because r is negative, we first find the limits as if it were positive; we temporarily use $r = +0.59$. Then, $1 + r = 1.59$, $1 - r = 0.41$, $n - 3 = 123$, and $\sqrt{123} = 11.0905$. The exponent is ± 0.3535. Substitution in Eq. (14.13) yields

$$P\left[\frac{1+r-(1-r)e^{\frac{2z_{1-\alpha/2}}{\sqrt{n-3}}}}{1+r+(1-r)e^{\frac{2z_{1-\alpha/2}}{\sqrt{n-3}}}} < \rho < \frac{1+r-(1-r)e^{-\frac{2z_{1-\alpha/2}}{\sqrt{n-3}}}}{1+r+(1-r)e^{-\frac{2z_{1-\alpha/2}}{\sqrt{n-3}}}}\right]$$

$$= P[0.4628 < \rho < 0.6934] = 0.95.$$

To convert the limits to bound $r = -0.59$, we change 0.4628 to -0.4628 and put it in the right (larger limit) position and then change 0.6934 to -0.6934, putting it in the left (smaller limit) position. The resulting confidence interval on ρ when $r = -0.59$ from a sample of 126 becomes

$$P[-0.6934 < \rho < -0.4628] = 0.95.$$

We are 95% confident that the population correlation coefficient between Hct and serum erythropoietin levels in patients with normal renal function lies between -0.69 and -0.46.

Exercise 14.5. An orthopedist is experimenting with the use of Nitronox as an anesthetic in the treatment of arm fractures in children.[26,27] The orthopedist finds that it provides a shorter duration of surgery and less pain than other procedures used. In addition, the orthopedist expects that the shorter the procedure, the less the perceived pain. Do the data bear out this expectation? More specifically, is the confidence interval on the population correlation coefficient between the two variables entirely within the positive range? The orthopedist treats $n = 50$ children, recording the treatment time in minutes and the score from a CHEOPS pain scale. The orthopedist finds $r = 0.267$. Find the 95% confidence interval on this r.

ANSWERS TO EXERCISES

14.1. To substitute in Eq. (14.3), the z-values must be found from Table 14.1 (or Table I). Under the column for "Two-tailed $1 - \alpha$," 0.900 (90%) yields $z = 1.645$ and 0.990 (99%) yields 2.576. Substituting $m = 47.84$ and $\sigma_m = 1.36$, we find $P[47.84 - 1.645 \times 1.36 < \mu < 47.84 + 1.645 \times 1.36] = P[45.60 < \mu < 50.08] = 0.90$ and $P[47.84 - 2.576 \times 1.36 < \mu < 47.84 + 2.576 \times 1.36] = P[44.34 < \mu < 51.34] = 0.99$.

14.2. The surgeon calculates $s_m = s/\sqrt{n} = 4.57/\sqrt{20} = 1.02$. From Table II, t for 95% "Two-tailed $1 - \alpha$ (except both tails)" for 19 df is 2.093. The surgeon substitutes these values in Eq. (14.7) to obtain $P[m - t_{1-\alpha/2}s_m < \mu < m + t_{1-\alpha/2}s_m] = P[6.56 - 2.093 \times 1.02 < \mu < 6.56 + 2.093 \times 1.02] = P[4.43 < \mu < 8.69] = 0.95$. The surgeon is 95% confident that mean time to full feeding will lie between 4.4 and 8.7 hours.

14.3. $s^2 = 36{,}000{,}000$ and $df = 18$. From Table 14.3, the intersection of the 0.025 column with 18 df is 31.53 for the right tail and 8.23 for the left tail. Substitution in Eq. (14.8) yields $P[20{,}551{,}855 < \sigma^2 < 78{,}736{,}330] = 0.95$, and taking square roots yields $P[4533 < \sigma < 8873] = 0.95$. Because σ for a healthy population would be about 1200–1300, the variability does indeed appear to be abnormally large.

14.4. p is not near 0 or 1, so $\sigma = \sqrt{p(1-p)/n} = \sqrt{0.261 \times 0.739/226} = 0.0292$. Substitute in Eq. (14.9) to obtain $P[p - 1.96\sigma - 1/2n < \pi < p + 1.96\sigma + 1/2n] = P[0.261 - 1.96 \times 0.029 - 1/452 < \pi < 0.261 + 1.96 \times 0.029 + 1/452] = P[0.202 < \pi < 0.320] = 0.95$.

14.5. $1 + r = 1.267$, $1 - r = 0.733$, and the exponential term is 0.5718. Substitution in Eq. (14.13) yields $P[-0.0123 < \rho < 0.5076] = 0.95$. The confidence interval crosses zero, indicating that the population correlation coefficient could be zero or negative and the 0.267 sample coefficient could have occurred by chance.

Chapter 15

Tests on Categorical Data

15.1. CATEGORICAL DATA SUMMARY

TERMS AND SYMBOLS

Section 6.2 provides a more complete treatment of the basics of categorical data. *Categorical* or *nominal* data are data placed into distinct categories rather than being measured as a point on a scale or ranked in order. The basic sample statistics are the *count*, obtained by counting the number of events per category, and the *proportion* of data in a category, the count divided by the total number in the variable. If n denotes the total number of events and n_a the number in category a, the proportion in a is $p_a = n_a/n$. Often percent, $100p_a$, is used. π denotes the population proportion, which is known from theory or closely approximated by a very large sample.

Example

A vascular surgeon treated 29 patients endoscopically for thrombosis in the leg,[55] after which thrombosis recurred in 6 patients. $n = 29$, $n_a = 6$, and $p_a = 6/29 = 0.207$. About 21% of the patients experienced recurrence of thrombosis.

ORGANIZING DATA FOR CATEGORICAL METHODS

If the relationship between two variables is of interest, such as whether they are independent or associated, the number of observations simultaneously falling into the categories of two variables is counted. For example, the vascular surgeon is interested in the relation between recurrence and ulcer condition. The first variable may be categorized as recurred or not recurred and the second as open ulcers or healed ulcers. This gives rise to the four categories of counts, which are shown together with example counts in Table 15.1. We can see that five is the number of recurrences, contingent on the ulcers being healed. Tables of this sort are called *contingency tables*, because the count in each cell is the number in that category of that variable contingent on also lying in a particular category of the other variable.

Table 15.1

Contingency Table of Thrombosis Recurrence and Ulcer Condition

		Ulcers		
		Open	Healed	Total
Recurred	Yes	$n_{11} = 1$	$n_{12} = 5$	$n_{1.} = 6$
	No	$n_{21} = 7$	$n_{22} = 16$	$n_{2.} = 23$
	Total	$n_{.1} = 8$	$n_{.2} = 21$	$n = 29$

CONTINGENCY TABLE SYMBOLS

Contingency tables often have more than four categories. For example, if the preulcerous condition of heavy pigmentation were added to the ulcer condition variable, there would be three conditions, resulting in a 2×3 table. In general, the number of rows is denoted by r and the number of columns by c, yielding an $r \times c$ table. Because the values of r and c are unknown until a particular case is specified, the cell numbers and the marginal row and column totals must be symbolized in a way that can apply to tables of any size. The usual symbolism is shown in Table 15.1: n, for "number," with two subscripts, the first indicating the row and the second indicating the column. Thus, the number of cases observed in row 1, column 2 is n_{12}. The row and column totals are denoted by inserting a dot in place of the numbers we summed to get the total. Thus, the total of the first row would be $n_{1.}$, and the total of the second column would be $n_{.2}$. The grand total can be denoted just n (or $n_{..}$).

CHOOSING CATEGORICAL METHODS

Methods to assess association between two categorical variables exist for both counts and proportions. When all cells of the table of counts can be filled in, use the methods for counts. Tests of proportions are appropriate for cases in which some table totals are missing but for which proportions can still be found. A first-step guide to choosing a test was introduced in Table 10.1. A more complete form of this table appears as Table 12.1, and we will refer to only Table 12.1 in the remainder of this book. The tests in this chapter address columns 3–5 in Table 12.1. First, decide if the question of your study asks about proportions (columns 3 and 4) or about counts, that is, number of cases appearing in the cells of a table (column 5). To select the row of Table 12.1, note if your sample is small or large and how many subsamples you have. Sections 15.2–15.5 address methods for count-type data, Sections 15.6 and 15.7 addresses proportions, and Section 15.8 addresses matched pairs.

DEGREES OF FREEDOM FOR CONTINGENCY TESTS

Contingency test degrees of freedom (*df*) are given by the minimum number of cells that must be filled to be able to calculate the remainder of cell entries using the totals at the side and bottom (often termed *margins*). For example, only one cell of a 2×2 table with the sums at the side and bottom needs to be filled, and the others can be found by subtraction; it has 1 *df*. A 2 × 3 table has 2 *df*. In general, an $r \times c$ table has $df = (r - 1)(c - 1)$.

CATEGORIES VERSUS RANKS

Cases for which counts or proportions for each sample group can be reordered without affecting the interpretation are constrained to categorical methods. An example would be group 1: high hematocrit (Hct); group 2: high white blood cell count; and group 3: high platelet count. These groups could be written in order as 2, 1, 3 or 3, 2, 1 without affecting our understanding of the clinical implications. However, if sample groups fall into a natural order in which the logical implication would change by altering this order—for example, group 1: low Hct; group 2: normal Hct; group 3: high Hct—rank methods are appropriate. Although these groups could be considered as categories and categorical methods could be used, rank methods give better results. Whereas rank methods lose power when there is a large number of ties, they still are more powerful than categorical methods, which are not very powerful. *When ranking is a natural option, rank methods of Chapter 16 should be used.*

METHODS ADDRESSED IN THIS CHAPTER

The basic question being tested asks about the relationship between two variables. Different tests exist to answer the question, including Fisher's exact test (FET) and the chi-square test of contingency (see Section 15.2) and, less commonly used, a test of logarithm of relative risk (log RR) and a test of logarithm of odds ratio (log OR) (see Section 15.5). RR and OR are measures of association, and they are discussed in Section 15.4. Although all of these tests give rather similar results, the choice of test can be aided by noting the question being asked. If we believe that two variables are independent and our goal is to assess the probability that this is true, we would prefer FET or its approximation the chi-square test of contingency, which approximates it in cases of large samples or several categories. If we have a measure of association that we believe is meaningful and our goal is to assess the statistical significance of this association, we would prefer to test the OR (or the RR, although the log OR test is a little more general in that it can be used in more situations than can the log RR test). If we know proportions but do not have the marginal totals required to use the FET or chi-square test, we can test the proportions directly (see Sections 15.6 and 15.7). Finally, if one of the variables includes matched pairs rather than counts from different samples, the McNemar test would be used (see Section 15.8).

15.2. 2 × 2 TABLES: CONTINGENCY TESTS

ARE THE ROW CATEGORIES INDEPENDENT OF THE COLUMN CATEGORIES IN A 2 × 2 TABLE?

A "yes" answer implies that the rows' variable has not been shown to be related to the columns' variable; knowing the outcome of one of the variables give us no information as to the associated outcome of the other variable.

A "no" answer tells us that *some* association exists, but it does *not* tell us what the strength of this association is. (Section 15.5 provides a test of the strength of association.)

THE CHI-SQUARE CONTINGENCY TEST FROM SECTION 6.3

The basic concepts and examples of application of the chi-square contingency test appear in Section 6.3. A reader new to these ideas should become familiar with Section 6.3 before continuing. Only a brief summary of that section appears later in the Methods section. The chi-square test of contingency is based on the differences between the observed values and those that would be expected if the variables were independent.

FISHER'S EXACT TEST IS ANOTHER TEST OF INDEPENDENCE BETWEEN TWO CATEGORICAL VARIABLES

If the counts contained in the contingency table cells are small, the probability of independence should be calculated directly by *Fisher's exact test* (FET). (The test could well have been named the Fisher–Irwin test; Irwin and Fisher developed the theory independently and their work appeared in print almost simultaneously.) Fisher's exact test can be found in statistical packages on modern computers. "Exact" implies that its results are calculated by methods from probability theory rather than approximated, as are results for the chi-square contingency test. The wide use of the chi-square test arose before there was adequate computing power for the exact test. In cases where there are cell counts less than 5 or there is an expected value less than 1, FET must be used. For larger cell counts and expected values, the two tests will give not very dissimilar results and either may be used, although preference for FET is increasing. If there are many cells in the contingency table, perhaps 10 or 15 (depending on the speed of the computer), the exact method may still be too computation intensive and the chi-square approximation will have to be used.

EXAMPLE: DEPENDENCE OF DIGITAL RECTAL EXAMINATION AND BIOPSY RESULTS

In Chapter 6, we investigate whether the digital rectal examination (DRE) is independent of the biopsy result (BIOP). For the 301 urologic patients, the 2×2 contingency (see Table 6.3) reappears in this chapter as Table 15.2. The chi-square statistic was calculated as $\chi^2 = 5.98$, which yields a significant p-value of 0.014, indicating dependence between DRE and biopsy. The FET p-value, calculated by a statistical software package, is similar, 0.016, yielding the same interpretation.

Table 15.2

Contingency Table of Simultaneous Biopsy and Digital Rectal Examination Results from 301 Patients

		DRE		Total
		1	0	
BIOP	1	$n_{11} = 68$	$n_{12} = 27$	$n_{1.} = 95$
	0	$n_{21} = 117$	$n_{22} = 89$	$n_{2.} = 206$
	Total	$n_{.1} = 185$	$n_{.2} = 116$	$n = 301$

0, no; 1, yes; BIOP, biopsy; DRE, digital rectal examination.

METHOD: 2 × 2 CONTINGENCY TABLE TESTS

Two categories of each variable yields a 2 × 2 contingency table. The two tests of independence are FET or its approximation, the chi-square test of contingency. Chapter 6 explains the basis of the chi-square test. In summary, the actual chi-square statistic is the sum of these differences squared in ratio to the expected value. The expected value for row i and column j is given by Eq. (6.1) as

$$e_{ij} = \frac{n_{i.}n_{.j}}{n}. \tag{15.1}$$

The chi-square statistic was given by Eq. (6.2) as

$$\chi^2 = \sum_i^2 \sum_j^2 \frac{(n_{ij} - e_{ij})^2}{e_{ij}}. \tag{15.2}$$

If this statistic is small, there is little dependence between the variables; a large statistic indicates dependence. If the chi-square statistic is large enough that it is unlikely to have occurred by chance, we say it is significant and conclude that the rows' variable is not totally independent of the columns' variable; however, it does *not* follow that one can be well predicted by the other. The reason for this is that significance is influenced by sample size and association. Section 6.3 describes the influence of sample size.

ASSUMPTIONS UNDERLYING CHI-SQUARE CONTINGENCY

The chi-square method should not be used for extremely small samples. In particular, it is valid *only* if the expected value (row sum × column sum ÷ total sum) of every cell is at least 1 and a minimum count of 5 appears in every cell. Even when valid, chi-square yields only an approximately correct p-value. When cell counts are other than large, the FET is preferred, because, as the name implies, it is exact.

THE BASIS OF FISHER'S EXACT TEST

Fisher's exact test is calculated by a statistical software package. Its *p*-value gives the probability that the observed deviation from independence would occur by chance alone. A small *p*-value indicates that causes other than chance are influencing the outcome, and therefore the two variables probably are not independent. Fisher's exact test uses a probability distribution known as the hypergeometric distribution for the observed counts, calculating the probability of all other 2 × 2 tables with the same marginal totals (*n* counts). The *p*-value is the sum of probabilities of all such outcomes with counts less likely than the observed counts. The interpretation of the FET *p*-value is the same as the chi-square *p*-value. A significant FET or chi-square result indicates that the variables are not independent, but it does not indicate how closely they are associated.

ADDITIONAL EXAMPLE

Improvement with Surgical Experience Tested by Fisher's Exact Test

An ophthalmologist investigated his learning curve in performing radial keratotomies.[5] He was able to perform a 1-month postoperative refraction on 78 of his first 100 eyes. He classified them as 20/20 or better versus worse than 20/20. The results appear in Table 15.3. The ophthalmologist notes that the percentage of eyes in the 20/20 or better group increased from 41% to 73% from earlier to later surgeries. His null hypothesis is that resultant visual acuity is independent of when in his sequence the surgery was done. Fisher's exact test yields $p = 0.006$. The ophthalmologist rejects the null hypothesis and concludes that his surgical skills have improved significantly.

Improvement with Surgical Experience Tested by Chi-square

The ophthalmologist could have approximated the FET result by a chi-square. He calculates his expected value as $e_{ij} = n_{i.} \times n_{.j}/n.e_{11} = 44 \times 41/78 = 23.13, e_{12} = 20.87, e_{21} = 17.87$, and $e_{22} = 16.13$. Substituting in Eq. (15.2), he finds

$$\chi^2 = \sum_i^2 \sum_j^2 \frac{(n_{ij} - e_{ij})^2}{e_{ij}} = \frac{(17 - 23.13)^2}{23.13} + \cdots + \frac{(10 - 16.13)^2}{16.13} = 7.86.$$

Table 15.3

Refraction of Postoperative Eyes by Position in Sequence of Surgery

	First 50 eyes treated	Second 50 eyes treated	Total
20/20 or better	17	27	44
Worse than 20/20	24	10	34
Total	41	37	78

Table 15.4

A Fragment of Table III, the Right Tail of the Chi-square Distribution

α (area in right tail)	0.10	0.05	0.025	0.01
$df = 1$	2.71	3.84	5.02	6.63

Chi-square (χ^2) values (distances to the right of zero) are tabulated for 1 df for four values of α. See Tables of Probability Distributions to review Table III.

Table 15.5

Data on Ankle Ligament Repair

		Treatment		
		MB	CS	Total
Success	Excellent	10	3	13
	Less	10	16	26
	Total	20	19	39

CS, Chrisman-Snook; MB, Modified Bostrom.

From Table III (see Tables of Probability Distributions), a fragment of which is reproduced as Table 15.4, the critical value of χ^2 for $\alpha = 0.05$ in one tail with 1 df is 3.84, which is much less than 7.86; again, the null hypothesis is rejected. Calculation by a software package yields $p = 0.005$, which is similar to the p-value for FET.

Exercise 15.1: Comparing Methods of Torn Ankle Ligament Repair. Two methods of torn ankle ligament repair, Modified Bostrom (MB) and Chrisman–Snook (CS), were compared in an orthopedics department.[28] The clinical success on 39 patients was rated as excellent or less than excellent. The contingency table is given in Table 15.5. Find the p-values using FET and the chi-square test. Are the two treatments significantly different?

15.3. r × c TABLES: CONTINGENCY TESTS

TABLES LARGER THAN 2 × 2

This section addresses the question: *Are the row categories independent of the column categories in tables with* r *rows and* c *columns (r and/or c > 2)?* A positive answer, that is, a nonsignificant p-value, implies that the rows' variable has not been shown to be related to the columns' variable.

EXAMPLE POSED: IS THE USE OF SMOKELESS TOBACCO RELATED TO ETHNIC ORIGIN?

With increased restriction on smoking in ships and shore stations, many U.S. Navy and Marine Corps personnel are changing to smokeless tobacco.[50] Navy medicine is concerned about the effect of this pattern on health and seeks

Contingency Table (2 × 3) of Smokeless Tobacco Use Associated with Categories of European, African, and Hispanic Ethnic Origin from 2004 Patients

		Ethnic background			
		European	African	Hispanic	Total
Smokeless tobacco	Use	424	13	28	465
	Do not use	1075	290	174	1539
	Total	1499	303	202	2004

causes for this change. Cultural influences may be involved. Is ethnic background related to smokeless tobacco use? A count of the numbers in categories of use and ethnicity yielded Table 15.6. With two rows and three columns, $r = 2$ and $c = 3$. Smokeless tobacco use was ascertained from 2004 sailors and marines of three ethnic backgrounds from a wide variety of unit assignments. A total of 74.8% were of European background but represented 91.2% of users, 15.1% were of African background but represented 2.8% of users, and 10.1% were of Hispanic background but represented 6.0% of users. It was suspected that use is not independent of ethnic background, and the relationship was tested.

METHOD: FISHER'S EXACT TEST FOR $r \times c$ CONTINGENCY TABLES

When the data are divided into r row and c column categories, an $r \times c$ contingency table is appropriate. Fisher's exact test uses a distribution named the hypergeometric for calculating the probability of occurrence of all other $r \times c$ tables with the same marginal totals. The p-value is the sum of probabilities of all such outcomes with counts less likely than the observed counts.

CHI-SQUARE APPROXIMATES THE RESULT FOR LARGER TABLES

For large r and c (perhaps $r + c > 7$, depending on computer speed), calculating p for FET becomes lengthy, often too lengthy for even a modern computer. In this case, the chi-square contingency calculation of Eq. (15.3), similar to Eq. (15.2), is usually used. By denoting the observed count in row i and column j as n_{ij} and calculating the corresponding expected entry e_{ij} by row i total × column j total ÷ grand total, chi-square may be written as

$$\chi^2 = \sum_i^r \sum_j^c \frac{(n_{ij} - e_{ij})^2}{e_{ij}}. \tag{15.3}$$

The difference from Eq. (15.2) is that, instead of summing over 2 rows and 2 columns, we sum over r rows and c columns. Table III gives the chi-square p-value (looking it up as if it were α), using df equal to 1 less than r multiplied by 1 less than c, or $df = (r - 1)(c - 1)$. For example, a 3×4 contingency table has $(3 - 1)(4 - 1) = (2)(3) = 6$ df. Choose the largest α having a table entry less than the calculated χ^2; the p-value of the test, that is, the chance of being wrong when a "dependent" decision is made, is less than that α. Alternatively, use a statistical package on a computer and calculate the p-value directly. (See Sections 5.2 and 5.3 for a discussion of the distinction between using a tabulated critical value and a calculated p-value.)

WHEN CHI-SQUARE ASSUMPTIONS ARE VIOLATED

For the chi-square approximation to be valid, all cell entries should be at least 5 and the expected value for all cells should be at least 1. If this assumption is violated, FET should be used. If the table is too large for FET to be used, cells with small entries or expectation may have to be amalgamated, which, of course, reduces the df and the information contained.

EXAMPLE CONTINUED: SMOKELESS TOBACCO AND ETHNIC ORIGIN—FISHER'S EXACT TEST

Here $r + c$ is 5, small enough for practical use of the FET when the user has access to a computer statistical package. The (FET) p-value is less than 0.001. (Statistical software packages may display p-values like 0.000, implying that the first significant digit lies more than three digits to the right of the decimal point. However, values less than 0.001 are not accurate, so $p < 0.001$ should be reported.)

EXAMPLE CONCLUDED: SMOKELESS TOBACCO AND ETHNIC ORIGIN—CHI-SQUARE TEST

The chi-square contingency approximation formula appears in Eq. (15.3); the critical value may be found in Table III for $(r - 1)(c - 1) = 2$ df. The observed count in row i and column j is n_{ij} (e.g., $n_{11} = 424$), and the corresponding expected entry e_{ij} is row i total $\times$ column j total/grand total (e.g., $e_{11} = 465 \times 1499/2004 = 347.82$). The chi-square component of the first cell is $(n_{11} - e_{11})^2/e_{11} = (424 - 347.82)^2/347.82 = 16.69$. Following this pattern and summing over the six cells, the chi-square statistic becomes

$$\chi^2 = 16.69 + 46.71 + 7.60 + 5.04 + 14.12 + 2.30 = 92.46.$$

In Table 15.7, a fragment of chi-square Table III, the entry for $\alpha = 0.0005$ and $df = 2$ is 15.21, which is much less than the calculated χ^2. Our p-value is

Table 15.7

A Fragment of Table III, the Right Tail of the Chi-square Distribution[a]

α (area in right tail)	0.10	0.05	0.025	0.01	0.005	0.001	0.0005
$df = 2$	4.61	5.99	7.38	9.21	10.60	13.80	15.21

[a] χ^2 values (distances to the right of zero on a χ^2 curve) are tabulated for 2 df for seven values of α. See Tables of Probability Distributions to review Table III.

then reported as less than 0.001. Smokeless tobacco use and ethnic background clearly are related.

ADDITIONAL EXAMPLE: WHO CHOOSES THE TYPE OF THERAPIST FOR PSYCHIATRY RESIDENTS?

In a study on the desirability of psychotherapy for psychiatry residents,[11] a question asked was whether the choice of therapist was independent of whomever paid for the therapy. The responses comprise Table 15.8. Fisher's exact test using a computer yielded $p = 0.001$. The choice of type of analyst clearly is related to who pays for the analyst. To perform the chi-square test, the expected values $e_{ij} = n_i.n_{.j}/n$, which are $e_{11} = 45 \times 38/155 = 11.032$, $e_{12} = 102 \times 38/155 = 25.006, \dots, e_{43} = 0.413$ are calculated first. (Note that the last expected value is less than 1, which violates the assumptions required for the chi-square approximation. However, the term involving that value contributes very little to the total sum, which is huge; therefore, we may continue. We must remember, however, that the resulting chi-square will not be exactly accurate.) We note that $df = (4-1)(3-1) = 6$. Substituting the values in Eq. (15.3) yields $\chi^2 = (15 - 11.032)^2/11.032 + \cdots = 24.45$. The α associated with 24.45 in the row for 6 df in Table III is a little less than 0.0005. We record $p < 0.001$, which is similar to the FET p-value. Psychiatrists and psychoanalysts were chosen by the overwhelming majority of residents and residency programs but not by insurance companies. Psychologists were chosen by 63% of insurance companies, although the costs, interestingly, were similar.

Table 15.8

Type of Therapist Chosen as Contingent on Who Chooses the Psychotherapy Given to Psychiatry Residents as Part of Their Training

	Who paid for therapy			
Type of therapist chosen	Resident	Residency program	Insurance company	Total
Psychoanalyst	15	22	1	38
Psychiatrist	18	64	1	83
Psychologist	7	14	5	26
Social worker	5	2	1	8
Total	45	102	8	155

Table 15.9

Data on Change in Blood Pressure in Response to Change in Position

	Increase in BP	No change in BP	Decrease in BP	Total
Systolic BP	8	20	72	100
Diastolic BP	11	30	54	95
Total	19	50	126	195

BP, blood pressure.

Exercise 15.2: Change in Blood Pressure Resulting from Change in Position.
The influence of patient position during the measurement of blood pressure
(BP) has long been questioned. A 1962 article[90] examined alterations in BP of
195 pregnant women resulting from a change from upright to supine positions.
We ask the following: Is the pattern of change the same or different for systolic
and diastolic BPs? The data are given in Table 15.9.

15.4. RISKS AND ODDS IN MEDICAL DECISIONS

EXAMPLE AND METHOD COMBINED: ACCURACY AND ERRORS IN THE DIAGNOSIS OF APPENDICITIS

Pain and tenderness in the lower right quadrant of the abdomen may or may
not be appendicitis. Clinical diagnoses are often wrong. Of 250,000 cases treated
yearly in the United States, 20% of appendicitis cases are missed, leading to
ruptures and complications, and the appendix is removed unnecessarily in 15%
to 40% of cases. In a recent study at Massachusetts General Hospital,[61] 100
cases of abdominal pain were diagnosed by computed tomography (CT) scans
and were followed for eventual verification of diagnosis. A total of 52 of 53 cases
diagnosed as appendicitis were correct, and 46 of 47 cases diagnosed as other
causes were correct. We can ask several questions about the risks of being wrong
and the odds of being right. The possible errors and their risks are discussed
in Section 5.2. A case may be diagnosed as appendicitis when it is not (false
positive), or it may not be diagnosed when it is (false negative).

TRUE AND FALSE POSITIVE AND NEGATIVE EVENTS

A special case of the contingency table is the case in which one variable represents a prediction that a condition will occur and the other represents the
outcome—that is, "truth." For example, the occurrence of appendicitis among
a sample of patients was predicted by a CT scan, and outcomes were known
from follow-up. (The appendicitis example is used in Exercise 15.3.) Section 6.4
discusses such a table and some of its implications. The frequencies of possible
prediction and outcome combinations can be counted and arrayed as $n_{11} - n_{22}$ in
a 2 × 2 *truth table*, as illustrated in Table 15.10. The true and false positives and
negatives are shown along with names for the associated risks (probabilities of
error). The same concepts are used in an epidemiologic context with somewhat

different terms. "Prediction" is "exposure to a putative causal agent," "truth" is "occurrence or nonoccurrence" of the malady, and the table might be thought of as an "occurrence table."

Table 15.10

Truth Table Showing Relationships and Counts of the Prediction of Presence or Absence of a Malady (or Exposure or Nonexposure of a Putative Causal Agent) as Related to the Truth (or Occurrence) of That Presence or Absence

		Prediction (e.g., test result) (or exposure to epidemiologic factor)		
		Have disease (or exposed)	Do not have disease (or not exposed)	Total
Truth	Have disease	True-positive correct decision (probability $1 - \beta$) frequency n_{11}	False-negative Type II error (probability β) frequency n_{12}	$n_1.$(yes)
	Do not have disease	False-positive Type I error (probability α) frequency n_{21}	True-negative correct decision (probability $1 - \alpha$) frequency n_{22}	$n_2.$(no)
	Total	n_1 (predict yes)	n_2 (predict no)	n (or $n..$)

Example Data

In Table 15.11, which reproduces the data of Table 6.7, the format of Table 15.10 is used to show DRE values of the 301 prostate patients in predicting biopsy results. The DREs are symbolized with a plus sign $(+)$ for predicted positive biopsy and a minus sign $(-)$ for predicted negative biopsy. In Table 15.12, the similar format shows some data on the relationship between exposure versus nonexposure to dust in a region believed to be heavily infused with *Coccidioides immitis* and the occurrence versus nonoccurrence of coccidioidomycosis in patients with fever, cough, and chest pains.

Table 15.11

Table of Biopsy Results Predicted by Digital Rectal Examination as Related to the True Outcomes of the Biopsies for 301 Patients

		Predicted by DRE		
		$+$	$-$	Total
Truth (biopsy)	$+$	68: n_{11}	27: n_{12}	95
	$-$	117: n_{21}	89: n_{22}	206
	Total	185	116	301

Table 15.12

Table of Occurrence or Nonoccurrence of Coccidioidomycosis in Patients with Certain Flulike Symptoms as Related to Exposure or Nonexposure to Dust in Outdoor Work in a Suspicious Region in the Southwestern United States

		Exposed to Dust		
		+	−	Total
Truth (coccidioidomycosis occurred or not)	+	$20: n_{11}$	$39: n_{12}$	59
	−	$35: n_{21}$	$362: n_{22}$	397
	Total	55	401	456

TRUTH TABLE STATISTICS ARE EXPLAINED IN TABLE 15.13

A number of truth table statistics introduced in Section 6.4 and several additional concepts are given in Table 15.13, an expansion of Table 6.8. Column 1 gives the names of the concepts, and column 2 gives the formulas for computing them from the sample truth table. Columns 3 and 4 give population definitions and the relationships these concepts are estimating, respectively, that is, what they would be if all data in the population were available. These are the values that relate to error probabilities in statistical inference, as discussed in Sections 5.1 and 5.2. As before, a vertical line (|) is read as "given." Columns 5 and 6 show numerical estimates of the probabilities and odds in predicting biopsy results calculated from DRE values (see counts in Table 15.12) and coccidioidomycosis occurrences (see Table 15.13), respectively. Interpretations of the statistics discussed in Chapter 6 are summarized next and new concepts are discussed in more detail.

SENSITIVITY, SPECIFICITY, ACCURACY, AND ODDS RATIO

Definitions of a number of concepts related to estimates of correct and erroneous predictions from the sample truth table are introduced in Section 6.4. These include sensitivity, the chance of detecting a disease when present; specificity, the chance of ruling out a disease when absent; accuracy, the chance of making a correct prediction; and the OR, the odds of being correct when the disease is predicted. It is important to note that the OR can be tested for statistical significance (see the next section). These values are best understood and are rather straightforward in the context of diagnosis. They are not easy to interpret in the context of the disease exposure occurrences of epidemiology.

POSITIVE AND NEGATIVE PREDICTIVE VALUES

Predictive values indicate the relative frequency of a predictor being correct or an exposure leading to infection. The positive predictive value (PPV) of 68/185 = 37% for DRE indicates that we will be correct 3 of 8 times when we

Table 15.13

Concepts of False Positive and Negative, Sensitivity, Specificity, Accuracy, Positive and Negative Predictive Values, Relative Risk, Odds Ratio, Positive and Negative Likelihood Ratios, and Attributable Risk Based on Sample Values[a]

Names of estimates	Values found from a sample truth table — Sample computing formulas	Population definitions of probabilities and odds of values being estimated — Definitions	Relationships	Examples — By DRE	Examples — By dust exposure
False-positive sample rate (*p*-value)	$n_{21}/n_{\cdot 2}$	Probability of a false positive (α)	$P(\text{predict yes} \mid \text{no})$	0.568	0.089
False-negative sample rate	$n_{12}/n_{\cdot 1}$	Probability of a false negative (β)	$P(\text{predict no} \mid \text{yes})$	0.284	0.661
Sensitivity	$n_{11}/n_{\cdot 1}$	Probability of a true positive: *power* $(1 - \beta)$	$P(\text{predict yes} \mid \text{yes})$	0.716	0.339
Specificity	$n_{22}/n_{\cdot 2}$	Probability of a true negative $(1 - \alpha)$	$P(\text{predict no} \mid \text{no})$	0.433	0.912
Accuracy	$(n_{11} + n_{22})/n$	Overall probability of a correct decision	$P(\text{predict no} \mid \text{no or yes} \mid \text{yes})$	0.522	0.838
Positive predictive value	$n_{11}/n_{1\cdot}$	Probability that a positive prediction is correct	$P(\text{yes} \mid \text{predicted yes})$	0.368	0.364
Negative predictive value	$n_{22}/n_{2\cdot}$	Probability that a negative prediction is correct	$P(\text{no} \mid \text{predicted no})$	0.767	0.903
Relative risk	$n_{11}n_{\cdot 2}/n_{12}n_{\cdot 1}$	Probability of a disease when predicted in ratio to probability of the disease when not predicted	$P(\text{yes} \mid \text{predicted yes})/P(\text{yes} \mid \text{predicted no})$ or PPV/(1 − NPV)	1.579	3.739
Odds ratio	$n_{11}n_{22}/n_{12}n_{21}$	Odds of a disease when predicted in ratio to odds of the disease when not predicted	ratio(yes/no | predicted yes)/ratio(yes/no | predicted no)	1.916	5.304
Likelihood ratio	$n_{11}n_{\cdot 2}/n_{21}n_{\cdot 1}$	Probability of correctly predicting a disease in ratio to probability of incorrectly predicting the disease	$P(\text{predict yes} \mid \text{yes})/P(\text{predict yes} \mid \text{no})$ or $(1 - \beta)/\alpha$ or sensitivity/(1 − specificity)	1.260	3.845
Attributable risk	$n_{11}/n_{\cdot 1} - n_{12}/n_{\cdot 2}$	Amount of incidence rate attributable to result of prediction or to exposure factor	$P(\text{yes} \mid \text{predicted yes}) - P(\text{yes} \mid \text{predicted no})$	0.135	0.266

[a]The third and fourth columns show the population entities these values are estimating. The fifth and sixth columns show biopsy outcomes and disease occurrence rates from the data of Tables 15.11 and 15.12.
"Prediction" may be thought of as "exposure" in the epidemiologic context.
DRE, digital rectal examination.

predict that the patient has cancer, and the negative predictive value (NPV) of $89/116 = 77\%$ for DRE indicates that we will be correct 3 of 4 times when we predict that a patient is free of cancer. These measures are not as useful clinically as sensitivity, the rate at which we detect cancer that is present, or specificity, the rate at which we rule out cancer that is absent. However, in an epidemiologic context, PPV and NPV are more useful than sensitivity or specificity. In the dust exposure example, the PPV tells us that 36% of the exposed subjects with the suspicious syndrome have contracted coccidioidomycosis, and the NPV tells us that 90% of the unexposed patients have not.

RELATIVE RISK

The RR gives the rate of disease given exposure in ratio to the rate of disease given no exposure. It could also represent the ratio of incidence rates of disease under two different conditions or strategies (including treatment as one strategy and nontreatment as the other strategy). In the DRE application, the RR gives the rate of cancer among *yes* predictions in ratio to the rate of cancer among *no* predictions; the 1.579 ratio tells us that prostate cancer is half as likely in a patient with a positive DRE as in one with a negative DRE. In the dust example, the RR gives the rate of coccidioidomycosis among exposed subjects in ratio to the rate among those not exposed; the 3.739 ratio tells us that the disease is between 3 and 4 times more likely to manifest in exposed subjects than in unexposed subjects. Relative risk sometimes appears under other names—for example, risk ratio or relative rate.

THE RELATIONSHIP BETWEEN RELATIVE RISK AND ODDS RATIO

The RR and OR seem rather similar in their definitions. The relationship between them is worth noting. If the disease or malfunction is rare, n_{11} and n_{12} are small so that their product $n_{11}n_{12}$ is almost zero and drops out of $\text{RR} = n_{11}(n_{12} + n_{22})/n_{12}(n_{11}+n_{21}) = (n_{11}n_{12}+n_{11}n_{22})/(n_{11}n_{12}+n_{12}n_{21}) \approx n_{11}n_{22}/n_{12}n_{21} = \text{OR}$. In the case of rare disease, the OR approximates the RR. As a simple example, suppose a subject exposed to a factor has a probability of 0.01 of contracting an associated disease, whereas a subject not exposed has a probability of only 0.001. The RR is 10.00, and the OR is 10.09.

LIKELIHOOD RATIO

The likelihood ratio (LR) gives the probability of correctly predicting cancer in ratio to probability of incorrectly predicting cancer. The LR indicates how much a diagnostic test result will raise or lower the pretest probability of the suspected disease. An LR of 1 indicates that no diagnostic information is added by the test. An LR greater than 1 indicates that the test increased the assessment of the disease probability; if less than 1, it decreased. The DRE LR of 1.260 indicates that the chance of correctly predicting a positive biopsy result is about one-fourth that of incorrectly predicting a positive biopsy result. The use of LR in epidemiology is less straightforward, but it has its uses. The LR of 3.845 in the dust example indicates that a patient with coccidioidomycosis is almost 4 times as likely to have been exposed as not exposed.

NEGATIVE LIKELIHOOD RATIO

Sometimes reference to a negative LR is seen. Whereas the positive LR is the probability of a positive test in a patient with the malady in question in ratio to a positive test in a patient without that malady, the negative LR is the ratio of a negative test in a patient without the malady in ratio to a negative test in a patient with the malady. In the DRE example, the negative LR is 1.520, indicating that a patient with a negative DRE is half as likely to be free of prostate cancer as to have it.

BE WARY OF THE TERM *LIKELIHOOD RATIO*

Likelihood ratio as applied here is a useful concept in describing and assessing diagnosis and treatment options, but the name is somewhat unfortunate. The term *likelihood ratio* (LR) has long been used in statistical theory for concepts other than that adopted in the medical community as described here. The user should be careful to differentiate interpreting results and presenting results so that the meaning will not be misunderstood by others.

ATTRIBUTABLE RISK

Attributable risk (AR) is the portion of disease rate attributable to the exposure factor in the epidemiologic context, the portion of correct diagnosis rate attributable to a positive predictive result (e.g., lab test) in the clinical context, or the portion of beneficial outcome rate attributable to a treatment. The AR of 0.266 in the dust example indicates that more than one fourth of the disease occurrences were caused by the exposure. The AR of 0.135 for DRE as a prediction of positive prostate biopsy indicates that nearly one-seventh more positive patients will be discovered by using the DRE. Attributable risk is sometimes seen in other forms; therefore, the reader must be wary. The form adopted here is perhaps its most useful form. A related statistic, sometimes erroneously called AR, is attributable fraction, the proportion that the occurrence would be reduced if the intervention (exposure, treatment, etc.) were removed. It can be calculated as $1 - 1/RR$.

THE RELATIONSHIP BETWEEN RELATIVE AND ATTRIBUTABLE RISKS

Relative risk is the ratio of rates of occurrence (among patients exposed, predicted, or treated to patients unexposed, not predicted, or untreated), whereas AR is the difference in those same rates of occurrence. Relative risk could be thought of as the incidence rate of exposed patients as a multiple of the incidence rate of unexposed patients. Attributable risk could be thought of as the amount the rate increases because of the exposure.

RECEIVER OPERATING CHARACTERISTIC CURVE

A receiver operating characteristic (ROC) curve is a graph displaying the relationship between the true-positive rate (on the vertical axis) and the false-positive rate (on the horizontal axis). Brought into the medical field from engineering usage, this curve is usually abbreviated as ROC. Figure 15.1 shows the ROC curve for prostate-specific antigen (PSA) levels. A ROC curve helps one choose the critical value at which a predictor best discriminates between choices, such as choosing the value of PSA that best predicts the presence or absence of a positive biopsy result. The critical value is found as the value at which the curve's deviation from the diagonal line from (0,0) (bottom left) to (1,1) (top right) is the greatest. However, there is an additional advantage in being able not just to maximize the correct response but to weight it for relative loss associated with the two types of error. For example, a missed cancer (false negative) is more detrimental to the patient than an unnecessary biopsy (false positive). If a false-negative error is assessed as 4 times as serious as a false-positive error, the corresponding position on the ROC curve can be located and the associated critical value determined.

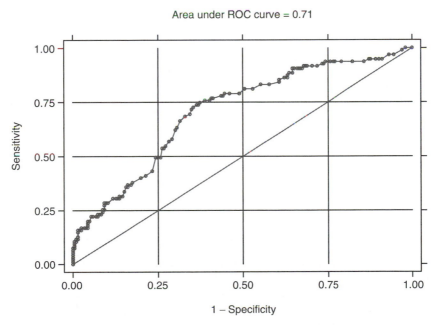

Figure 15.1   Receiver operating characteristic curve for biopsy result as predicted by prostate-specific antigen level.

CALCULATING A RECEIVER OPERATING CHARACTERISTIC CURVE

In our urologic data, a DRE is either positive or negative. There is no ROC curve for it. (More exactly, its ROC curve would consist of a single point.) However, for PSA level, we could choose any critical value. As one error type grows smaller, the other grows larger. The issue is choosing the trade-off of the two error values (or their complements, sensitivity and specificity).

In Fig. 15.1, calculation started with the critical value of PSA at zero, where all patients are classed as cancerous, and the error rates were counted. The critical value was taken as 0.1, then 0.2, and so on, and the error rates for each critical value were calculated; the points were plotted on the evolving ROC curve. Such a procedure is time consuming and is best done with computer software.

CHOOSING THE BEST CRITICAL VALUE

The strongest predictor of biopsy result will be indicated by where the perpendicular distance from (and above) the (45°) line of equality is a maximum, which can be seen by inspection to be about at the point (0.36, 0.75). We use these coordinates to select the number of negative biopsy results predicted that led to that point, and from that number we find the critical value of PSA. If we calculated the ROC values ourselves, we just select it from the list of calculations. If the ROC curve was done on a computer, we have to "back calculate" it. We know that 95 patients have a positive biopsy result (i.e., $n_1. = 95$) and that 206 have a negative biopsy result (i.e., $n_2. = 206$). We use this information to find the $n_{11}...n_{22}$ elements of a new 2×2 table. The user should draw such a table and fill it in step by step during the following calculations. Sensitivity $\times n_1. = n_{11}$ or $0.75 \times 95 = 71$. Because $n_1. = n_{11} + n_{12}$, $n_{12} = 95 - 71 = 24$. Similarly, $n_{21} = (1 - \text{specificity}) \times n_2. = 0.36 \times 206 = 74.16$, which we may round to 74. $n_{22} = n_2. - n_{21} = 206 - 74 = 132$. Finally, $n_{12} + n_{22} = 23 + 132 = 155$. Predicting the smallest 155 values of PSA to have a negative biopsy result would lead to $1 - \text{specificity} = 0.36$ and sensitivity $= 0.73$. We rank the PSA values in order from smallest to largest and note that the PSA between the 155th and 156th values is 5.95, the required PSA critical value. For clinical convenience, we may denote the cut point as a PSA value of 6.

CHOOSING A BEST WEIGHTED CRITICAL VALUE

Suppose we were to require that the rate of false-positive errors (the number of unnecessary biopsies) be 4 times the rate of false-negative errors (the number of missed cancers), or $1 - \text{specificity} = 4(1 - \text{sensitivity})$. What critical value of PSA would provide this result? By inspection, $1 - \text{specificity}$ would be about 0.60 and $1 - \text{sensitivity}$ about 0.15. Following a procedure similar to that in the preceding paragraph, we find $n_{21} = 124$, $n_{12} = 14$, $n_{22} = 82$, and negative predictions total 96, which lead to a critical PSA value of 4.25.

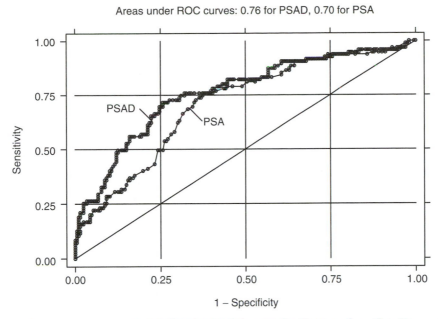

Figure 15.2 Superposed receiver operating characteristic curves for biopsy result predicted by prostate-specific antigen density and prostate-specific antigen level.

USING THE RECEIVER OPERATING CHARACTERISTIC CURVE TO CHOOSE THE BETTER OF TWO INDICATORS (RISK FACTORS)

Another use of ROC curves is to compare two indicators, for example, PSA and PSAD, the ROC curves that are shown superposed in Fig. 15.2. An ROC curve that contains a larger area below it is a better predictor than one with a smaller area. However, if the curves cross, an indicator with a smaller area can be better in some regions of the indicator. The area under the PSAD curve is greater than that under the PSA curve, and the PSAD curve is uniformly greater than or equal to the PSA curve.

ADDITIONAL EXAMPLE: TRUTH TABLES AS RELATED TO SURVIVAL RATES

The risks and odds related to various medical occurrences follow the same numerical patterns but must be interpreted according to their own definitions. Survival as related to treatment represents an interpretation quite different from the results of a lab test.

Pancreatic Cancer Data

Oncologists examined the effect of reoperation for pancreatic cancer on patient survival 1 year after surgery.[37] Of 28 patients, 16 were resectable and 12 were not. All were followed for 1 year or until death, whichever came first.

Table 15.14

Data on Survival as Related to Treatment for Pancreatic Cancer

	Resection	No resection	Total
Survived at 1 year	8 (n_{11})	2 (n_{12})	10 ($n_{1.}$)
Did not survive at 1 year	8 (n_{21})	10 (n_{22})	18 ($n_{2.}$)
Total	16 ($n_{.1}$)	12 ($n_{.2}$)	28 (n)

We know which patients survived for 1 year and which did not; we are interested in what resection portends about that survival. Results are given in Table 15.14.

False-Positive and False-Negative Rates

The false-positive rate is the chance of having been resected given no survival for 1 year, or $n_{21}/n_{2.} = 8/18 = 44\%$. The false-negative rate is the chance of not having been resected given survival, or $n_{12}/n_{1.} = 2/10 = 20\%$. In the prostate example, the values for false-positive and -negative rates, sensitivity, and specificity dealt with the rates at which our predictions came true and had considerable meaning. In this example, they have less meaning, but the PPVs and NPVs are more relevant to what interests us.

Sensitivity, Specificity, and Accuracy

Sensitivity is the chance of having had a resection, given survival for 1 year, or $n_{11}/n_{1.} = 8/10 = 80\%$. The specificity is the chance of having had no resection given no survival, or $n_{22}/n_{2.} = 10/18 = 56\%$. The accuracy, that is, the overall chance of an appropriate resection, is $(n_{11} + n_{22})/n = (8 + 10)/28 = 64\%$.

Positive and Negative Predictive Values

The PPV is the chance of survival given a resection, or $n_{11}/n_{.1} = 8/16 = 50\%$. This useful information is the relative frequency with which resected patients survive. Similarly, NPV, the chance of death given no resection, or $n_{22}/n_{.2} = 10/12 = 83\%$, is the relative frequency with which patients without resection die.

Relative Risk

The RR provides the rate of survival given resection in ratio to the rate of survival given no resection, which is essentially the improvement in survival rate because of resection. Relative risk $= n_{11}n_{.2}/n_{12}n_{.1} = 8 \times 12/2 \times 16 = 3$ indicates that the resected patient's chance of survival to 1 year is 3 times that of the patient who did not undergo resection.

Odds Ratio

The OR gives the odds of resected patients' survival in ratio to the odds of survival for the patients without resection. The OR $= n_{11}n_{22}/n_{12}n_{21} = 8 \times 10/2 \times 8 = 5$. The improvement in survival odds because of resection is 5 times that for no resection.

Table 15.15

Truth Table for Appendicitis Data

		Appendicitis predicted		Total
		Yes	No	
Appendicitis verified	Yes	52 (n_{11})	1 (n_{12})	53 ($n_{1.}$)
	No	1 (n_{21})	46 (n_{22})	47 ($n_{2.}$)
	Total	53 ($n_{.1}$)	47 ($n_{.2}$)	100 (n)

Likelihood Ratio

The LR gives the ratio of appropriate resections to inappropriate resections. The LR $= n_{11}n_{2.}/n_{21}n_{1.} = 1.8$. The rate of appropriate resections is nearly double the rate of inappropriate ones.

Attributable Risk

Attributable risk gives the difference in rates of survival with and without resection. The AR $= n_{11}/n_{.1} - n_{12}/n_{.2} = 1/3$. Resection increases survival rate by one third.

Exercise 15.3: Appendicitis Diagnosis. Table 15.15 displays the Massachusetts General Hospital appendicitis data[61] as a truth table. Calculate and interpret sensitivity, specificity, accuracy, PPV, NPV, RR, OR, LR, and AR.

15.5. 2 × 2 TABLES: TESTS OF ASSOCIATION

THE REASON FOR THIS SECTION AND WHAT IS IN IT

The chi-square and Fisher's exact tests of contingency are so influenced by sample size that they are poor indicators of association. A small sample is unlikely to indicate a significant relationship between variables, however strong the association may be. However, take a large enough sample and you will find evidence of association, no matter how weak is the influence of that association. A better test of the level of association is needed. Two such tests are the log OR and the log RR tests. This section discusses only the log OR test because it is more general and has a better mathematical foundation. In addition, AR often may be used as a clinical indicator of the relationship between two variables. This indicator has inadequate mathematical foundation to relate it to probabilistic tests, but it adds another dimension to the understanding of interrelationships on just a clinical basis.

THE LOG ODDS RATIO TEST

Is the OR large enough to indicate that the association between the two types of categories is significant in probability? This is the first question addressed in this section. If we have estimated the level of association by the OR, we can go directly to a test using that value. The OR has a difficult, asymmetric distribution (as does the RR). To put it into a form with a known and usable probability distribution, the *log OR*, which we denote by *L*, is used. *L* is symmetric about zero, unlike the OR. The square of *L* divided by its standard deviation is distributed as chi-square with 1 *df*. (The chi-square test of log OR is not the same test as the chi-square test of contingency; both just use the chi-square probability distribution.)

EXAMPLE POSED: IS AN ODDS RATIO FOR PREDICTING A POSITIVE BIOPSY RESULT FROM DIGITAL RECTAL EXAMINATION SIGNIFICANT?

From Table 15.2, which provides DRE and biopsy counts, the OR is $(68 \times 89)/(117 \times 27) = 1.92$. This OR is greater than the value of 1, which would indicate no association, but is 1.9 significant in probability? (Because the FET showed dependence in the example in Section 15.2, we expect the log OR to be significant.)

METHOD: 2 × 2 TABLE TEST OF ASSOCIATION

A test of the significance of 2×2 contingency requires the probability distribution of the measure. The *log OR* (see Section 6.4 for OR) is one that can be put into a form having a chi-square distribution. (The log RR might also be used, but it is slightly less general than the log OR. Others exist, for example, the Mantel–Haenszel chi-square test, but are more complicated and no better for the simpler cases examined in this book.) Using the *n* notation introduced in Section 6.2, log OR is given by

$$L = \ln\left[\frac{(n_{11} + 0.5) \times (n_{22} + 0.5)}{(n_{12} + 0.5) \times (n_{21} + 0.5)}\right], \tag{15.4}$$

where *ln* denotes natural logarithm. The 0.5 values are a continuity correction added to improve the approximation. The standard error of log OR (SEL) is

$$SEL = \sqrt{\left(\frac{1}{n_{11} + 0.5}\right) + \left(\frac{1}{n_{12} + 0.5}\right) + \left(\frac{1}{n_{21} + 0.5}\right) + \left(\frac{1}{n_{22} + 0.5}\right)}. \tag{15.5}$$

To test *L* against a hypothesized log OR λ (e.g., H_0: $\lambda = 0$ if the category types are independent), the quantity

$$\chi^2 = \left(\frac{L - \lambda}{SEL}\right)^2 \tag{15.6}$$

may be looked up in the chi-square table (see Table III) for 1 *df*. The resulting *p*-value is the probability that such an association would occur by chance alone. (Note: log OR $= 0$ is equivalent to OR $= 1$, which implies no association.)

EXAMPLE COMPLETED: IS AN ODDS RATIO FOR PREDICTING A POSITIVE BIOPSY RESULT FROM DIGITAL RECTAL EXAMINATION SIGNIFICANT?

$$L = \ln\left[\frac{(n_{11} + 0.5) \times (n_{22} + 0.5)}{(n_{12} + 0.5) \times (n_{21} + 0.5)}\right] = \ln\left[\frac{68.5 \times 89.5}{27.5 \times 117.5}\right] = 0.6404.$$

Its standard error is

$$SEL = \sqrt{\left(\frac{1}{n_{11} + 0.5}\right) + \left(\frac{1}{n_{12} + 0.5}\right) + \left(\frac{1}{n_{21} + 0.5}\right) + \left(\frac{1}{n_{22} + 0.5}\right)}$$

$$= \sqrt{\left(\frac{1}{68.5}\right) + \left(\frac{1}{27.5}\right) + \left(\frac{1}{117.5}\right) + \left(\frac{1}{89.5}\right)} = 0.2658.$$

The chi-square statistic to test against $\lambda = 0$ is:

$$\chi^2 = \left(\frac{L - \lambda}{SEL}\right)^2 = \left(\frac{0.6404 - 0}{0.2658}\right)^2 = 5.8049$$

with 1 *df*. This chi-square value falls between 0.025 and 0.01, as seen in Table 15.4 (or Table III); therefore, we may state that the *p*-value is significant, confirming our initial belief. The actual *p*-value for chi-square with 1 *df* calculated from a software package is 0.016, which is close to the FET result. We have adequate evidence from the sample to conclude that DRE result is associated with biopsy result.

ADDITIONAL EXAMPLE: DOES A SURGEON'S RADIAL KERATOTOMY EXPERIENCE AFFECT THE VISUAL ACUITY OF THE PATIENTS?

In the additional example on radial keratotomy (RK) surgery discussed in Section 15.2, the ophthalmologist found that postoperative visual acuity and position in his surgical sequence were dependent. The OR, calculated from the formula in Table 15.13, is 0.26, which is much less than 1, suggesting a negative relationship: the greater the number of surgeries, the fewer the eyes with poor refraction. Is the association as indicated by the OR significant? The ophthalmologist substitutes data from Table 15.3 in Eqs. (15.4), (15.5), and finally (15.6) to calculate

$$L = \ln\left[\frac{(n_{11} + 0.5) \times (n_{22} + 0.5)}{(n_{12} + 0.5) \times (n_{21} + 0.5)}\right] = \ln\left[\frac{17.5 \times 10.5}{27.5 \times 24.5}\right] = 1.2993$$

$$SEL = \sqrt{\frac{1}{n_{11} + 0.5} + \frac{1}{n_{12} + 0.5} + \frac{1}{n_{21} + 0.5} + \frac{1}{n_{22} + 0.5}}$$

$$= \sqrt{\frac{1}{17.5} + \frac{1}{27.5} + \frac{1}{24.5} + \frac{1}{10.5}} = 0.4791.$$

$$\chi^2 = \left(\frac{L - \lambda}{SEL}\right)^2 = \left(\frac{-1.2993 - 0}{0.4791}\right)^2 = 7.35.$$

From Table 15.4 (or Table III), the critical value of χ^2 for $\alpha = 0.05$ for 1 *df* is 3.84. Because 7.35 is much larger than 3.84, the ophthalmologist concludes that there is a significant association between the two factors. (The actual $p = 0.007$.)

Exercise 15.4: Is the Quality of Ligament Repair Significantly Associated with the Method of Repair? Use the data in Exercise 15.1 (on ligament repair) to calculate L and SEL and test the hypothesis that $\lambda = 0$, that is, that there is no difference between the MB and CS methods of repair.

CLINICAL RELEVANCE OF THE DEPENDENCE BETWEEN CATEGORIES IN A 2 × 2 TABLE

The clinical relevance of the dependence is another issue. The RR and OR do not answer this question well, because the actual rates of occurrence have been factored out. The clinical relevance may be indicated by the AR, the difference in rates of occurrence. The importance and interpretation of this statistic lies in clinical judgment rather than in a probability statement. For example, from Table 15.12, the AR for the data of Table 15.2 is given as 0.135. We get roughly a one-seventh greater chance of correctly predicting an existing positive biopsy result from a positive DRE than from a negative DRE, which is not a startling improvement. As another example, we may examine the AR for the data of the learning advantage of repeated surgery from Table 15.3. What is the clinical importance of performing eye surgery later in the learning curve rather than earlier? The rate of a good result later in the curve is $27/37 = 0.73$, whereas the rate of a good result earlier is $17/41 = 0.41$. By the formula from Table 15.12, the AR is $0.73 - 0.41 = 0.32$. The rate of a good surgical outcome is increased nearly one-third by the additional experience.

15.6. TESTS OF PROPORTION

WHAT A TEST OF PROPORTION WILL DO AND WHY IT IS NEEDED

Suppose we are concerned with the efficacy of an antibiotic in treating pediatric otitis media accompanied by fever. We know that, say, 85% of patients get well within 7 days without treatment. We learn that, of 100 patients, 95% of those treated with the antibiotic got well within 7 days. Is this improvement statistically significant? Note that we cannot fill in a contingency table because we do not have the number of untreated patients or the total number of patients. However, we do have the theoretical proportion, so we can answer the question using a test of proportions. Testing a *proportion* will answer one of two questions: *Is a sample proportion different from a theoretical or population proportion?* (difference test) or *Is a sample proportion the same as a theoretical or population proportion?* (equivalence test). The difference test is addressed here. The equivalence test requires further development and is addressed in Chapter 21.

When a Proportion Is Very Small

Suppose you are concerned with a rare disease having a prevalence of 0.002. This section deals with proportions not very close to 0 (or to 1). Slightly different methods should be used if the proportion being examined is very close to 0 (see Section 15.7 for a discussion of these methods).

BASIS OF THE TEST

Proportions for binary data (e.g., yes/no, male/female, survive/die) follow the *binomial distribution*, because outcomes from a sampling must fall into one category or the other. Probabilities for the binomial distribution appear in Table VI (see Tables of Probability Distributions). Suppose we wanted to compare our 30% positive biopsy rate from Table DB1.1 with a theoretical result of long standing in which 25% of patients older than 50 years presenting to urology clinics had positive biopsy results. We would use the binomial distribution with theoretical proportion $\pi = 0.25$. (The Greek symbol π is used for the binomial population proportion and the Roman p is used for the sample proportion. This p does not represent the common usage designating probability [e.g., p-value]; the use of the same letter is coincidental. We must keep track of the usage by context.) We ask, "Could a rate as large as or larger than our $p = 0.30$ rate have occurred by chance alone?" (The "larger than" portion arises because we are really considering 1 minus the probability that the rate is less than 0.30.) The probability of obtaining a proportion $p \geq 0.30$ by chance alone, given $\pi = 0.25$, can be calculated or looked up in a table. The resulting probability is the p-value of the binomial test of proportion. In Table VI, the binomial probabilities are given for π no larger than 0.50; for larger values of π, just reverse the tails of the distribution. For example, for a particular disease occurring in an undeveloped country, the mortality rate is 80%. Instead of examining the chance of death, examine the chance of survival, having $\pi = 20\%$.

A Different Method Is Used if *n* Is Large

For a large sample n, exact probabilities do not appear in the tables and they would be very time-consuming to calculate. Fortunately, good approximations have been found for large samples. For the binomial distribution, when both $n\pi$ and $n(1 - \pi)$ are greater than 5, the sample proportion is distributed approximately *normal* with mean equal to the theoretical proportion, that is, normal $\mu = \pi$, and standard deviation also depending on π. The null hypothesis is that the π for the distribution from which the sample was drawn is the same as the theoretical π. Knowing μ and σ, we may calculate a z-score as the difference between the sample and theoretical proportions divided by the standard deviation in much the same conceptual form as described in Section 2.8. We then use normal tables to find p-values.

Test of Two Proportions

A generalization of the large sample test of a proportion exists to test for a significant difference between two sample proportions. We sample n_1 and n_2 readings from two populations having proportions π_1 and π_2. The null hypothesis is H_0: $\pi_1 = \pi_2$. We estimate π_1 and π_2 by p_1 and p_2. We calculate a combined p_c pooled over the two samples and a standard deviation depending on p_c. The difference between p_1 and p_2 divided by the standard deviation is again normal in form, and we find p-values from the normal table.

For small samples, the sample sizes are usually known. From these and the sample proportion values, the actual counts can be found, which will permit the data to be posed in the form of a contingency table. Use the methods explained in Section 15.2.

Test of Three or More Proportions

When three or more proportions are to be compared, small samples can be treated by posing the data in a contingency table larger than 2×2, as in Section 15.3. For large samples, the associated distribution is the multinomial distribution, a generalization of the binomial. This problem is not met frequently enough to pursue here. However, in some cases in which it is met, a satisfactory answer can be found without the multinomial by combining categories and reducing the question to two proportions.

METHODS OF TESTS OF PROPORTION

Test of a Sample Proportion Against a Theoretical Proportion

This method tests a sample proportion against a theoretical or (previously known) population proportion, π. We sample n observations randomly from a population with probability π_s of a "success." (A "success" is the occurrence of the event we are testing, not necessarily what we would wish clinically.) We want to know if π_s, the proportion for the population we sampled from, is the same as π, the proportion for the theoretical population. Thus, H_0 is $\pi_s = \pi$. We obtain n_s successes, estimating π_s by the sample proportion $p_s = n_s/n$.

Case of Small n

If n is less than 17, Table VI gives the p-value, the probability that, for expected chance of a success π, we would find n_s successes or p_s proportion successes by chance alone. More extensive tabulation may be found in books on mathematical or statistical tables.

Case of Large n

For larger n, a normal approximation to the binomial distribution is adequate, so binomial tables are not required. "Large" may be thought of in the context of the number necessary for the binomial to adequately approximate the normal, specifically $n\pi > 5$. The normal mean μ is π, and the standard deviation is

$$\sigma = \sqrt{\frac{\pi(1-\pi)}{n}}. \qquad (15.7)$$

The difference between the sample and theoretical proportions is

---

transformed to a normal variate z by dividing by σ, or

$$z = \frac{p_s - \pi}{\sqrt{\dfrac{\pi(1 - \pi)}{n}}}. \tag{15.8}$$

The two-tailed p-value corresponding to the calculated z is then found from the normal table (see Table I). Because a normal variable is continuous and a binomial variable is discrete, the quantity $1/2n$ may be subtracted from the absolute value of the numerator to increase the accuracy of the approximation. If n is small, the exact form should be used; if n is large, the adjustment adds little. There may be cases of moderate sized n in which the adjustment is beneficial.

Case of Two Proportions

The smaller sample form can usually be treated as a contingency table. If $n_1\pi_1$, $n_1(1 - \pi_1)$, $n_2\pi_2$, and $n_2(1 - \pi_2)$ are all greater than 5, the larger sample form may be used. Here, we test $H_0 \colon \pi_1 = \pi_2$ by contrasting $p_1 - p_2$. The standard deviation for the combined samples, σ_c, is a little more complicated than Eq. (15.7). First, we find a combined sample proportion estimate, p_c, as

$$p_c = \frac{n_1 p_1 + n_2 p_2}{n_1 + n_2}, \tag{15.9}$$

and then the combined standard deviation as

$$\sigma_c = \sqrt{p_c(1 - p_c)\left(\frac{1}{n_1} + \frac{1}{n_2}\right)}. \tag{15.10}$$

The normal z statistic is given by

$$z = \frac{p_1 - p_2}{\sqrt{p_c(1 - p_c)\left(\dfrac{1}{n_1} + \dfrac{1}{n_2}\right)}}. \tag{15.11}$$

The critical value(s) to test H_0 can be found as area(s) under the normal curve. For example, a two-tailed test has critical values ± 1.96, corresponding to 2.5% of the curve under each tail.

EXAMPLES OF TESTS OF PROPORTION

Example: Sample Proportion Against Theoretical Proportion, Small n—Is a Rate of Positive Biopsy Results Consistent with a Theoretical Rate?

We want to compare the positive biopsy result rate $p_s = 0.30$ (30%) from a sample size of 10 patients (see Table DB1.1) with a theoretical positive biopsy result rate defined as 0.25 (25%) of patients older than 50 years presenting to

Table 15.16

A Portion of Table VI^a

						π					
n	n_o	0.05	0.10	0.15	0.20	0.25	0.30	0.35	0.40	0.45	0.50
9	4	0.001	0.008	0.034	0.086	0.166	0.270	0.391	0.517	0.639	0.746
10	3	0.012	0.070	0.180	0.322	0.474	0.617	0.738	0.833	0.900	0.945

^{a}Values of cumulative binomial distribution, depending on π, n, and n_o (number of occurrences observed).
The corresponding entry in the table body represents the probability that n_o or more occurrences (or n_o/n proportion) would have been observed by chance alone.
See Tables of Probability Distributions to review Table VI.

a urology clinic. We use the binomial distribution with theoretical proportion $\pi = 0.25$. To find the probability of drawing $n_s \geq 3$ of $n = 10$ by chance alone, given $\pi = 0.25$, we go to Table VI. A portion of Table VI appears as Table 15.16. The table entry is 0.474; the probability of finding such a result is nearly half of that we would find by chance alone (the p-value). Thus, we conclude that there is insufficient evidence to say that our proportion of positive biopsy results is larger than that found in the clinic.

Example: Sample Proportion Against Theoretical Proportion, Large n—The Issue of Physician Cleanliness 150 Years Ago

In 1847 in Vienna, Ignaz Semmelweis experimented with the effect on patient mortality of physician hand cleansing between patients.[65] Original data are not available (if they ever were), but a reconstruction from medical history accounts yields the following results: The long-term mortality rate for patients treated without the cleansing had been about 18%. Subsequently, for 500 patients treated with the cleansing, the mortality rate was 1.2%. The requirement $n\pi > 5 (90 > 5)$ is satisfied. Because hand cleaning could not increase the mortality, a one-tailed test is appropriate. We substitute in the formula for a normal z-value, obtaining

$$z = \frac{p_s - \pi}{\sqrt{\pi(1-\pi)/n}} = \frac{0.012 - 0.18}{\sqrt{0.18 \times 0.82/500}} = -9.78.$$

By symmetry in the normal curve, we may use the upper tail, that is, $z = 9.78$. The area in the tail is far less than the smallest tabulated value in Table I for one-tailed α, which yields $p < 0.0002$ (reported as $p < 0.001$) for $z = 3.5$. The cleansing significantly reduced the mortality rate. Nonetheless, as a matter of historical interest, the local medical society attacked Semmelweis' findings viciously. He was reduced in rank and his practicing privileges were curtailed. He was broken and died in an asylum.

Example: Test of Two Sample Proportions

In DB13, there are 104 International Normalized Ratio (INR) readings from each of the Coumadin Clinic and the hospital laboratory that are classed as below, in, or above range. If we combine the above and below ranges into an out-of-range category, we have in range versus out of range. We find 33 out of range readings from the clinic for a rate of 31.7%, and 57 from the laboratory for a rate of 54.8%. Are the rates from the clinic and the laboratory significantly different? We can test the two rates by a test of two proportions. (That said, we should note that we are conducting a test of two independent rates and the readings in this database are paired. We will carry on for the purpose of example, but McNemar's test in Section 15.8 is the proper form of analysis. These data are exemplified using McNemar's test in that section.) $n_1 = n_2 = 104$, $p_1 = 0.317$, and $p_2 = 0.548$. We calculate $p_c = 0.4325$ from Eq. (15.9) and $z = 3.362$ from Eq. (15.11). From Table I, we see that a z of 3.36 provides a two-sided area under the tails between 0.0010 and 0.0006; therefore, we would report $p < 0.001$. The out-of-range rates are different for the clinic and the laboratory.

ADDITIONAL EXAMPLES

Small *n*: Does Radiation Therapy Improve the Proportion Survival?

The 2-year survival rate of an unusual form of malignant cancer has been $\pi = 0.20$. A radiation oncology department treated nine patients.[36] Two years later, four patients had survived. Does the treatment improve survival? From Table 15.16, which is a segment of Table VI, the column for $\pi = 0.20$, the row for $n = 9$, and $n_o = 4$ yield a probability of 0.086 that four or more patients with the 20% untreated survival rate would survive by chance alone. There is not enough evidence to infer treatment effectiveness on the basis of this small a sample.

Large *n*: Does Absorbable Mesh in Closure of Large Abdominal Wall Openings Reduce the Proportion of Dense Adhesion Formation?

The use of permanent prosthetic mesh has led to formation of dense adhesions with later obstructions and fistula formations. We ask whether the population proportion π_a of these problems is less using absorbable mesh. $H_0: \pi_a = \pi$, and $H_1: \pi_a < \pi$. A number of studies in the literature report the proportion of dense adhesion formation from permanent mesh to be about 0.63, which is taken as π. An experiment[50] on 40 rats finds the proportion of dense adhesions from absorbable mesh to be $p_a = 0.125$ (5 rats), which estimates π_a. The clinical difference is obvious, but we need to state the risk for error. From Eq. (15.7), $\sigma = 0.0763$; from Eq. (15.8), $z = (|p_a - \pi|)/\sigma = (|0.125 - 0.63|)/0.0763 = 6.62$. The area under the tail of a normal distribution more than 6.62 standard deviations away from the mean is smaller than any tabulated value, certainly less than 0.001, which is the smallest acceptable value to quote. (The discrepancy between the approximation and the true probability curve becomes more than the curve height in the far tails; therefore, p-values smaller than 0.001 should not be listed, even though they can be calculated.)

Exercise 15.5: Small *n*—Are More Boys Than Girls Born in This Family?   A small town obstetrician has been delivering babies from a particular family for three generations. He notes a high proportion of male babies and suspects he may be seeing a genetic phenomenon. Of 16 babies born during his experience,

11 are boys. Does he have evidence of a nonrandom causative factor, or could it have occurred by chance?

Exercise 15.6: Large *n*—Do Boys and Girls Have Limb Fractures in Equal Numbers? Theoretically, the population proportion would be $\pi = 0.5$ if the proportion were sex independent. We recorded the sex of the next 50 children presenting to the orthopedic clinic with limb fractures and found 34 boys. What do we conclude?

15.7. TESTS OF A SMALL PROPORTION (CLOSE TO ZERO)

WHAT A TEST OF A SMALL PROPORTION WILL DO AND WHY IT IS NEEDED

Suppose we suspect a sudden increase in the occurrence rate of a rare disease, with prevalence known to be 0.002 under ordinary conditions. Testing our sample proportion will answer the question: *Is a sample proportion different from the theoretical or population proportion?* The proportion 0.002 is too small to use the binomial methods discussed in Section 15.6. If π is near zero, and if the relationship $n \gg n\pi \gg \pi$ (recall that $\gg$ denotes "much greater than") may be assumed, the result can be well approximated using the *Poisson distribution*, for which probabilities appear in Table VII (see Tables of Probability Distributions). The tabulation of the Poisson probabilities uses a parameter $\lambda = n\pi$ rather than π alone, which is used for binomial probabilities. Suppose our sample shows a prevalence of 0.006, which is 3 times the expected prevalence. The probability of obtaining a proportion $p \geq 0.006$ by chance alone, given $\pi = 0.002$, can be calculated or looked up in a table, as detailed in the methods and examples paragraphs that appear later in this section. The resulting probability is the *p*-value of the binomial test of proportion.

WHAT IS DONE IF π IS VERY CLOSE TO 1 RATHER THAN 0

If π is near 1, we can use the complementary proportion, $1 - \pi$, because the distribution is symmetric. If survival from a disease is 0.998, we examine the prevalence of death as 0.002.

A DIFFERENT METHOD IS USED FOR LARGE SAMPLES

If *n* is large enough that the input value $\lambda = n\pi$ does not appear in Table VII (i.e., $\lambda > 9$), a large-sample approximation is used. It has been shown that, for the Poisson distribution, the sample proportion is distributed approximately *normal* with mean equal to the theoretical proportion π, as for the binomial, but with a slightly different standard deviation σ. By knowing μ and σ, we may calculate a *z*-score and use a normal table to find *p*-values, as detailed in the methods and examples paragraphs that follow.

EXAMPLES POSED

Small λ: Are There Too Many Large Prostates in a Distribution?

The sample of $n = 301$ urologic patients contains 3 men with prostate volumes in excess of 90 ml. We might hypothesize that the distribution of volumes in the general population is normal. If so, would we be likely to find $x = 3$ volumes greater than 90 ml by chance alone, or should we suspect that our sample is not normal in form? In a normal curve with $\mu = 36.47$ and $\sigma = 18.04$ ml, 90 ml would lie 2.967σ above the mean. From Table I, the area in the right tail would be about $0.002 = \pi$. There is a large number of opportunities for a volume to be greater than 90 but only a small chance that any one randomly chosen would be, suggesting the Poisson process. The Poisson constant is $\lambda = n\pi = 301 \times 0.002 = 0.602$.

Large λ: Does a New Drug Cause Birth Defects?

It has been established that 1.75% ($\pi = 0.0175$) of babies born to women exposed during pregnancy to a certain drug have birth defects. It is believed that an adjuvant drug will reduce the rate of a birth defect. The two drugs are used together on 638 pregnant women, and 7 of their babies have birth defects.[50] Did the adjuvant drug reduce the rate of defects? $\lambda = n\pi = 638 \times 0.0175 = 11.165$, which is larger than the values of λ in Table VII, so the normal approximation is used.

METHOD: TEST OF A VERY SMALL PROPORTION (NEAR ZERO)

This method tests a sample proportion against a theoretical (or previously known) population proportion, π, where π is close to 0 (or to 1) and where the Poisson assumption $n \gg n\pi \gg \pi$ is satisfied. We sample n observations randomly from a population with probability π_s of a "success." We want to know if π_s, the proportion for the population sampled, is the same as π. Thus, H_0 is $\pi_s = \pi$. We obtain n_s successes, estimating π_s by the sample proportion $p_s = n_s/n$. The Poisson constant is $\lambda = n\pi$, which is also the number of occurrences expected on average. Calculate $\lambda = n\pi$ and verify the assumption.

Case of Small λ

Table VII gives the probability of n_s or more successes by chance alone, which is also the *p*-value for the test that the sample observations have π proportion success.

Case of Large λ

If $\lambda > 9$ (e.g., $n = 500$ and $\pi = 0.02$), a normal approximation to the Poisson distribution is adequate. The normal mean $\mu = \pi$, and the standard deviation σ is

$$\sigma = \sqrt{\frac{\pi}{n}}. \tag{15.12}$$

[Note the similarity to and the difference from Eq. (15.7).] Calculate usual standard normal form, here the difference between the observed and theoretical proportions divided by the standard deviation,

$$z = \frac{p_s - \pi}{\sqrt{\pi/n}} = (p_s - \pi)\sqrt{n/\pi} \qquad (15.13)$$

and test it using the normal Table I. (Either the expression in the middle or on the right may be used as convenient.) The "continuity correction" $1/2n$ subtracted from the absolute value of the numerator may be beneficial in cases of moderate sized samples, as noted in Section 15.6.

Examples Completed

Small λ

The Poisson assumption $n \gg n\pi \gg \pi$ is satisfied: $301 \gg 0.6 \gg 0.002$. From Table 15.17, a portion of Table VII, the Poisson probability of observing 3 or more volumes greater than 90 ml, given a Poisson constant of 0.6, is 0.023. With a p-value this small, we reject the hypothesis and conclude that our sample is not normal in form.

Large λ

A large number of opportunities to occur (638) but a small chance of occurring on any one opportunity (0.0175) indicates Poisson. Because the adjuvant drug would be used clinically if defects reduce but not if they stay the same or increase, a one-tailed test is appropriate. $\lambda = n\pi = 638 \times 0.0175 = 11.165$, which is larger than the value in Table VII, so the normal approximation is used. $p_s = 7/638 = 0.0110$. Substitution in Eq. (15.13) yields $z = -1.2411$. By normal curve symmetry, we can look up 1.2411 in Table 15.18. By interpolation, we can see that the p-value, that is, the α that would have given this result, is about 0.108. A reduction in defect rate has not been shown.

Table 15.17

A Portion of Table VII[a]

n_o						$\lambda(= n\pi)$							
	0.1	0.2	0.3	0.4	0.5	0.6	0.7	0.8	0.9	1.0	1.1	1.2	1.3
3	0.000	0.001	0.004	0.008	0.014	0.023	0.034	0.047	0.063	0.080	0.100	0.121	0.143

[a]Values of the cumulative Poisson distribution, depending on $\lambda = n\pi$ and n_o (number of occurrences observed).

Given λ and n_o, the table entry represents the probability that n_o or more occurrences would have been observed by chance alone.

See Tables of Probability Distributions to review Table VII.

Table 15.18

A Fragment of Table I, Normal Distribution[a]

z (no. standard deviations to right of mean)	One-tailed α (area in right tail)	Two-tailed α (area in both tails)
1.20	0.115	0.230
1.30	0.097	0.194
1.40	0.081	0.162
1.50	0.067	0.134

[a]For selected distances (z) to the right of the mean, given are one-tailed α, the area under the curve in the positive tail, and two-tailed α, the areas combined for both tails. See Tables of Probability Distributions to review Table I.

ADDITIONAL EXAMPLES

Small λ: Does The Site of Intramuscular Injection Affect the Incidence of Sarcoma?

A veterinarian[50] finds that cats contract vaccine-associated feline sarcoma when vaccinated in the back with a rate of 2 in 10,000 ($\pi = 0.0002$). She initiates a study in which 5000 cats are given their rabies and feline leukemia vaccinations in the rear leg. The veterinarian finds $n_o = 3$ cases of the sarcoma. Is the incidence of sarcoma for leg vaccination different from that for the back? There are many opportunities for the sarcoma to occur, but the probability of it occurring in any randomly chosen case is near zero, indicating Poisson methods. If the incidences at the two sites are different, either one could be greater; therefore, a two-tailed test is appropriate. In Table VII, the table entry is the probability that n_o or more sarcomas would have occurred by chance alone; therefore, the critical values for $\alpha = 5\%$ will be 0.975 (i.e., $1 - 0.025$) for the case of fewer sarcomas in the new site and 0.025 for the case of more sarcomas in the new site. $\lambda = n\pi = 5000 \times 0.0002 = 1$. The assumption $n \gg n\pi \gg \pi$ is satisfied ($5000 \gg 1 \gg 0.0002$). In Table VII, the probability is 0.080 between the two critical values. The veterinarian concludes that the data did not provide sufficient evidence to conclude that leg vaccination gives a sarcoma rate different from back vaccination.

Large λ: Is Anorexia in the Military Different from That in the General Public?

Anorexia among the general U.S. population of women is known to be $\pi = 0.020$ (2%). We want to know whether this rate is also true for female military officers.[48] Let us use π_m to represent the true but unknown female military officer population proportion, which we will estimate by p_m. $H_0: \pi_m = \pi$, and $H_1: \pi_m \neq \pi$. We take a sample of 539 female Navy officers and find $p_m = 0.011$. Do we reject H_0? Because $\lambda = 0.02 \times 539 = 10.78$ is larger than the tabulated values, we use the normal approximation to the Poisson distribution. By using Eq. (15.13), we find

$$z = \frac{p_s - \pi}{\sqrt{\pi/n}} = \frac{0.011 - 0.020}{\sqrt{0.020/539}} = -1.48.$$

By symmetry of the normal curve, the result is the same as for $z = 1.48$, which is between $z = 1.40$ and $z = 1.50$ in Table 15.18 (or Table I). The two-tailed p-values (found in the table as if they were α's) for these z-values are 0.162 and 0.135. Interpolation to $z = 1.48$ yields $p = 0.157$, a result not statistically significant. We do not reject H_0 and conclude that we have no evidence that the rate of anorexia in military officers is different from the rate of the general population.

Exercise 15.7: Does Blood-Bank Screening Reduce Hepatitis-Infected Blood? Studies through 1992 found the rate of viral hepatitis infection caused by red blood cell transfusions in the United States to be 1 per 3000 units transfused, or $\pi = 0.000333$. Introduction of second- and third-generation hepatitis C screening tests is expected to reduce this rate.[39] Test the significance of results from 15,000 transfused units reducing the rate to 1:5000, or $p = 0.000200$.

15.8. MATCHED PAIR TEST (McNEMAR'S TEST)

McNemar's test assesses the dependence of categorical data that are matched or paired. In DB13, the INR readings of 104 diabetic patients were assessed by the Coumadin Clinic and the hospital laboratory. Is the rate of out-of-range readings different for the two assessors? We are not comparing a sample of clinic patients with a sample of laboratory patients; all patients have readings from both assessors, so the data are paired. This pairing was not the case for contingency tables and their methods. McNemar's test fills this methodologic gap.

EXAMPLES POSED

Is the Rate of Out-of-Range Readings Different for Clinic Versus Laboratory?

The INR readings of 104 diabetic patients were assessed by both the Coumadin Clinic and the hospital laboratory and thus are paired. Is the out-of-range rate different?

Is Smoking Associated with Lung Cancer?

We randomly choose the records of 10 patients with lung cancer and find 10 control patients without lung cancer with a match one-for-one in age, sex, health history, socioeconomic level, and air-quality of living region.[50] Then we record whether each patient smokes (yes) or not (no). Are the paired rates different?

METHOD: MCNEMAR'S TEST

McNemar's test assesses the dependence of categorical data that are matched or paired. We want to determine whether a certain characteristic is associated with a disease (or other malady). We identify n patients having the disease and pair them one-for-one with control patients without that disease, but with other possibly relevant characteristics being the same.

Table 15.19

Recording Format for Patients Paired for Disease and Nondisease

	Factor present for:	
Pair number	Diseased member of pair	Nondiseased member of pair
1	Yes	No
2	Yes	Yes
3	No	No
etc.	etc.	etc.

Table 15.20

Recording Format for Tallying the Data of Table 15.18

		Control members of pair	
		Yes	No
Diseased members of pair	Yes	a	b
	No	c	d

Then we record the presence or absence of the characteristic in each patient and summarize the result in a two-way table. The data are listed in the format of Table 15.19, where "yes" and "no" indicate the presence or absence, respectively, of the characteristic. A tally of the results can be recorded in a two-way table (Table 15.20), where a is the number of pairs with a "yes-yes" sequence (as in patient 2), b is the number of pairs with a "yes-no" sequence (as in patient 1), and so on.

The difference in proportions of cases having and not having the characteristic under study is given by $(b - c)/n$ and its standard error by $\sqrt{b + c}/n$. Chi-square with 1 df is calculated by Eq. (15.14), and the p-value is obtained from Table III.

$$\chi^2_{1df} = \frac{(|b - c| - 1)^2}{b + c} \tag{15.14}$$

The p-value is the probability of being wrong if we conclude that the characteristic is associated with the disease.

EXAMPLES COMPLETED

Is the Rate of Out-of-Range Readings Different for Clinic Versus Laboratory?

We tally the in-range and out-of-range INR readings for the clinic and the laboratory in the format of Table 15.20 to obtain Table 15.21. $b = 27$, and $c = 3$. By substituting in Eq. (15.14), we find $\chi^2_{1df} = 23^2/30 = 17.633$, which is much

Table 15.21

**In-Range and Out-of-Range International Normalized Ratio
Readings for the Clinic and the Laboratory**

| | | Laboratory | |
		In Range	Out of Range
Clinic	In range	44	27
	Out of range	3	30

larger than the 5% critical value of 3.84. The largest tabulated value for 1 *df* in Table III is 12.13, which is associated with $p = 0.0005$. Clearly, $p < 0.001$. The out-of range rates are very different for the clinic versus the laboratory.

Is Smoking Associated with Lung Cancer?

We list the relevant data in the format of Table 15.19 as Table 15.22 and then tally the numbers of pairs in the "yes-yes," "yes-no," and so on categories in the two-way table (Table 15.23). The 5% critical value of chi-square with 1 *df* is 3.84. For the lung cancer data, McNemar's test statistic is

$$\chi^2_{1df} = \frac{(|b - c| - 1)^2}{b + c} = \frac{(5 - 1)^2}{5} = 3.2.$$

From Table III, a chi-square of 3.2 with 1 *df* lies between 5% and 10%; for an α of 5%, we do not have quite enough evidence to say that smoking is related to

Table 15.22

Smoking Records of Patients Paired for Lung Cancer and Control

Pair no.	Cancer	Control
1	Yes	No
2	No	No
3	Yes	Yes
4	Yes	No
5	Yes	No
6	No	No
7	No	No
8	Yes	Yes
9	Yes	No
10	Yes	No

Table 15.23

Tally of Counts from Table 15.22

| | | Control | |
		Yes	No
Lung cancer	Yes	$a = 2$	$b = 5$
	No	$c = 0$	$d = 3$

lung cancer. The actual p-value, calculated using a statistical software package, is 0.074.

ADDITIONAL EXAMPLE: IS THERE A GENETIC PREDISPOSITION TO STOMACH CANCER?

A company physician suspects a genetic predisposition to stomach ulcers and targets a specific gene.[50] She chooses workers in the company who have ulcers. The physician matches each subject with a nonulcerous coworker in the same job (same work environment) with similar personal characteristics (sex, marital state, etc.). She then checks all subjects for the presence of the suspect gene, obtaining the data displayed in Table 15.24, and then tallies these data in the format of Table 15.20 to obtain Table 15.25. McNemar's chi-square statistic is $\chi^2_{1df} = (|b - c| - 1)^2/(b + c) = (7 - 1)^2/9 = 4$. From Table III, the critical value of χ^2 for $\alpha = 0.05$ with 1 df is 3.84. Because $4.0 > 3.84$, $p < 0.05$; therefore, the

Table 15.24

Presence of a Suspect Gene in Patients Paired for Ulcers and Healthy Subjects

	Gene present	
Pair no.	Ulcer patient	Healthy subject
1	Yes	No
2	Yes	Yes
3	No	No
4	No	No
5	No	Yes
6	Yes	Yes
7	Yes	No
8	Yes	No
9	Yes	No
10	Yes	No
11	No	No
12	Yes	Yes
13	Yes	No
14	Yes	No
15	Yes	No

Table 15.25

Tally of Counts from Table 15.24

		Gene in healthy member of pair	
		Yes	No
Gene in member of pair with ulcer	Yes	$a = 3$	$b = 8$
	No	$c = 1$	$d = 3$

Table 15.26

Accident Occurrence in Children Paired for Boy and Girl

Pair no.	Had accident(s) Boy	Had accident(s) Girl
1	Yes	No
2	Yes	Yes
3	No	No
4	Yes	No
5	No	Yes
6	Yes	No
7	Yes	yes
8	No	No
9	Yes	No
10	Yes	No
11	Yes	No
12	No	Yes

physician concludes that evidence is present for a genetic predisposition. The actual p-value is 0.046.

Exercise 15.8: Are Boys More Accident Prone Than Girls?   You suspect that boys between the ages of 3 and 5 years are more accident prone than girls. You identify families with a boy and a girl and check their medical records for accidents during those ages.[50] Your data are displayed in Table 15.26. Use McNemar's test to decide if boys have significantly more accidents.

ANSWERS TO EXERCISES

15.1. $p = 0.041$ for the FET, indicating a significant difference between the success of the two treatments with a risk for error less than 5%; MB is better than CS. $\chi^2 = 5.132$, yielding $p = 0.023$, also showing a significant difference. A cell entry of less than 5 indicates that the chi-square approximation may not be adequate. Fisher's exact test is the appropriate test.

15.2. $e_{11} = 19 \times 100/195 = 9.7436$, and so on. $\chi^2 = (8 - 9.7436)^2/9.7436 + \cdots = 4.92$. $df = 2$. From Table III, 4.92 for 2 df lies between 0.05 and 0.10. A computer evaluates the p as 0.085. Because $p > 0.05$, there is inadequate evidence to claim a systolic and diastolic difference.

15.3. Sensitivity $= 0.981$, the chance of detecting appendicitis using magnetic resonance imaging. Specificity $= 0.979$, the chance of correctly not diagnosing appendicitis. Accuracy $= 0.980$, the overall chance of making a correct diagnosis. The PPV $= 0.981$, the chance that a diagnosis of appendicitis is correct. The NPV $= 0.979$, the chance that a diagnosis ruling out appendicitis is correct. That these values are the same as sensitivity and specificity is an unusual coincidence because of the equality of the marginal totals. The RR $= 46.11$, the chance of appendicitis when predicted in ratio the chance of appendicitis when not predicted. The OR $= 2392$, the odds of correctly diagnosing appendicitis in ratio to the odds of incorrectly diagnosing appendicitis. The LR = 46.11, the chance of correctly diagnosing appendicitis in ratio to the chance of incorrectly diagnosing appendicitis, which is coincidentally the same as RR because of identical marginal totals. The AR = 0.96, the proportion of appendicitis cases discovered by the clinical diagnostic procedure. Some of these figures, lacking in clinical meaning or usefulness, were calculated for the exercise.

15.4. $L = 1.5506$, $SEL = 0.7327$, and $\chi^2 = (1.5506/0.7327)^2 = 4.4787$. From Table III, the p-value for this result is between 0.025 and 0.05, a significant result. The actual p-value is 0.034, which is similar to the 0.041 FET result in Exercise 15.1. We have shown that quality of result is significantly associated with method of ligament repair.

15.5. $n = 16$, $n_o = 11$, and $\pi = 0.50$. From Table VI, the probability of 11/16 occurring by chance alone is 0.105, which is too large to infer evidence of a nonrandom phenomenon.

15.6. If π_b is the true population proportion of boys presenting, H_0: $\pi_b = \pi$ and H_1: $\pi_b \neq \pi$. Then, $p_b = 0.68$. Using the normal approximation, $\sigma = 0.0707$. From Eq. (15.8), $z = (|p_b - \pi|)/\sigma = 0.18/0.0707 = 2.55$. From Table I, $z = 2.55$ yields two-tailed $p = 0.010$. We reject H_0 and conclude that more boys than girls present with limb fractures.

15.7. We judge that introducing better testing can only reduce the rate; therefore, the test is one-sided. From Table I, a critical value of z at $\alpha = 0.05$ will be -1.645. We note that $\lambda = n\pi = 5$, so we may use Table VII. We want the probability of three or fewer cases [$P(3$ or fewer$)$], which is $1 - P(4$ or more$)$, occurring by chance alone. Using $n_o = 4$, $P(3$ or fewer$) = 1 - 0.735 = 0.265$. There is inadequate evidence to infer an improvement due to the additional screening tests. For illustration, let us calculate the normal approximation, although we know the result will be erroneously different from the Table VII result because of the small λ. From Eq. (15.13),

$$z = (p_s - \pi)\sqrt{n/\pi} = (0.000200 - 0.000333) \times \sqrt{15000/0.000333} = -0.893,$$

which is less distant from zero than the critical value; therefore, there is inadequate evidence to infer an improvement. From a statistical software package, $p = 0.186$.

15.8. The data tally appears in Table 15.27. $\chi^2_{1df} = (|b - c| - 1)^2/(b + c) = (4 - 1)^2/8 = 1.125$. From Table III, the critical value of χ^2 for $\alpha = 0.05$ with 1 df is 3.84. Because $1.125 < 3.84$, you conclude that boys are not shown to have a greater accident rate. The actual p-value is 0.289.

Table 15.27

Tally of Counts from Table 15.26

		Female member of pair had accident(s)	
		Yes	No
Male member of pair had accident(s)	Yes	$a = 2$	$b = 6$
	No	$c = 2$	$d = 2$

Chapter 16

Tests on Ranked Data

16.1. BASICS OF RANKS

WHAT RANKS ARE

This section summarizes the basics of ranked data as seen in Section 6.5. Ranked data are values put in order according to some criterion (e.g., smallest to largest). Rank-order data may arise from ranking events directly or from putting already recorded continuous-type quantities in order. If categories occur in a natural ordering (e.g., most to least severe skin lesions, but not ethnic groups), they may be treated as rank-order data.

WHEN AND WHY DO WE USE RANKS?

Continuous measurements contain more information than do ranks, and ranks contain more information than do counts. When the user can rank events but cannot measure them on a scale, rank-order statistical methods will give the best results. When the sampling distribution of continuous data is skewed or otherwise poorly behaved, the assumptions underlying continuous-based methods may be violated, giving rise to spurious results. Ranks do not violate these assumptions. When sample sizes are too small to verify the satisfaction of these assumptions, rank-order methods are safer. In summary, rank-order methods should be used (1) when the primary data consist of ranks, (2) when samples are too small to form conclusions about the underlying probability distributions, or (3) when data indicate that necessary assumptions about the distributions are violated.

TIES IN RANKED DATA

If ranks are tied, the usual procedure is to average the tied ranks. If a set of heart rates (HRs) per minute includes 64, 67, 67, and 71, they would have ranks of 1, 2.5, 2.5, and 4. Ties reduce the quality of rank methods somewhat, but they are still stronger than categorical methods.

16.2. SINGLE OR PAIRED SMALL SAMPLES: THE SIGNED-RANK TEST

EXAMPLES POSED

Single Sample: Are Our First 10 Prostate-Specific Antigen Levels from the Same Population as the Remainder of Our Sample?

Table DB1.1 gives prostate-specific antigen (PSA) levels for the first 10 of the 301 urology patients. The median PSA of the next 291 patients is 5.7. We ask if the first 10 patients have an average (more exactly, a median) similar to the remaining 291. Because the second median could be either more or less than the first, we use a two-tailed test. If the averages are similar, the ranks of the deviations of the first 10 from the median of the remainder should vary approximately equally (about zero).

Paired (Matched) Observations: Does a Drug Change the Heart Rate?

Anecdotal information suggests that HR changes in response to a certain ophthalmologic drug designed to reduce intraocular pressure. For a sample of eight patients,[50] the HRs (measured in beats per minute) before beginning treatment, 24 hours after the first dose, and the change are as follows:

Before:	64,	66,	67,	71,	72,	76,	78,	89
After:	66,	58,	68,	65,	75,	67,	59,	74
Difference:	−2,	8,	−1,	6,	−3,	9,	19,	15

Are the before and after readings different?

METHOD: THE SIGNED-RANK TEST

Often referred to as the Wilcoxon signed-rank test, this method may be thought of as testing the hypothesis that the distribution of differences has a median equal to zero. It may test:

(1) A set of observations deviating from a hypothesized common value or
(2) Pairs of observations on the same individuals, such as before and after data.

Steps to perform the test are as follows:

1. Calculate the differences of the observations as in (1) or (2) just listed (zero differences are just omitted from the calculation, and the sample size is reduced accordingly).
2. Rank the magnitudes, that is, the differences without signs, the smallest being rank 1. (For ties, use the average of the ranks that would have occurred without ties in the same positions.)
3. Reattach the signs to the ranks.

4. Add up the positive and negative ranks.
5. Denote by T the unsigned value of the smaller sum.
6. Look up the p-value for the test in Table VIII (see Tables of Probability Distributions). If $n > 12$, go to Section 16.6.

ONE TAIL OR TWO?

If we have some sound nonstatistical reason why the result of the test must lie in only one tail, such as a physiologic impossibility to occur in the other tail, we can halve the tabulated value to obtain a one-tail error probability.

EXAMPLES COMPLETED

Single Sample: Are Our First 10 Prostate-Specific Antigen Levels Similar to the Remainder?

We rank these deviations of our first 10 PSAs from the median of the remainder without regard to the sign ($+$ or $-$), then reattach these signs to the ranks, and add the positive and the negative ranks. If the first 10 observations are similar to the remaining observations, the positive rank sum will be similar to the negative rank sum. These deviations and their ranks are as follows:

Deviations	1.9	−1.6	0.2	3.3	1.1	2.3	2.0	−1.3	0.4	2.2
Signless ranks	6	5	1	10	3	9	7	4	2	8
Signed ranks	6	−5	1	10	3	9	7	−4	2	8

The sum of positive ranks is 46 and that of negative ranks is −9, which are quite different. Denote by T the unsigned value of the smaller sum; in this case, $T = 9$. Table 16.1 shows a portion of Table VIII (see Tables of Probability Distributions). In Tables 16.1 and VIII, the intersection of the row for $T = 9$ and the column for $n = 10$ yields $p = 0.065$. For a risk for error exceeding 0.05, we conclude that the median of the first 10 PSA readings is not different from that of the next 291 readings.

Table 16.1

A Segment of Signed-Rank Probabilities, Two-Tailed Probabilities of T[a]

T	Sample size n								
	4	5	6	7	8	9	10	11	12
3	0.625	0.313	0.156	0.078	0.039	0.020	0.010	0.005	0.003
4	0.875	0.438	0.219	0.109	0.055	0.027	0.014	0.007	0.004
6		0.801	0.438	0.219	0.109	0.055	0.027	0.014	0.007
9			0.844	0.469	0.250	0.129	0.065	0.032	0.016

[a]For a sample of size n and value of T, the entry gives the p-value.
Table VIII in Tables of Probability Distributions provides an expanded version of this table.

Paired Observations: Does a Drug Change the Heart Rate?

The differences in heart rates (measured in beats per minute), before medication minus after medication, are ranked by magnitude, that is, with their signs removed. The data, differences, and rank entries are as follows:

Before	After	Difference	Magnitude rank	Signed rank
64	66	−2	2	−2
66	58	8	5	5
67	68	−1	1	−1
71	65	6	4	4
72	75	−3	3	−3
76	67	9	6	6
78	59	19	8	8
89	74	15	7	7

The two unsigned rank sums are 6 for negative and 30 for positive; $T = 6$. In Table 16.1, the intersection of the $n = 8$ column and the $T = 6$ row yields $p = 0.109$, which is too large a p-value to conclude that the drug changed the intraocular pressure.

ADDITIONAL EXAMPLE: DOES CERTAIN HARDWARE RESTORE ADEQUATE FUNCTIONALITY IN ANKLE REPAIR?

An orthopedist installs the hardware in the broken ankles of nine patients.[28] He scores the percent functionality of the joint. The orthopedist asks: "Is the average functionality percentage less than 90% that of normal functionality?" These data are 75%, 65%, 100%, 90%, 35%, 63%, 78%, 70%, and 80%. A quick frequency plot of the data shows they are far from a normal distribution; therefore, the orthopedist uses a rank-based test. He subtracts 90% from each (to provide a base of zero) and ranks them, ignoring signs. Then he attaches the signs to the ranks to obtain the signed ranks. These results are as follows:

Deviation from 90%	−15	−25	10	0	−55	−27	−12	−20	−10
Unsigned ranks	5	7	2.5	1	9	8	4	6	2.5
Signed ranks	−5	−7	2.5	1	−9	−8	−4	−6	−2.5

The sum of positive signs will obviously be the smaller sum, namely, $3.5 = T$. From Table 16.1 (or Table VIII), the p-value for $n = 9$ with $T = 3.5$ for a two-tailed test is between 0.020 and 0.027, about 0.024. Because the orthopedist chose a one-tailed test, the tabulated value may be halved, or $p = 0.012$, approximately, which is clearly significant. He concludes that the patients' average functionality is significantly less than 90%.

Exercise 16.1: Are the Readings of a Certain Thermometer on a Healthy Patient Normal?

We are investigating the reliability of a certain brand of tympanic thermometer (temperature was measured by a sensor inserted into the patient's ear).[63] Eight readings were taken in the right ear of a healthy patient at 2-minute intervals. Data were 98.1°, 95.8°, 97.5°, 97.2°, 97.7°, 99.3°, 99.2°, and 98.1°F. Is the median different from the population average of 98.6°F?

16.3. TWO SMALL SAMPLES: THE RANK-SUM TEST

Section 6.6 introduces the method of the rank-sum test. This test uses sample ranks to compare two samples in which the observations are not paired; n_1 observations are drawn from one sample, and n_2 from the other. (n_1 is always assigned to the smaller of the two n's.)

EXAMPLE: REVIEW OF THE EXAMPLE FROM SECTION 6.6

We compared PSA levels from Table DB1.1 for the $n_1 = 3$ patients having positive biopsy results with the $n_2 = 7$ patients having negative biopsy results. We ranked the PSA levels, then added up the ranks for the smaller sample, naming this rank sum T. We converted T to the Mann–Whitney U statistic by the formula $U = n_1 n_2 + n_1(n_1 + 1)/2 - T$, and looked up the probability in Table IX (see Tables of Probability Distribution) that a U-value this small would have arisen by chance alone. $n_2 = 7$, $n_1 = 3$, $T = 18$, $U = 9$, and the p-value associated with $U = 9$ was 0.834. We concluded that we had no evidence from the Table DB1.1 data that PSA levels tend to be different for positive and negative biopsy results.

METHOD: THE RANK-SUM TEST

Given two samples, the hypothesis being tested is whether the value for a randomly chosen member of the first sample is probably smaller than one of the second sample, a slightly technical concept. For practical purposes, the user may think of it informally as testing whether the two distributions have the same median. Section 6.6. discusses the sampling, choosing hypotheses, and α. The steps to conduct the test may be summarized as follows:

1. Name the sizes of the samples n_1 and n_2; n_1 is the smaller sample. If $n_2 > 8$, go to Section 16.7.
2. Combine the data, keeping track of the sample from which each datum arose.
3. Rank the data.
4. Add up the ranks of the data from the smaller sample and name it T.
5. Calculate $U = n_1 n_2 + n_1(n_1 + 1)/2 - T$.
6. Look up p-value from Table IX and use it to decide whether or not to reject the null hypothesis.

One-Tailed Test

Table IX gives a two-tailed p-value. If we have a sound nonstatistical reason why the result of the test must lie in only one tail, we can halve the tabulated value to obtain a one-tailed p-value.

Other Names for the Rank-Sum Test

The rank-sum test may be referred to in the literature also as the Mann–Whitney U test, Wilcoxon rank-sum test, or Wilcoxon–Mann–Whitney test. Mann and Whitney published what was thought to be one test and Wilcoxon published another test. Eventually, these tests were shown to be different forms of the same test.

ADDITIONAL EXAMPLE: HEMATOCRIT COMPARED FOR LAPAROSCOPIC VERSUS OPEN PYLOROMYOTOMIES

Among the indicators of patient condition following pyloromyotomy (correction of stenotic pylorus) in neonates is hematocrit (Hct) percentage. A surgeon[21] wants to compare average Hct value on 16 randomly allocated laparoscopic versus open pyloromyotomies. ($n_1 = n_2 = 8$.) A quick frequency plot shows the distributions to be far from normal; a rank method is appropriate. Data are as follows: open—46.7%, 38.8%, 32.7%, 32.0%, 42.0%, 39.0%, 33.9%, and 43.3%; laparoscopy—29.7%, 38.3%, 32.0%, 52.0%, 43.9%, 32.1%, 34.0%, and 25.6%. The surgeon puts the data in order and assigns ranks, obtaining the following data:

Hct (%)	Rank	Open versus laparoscopy
25.6	1	1
29.7	2	1
32.0	3.5	0
32.0	3.5	1
32.1	5	1
32.7	6	0
33.9	7	0
34.0	8	1
38.3	9	1
38.8	10	0
39.0	11	0
42.0	12	0
43.3	13	0
43.9	14	1
46.7	15	0
52.0	16	1

0, open; 1, laparoscopy.

The surgeon finds the rank sums to be 77.5 for open pyloromyotomy and 58.5 for laparoscopy. Because of symmetry, either may be chosen; $T = 77.5$. He calculates $U = n_1 n_2 + n_1(n_1 + 1)/2 - T = 8 \times 8 + 8(9)/2 - 77.5 = 22.5$. Table 16.2 is a segment of Table IX. From Table 16.2 (or Table IX), the p-value for $n_1 = n_2 = 8$ and $U = 22.5$ falls between 0.33 and 0.38, or about 0.335. The surgeon concludes that postoperative Hct percentage has not been shown to be different for open versus laparoscopic pyloromyotomy.

Table 16.2

Table 16.2

A Segment of Rank-Sum *U* Probabilities[a]

	n_2:					8			
	n_1:	1	2	3	4	5	6	7	8
>*U*	22						0.852	0.536	0.328
	23						0.950	0.612	0.382

[a]Two-tailed probabilities for the distribution of *U*, the rank-sum statistic. For two samples of size n_1 and $n_2 (n_2 > n_1)$ and the value of *U*, the entry gives the *p*-value.
Table IX in Tables of Probability Distributions provides an expanded version of this table.

Exercise 16.2: Does a Certain Tympanic Thermometer Measure the Same in Both Ears? We are investigating[63] the reliability of a certain brand of tympanic thermometer (temperature measured by a sensor inserted into the patient's ear). Sixteen readings (measured in degrees Fahrenheit [°F]), eight per ear, were taken on a healthy subject at intervals of 1 minute, alternating ears. Data were as follows:

Left ear:	95.8	95.4	95.3	96.0	96.9	97.4	97.4	97.1
Right ear:	98.1	95.8	97.5	97.2	97.7	99.3	99.2	98.1

Are the temperature readings for the two ears different?

16.4. THREE OR MORE INDEPENDENT SAMPLES: THE KRUSKAL–WALLIS TEST

EXAMPLE POSED: IS THE PROSTATE-SPECIFIC ANTIGEN LEVEL THE SAME AMONG PATIENTS WITH BENIGN PROSTATIC HYPERTROPHY, PROSTATE CANCER, AND NO EVIDENCE OF DISEASE?

A urologist asks whether the PSA level is different among three patient groups: six patients with benign prostatic hypertrophy (BPH), eight patients with positive biopsy results for prostate cancer, and eight patients with negative biopsy results and no evidence of disease.[33] The PSA levels (measured in nanogram per milliliter [ng/ml]) are as follows:

BPH	5.3	7.9	8.7	4.3	6.6	6.4		
Positive biopsy	7.1	6.6	6.5	14.8	17.3	3.4	13.4	7.6
Negative biopsy	11.4	0.5	1.6	2.3	3.1	1.4	4.4	5.1

A quick data plot shows the distributions to be nonnormal. We want to compare the three groups with a rank-order–based test.

METHOD: THE KRUSKAL–WALLIS TEST

Methodologically, the Kruskal–Wallis test is just the rank-sum test extended to three or more samples. It is used when rank-order data arise naturally in three or more groups or if the assumptions underlying the one-way analysis of variance test are not satisfied. The hypothesis being tested is whether or not the value for a randomly chosen member of one sample is probably smaller than one of another sample. For practical purposes, the user may think of it informally as testing whether the several distributions have the same median. The chi-square (χ^2) approximation is valid only if there are five or more members in each sample. The method is as follows:

1. Name the number of samples k $(3, 4, \ldots)$.
2. Name the sizes of each of the samples $n_1, n_2, \ldots, n_k$; n is the grand total.
3. Combine the data, keeping track of the sample from which each datum arose.
4. Rank the data.
5. Add the ranks from each sample separately, naming the sums $T_1, T_2, \ldots, T_k$.
6. Calculate the Kruskal–Wallis H statistic, which is distributed as chi-square, by

$$H = \frac{12}{n(n+1)} \left(\frac{T_1^2}{n_1} + \frac{T_2^2}{n_2} + \cdots + \frac{T_k^2}{n_k} \right) - 3(n+1). \qquad (16.1)$$

7. Obtain the p-value (looked up as if it were α) from Table III (χ^2 right tail) for $k - 1$ degree of freedom (df).

EXAMPLE COMPLETED: COMPARE PROSTATE-SPECIFIC ANTIGEN LEVELS FOR THREE DISEASE GROUPS

The number of groups and number of members within groups are $k = 3$, $n_1 = 6$, $n_2 = 8$, $n_3 = 8$, and $n = 22$. We rank Fall 22 data in order, keeping track of the sample each came from, as shown following this paragraph. The sums (totals) of ranks by group are $T_1 = 76.5$, $T_2 = 125.5$, and $T_3 = 51$. We calculate the Kruskal–Wallis H statistic as

$$H = \frac{12}{n(n+1)} \left(\frac{T_1^2}{n_1} + \frac{T_2^2}{n_2} + \frac{T_3^2}{n_3} \right) - 3(n+1)$$

$$= \frac{12}{22(23)} \left(\frac{5852.25}{6} + \frac{15750.25}{8} + \frac{2601}{8} \right) - 3(23) = 8.53.$$

Table 16.3

A Segment of Chi-square Distribution, Right Tail[a]

α (area in right tail)	0.10	0.05	0.025	0.01	0.005	0.001
$df = 2$	4.61	5.99	7.38	9.21	10.60	13.80
3	6.25	7.81	9.35	11.34	12.84	16.26
4	7.78	9.49	11.14	13.28	14.86	18.46

[a]Selected χ^2 values (distances above zero) for various df and for α, the area under the curve in the right tail.
Table III in Tables of Probability Distributions provides an expanded version of this table.

Table 16.3 is a segment of Table III (see Tables of Probability Distributions). In Table 16.3 (or Table III), the critical chi-square for $k - 1 = 2$ df for $\alpha = 0.05$ is 5.99. The p-value for $H = 8.53$ is calculated as 0.014. Because H is greater than the critical χ^2, and indeed the p-value is quite small, the urologist has sufficient evidence to conclude that the PSA levels are different among patients with BPH, patients with biopsy results positive for prostate cancer, and patients without BPH with negative biopsy results.

PSA (ng/ml)	Rank for patients with BPH	Rank for patients with positive biopsy results	Rank for patients with negative biopsy results
0.5			1
1.4			2
1.6			3
2.3			4
3.1			5
3.4		6	
4.3	7		
4.4			8
5.1			9
5.3	10		
6.4	11		
6.5		12	
6.6		13.5	
6.6	13.5		
7.1		15	
7.6		16	
7.9	17		
8.7	18		
11.4			19
13.4		20	
14.8		21	
17.3		22	
Total (T)	76.5	125.5	51

BPH, benign prostatic hypertrophy; PSA, prostate-specific antigen.

ADDITIONAL EXAMPLE: DO SURGICAL INSTRUMENTS FROM FIVE DIFFERENT MANUFACTURERS PERFORM DIFFERENTLY?

As part of an instrument calibration, we wish to compare $k = 5$ disposable current-generating instruments used in surgery to stimulate (and thereby help locate) facial nerves.[56] Among the variables recorded is current (measured in milliamperes). $n_1 = n_2 = n_3 = n_4 = n_5 = 10$; therefore, 10 readings are taken from each machine for a total of $n = 50$. A quick plot shows the data are clearly not normal, so a rank method must be used. The ordered current readings, separated by instrument from which each datum arose, and their corresponding ranks are as follows:

Instr- ument 1	Ranks 1	Instrument 2	Ranks 2	Instrument 3	Ranks 3	Instrument 4	Ranks 4	Instrument 5	Ranks 5
		1.90	1						
		1.95	2.5	1.95	2.5				
		1.96	6	1.96, 1.96	6×2	1.96	6	1.96	6
		1.97	9						
				1.98	12	1.98, 1.98	12×2	1.98, 1.98	12×2
2.00, 2.00	16.5×2			2.00	16.5			2.00	16.5
2.01	20.5	2.01	20.5	2.01	20.5	2.01	20.5		
2.03, 2.03	25×2	2.03	25					2.03, 2.03	25×2
2.04	29.5	2.04	29.5	2.04	29.5	2.04	29.5		
		2.06	33.5	2.06	33.5	2.06	33.5	2.06	33.5
2.07, 2.07	38×2	2.07	38	2.07	38			2.07	38
						2.08	41		
20.9	42								
						2.10	43		
						2.11	44.5	2.11	44.5
2.12	46								
		2.31	47						
				2.76	48				
						3.02	49		
								3.03	50
$T =$	297		212		212.5		291		262.5

We calculate the H statistic to obtain

$$
H = \frac{12}{n(n+1)} \left(\frac{T_1^2}{n_1} + \frac{T_2^2}{n_2} + \cdots + \frac{T_5^2}{n_5} \right) - 3(n+1)
$$

$$
= \frac{12}{50(51)} \left(\frac{88209}{10} + \frac{44944}{10} + \frac{45156.25}{10} + \frac{84681}{10} + \frac{68906.25}{10} \right) - 3(51)
$$

$$
= 3.187.
$$

From Table 16.3 (or Table III), the critical value of χ^2 for $\alpha = 0.05$ with 4 df is 9.49, which is much larger than 3.187; therefore, H_0 is not rejected. There is inadequate evidence to show the current to be different from instrument to instrument. Using a statistical software package, we find $p = 0.527$.

Exercise 16.3: Are Three Methods of Removing Excess Fluid in Infected Bursitis Different? An orthopedist compares an inflow–outflow catheter (treatment 1), needle aspiration (treatment 2), and incision-and-drain surgery (treatment 3).[91] Among the measures of effectiveness is posttreatment erythrocyte sedimentation rate (ESR; measured in millimeters per hour). The orthopedist uses each method on $n_1 = n_2 = n_3 = 10$ patients for a total of $n = 30$. Data are as follows:

Instrument 1	28	10	94	29	2	27	1	58	25	26
Instrument 2	64	30	30	11	30	9	9	30	9	25
Instrument 3	16	22	108	18	45	10	15	40	15	24

Draw rough frequency plots of ESRs for groups of 10 to show that they are not normal and that a rank method is appropriate. Use the Kruskal–Wallis method to test the null hypothesis that ESR is the same for all treatments.

16.5. THREE OR MORE MATCHED SAMPLES: THE FRIEDMAN TEST

EXAMPLE POSED: ARE TWO SKIN TESTS FOR SENSITIVITY TO AN ALLERGEN DIFFERENT?

An allergist applies three prick skin tests (two competing stimuli and a control) to the inside forearms, randomizing the order proximal to distal, to each of eight patients.[38] After 15 minutes, the allergist ranks the wheal-and-erythema reactions for each patient by severity (1 = most severe). The results (ranks) were as follows:

Patient no.	Test 1	Test 2	Control
1	1	2	3
2	2	1	3
3	1	3	2
4	3	2	1
5	1	2	3
6	1	2	3
7	2	1	3
8	1	2	3

Are the outcomes different?

METHOD: THE FRIEDMAN TEST

If three or more treatments are given to each patient of a sample, we have an extension of the paired-data concept. More exactly, it is called a randomized block design. In the example, a dermatologist applied three skin patches to each of eight patients to test for an allergy. The three skin test results for each patient is called a *block*. (In contrast, three groups of patients with a different skin test being used in each group would require the Kruskal–Wallis test.) The hypothesis being tested is that the several treatments have the same distributions. The chi-square approximation is valid only if there are five or more blocks (e.g., patients) in the sample. The method is as follows:

1. Name the number of treatments (3, 4, …) k and name the blocks (e.g., patients) n.
2. Rank the data within each block (e.g., rank the treatments for each patient).
3. Add the ranks for each treatment separately; name the sums T_1, T_2, …, T_k.
4. Calculate the Friedman F_r statistic, which is distributed as chi-square, by

$$F_r = \frac{12}{nk(k+1)} \left(T_1^2 + T_2^2 + \cdots + T_k^2 \right) - 3n(k+1). \qquad (16.2)$$

5. Obtain the p-value (as if it were α) from Table III (χ^2 right tail) for $k - 1$ *df.*

EXAMPLE COMPLETED: SKIN SENSITIVITY

The number of treatments is $k = 3$, and the number of blocks is $n = 8$. The rank sums for the three treatments are 12, 15, and 21. We calculate the Friedman statistic as

$$F_r = \frac{12}{nk(k+1)}(T_1^2 + T_2^2 + \cdots + T_k^2) - 3n(k+1)$$

$$= \frac{12}{8(3)(4)}(144 + 225 + 441) - 3(8)(4) = 5.25.$$

In Table 16.3 (or Table III), the critical chi-square for $k - 1 = 2$ *df* for $\alpha = 0.05$ is 5.99. Because F_r is less than critical χ^2, the allergist has insufficient evidence to infer any difference in efficacy between the two tests or that the tests are more efficacious than the control. The p-value is calculated as 0.072.

ADDITIONAL EXAMPLE: DOES THE SYSTEMIC LEVEL OF GENTAMICIN TREATMENT DECLINE OVER TIME?

In treating an infected ear, gentamicin, suspended in a fibrin glue, can be inserted into the middle ear, and some enters into the system. The time after application that it resides in the system is unknown. $n = 8$ chinchillas, which have large ears anatomically similar to those of humans, were used.[29] The serum gentamicin level (microgram per milliliter [µg/ml]) was measured at 8 hours, 24 hours, 72 hours, and 7 days after the administration. If the declining levels at these times test significantly different, an approximate "fade-out" time can be inferred. A quick plot showed the data per time grouping were far from normal; therefore, a rank method must be used. Because the data arose through time from the same animal, Friedman's test was appropriate. The data, listed and then ranked for each animal, were as follows:

Animal no.	8 hours Level	8 hours Rank	24 hours Level	24 hours Rank	72 hours Level	72 hours Rank	7 days Level	7 days Rank
1	482	3	877	4	0	1.5	0	1.5
2	124	3	363	4	66	2	0	1
3	280	3	1730	4	50	2	13	1
4	426	4	102	3	0	1.5	0	1.5
5	608	4	2	3	0	1.5	0	1.5
6	161	4	0	1.5	0	1.5	23	3
7	456	3	1285	4	48	2	0	1
8	989	4	189	2	378	3	0	1
$T =$		28		25.5		15		11.5
$T^2 =$		784		650.25		225		132.25

The results were substituted in Eq. (16.2) to obtain

$$F_r = \frac{12}{nk(k+1)}(T_1^2 + T_2^2 + \cdots + T_k^2) - 3n(k+1)$$

$$= \frac{12}{8 \times 4 \times 5}(1791.5) - 3 \times 8 \times 5 = 14.36.$$

From Table 16.3 (or Table III), the critical value of χ^2 for $\alpha = 0.05$ with $k - 1 = 3$ df is 7.81. F_r is much greater than 7.81, so the null hypothesis of equal levels over time was rejected. The serum gentamicin level declines over time. The p-value calculated from a statistical software package is 0.002.

Exercise 16.4: Do Posttherapy Prostate-Specific Antigen Levels in Patients with Prostate Cancer Remain Stable? $n = 9$ patients with prostate cancer had been without clinical evidence of disease 10 years after a negative staging pelvic lymphadenectomy and definitive radiation therapy.[36] Prostate-specific antigen levels were then measured in three successive intervals about a year apart. Data were as follows:

Patient no.	First PSA (ng/ml)	Second PSA (ng/ml)	Third PSA (ng/ml)
1	3.90	3.95	4.95
2	3.70	3.80	0.10
3	1.80	1.86	3.03
4	0.80	0.30	0.30
5	3.80	7.68	13.5
6	1.80	2.10	2.54
7	1.80	1.67	0.80
8	3.60	4.51	6.80
9	0.62	0.65	0.42

PSA, prostate-specific antigen level.

The data distributions are not normal; therefore, a rank test is appropriate. Test the hypothesis that PSA level is not changing through time.

16.6. SINGLE LARGE SAMPLES: NORMAL APPROXIMATION TO SIGNED-RANK TEST

EXAMPLE POSED: IS THE EARLY SAMPLING OF OUR 301 UROLOGIC PATIENTS BIASED RELATIVE TO THE LATER SAMPLING?

We ask if the PSA levels of the first 20 patients differ significantly from 8.96 ng/ml (the average of the remaining 281 patients). When the sample is large, it exceeds the tabulated probabilities, but may be approximated by the normal distribution, with μ and σ calculated by simple formulas.

METHOD: THE SIGNED-RANK TEST FOR LARGE SAMPLES

The normal approximation to the Wilcoxon signed-rank test tests the hypothesis that the distribution of differences has a median of zero. (The median and mean are the same in the normal distribution.) It may test (1) a set of observations deviating from a hypothesized common value or (2) pairs of observations on the same individuals, such as before and after data. The approximation is valid only if the sample is large enough, and the size of this "large enough" is not established. Like all large-sample approximations, the larger the sample is, the better is the agreement with exact tests. If reliable statistical software is available, it should be used to calculate the exact test. If such software is not available and the approximation is calculated manually, a minimum sample size may be taken as 16, a number stated in the classic textbook *Statistical Methods* by George Snedecor.[78] The *p*-value will not be identical with the exact method, but only rarely will this difference change the outcome decision. The steps in performing the test manually are as follows:

1. Calculate the differences between pairs or from a hypothesized central value.
2. Rank the magnitudes (i.e., the differences without signs).
3. Reattach the signs to the ranks.
4. Add up the positive and negative ranks.
5. Denote by T the unsigned value of the smaller sum of ranks; n is sample size (number of ranks).
6. Calculate $\mu = n(n + 1)/4$, $\sigma^2 = (2n + 1)\mu/6$, and then $z = (T - \mu)/\sigma$.
7. Obtain the *p*-value (as if it were α) from Table I (see Tables of Probability Distributions) for a two- or one-tailed test as appropriate.

EXAMPLE COMPLETED: FIRST 20 PROSTATE-SPECIFIC ANTIGEN LEVELS VERSUS REMAINING LEVELS

We compute the differences of the first 20 readings from 8.96, rank them by magnitude (i.e., regardless of sign), and find the sums of positive and negative ranks as in the following table. $T = 1$, the unsigned value of the smaller sum of ranks; $\mu = n(n + 1)/4 = 20(21)/4 = 105$; $\sigma^2 = (2n + 1)\mu/6 = 41(105)/6 = 717.5$; $\sigma = \sqrt{717.5} = 26.7862$; $z = (T - \mu)/\sigma = (1 - 105)/26.7862 = -3.8826$. A value of z this large is off the scale in Table I, indicating that $p < 0.001$; the early values are different. (Calculated exactly, a two-tailed $p = 0.0001$.) We have strong evidence that PSA levels from early sampling were different from later sampling. The claim of bias is credible.

PSA difference	Rank	Negative ranks	Positive ranks
−0.04	1	1	
0.06	2		2
0.96	3		3
1.06	4		4
1.26	5.5		5.5
1.26	5.5		5.5
1.36	7.5		7.5
1.36	7.5		7.5
2.16	9		9
2.63	10		10
2.86	11		11
3.06	12		12
3.26	13		13
3.36	14		14
3.66	15		15
4.16	16		16
4.36	17		17
4.56	18		18
4.86	19		19
7.66	20		20
Sums of ranks		1	209

ADDITIONAL EXAMPLE: DOES A CERTAIN TYPE OF HARDWARE INSTALLED IN BROKEN ANKLES RESTORE FUNCTIONALITY?

An orthopedist installs a certain type of hardware in ankle repair in $n = 19$ patients.[28] (This example is similar to the Additional Example in Section 16.2 but with a larger sample size.) The orthopedist scores the percent functionality of the joint. He asks: Is the average functionality as good as 90% of normal? Because it could be more or less, his test is two-sided. His data are 75%, 65%, 100%, 90%, 35%, 63%, 78%, 70%, 80%, 98%, 95%, 45%, 90%, 93%, 85%, 100%, 72%, 99%, and 95%. A quick frequency plot of the data show that they are far from normal; therefore, the orthopedist uses a rank test. Because he has more than 16 data, the normal approximation is adequate. He subtracts

90% from each (to test against a basis of zero), obtaining −15, −25, 10, 0, −55, −27, −12, −20, −10, 8, 5, −45, 0, 3, −5, 10, −18, 9, 5. He puts them in order and assigns ranks to them, ignoring signs. Then he attaches the signs to the ranks to obtain the signed ranks. These rankings may be seen in the following table. The sum of negative signed ranks (−55) is the smaller value; therefore, $T = 55. \mu = n(n+1)/4 = 19 \times 20/4 = 95; \sigma^2 = (2n+1)\mu/6 = 39 \times 95/6 = 617.5$; and $\sigma = 24.85$. Then, $z = (T - \mu)/\sigma = (55 - 95)/24.85 = -1.61$. The critical value of a 5% two-tailed α is the usual ± 1.96. Because the calculated z is inside these bounds, the orthopedist concludes that the postoperative patients' average ankle functionality is not different from 90%. Table 16.4 shows the relevant portion of Table I, which presents the normal distribution. From Table 16.4 (or Table I), interpolation gives $p = 0.108$.

Deviations ranked	Unsigned ranks	Signed ranks
−55	1	−1
−45	2	−2
−27	3	−3
−25	4	−4
−20	5	−5
−18	6	−6
−15	7	−7
−12	8	−8
−10	9	−9
−5	10	−10
0	11.5	11.5
0	11.5	11.5
3	13	13
5	14.5	14.5
5	14.5	14.5
8	16	16
9	17	17
10	18.5	18.5
10	18.5	18.5

Table 16.4

A Segment of Normal Distribution[a]

z (no. standard deviations to right of mean)	Two-tailed α (area in both tails)
0.60	0.548
0.70	0.484
1.60	0.110
1.70	0.090
1.80	0.072
1.90	0.054

[a]For selected distances (z) to the right of the mean, given are two-tailed α, the areas combined for both positive and negative tails.
Table I in Tables of Probability Distributions provides an expanded version of this table.

Exercise 16.5. In Exercise 16.2, $n = 8$ temperature readings were taken on a patient. The question was: Is the median different (implying a two-tailed test) from the population average of 98.6°F? As an exercise, answer the question using the normal approximation to the signed-rank test, even though the sample is smaller than appropriate. Data were 98.1°, 95.8°, 97.5°, 97.2°, 97.7°, 99.3°, 99.2°, and 98.1°F.

16.7. TWO LARGE SAMPLES: NORMAL APPROXIMATION TO RANK-SUM TEST

EXAMPLE POSED: IS THE PROSTATE-SPECIFIC ANTIGEN LEVEL FOR THE 10 PATIENTS REPORTED IN TABLE DB1.1 THE SAME AS THAT FOR PATIENTS WITH BENIGN PROSTATIC HYPERTROPHY?

We obtain a sample of 12 patients with BPH. We want to compare PSA levels for these patients with that for the 10 patients reported in Table DB1.1. $n_1 = 10$, and $n_2 = 12$. The data are as follows:

PSA from Table DB1.1 (ng/ml)	7.6	4.1	5.9	9.0	6.8	8.0	7.7	4.4	6.1	7.9		
PSA for patients with BPH (ng/ml)	5.3	7.9	8.7	4.3	6.6	6.4	20.2	8.5	6.5	6.5	12.5	7.1

BPH, benign prostatic hypertrophy; PSA, prostate-specific antigen level.

METHOD: THE RANK-SUM TEST FOR LARGE SAMPLES

This method is a normal approximation to the rank-sum test for larger sample sizes, in which "larger" is not well established. The larger the sample is, the better is the agreement with exact tests. If reliable statistical software is available, it should be used to calculate the exact test. If such software is not available and the approximation is calculated manually, a minimum sample size may be taken as 15, a number used in *Statistical Methods*.[78]

Given two samples, the hypothesis being tested is whether or not the value for a randomly chosen member of the first sample is probably smaller than one of the second sample. For practical purposes, the user may think of it informally as testing if the two distributions have similar centers. Steps to perform the test are as follows:

1. Name the sizes of the two samples n_1 and n_2; n_1 is the smaller sample.
2. Combine the data, keeping track of the sample from which each datum arose.
3. Rank the data.
4. Add up the ranks of the data from each sample separately.

> 5. Denote as T the sum associated with n_1.
> 6. Calculate $\mu = n_1(n_1 + n_2 + 1)/2$, $\sigma^2 = n_1 n_2(n_1 + n_2 + 1)/12$, and $z = (T - \mu)/\sigma$.
> 7. Obtain the p-value (as if it were α) from Table I for a two- or one-tailed test as appropriate.

EXAMPLE COMPLETED: PROSTATE-SPECIFIC ANTIGEN LEVELS FOR TWO GROUPS OF PATIENTS

Keeping track of the sample each datum came from, we rank all 22 data in order as follows:

PSA (ng/ml)	Rank for patients with BPH	Rank for patients reported in Table DB1.1
4.1		1
4.3	2	
4.4		3
5.3	4	
5.9		5
6.1		6
6.4	7	
6.5	8.5	
6.5		8.5
6.6	10	
6.8		11
7.1	12	
7.6		13
7.7		14
7.9	15.5	
7.9		15.5
8.0		17
8.5	18	
8.7	19	
9.0		20
12.5	21	
20.2	22	
Sum of ranks	147.5	105.5

BPH, benign prostatic hypertrophy; PSA, prostate-specific antigen level.

$T = 105.5$, the smaller rank sum; $\mu = n_1(n_1+n_2+1)/2 = 10(10+12+1)/2 = 115$; $\sigma^2 = n_1 n_2(n_1+n_2+1)/12 = 10(12)(10+12+1)/12 = 230$; $\sigma = \sqrt{230} = 15.1658$; $z = (T - \mu)/\sigma = (105.5 - 115)/15.1658 = -0.626$. From Table 16.4 (or Table I), $z = 0.60$ yields an area in both tails of 0.548, which is slightly larger than the p-value that would arise from a z of 0.626. (An exact calculation gives an area, i.e., p-value, of 0.532.) There is no evidence that the two samples have differing PSA levels.

ADDITIONAL EXAMPLE: DOES REMOVAL OF HARDWARE IN BROKEN ANKLE REPAIR ALTER JOINT FUNCTIONALITY?

An orthopedist installs hardware in in $n = 19$ broken ankles.[28] After healing, he removes the hardware in $n_1 = 9$ of the patients, randomly selected, and leaves it permanently in place in the remaining $n_2 = 10$. He judges the postoperative percent of functionality of the ankle joint and asks: Is the average functionality different for the two treatments? Because either one might be the better, his test is two-sided. His data are as follows: remove—75%, 65%, 100%, 90%, 35%, 63%, 78%, 70%, and 80%; retain—98%, 95%, 45%, 90%, 93%, 85%, 100%, 72%, 99%, and 95%. A quick frequency plot of the data shows that they are far from normal; therefore, he uses a rank test. Although the sample sizes are rather small for the approximation, we will use the large sample method and note that its outcome agrees with the small sample method. His data, put in order with ranks assigned, become

Functionality (%)	Rank	Removed
35	1	Yes
45	2	No
63	3	Yes
65	4	Yes
70	5	Yes
72	6	No
75	7	Yes
78	8	Yes
80	9	Yes
85	10	No
90	11.5	Yes
90	11.5	No
93	13	No
95	14.5	No
95	14.5	No
98	16	No
99	17	No
100	18.5	Yes
100	18.5	No

By summing the smaller group ("yes"), he finds $T = 67$. $\mu = n_1(n_1 + n_2 + 1)/2 = 9(9 + 10 + 1)/2 = 90$; $\sigma^2 = n_1 n_2(n_1 + n_2 + 1)/12 = 150$; and $\sigma = 12.25$. Then, $z = (67 - 90)/12.25 = -1.88$. From Table 16.4 (or Table I), the critical value of a 5% two-tailed test is ± 1.96. Because the calculated z is just inside these bounds, the orthopedist must conclude that there is inadequate evidence to show that leaving or removing the hardware provides a different percent of functionality. From a computer package, the normal p-value and that of the software-based rank-sum test are both 0.060.

Exercise 16.6. In Exercise 16.2, $n_1 = n_2 = 8$ temperature readings; each was taken from a patient's left and right ears.[63] The question was: Are the distribution centers for the two ears different (implying a two-tailed test)?

As an exercise, answer the question using the normal approximation to the rank-sum test, even though the sample size is small for the approximation. Is the resulting *p*-value close to that for the exact method? Data were as follows:

Left ear	95.8	95.4	95.3	96.0	96.9	97.4	97.4	97.1
Right ear	98.1	95.8	97.5	97.2	97.7	99.3	99.2	98.1

ANSWERS TO EXERCISES

16.1. The data differences (98.6 − each datum) are 0.5, 2.8, 1.1, 1.4, 0.9, −0.7, −0.6, and 0.5. The unsigned ranks are 1.5, 8, 6, 7, 5, 4, 3, and 1.5; the signed ranks are 1.5, 8, 6, 7, 5, −4, −3, and 1.5. The smaller rank sum is 7. From Table VIII, the probability that a signed rank sum is 7, given $n = 8$, is 0.149. This *p*-value is too large to infer a difference; the patient's median temperature is not shown to be different from 98.6°F.

16.2. Rank the combined data, keeping track of which ear gave rise to each datum. The ranks are 3.5, 2, 1, 5, 6, 9.5, 9.5, 7, 13.5, 3.5, 11, 8, 12, 16, 15, and 13.5. The sum of the first eight ranks for the left ear is 43.5; the sum of the remainder for the right ear is 92.5. Because the *n*'s are the same, either *n* could be designated n_1. The associated *U*-values are $U = n_1 n_2 + n_1(n_1+1)/2 - 43.5 = 56.5$ and $U = n_1 n_2 + n_1(n_1+1)/2 - 92.5 = 7.5$. Because 56.5 is off the table, we choose $U = 7.5$. (The test is symmetric.) In Table IX, for $n_1 = n_2 = 8$, $U = 7.5$, designates a *p*-value lying halfway between 0.006 and 0.010 (i.e., 0.008). The data provide strong evidence that the temperature readings for the two ears tend to be different.

16.3. Frequency plots per group are as follows:

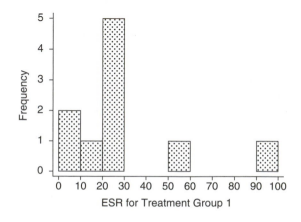

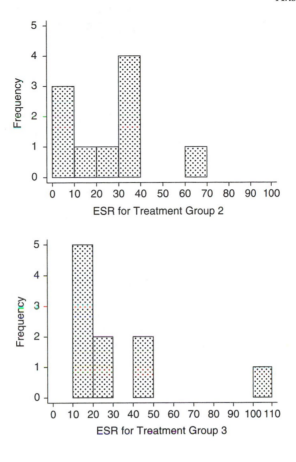

Erythrocyte sedimentation rate values, separated by treatment from which each datum arose, and their corresponding ranks are as follows:

ESR 1	Ranks 1	ESR 2	Ranks 2	ESR 3	Ranks 3
1	1				
2	2				
		9, 9, 9	4 × 3		
10	6.5			10	6.5
		11	8		
				15, 15	9.5 × 2
				16	11
				18	12
				22	13
				24	14
25	15.5	25	15.5		
26	17				
27	18				
28	19				
29	20				

(Continued)

301

ESR 1	Ranks 1	ESR 2	Ranks 2	ESR 3	Ranks 3
		30, 30, 30, 30	22.5×4		
				40	25
				45	26
58	27				
		64	28		
94	29				
				108	30
$T =$	155		153.5		156.5
$T^2 =$	24025		23562.25		24492.25

$$H = \frac{12}{n(n+1)} \left(\frac{T_1^2}{n_1} + \frac{T_2^2}{n_2} + \frac{T_3^2}{n_3} \right) - 3(n+1)$$

$$= \frac{12}{30 \times 31} \left(\frac{24025}{10} + \frac{23562.25}{10} + \frac{24492.25}{10} \right) - 3 \times 31 = 0.0058$$

From Table III, the 5% critical value for 2 df is 5.99, which is much larger than H; ESR is not different for the three treatments. From a statistical software package, $p = 0.997$.

16.4. The ranks for each patient according to PSA levels, with rank sums at the bottom, are as follows:

	Rank		
Patient no.	First PSA	Second PSA	Third PSA
1	1	2	3
2	2	3	1
3	1	2	3
4	3	1.5	1.5
5	1	2	3
6	1	2	3
7	3	2	1
8	1	2	3
9	2	3	1
T	15	19.5	19.5
T^2	225	380.25	380.25

$$F_r = \frac{12}{nk(k+1)}(T_1^2 + T_2^2 + \cdots + T_k^2) - 3n(k+1)$$

$$= \frac{12}{9(3)(4)}(225 + 380.25 + 380.25) - 3(9)(4) = 1.50$$

From Table III, the critical chi-square for $k - 1 = 2$ df for $\alpha = 0.05$ is 5.99. Because $1.50 < 5.99$, there is no evidence that PSA level is increasing over time in this population. $p = 0.472$.

16.5. Follow the answer to Exercise 16.1 until the smaller rank sum $T = 7$ is found. Calculate $\mu = n(n + 1)/4 = 8 \times 9/4 = 18$; $\sigma^2 = (2n + 1)\mu/6 = 17 \times 18/6 = 51$; and then $z = (T - \mu)/\sigma = (7 - 18)/\sqrt{51} = -1.54$. From Table I, the two-tailed α lies between 0.134 and 0.110, or 0.124 by interpolation. We do not have enough evidence to reject the null hypothesis of no difference. Note that use of Table VIII yielded $p = 0.149$. The discrepancy is due to too small a sample for a good approximation.

16.6. Follow the answer to Exercise 16.2 until the rank sums $T = 43.5$ or 92.5 are found. Calculate $\mu = n_1(n_1 + n_2 + 1)/2 = 8 \times 17/2 = 68$; $\sigma^2 = n_1 n_2(n_1 + n_2 + 1)/12 = 8 \times 8 \times 17/12 = 90.67$; $\sigma = 9.52$; and then $z = (T - \mu)/\sigma = (43.5 - 68)/9.52 = -2.57$. (The other $T = 92.5$ yields $+2.57$; note the symmetry.) From Table I, $p = 0.010$. The result differs little from $p = 0.008$ in Table IX.

Chapter 17

Tests on Means of Continuous Data

17.1. SUMMARY OF MEANS TESTING

THE BASIC QUESTION BEING ASKED

Is a sample mean different from a population mean, or, alternatively, do two or more samples have different population means? The question is answered by testing the null hypothesis that the means are equal and then rejecting or failing to reject this hypothesis. Sections 5.1 and 5.2 discuss the concept of a hypothesis test. We also could ask if the means are the same, rather than are they different. This question falls under the heading of equivalence testing and is addressed in Chapter 21.

ASSUMPTIONS UNDERLYING A HYPOTHESIS TEST

Tests of means using continuous data were developed under the assumptions that (1) the sample observations are independent from each other and (2) they were drawn from a normal distribution. In addition, when two sample means are being tested, the hypothesis that they arose from the same distribution implies (3) equal standard deviations. These assumptions are discussed in Section 6.7. Also discussed in Section 6.7 is the property of *robustness*, that is, how impervious the test is to violations of the assumptions. Although these assumptions usually are not satisfied exactly, the robustness of the test allows it to be valid if the assumptions are roughly approximated. Assumption (1) is essential for all types of test; if it is badly violated, the results of any test will be spurious. If assumption (2) is badly violated, a valid test may still be made using the rank methods of Chapter 16. If assumption (3) is violated in a test of two means, the user should apply either an unequal-variance form of the means test or a rank test. Section 17.4 provides further guidance on this topic.

WE MUST SPECIFY THE NULL AND ALTERNATE HYPOTHESES

The null hypothesis states that the mean of the population from which the sample is drawn is not different from a theorized mean or from the population mean of another sample. The alternate hypothesis may claim that the two means are not equal or that one is greater than the other. The form of the alternate hypothesis is discussed in Section 6.7. We should specify the hypotheses before seeing the data so that our choice will not be influenced by the outcome.

17.2. NORMAL (*z*) AND *t* TESTS FOR SINGLE OR PAIRED MEANS

SINGLE SAMPLES AND PAIRED SAMPLES ARE TREATED THE SAME

Single and paired samples are treated by the same method, because we may create a single observation from a pair by subtraction, for example, the after minus the before measurement, or a patient's response to drug 1 minus the response to drug 2.

EXAMPLES POSED

Normal Test: Does the Average Beginning Member of Our Prostate-Specific Antigen Sample Differ from the Average Later Member?

We ask if the mean prostate-specific antigen (PSA) level ($m = 6.75$) of the 10 patients of Table DB1.1 is different from that of the remaining 291 patients, a sample large enough for its sample mean and standard deviation to be treated as if they were population mean and standard deviation. Thus, $\mu = 8.85$ and $\sigma_m = \sigma/\sqrt{n} = 17.19/\sqrt{291} = 1.01$. Is m different from μ?

t Test: Does Asthma Training of Pediatric Patients Reduce Acute Care Visits?

We include all asthma patients satisfying inclusion criteria presenting over a period of time (in this case, $n = 32$) and record the number of acute care visits during a year.[20] We then provide them a standardized course of asthma training and record the number of acute care visits for the following year. These "before and after" data allow us to analyze the change per patient: d (named for "difference" or "delta") designates the number of visits before minus the number of visits after training. H_0: $\mu_d = 0$. What is the alternate hypothesis? We would *expect* the training to reduce the number of visits, but we are not *certain*. Perhaps the training will increase the child's awareness and fear, causing an increase in visits; this is unlikely, but we cannot rule it out. Therefore, we cannot in good faith claim that the error can lie in only one direction, so we must use a two-tailed test. H_1: $\mu_d \neq 0$.

METHOD: THE ONE-SAMPLE/PAIRED-SAMPLE *z* AND *t* TESTS

We want to test the hypothesis that the mean of the distribution from which we draw our sample, denoted μ_0, is the same as the known (theoretical) mean μ; or H_0: $\mu_0 = \mu$. (If we are dealing with differences,

---

e.g., before minus after, μ_0 may be denoted μ_d.) The alternate hypothesis may be that μ_0 is greater than or is less than μ, giving a one-tailed test, or that μ_0 is either, giving a two-tailed test; these options are, respectively, H_1: $\mu_0 > \mu$, H_1: $\mu_0 < \mu$, or H_1: $\mu_0 \neq \mu$. We must assume that the basic data are distributed in a roughly normal distribution. Choose an appropriate α. The test is a normal z test or a t test in the form of a standardized mean. When the standard error of the mean σ_m is known theoretically or the sample is large enough that s_m is close to σ_m (e.g., >50 or >100), use z, calculated as in Eq. (17.1) and Table I (see Tables of Probability Distributions):

$$z = \frac{m - \mu}{\sigma_m} = \frac{m - \mu}{\sigma/\sqrt{n}}. \tag{17.1}$$

When σ_m is unknown and therefore is estimated by s_m, and n is smaller than the preceding guidance, use t, calculated as in Eq. (17.2) and Table II (see Tables of Probability Distributions), with $n - 1$ degrees of freedom (df):

$$t = \frac{m - \mu}{s_m} = \frac{m - \mu}{s/\sqrt{n}}. \tag{17.2}$$

Follow these steps for either test:

1. Specify null and alternate hypotheses and choose α.
2. Make a quick, informal frequency plot of the basic data to check for normal shape.
3. Look up the critical value in the appropriate table for the chosen α.
4. Calculate the appropriate statistic from Eq. (17.1) or (17.2).
5. Make the decision to reject or not reject the null hypothesis.

EXAMPLES COMPLETED

Normal Test: Average Beginning Prostate-Specific Antigen Versus Average Later Prostate-Specific Antigen Level

The distribution is similar to that shown in Fig. 2.3 for all 301 patients, which is close enough to normal in shape for the test to be valid. The null hypothesis is that m is drawn from the population having mean μ, so m and μ should be different only by random influence. We have no reason to anticipate whether m should be larger or smaller than μ; therefore, we use a two-tailed test. The null hypothesis is tested by the normal statistic $z = (m - \mu)/\sigma_m = (6.75 - 8.85)/1.01 = -2.08$. Because we are concerned with how far z is from zero regardless of the direction, normal curve symmetry allows us to look up +2.08 in the table. Table 17.1 is a portion of Table I. In either table, the two-tailed p-value (looked up as if it were α) for this z is slightly greater than 0.036 (0.037 from exact calculation). The chance that the difference between m and μ occurred by a random influence is less than 1 in 25; we have evidence to conclude that they are different.

Table 17.1

A Segment of the Normal Distribution[a]

z (no. standard) deviations to right of mean	Two-tailed α (area in both tails)
1.90	0.054
1.960	*0.050*
2.00	0.046
2.10	0.036
2.30	0.022
2.326	*0.020*
2.40	0.016

[a]For selected distances (z) to the right of the mean, given are two-tailed α values, the areas combined for both tails. Entries for the most commonly used areas are in italics.
Table I in Tables of Probability Distributions provides an expanded version of this table.

Table 17.2

A Segment of the *t* Distribution[a]

Two-tailed α	0.10	0.05	0.02	0.01	0.002	0.001
df = 11	1.796	2.201	2.718	3.106	4.025	4.437
14	1.761	2.145	2.624	2.977	3.787	4.140
26	1.706	2.056	2.479	2.779	3.435	3.707
30	1.697	2.042	2.457	2.750	3.385	3.646
40	1.684	2.021	2.423	2.704	3.307	3.551
100	1.660	1.984	2.364	2.626	3.174	3.390
∞	1.645	1.960	2.326	2.576	3.090	3.291

[a]Selected distances (t) to the right of the mean are given for various degrees of freedom (*df*) and for two-tailed α, areas combined for both positive and negative tails.
Table II in Tables of Probability Distributions provides an expanded version of this table.

t Test: Does Asthma Training of Pediatric Patients Reduce Acute Care Visits?

A plot of the data shows an approximately normal shape, satisfying that assumption. The 32 *d*-values were: 1, 1, 2, 4, 0, 5, −3, 0, 4, 2, 8, 1, 1, 0, −1, 3, 6, 3, 1, 2, 0, −1, 0, 3, 2, 1, 3, −1, −1, 1, 1, and 5. $m_d = 1.66$, and $s_d = 2.32$. $t = (m_d - \mu_d)/s_m = (1.66 - 0)/(2.32/\sqrt{32}) = 1.66/0.41 = 4.05$. Table 17.2 is a portion of Table II (see Tables of Probability Distributions). In either table, look under the two-tailed α column for $df = 30$ to find the critical value of t; our $df = 31$ will yield a probability just less than the tabulated value, or about 2.04. Because 4.05 is greater than 2.04, and in fact greater than the critical 3.64 for $\alpha = 0.001$, we can state that there is less than a 1 in 1000 chance of being wrong if we conclude that the asthma training was efficacious.

ADDITIONAL EXAMPLE: IS A NEW DYSPEPSIA TREATMENT EFFECTIVE IN THE EMERGENCY DEPARTMENT?

Normal (z) Test

An emergency medicine physician wants to test the effectiveness of a "GI cocktail" (antacid plus viscous lidocaine) to treat emergency dyspeptic symptoms as measured on a 1–10 pain scale.[50] H_0: $\mu = 0$. She suspects that the treatment will not worsen the symptoms, but is not totally sure, so she uses H: $\mu \neq 0$, implying a two-sided test. She chooses $\alpha = 0.05$. She decides to accept as the population standard deviation that for scoring of a large number of patients without treatment, $\sigma = 1.73$. She samples $n = 15$ patients, measuring the difference in pain before and after treatment. Data are 6, 7, 2, 5, 3, 0, 3, 4, 5, 6, 1, 1, 1, 8, and 6. $m = 3.87$. The physician substitutes in Eq. (17.1) to find $z = (m - \mu)/\sigma_m = (3.87 - 0)/(1.73/\sqrt{15}) = 8.66$. Because the critical value from Table 17.1 (or Table I) is 1.96, which is much less than 8.66, she rejects H_0 and concludes that the treatment is effective. The actual p-value is 0 to more than three decimal places and thus is stated as $p < 0.001$.

t Test

Suppose the standard deviation came from a small sample instead of a large sample. If the physician had decided not to use $\sigma = 1.73$ because it arose from untreated patients, she would have estimated the standard deviation from the data as $s = 2.50$ and would have used the t test. From Table 17.2 (or Table II), the critical value for a two-tailed t with 14 df is 2.145. Substituting in Eq. (17.2), she finds $t = (m - \mu)/s_m = (3.87 - 0)/(2.50/\sqrt{15}) = 6.00$. The conclusion is the same as that for large-sample test statistics.

Exercise 17.1: The Original Example That "Student" Used for His t In W. S. Gossett's classic 1908 article[80] that introduced the t test, he used the following data: Two soporific drugs were administered in turn to 10 patients with insomnia, and the number of hours of additional sleep each drug provided were recorded. Data were as follows:

Patient no.	Dextro	Laevo	Difference (d)
1	0.7	1.9	1.2
2	−1.6	0.8	2.4
3	−0.2	1.1	1.3
:	:	:	:
10	2.0	4.3	1.4
Mean	0.75	2.33	1.58
Standard deviation			1.23

Values in the second and third columns represent the amount (measured in hours) of additional or less sleep that the drug provided.

We denote the unknown population's mean of differences as δ, estimated from the sample by the mean of d. At the $\alpha = 0.05$ level of significance, test H_0: $\delta = 0$ against H_1: $\delta \neq 0$.

17.3. *POST HOC* CONFIDENCE AND POWER

POSTTEST CONFIDENCE IN THE ONE-SAMPLE MEAN

We have done our test on a mean and reached a conclusion. What is our confidence interval on this mean? It does not affect the test result, but it may provide additional understanding of the sizes of parameters involved. For known standard error of the mean difference σ_m, our test statistic $z = (\mu - m)/\sigma_m$ (or its negative; the distribution is symmetric) is the number of standard errors apart that the unknown theoretical and the observed means are. Using the logic of Chapter 14, we may write this relationship as $P[-z_{1-\alpha/2} < (\mu - m)/\sigma_m < z_{1-\alpha/2}] = 1 - \alpha$. (For example, $P[-1.96 < (\mu - m)/\sigma_m < 1.96] = 0.95$.) By adding m and then multiplying by σ_m throughout within the brackets, we achieve a form similar to Eq. (14.3):

$$P[m - z_{1-\alpha/2}\sigma_m < \mu < m + z_{1-\alpha/2}\sigma_m] = 1 - \alpha \qquad (17.3)$$

Equation 17.3 provides the confidence interval on the unknown theoretical mean. If we estimate σ_m by s_m, we can use similar logic to find the confidence interval on μ for that case as

$$P[m - t_{1-\alpha/2}s_m < \mu < m + t_{1-\alpha/2}s_m] = 1 - \alpha. \qquad (17.4)$$

For example, in the asthma training example, by substitution in Eq. (17.4), we find

$$P[1.66 - 2.04 \times 0.41 < \mu_d < 1.66 + 2.04 \times 0.41] = P[0.82 < \mu_d < 2.50] = 0.95.$$

We are 95% sure that training will reduce the average number of acute care visits for asthma by a little less than 1 visit to 2.5 visits. (We knew the confidence interval would not cross zero because the test showed significance.)

POWER OF THE TEST

The probability of a false-positive outcome (concluding significance when it is not there), or α, selects the critical value demarking rejection versus nonrejection regions for our test statistic. For example, a two-tailed test of a normally distributed variable yields critical values of ± 1.96 for $\alpha = 0.05$. This selection occurs *a priori*, that is, before the test is made (and we hope before the data are even gathered). At that point, the power of the test, that is, $1 - \beta$, is unknown, because β depends on the value of the alternate mean, which is unknown. However, after a significant test result, the observed mean is known and may be taken as the alternate-hypothesis mean. A power calculated *a posteriori*, more frequently referred to as *post hoc*, can be obtained.

CALCULATING THE *POST HOC* POWER

Recall that α was obtained as the area under the probability curve defined by the null hypothesis to the right of the critical value. To calculate the *post hoc* power, we just substitute the observed mean for the null-hypothesized mean, standardize the distribution, and find the area under the curve to the right of the critical value. For example, continuing the asthma training test, our t statistic was calculated to be 4.05. To standardize the distribution, we move the center, or mean, 4.05 units to the left to make it zero. This, of course, also moves the critical value, previously at 2.04, 4.05 units to the left, so it becomes -2.01. The power of the test is the area under a standard t with 31 df to the right of -2.01. Recall that power $= 1 - \beta$; it is easier to find β, which is the area to the left of -2.01, or, by symmetry, to the right of $+2.01$. The closest right tail area in Table II for the right tail area (top row) is 0.025, which is associated with t-value 2.04 for 30 df. A t of 2.01 for 31 df would be fairly close. Therefore, the power is close to $1 - 0.025 = 0.975$. From a statistical software package, the *post hoc* power $= 0.973$.

INTERPRETING THE *POST HOC* POWER

What does this *post hoc* power tell us? It adds nothing to the test, for that is complete. Because the power increases with sample size, it does add to our intuitive assurance that the sample is large enough for us to rely on the test result.

APPLYING TO OTHER TYPES OF TEST

The ideas of a *post hoc* confidence interval and *post hoc* power extend generally to other tests. These extensions are obvious in some cases, for example, to the tests addressed in this section, but they are not always so easy to identify. In some cases, the theory has never been detailed.

Exercise 17.2: What Was the Power of "Student's" Original t Test?   Follow the steps to calculate the *post hoc* power for the result of Exercise 17.1.

17.4. NORMAL (z) AND t TESTS FOR TWO MEANS

A TEST OF MEANS OF TWO SAMPLES

The particular form of the test to contrast two means depends on the relationship of the variances, which are pooled to obtain an overall estimate of variability. Unequal variances are pooled differently from equal variances. Fortunately, the tests are rather robust against differing variances, provided the sizes of the two samples are about the same. If sample sizes are quite different or if one variance is more than double the other, the variances should be tested for equality before making the means test. Section 19.3 describes methods for testing the equality

Table 17.3

Guide for Selecting an Appropriate Two-Sample Means Test

Total sample size	Subgroup sample size	Variances about equal	Variances somewhat different	Variances extremely different
Large size *or σ's* known	About equal	Normal (z) test, equal variances	Normal (z) test, equal variances	Rank-sum test
	Very different		Normal (z) test, unequal variances	Rank-sum test
Small size	About equal	t test, equal variances	t test, equal variances	Rank-sum test
	Very different		t test, unequal variances	Rank-sum test

of variances. The logic for selecting which form of means test to use appears in Table 17.3.

FORMAT OF THIS SECTION

This section presents the tests of Table 17.3, except for the rank-sum test, in the following order: normal (z) test, equal variances; t test, equal variances; normal (z) test, unequal variances; and t test, unequal variances. The rank-sum test is also discussed in detail in Sections 6.6 and 16.3.

DISAGREEMENT REGARDING THE LOGIC OF TABLE 17.3

Some controversy exists in the field of statistics about using unequal-variance normal and t methods. Although they are approximately correct and acceptably usable, some statisticians would prefer to use the rank-sum test whenever variances are unequal.

ASSUMPTIONS REQUIRED

As discussed at the beginning of this chapter, the two-sample means test requires the assumptions that data sampled are independent one from another and that their frequency distributions are approximately normal, that is, roughly bell-shaped.

EXAMPLES POSED

Normal (z) Test, Equal Variances: Is Age a Risk Factor for Prostate Cancer?

Whether or not age is a risk factor for prostate cancer is addressed in Section 6.8, which exemplifies the method of the z test with equal variances. The user is encouraged to review this example.

t Test, Equal Variances: Continuation of This Example

Also addressed in Section 6.8 is the effect of using the *t* test rather than the *z* test of means. The user should continue the review through this example.

Normal (z) Test, Unequal Variances: Is Prostate Volume a Risk Factor for Prostate Cancer?

We omit the five patients with known benign prostatic hypertrophy, because we want a population at risk for prostate cancer without other prostate disease. We ask if the mean volume for $n_1 = 201$ patients with negative biopsy results is different from that for $n_2 = 95$ patients with positive biopsy results. The sample sizes are large enough to use methods for normal instead of *t*. H_0: $\mu_1 = \mu_2$. Because it is theoretically possible for either mean to be the larger, H_1: $\mu_1 \neq \mu_2$. Plots of the frequency distributions appear in Fig. 17.1 with normal curves fitted. The shape for negative biopsy results (see Fig. 17.1A) is a little skewed to the right, whereas that for positive biopsy results seems adequately symmetric. However, let us assume for the moment that they are not unreasonably deviant from normal. The descriptive data are as follows:

Biopsy results	Sample size	Mean volume (ml)	Standard deviation (ml)
Negative	201	36.85	17.33
Positive	95	32.51	13.68

The sample sizes are quite different, and standard deviations are somewhat different. Using the methods of Section 19.3, the variances test significantly different with $p = 0.008$. The method for unequal variances is appropriate.

t Test, Unequal Variances: Using the *t* Test on the Question of Prostate Volume as a Risk Factor in Prostate Cancer

Would the *t* test of means have been appropriate in the previous example? Yes; for large samples, the normal and *t* tests are negligibly different (see Section 6.8 for a more detailed discussion of this topic). We have the same hypotheses and normality assumptions.

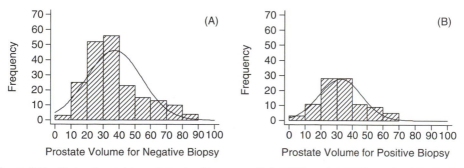

Figure 17.1 Plots of prostate volume by biopsy result for 296 patients without benign prostatic hypertrophy.

METHOD: TESTS BETWEEN MEANS FROM TWO SAMPLES

We have two samples of size n_1 and n_2. H_0: $\mu_1 = \mu_2$. Select the alternate as H_1: $\mu_1 \neq \mu_2$ (two-tailed α), or H_1: $\mu_1 < \mu_2$ or H_1: $\mu_1 > \mu_2$ (one-tailed α). We must assume that the basic data are independent from each other and are distributed roughly normal in shape. Choose an appropriate α. The test is a normal (z) test or a t test in the form of a standardized difference between means, that is, $m_1 - m_2$ divided by the standard error of this difference between means. It is this standard error that differs in the equal versus unequal standard deviation cases.

Normal (z) Test

m_1 and m_2 are the sample means. The samples have known σ's or are large enough for the s's to be close to the σ's. Calculate z using Eq. (17.5) with σ_d found from Eq. (17.6) or (17.7), as appropriate, and find the p-value from Table I.

$$z = \frac{m_1 - m_2}{\sigma_d} \tag{17.5}$$

When variances are taken as equal, use:

$$\sigma_d = \sigma\sqrt{\frac{1}{n_1} + \frac{1}{n_2}}. \tag{17.6}$$

When variances are taken as unequal, use:

$$\sigma_d = \sqrt{\frac{\sigma_1^2}{n_1} + \frac{\sigma_2^2}{n_2}}. \tag{17.7}$$

t Test

When the σ's are unknown and the n values are small (e.g., <50 or <100), calculate t using Eq. (17.8) with s_d found from Eq. (17.9) or (17.10), as appropriate, and find the p-value from Table II.

$$t = \frac{m_1 - m_2}{s_d} \tag{17.8}$$

When variances are taken as equal, use:

$$s_d = \sqrt{\left(\frac{1}{n_1} + \frac{1}{n_2}\right)\left[\frac{(n_1 - 1)s_1^2 + (n_2 - 1)s_2^2}{n_1 + n_2 - 2}\right]}, \tag{17.9}$$

with $df = n_1 + n_2 - 2$. When variances are taken as unequal, use:

$$s_d = \sqrt{\frac{s_1^2}{n_1} + \frac{s_2^2}{n_2}}, \tag{17.10}$$

with *df* rounded to the next smallest integer less than the rather peculiar expression of Eq. (17.11):

$$\text{approx. } (df) = \frac{\left(\dfrac{s_1^2}{n_1} + \dfrac{s_2^2}{n_2}\right)^2}{\dfrac{\left(\dfrac{s_1^2}{n_1}\right)^2}{n_1 - 1} + \dfrac{\left(\dfrac{s_2^2}{n_2}\right)^2}{n_2 - 1}} \tag{17.11}$$

STEPS TO FOLLOW IN TESTING

Follow these steps for either the *z* or the *t* test:

1. Specify null and alternate hypotheses and choose α.
2. Plot frequencies of the two samples' basic data to check for normality.
3. In light of assumption about variances, choose the appropriate form from Table 17.3. If using *t*, calculate *df*.
4. Look up the critical value in the appropriate table for the chosen α.
5. Calculate the statistic from Eq. (17.5) or (17.8), as appropriate, that will include the appropriate standard error calculation.
6. Make the decision to reject or not reject the null hypothesis.

EXAMPLES COMPLETED

z and *t*, Equal Variances: Age as a Risk Factor for Prostate Cancer

The examples testing age as a risk factor for prostate cancer have been completed in Section 6.8.

z, Unequal Variances: Prostate Volume as a Risk Factor for Prostate Cancer

From Eq. (17.7),

$$\sigma_d = \sqrt{\frac{\sigma_1^2}{n_1} + \frac{\sigma_2^2}{n_2}} = \sqrt{\frac{17.33^2}{201} + \frac{13.68^2}{95}} = 1.86,$$

so

$$z = (m_1 - m_2)/\sigma_d = (36.85 - 32.51)/1.86 = 2.33.$$

From Table 17.1 (or Table I), $z = 2.33$ corresponds to $p = 0.020$, which is statistically significant. We have evidence that the mean prostate volume for patients with negative biopsy results is larger than that for those with positive biopsy results.

t, Unequal Variances: Prostate Volume as a Risk Factor for Prostate Cancer

From Eq. (17.9),

$$s_d = \sqrt{\frac{s_1^2}{n_1} + \frac{s_2^2}{n_2}} = \sqrt{\frac{17.33^2}{201} + \frac{13.68^2}{95}} = 1.86,$$

so

$$t = (m_1 - m_2)/s_d = (36.85 - 32.51)/1.86 = 2.33,$$

which are the same standard error and standardized difference as for z. However, we must calculate df to find the p-value. From Eq. (17.11),

$$df \approx \frac{(s_1^2/n_1 + s_2^2/n_2)^2}{\dfrac{(s_1^2/n_1)^2}{n_1 - 1} + \dfrac{(s_2^2/n_2)^2}{n_2 - 1}} = \frac{(17.33^2/201 + 13.68^2/95)^2}{\dfrac{(17.33^2/201)^2}{200} + \dfrac{(13.68^2/95)^2}{94}} = 228.8,$$

which rounds to $df = 228$. From Table 17.2 (or Table II), $t = 2.33$ for 228 df corresponds to $p = 0.02$, as with the z test. A calculation on a computer to three decimal places yields $p = 0.021$, which is negligibly larger than that from the z test.

Suppose We Had Used the Rank-Sum Test to Assess Prostate Volume

Because the positive biopsy distribution was somewhat skewed, our underlying assumptions are questionable. We might better have used the rank-sum test from Section 6.6 or 16.3, which does not require assumptions of symmetry (or for that matter of equal variances). On doing so, the p-value turns out to be 0.097, which is *not* significant, changing our conclusion. Apparently, the positive biopsy distribution was too skewed for our assumption of symmetry. The median is a better measure of central tendency in skewed distributions, and the rank-sum test is closer to a test of medians than of means. Recall that, in a skewed distribution, the mean is "pulled" toward the skew. The mean of the patients with skewed negative biopsy results is pulled farther to the right than is the mean of the patients with the less-skewed positive biopsy results, exaggerating the difference between the means. In this example, the means were 36.85 and 32.51 for a difference of 4.34, whereas the medians were 33.20 and 30.87 for a difference of 2.33, which is slightly more than half the difference of the means. This is an example of the effect of violated assumptions on the z and t tests. The rank-sum result is a more conservative (safer) result.

ADDITIONAL EXAMPLES

z and *t* Tests, Equal Variances: Comparing the Effectiveness of Two Treatments

An emergency medicine physician wants to compare the relative effectiveness of a "GI cocktail" (antacid plus viscous lidocaine; treatment 1) versus intravenous ranitidine hydrochloride (treatment 2) to treat emergency dyspeptic symptoms as measured on a 1–10 pain scale.[50] The physician records data as pain before treatment minus pain 45 minutes after treatment for $n = 28$ patients randomly assigned to the two treatments; 15 patients are in the first group, and 13 are in the second group. In the medical literature, the physician finds a large-sample standard deviation of pain difference from untreated patients, $\sigma = 1.73$, and uses it as the standard deviation for both treatments. $H_0: \mu_1 = \mu_2$. He takes $\alpha = 0.05$. The test is two-tailed, because either treatment could be the more effective one. $H_1: \mu_1 \neq \mu_2$. The critical values of z from Table 17.1 (or Table I) are ± 1.96. The data are as follows:

| Treatment 1 | 6 | 7 | 2 | 5 | 3 | 0 | 3 | 4 | 5 | 6 | 1 | 1 | 1 | 8 | 6 |
| Treatment 2 | 0 | 1 | 8 | 4 | 7 | 4 | 7 | 7 | 6 | 1 | 0 | 4 | 4 | | |

$m_1 = 3.87$ and $m_2 = 4.08$. The physician substitutes in Eqs. (17.6) and (17.5) to find

$$\sigma_d = \sigma \sqrt{\frac{1}{n_1} + \frac{1}{n_2}} = 1.73\sqrt{\frac{1}{15} + \frac{1}{13}} = 0.6556$$

and

$$z = \frac{m_1 - m_2}{\sigma_d} = \frac{3.87 - 4.08}{0.6556} = -0.32.$$

z is within ± 1.96, so the physician fails to reject H_0; he has no evidence of a difference between the treatments. He interpolates from Table I to find $p = 0.75$.

A critic comments that the population σ of untreated patients is not an appropriate estimate for treated patients. The physician reanalyzes his results, using standard deviations estimated by his data. He finds $s_1 = 2.50$ and $s_2 = 2.84$. He substitutes in Eq. (17.9) and then Eq. (17.8) to find

$$s_d = \sqrt{\left(\frac{1}{n_1} + \frac{1}{n_2}\right)\left[\frac{(n_1 - 1)s_1^2 + (n_2 - 1)s_2^2}{n_1 + n_2 - 2}\right]}$$

$$= \sqrt{\left(\frac{1}{15} + \frac{1}{13}\right)\left[\frac{14 \times 2.5^2 + 12 \times 2.84^2}{15 + 13 - 2}\right]}$$

$$= 1.0088$$

and

$$t = \frac{3.87 - 4.08}{1.0088} = -0.21.$$

From Table 17.2 (or Table II), the critical t-value for 26 df is 2.056. t is well within ± 2.056, so he fails to reject H_0; no difference between treatments has been shown. From statistical software, $p = 0.83$.

t Test, Unequal Variances: Comparing Pain Relief from Two Drugs

In 1958, a study[50] compared a new postoperative pain relief drug (treatment 1) with the established Demerol (treatment 2). Data, consisting of reduction in pain measured on a 1–12 rating scale, for 23 patients who completed the protocol were as follows:

Treatment 1	2	0	3	3	0	0	7	1	4	2	2	1	3
Treatment 2	2	6	4	12	5	8	4	0	10	0			

$n_1 = 13$, $m_1 = 2.15$, $s_1 = 1.9513$, $s_1^2 = 3.8077$, $n_2 = 10$, $m_2 = 5.10$, $s_2 = 4.0125$, and $s_2^2 = 16.1000$. The variances seem quite different. (For a test of these variances, see Additional Example in Section 19.3.) We substitute in Eq. (17.10) and then Eq. (17.8) to find

$$s_d = \sqrt{\frac{s_1^2}{n_1} + \frac{s_2^2}{n_2}} = \sqrt{\frac{3.8077}{13} + \frac{16.1000}{10}} = 1.3795$$

$$t = \frac{m_1 - m_2}{s_d} = \frac{2.15 - 5.10}{1.3795} = -2.14.$$

The *df* calculation is a bother in this case. From Eq. (17.11),

$$\text{approx. } (df) = \frac{\left(\frac{s_1^2}{n_1} + \frac{s_2^2}{n_2}\right)^2}{\frac{\left(\frac{s_1^2}{n_1}\right)^2}{n_1 - 1} + \frac{\left(\frac{s_2^2}{n_2}\right)^2}{n_2 - 1}} = \frac{\left(\frac{3.0877}{13} + \frac{16.1000}{10}\right)^2}{\frac{\left(\frac{3.0877}{13}\right)^2}{12} + \frac{\left(\frac{16.1000}{10}\right)^2}{9}} = 11.66.$$

The next smallest integer gives $df = 11$. From Table 17.2 (or Table II), the critical value for two-tailed $\alpha = 0.05$ for 11 *df* is 2.201. Our calculated *t* is just shy of the critical value, so we do not have quite enough evidence to conclude that the new drug is better than Demerol. Statistical software gives $p = 0.056$.

Exercise 17.3 (Equal Variances): Are Tympanic Temperatures the Same for the Left and the Right Ears?

We are investigating the reliability of a certain brand of tympanic thermometer (temperature measured by a sensor inserted into the patient's ear). Sixteen readings (measured in degrees Fahrenheit [°F]), eight per ear, were taken on a healthy subject at intervals of 1 minute, alternating ears.[63] Data are given in Exercise 16.2. Because readings in either ear may be higher, the alternate hypothesis is two-sided. *L* denotes left; *R* denotes right. $m_L = 96.41°F$, $s_L = 0.88°F$, $m_R = 97.86°F$, and $s_R = 1.12°F$. At the $\alpha = 0.05$ level of significance, are the means of the two ears different?

Exercise 17.4 (Unequal Variances): Testing the Effectiveness of a Vaccine to Inhibit Human Immunodeficiency Virus.

The vaccine was tested by randomizing patients infected with human immunodeficiency virus (HIV) into unvaccinated (treatment 1) and vaccinated (treatment 2) groups and comparing number of HIV per milliliter blood.[85] The data format was as follows:

Patient no.	1	2	3	...	24	25	...	44	45	46
No. of virus/ml blood	134	19,825	38,068	...	4,315	8,677	...	292	67,638	4,811
Treatment	1	1	1	...	1	2	...	2	2	2

$n_1 = 24$, $m_1 = 21{,}457.3$, $s_1 = 29{,}451.98$, $n_2 = 22$, $m_2 = 32{,}174.5$, and $s_2 = 50{,}116.50$. The standard deviations appear quite different, so an unequal-variances t test was chosen. (An F test of the variances as in Section 19.3 yielded $p < 0.001$.) At the $\alpha = 0.05$ level, test H_0: $\mu_1 = \mu_2$ against H_1: $\mu_1 \neq \mu_2$.

17.5. THREE OR MORE MEANS: ONE-WAY ANALYSIS OF VARIANCE

EXAMPLE POSED: ARE THE LOW, UNCERTAIN, AND HIGH RISKS FOR PROSTATE CANCER RELATED TO AGE?

Prostate-specific antigen level is often used to screen patients for cancer risk. Prostate-specific antigen levels less than 4 ng/ml often represent low risk, levels between 4 and 10 ng/ml represent uncertain risk, and levels greater than 10 ng/ml represent high risk. We label the categories $i = 1, 2,$ and 3, respectively. Could age be associated with these PSA levels? We want to know if average ages (m_1, m_2, and m_3) are different for these groups. H_0: There are no differences among the m_i. The alternate hypothesis is H_1: A difference exists somewhere among the groups.

METHOD: ONE-WAY ANALYSIS OF VARIANCE

We want to know whether or not the means from k groups are the same. (For first-time reading, just replace k by 3 until the material is familiar.) Should we just make t tests for each possible pair? No. Each t test would increase the size of the error. (For two 5% tests, we would have the chance of error on the first, *or* the second, *or* both, which is 1 − the chance of no error on any, or $1 - [0.95]^2 = 0.0975$, nearly double.) For three groups, our 5% risk for error would become nearly 15% ($1 - [1 - \alpha]^3 = 14.3\%$); the risk for error would be 18.5% for four groups and 22.6% for five groups. We need to use a single test to detect any overall difference and, if one or more differences are embedded, find a way to detect which ones are significant without increasing the risk for error. This single overall test is an analysis of variance (ANOVA) because it analyzes or separates the variance components into that caused by mean differences and that caused by random influences. To distinguish this form of ANOVA from the analysis of two or more factors simultaneously, we call it a *one-factor ANOVA*, or often simply a *one-way ANOVA*.

One-Way Analysis of Variance Is Similar to the *t* Test Generalized to Three or More Means

Our total sample has n observations, mean m, and variance s^2. This sample is divided into k groups having $n_1, n_2, \ldots, n_k$ observations with group means $m_1, m_2, \ldots, m_k$. The method of one-way ANOVA may be thought of as a three-or-more-mean extension of the two-mean *t* test. It looks very different, but underneath, the mechanics are similar. If we took k to be 2, using ANOVA to test two means, we would find that the F statistic of the ANOVA is just the square of *t*; thus the *p*-values are the same. (The one-way ANOVA bears a relationship to the *t* test much as the Kruskal–Wallis bears to the rank-sum test, as discussed in Chapter 16.)

Assumptions Required for Analysis of Variance

A legitimate ANOVA, as would be expected from a generalization of the two-sample *t* test, requires the same three assumptions: (1) The data are independent from each other, (2) the distribution of each group in the original data is normal, and (3) the variances (or standard deviations) are the same for all groups. Analysis of variance is fairly robust against these assumptions, so we need not be stringent about them, but the data should not be extremely far off.

New Terms in Analysis of Variance: Mean Square and Sum of Squares

Mean square and sum of squares (usually abbreviated MS and SS, respectively) are only new names for familiar concepts. The mean square is just the sample variance, and the sum of squares is the numerator in the sample variance calculation. To analyze the variance (in the classic sense of separating it into components), the total sum of squares is separated into its component because of variability among the means and the remaining (*residual* or *error*) component. These component sums of squares are divided by their *df* to obtain variances (or mean squares). The differences-among-means variance (mean square of means [MSM]) divided by the error variance (mean square of error [MSE]) yields the F statistic. This is because, conceptually, the differences-among-means component is a variance caused by an identified factor, and the error variance is a variance assumed to be caused by random influences alone. If the differences-among-means variance is sufficiently larger than the error variance that it is unlikely to have happened by chance, we believe a significant difference exists among the means. The relationships among the various components in the ANOVA are shown in Table 17.4.

Interpretation of Analysis of Variance

In actuality, the error variance is caused by random influences *plus* unidentified causal influences. Part of a good study design is to control the variability so that the influence of unidentified causes is small. The test is conservative in that a significant outcome implies that the numerator variance (MSM) is larger than random *plus* other causal variability, so it certainly is larger than random variability alone. If the design is not controlled carefully, large unidentified causes may enter and an outcome would have been significant when tested against random variability alone will not show its significance. Thus, a significant result allows the investigator to conclude a difference, but a nonsignificant result does not allow the investigator to conclude no difference; it may only be said that significance *has not been shown*.

Most statistical software packages will perform a one-way ANOVA on command. The logic of a one-way ANOVA follows these steps:

1. Pose H_0: There are no differences among the means; and H_1: There are one or more differences somewhere among the means.
2. Verify approximate satisfaction of the assumptions: (a) normal distributions and (b) equal variances in the original data of the k groups.
3. Choose the error probability α, such as 5%, you are willing to accept, and look up the associated critical value of F, named F_0, in Table V (see Tables of Probability Distributions) for $k - 1$ (numerator) and $n - k$ (denominator) *df*.
4. Calculate m and s^2 for the total sample as usual, recording the sum of squares for the total (SST) before dividing by $n - 1$.
5. Calculate the means m_i for the k groups.
6. Calculate the sum of squares for means (SSM) by squaring each difference $m_i - m$ and adding them.
7. Calculate the sum of squares for error (SSE) by SST $-$ SSM.
8. Calculate s_m^2 (or MSM) $=$ SSM/$(k - 1)$ and s_e^2 (or MSE) $=$ SSE/$(n - k)$.
9. Calculate F $=$ MSM/MSE $= s_m^2/s_e^2$.
10. Compare the calculated F with F_0. If F $> F_0$, reject H_0; if F $< F_0$, do not reject H_0.

Table 17.4

Relationship of Components in a One-Way Analysis of Variance[a]

| Source of variability | Sum of squares | | | Variances, or mean squares | |
	Designation	Formula	*df*	Designation	Formula
Mean	SSM	$\sum^k n_i(m_i - m)^2$	$k - 1$	s_m^2 (or MSM)	SSM/$(k-1)$
Error	SSE	SST $-$ SSM	$n - k$	s_e^2 (or MSE)	SSE/$(n-k)$
Total	SST	$\sum^n (x_i - m)^2$	$n - 1$	s^2 (or MST)	SST/$(n-1)$

[a]The symbolism $\sum^k$ implies "sum (from 1) up to k"; likewise, $\sum^n$ implies "sum up to n."
df, degrees of freedom; MSE, mean square of error; MSM, mean square of means; MST, mean square of the total; SSE, sum of squares for error; SSM, sum of squares for means; SST, sum of squares for the total.

Table 17.5

A Comparison of Five Methods of Multiple Comparisons[a]

Method	Comment
Tukey HSD	Exact *p*-values if sample sizes are equal, conservative if they are different
Fisher LSD	Least conservative method; will find significance more often than others
Scheffé	Conservative method; finds no significance if ANOVA not significant
Bonferroni	Significance is α divided by number of pairs; conservative method
Sidák	Slight theoretical improvement on Bonferroni's method

[a]The term *conservative* implies high specificity, that is, a small chance of showing a difference when there is none.
ANOVA, analysis of variance; HSD, honestly significant differences; LSD, least significant differences.

IDENTIFYING THE MEAN DIFFERENCE(S) THAT CAUSED THE SIGNIFICANCE: MULTIPLE COMPARISONS TESTS

Suppose we reject H_0 and conclude that differences exist among the group means. Which pairs of means are different and which are not? Recall that we cannot just make *t* tests of all possible pairings, because the overall *p*-value accumulates. Methods testing the possible pairings without increasing α are called *multiple comparisons* tests (sometimes also termed *post hoc* comparisons). Several tests, named after their developers, have been devised, including Tukey's HSD (honestly significant differences), Fisher's LSD (least significant differences), Scheffé, Bonferroni, and Sidák that test all possible pairs. These are listed, together with comments comparing them, in Table 17.5. Although the comments show some theoretical differences, in practice, the tests do not give very different results. The Bonferroni method is the simplest to use, but it is also the most primitive theoretically. Use it only when you have no better method available. Having no better method available, however, includes a number of frequently used methods, for example, some rank and multivariate methods.

Two other tests compare subsets of pairs, including Dunnett's test, which pairs only the control mean with the treatment means, and Hsu's multiple comparison bounds (MCB), which pairs the strongest result with the others. In addition, the same goal can be accomplished by what are called *multiple range* tests, including those by Duncan and Newman–Keuls. However, they are not as easy to use and will not be examined further in this chapter.

Multiple Comparisons Using Statistical Software

The user will find some methods in one statistical software package and others in another package. Because of the similarity of results from the different methods, the user can review Table 17.5 to choose from among the options given. There will be little loss in using whatever is available.

When statistical software is used to make a multiple comparison test, the outcome is given as *p*-values adjusted so that each may be compared with our chosen overall α. Different software packages display the results using various schemes. One common display type is presented here; the user who understands it can follow other schemes with little difficulty. The display forms a table in which the group names are shown as both columns and rows, providing a position for every possible pair. (Of course, positions in which row and column show the same group are omitted.) If we denote the pairwise adjusted p-value for the pair i and j as p_{ij}, it appears as follows:

	Mean 1	Mean 2	...	Mean $k-1$
Mean 2	p_{12}			
Mean 3	p_{13}	p_{23}		
$\vdots$	$\vdots$	$\vdots$		
Mean k	p_{1k}	p_{2k}	...	$p_{k,k-1}$

We then identify the pairs for which the displayed p-value exceeds the overall α.

Multiple Comparisons Performed Manually

Sometimes a requirement for multiple comparisons arises when no statistical software is available or when using methods for which the software does not include it. Without much effort, we can use the Bonferroni method to contrast the pairs. We make two-sample t tests on each pair but choose the critical t from an adjusted α rather than $\alpha = 5\%$. The number of pairs is $q = k(k - 1)/2$, and the Bonferroni adjusted α is α/q. We find the df for each pair's t test. From Table II, we obtain a critical t-value for that df, using α/q instead of α and interpolating as required. The calculated t's in the test of the pair's mean must exceed that critical t-value. The t-values obtained from the q t tests may be displayed in a format similar to the preceding p-value display, and each can be compared with its critical t-value one by one.

EXAMPLE COMPLETED: AGE AS RELATED TO RISKS FOR PROSTATE CANCER

Descriptive Statistics and Assumptions

We have the ages of patients in three risk groups. The descriptive statistics are as follows:

PSA group	CaP risk group	Sample size	Mean	SD
PSA < 4 ng/ml	Low	89	66.1	9.1
PSA = 4–10 ng/ml	Uncertain	164	66.3	7.8
PSA > 10 ng/ml	High	48	69.6	6.4

CaP, prostate cancer; PSA, prostate-specific antigen level; SD, standard deviation.

Are the assumptions of normality and equal variance (or standard deviation) in the original data satisfied? We make a quick plot of the distributions and see that they are approximately normal. We examine the three standard deviations and decide that they are not strikingly different. (If we wished to be more formal, e.g., in preparing for journal publication, we could add evidence to our procedure by testing normality group by group using the methods of Section 20.2 and testing equality of variances using the methods of Section 19.4.)

Analysis of Variance Calculations

We choose $\alpha = 0.05$. Degrees of freedom are $k - 1 = 3 - 1 = 2$ and $n - k = 301 - 3 = 298$. For numerator $df = 2$ and denominator $df = 298$, Table V gives a critical value of F of between 3.06 and 3.00, approximately 3.03. We calculate (carrying six significant digits) the additional statistics needed for a one-way ANOVA as follows:

$$m = 66.7641$$
$$\text{SST} = 19670.3$$
$$s^2 = 65.5675$$
$$\text{SSM} = \sum n_i(m_i - m)^2 = 89(66.1124 - 66.7641)^2 + \cdots = 449.196$$
$$\text{SSE} = \text{SST} - \text{SSM} = 19221.104$$
$$s_m^2 \text{ (or MSM)} = \text{SSM}/(k - 1) = 449.196/2 = 224.598$$
$$s_e^2 \text{ (or MSE)} = \text{SSE}/(n - k) = 19221.104/298 = 64.5003$$
$$\text{F} = s_m^2/s_e^2 \text{ (or MSM/MSE)} = 224.598/64.5003 = 3.48$$

An ANOVA result is usually displayed in a table, giving the outcomes of the calculations of interest. Table 17.6 is such a table for this example. Because 3.48 > critical 3.03, we have evidence that a difference among means exists. From a computer calculation, $p = 0.032$.

Which Among the Possible Mean Differences Account(s) for the Significance?

Does the significance arise from low to uncertain risk (m_1 vs. m_2), low to high risk (m_1 vs. m_3), uncertain to high risk (m_2 vs. m_3), or some combination of these? For the example, Table 17.7 shows multiple comparisons calculated for all five methods listed in Table 17.5. It can be seen that the methods do not give very different results. In most cases, they will lead to the same conclusions. The mean ages between the low and the uncertain risk groups are far from significantly

Table 17.6

One-Way Analysis of Variance

Source	Sum of squares	df	Mean square	F	p
Among means	449.196	2	224.598	3.48	0.032
Within groups (error)	19221.104	298	64.5003		
Total	19670.300	300			

df, degrees of freedom.

p-Values for Age by Prostate Cancer Risk Group for Five Multiple Comparison Methods

Method	Low vs. uncertain	Low vs. high	Uncertain vs. high
Tukey HSD	0.983	0.043[a]	0.035[a]
Fisher LSD	0.860	0.017[a]	0.014[a]
Scheffé	0.985	0.058	0.048[a]
Bonferroni	1.000	0.051	0.041[a]
Sidák	0.997	0.050	0.041[a]

[a]$p < 0.05$, significant.
HSD, honestly significant differences; LSD, least significant differences.

different for all methods, and the mean ages between the uncertain and high-risk groups are significant for all methods. The p-values for the low- versus the high-risk groups do not vary greatly, but straddle 0.05. Fisher's LSD, the least conservative (higher sensitivity and lower specificity), and Tukey's HSD, a "middle-of-the-road" method, show significance, whereas the remaining, more conservative methods do not. We can safely conclude that greater PSA values occur in older patients.

Suppose We Do Not Have a Statistical Software Package

If we do not have access to statistical software, we can use Bonferroni's method to contrast the pairs. We make two-sample t tests on each pair but choose the critical t from an adjusted α rather than $\alpha = 5\%$. Because the number of possible pairings is $q = 3$, the Bonferroni adjusted $\alpha/q = 0.05/3 = 0.016$. For the different pairings, df varies from about 50 to about 150. In Table 17.2 (or Table II), we see that the t-value for df around 100 lying a third of the way from two-tailed $\alpha = 0.02$ to two-tailed $\alpha = 0.01$ is in the vicinity of 2.4. (A computer tells us that it is exactly 2.393.) Thus, the calculated t's in the tests of mean pairs must exceed 2.4 to drop below the overall 5% level. The t-values obtained from the three t tests, using a format similar to the previous p-value display, is:

	Low risk	Uncertain risk
Uncertain risk	0.17	
High risk	2.33	2.64

We note that the t for low to uncertain risk is far from significant, whereas the other two values are a little below and a little above the critical value, respectively. The results agree with that presented for Bonferroni's method in Table 17.7.

ADDITIONAL EXAMPLE: DOES STEROID DECREASE EDEMA AFTER RHINOPLASTY? IF SO, WHAT LEVEL SHOULD BE USED?

Following rhinoplasty, swelling may cause deformity during healing. Steroids may decrease the swelling, but the required level of steroid has not been known. $n = 50$ rhinoplasty patients were randomized into $k = 5$ groups of increasing

Table 17.8

Data on Edema Following Rhinoplasty

Swelling reduction (ml)	Level 1	Level 2	Level 3	Level 4	Level 5
Patients 1–5	1.6	4.4	5.5	4.0	8.0
Patients 6–10	2.3	5.8	6.4	8.3	6.2
Patients 11–15	2.3	6.4	6.9	5.8	3.1
:	:	:	:	:	:
Mean m_i	3.77	5.00	5.67	6.79	6.35

Table 17.9

Analysis of Variance for Rhinoplasty Data

Source	SS	df	MS	Calculated F	p
Mean	SSM = 56.57	4	MSM = 56.57/4 = 14.14	MSM/MSE = 4.26	0.005
Error	SSE = 149.57	45	MSE = 149.57/45 = 3.32		
Total	SST = 206.16	49	$(s^2 = 4.2074)$		

df, degrees of freedom; MS, mean square; MSE, mean square of error; MSM, mean square of means; SS, sum of squares; SSE, sum of squares for error; SSM, sum of squares for means; SST, sum of squares for the total.

steroid level having $n_1 = \cdots = n_5 = 10$. Swelling reduction was measured by magnetic resonance images before and after administration of the steroid.[9] H_0: $\mu_1 = \mu_2 = \mu_3 = \mu_4 = \mu_5$ was to be tested using $\alpha = 0.05$. The first few data are given in Table 17.8. The total mean (all data pooled) was $m = 5.52$, and total variance $s^2 = 4.2074$. The following calculations are needed:

$$SST = (n - 1)s^2 = 49 \times 4.2074 = 206.16,$$

$$SSM = 10[(3.77 - 5.52)^2 + (5 - 5.52)^2 + \cdots + (6.35 - 5.52)^2] = 56.57, \text{ and}$$

$$SSE = SST - SSM = 206.16 - 56.57 = 149.59.$$

Table 17.9 gives the ANOVA table. From Table V, the critical $F_{4,45df} = 2.58$. The calculated $F = 4.26$ is much larger, so H_0 is rejected; steroid level does affect swelling. From a statistical software package, the calculated $p = 0.005$.

Identifying the Steroid Dosage

Steroid level affects swelling, but which level should be selected for clinical use? Bonferroni's multiple comparisons procedure using a statistical software package yields the following significance levels (*p*-values), adjusted to be interpreted according to the usual 5% α, although the computer calculations were made so that the true α's used accumulated to 5%. (Bonferroni's method was used so that the reader can reproduce the outcomes without statistical software. The *p*-values showing 1.000 are really slightly less but round to 1.000.)

	Level 1	Level 2	Level 3	Level 4
Level 2	1.000			
Level 3	0.243	1.000		
Level 4	0.006	0.333	1.000	
Level 5	0.028	1.000	1.000	1.000

Level 1 is significantly worse than levels 4 or 5; thus, we reject level 1 as an acceptable steroid level. The other levels are not significantly different. Any one of levels 2 through 5 may be chosen for clinical use. Examining the means, we see that level 4 gives the greatest reduction in swelling, so one would tend to choose level 4 pending additional information.

Exercise 17.5: Are Experimental Animals Equally Resistant to Parasites? In a study on the control of parasites,[49] rats were injected with 500 larvae each of the parasitic worm *Nippostrongylus muris*. Ten days later, the rats were euthanized, and the number of adult worms were counted. The question arose: Is there a batch-to-batch difference in resistance to parasite infestation by groups of rats received from the supplier? $k = 4$ batches of $n_i = 5$ rats each were tested. Data were as follows:

	Group 1	Group 2	Group 3	Group 4
	279	378	172	381
	338	275	335	346
	334	412	335	340
	198	265	282	471
	303	286	250	318
Mean m_i	290.4	323.2	274.8	371.2

Total mean $m = 314.9$, and total variance $s^2 = 4799.358$. At the $\alpha = 0.05$ level of significance, test H_0: $\mu_1 = \mu_2 = \mu_3 = \mu_4$.

ANSWERS TO EXERCISES

17.1. Substitution in Eq. (17.2) (with renamed elements) yields $t = (d - 0)/s_d = 1.58/(1.23/\sqrt{10}) = 4.06$. From Table II, the critical value for a 5% two-tailed *t* with 9 *df* is 2.26. Because 4.06 is much larger, we reject H_0. Laevo, which has the larger mean, is more effective. From a statistical software package, $p = 0.003$.

17.2. The critical value of 9 *df* was 2.26, and the *t* statistic was 4.06. Moving values to the left 4.06 units to standardize the curve resets the critical value at -1.80. β is the area left of -1.80, or, by symmetry, right of 1.80. The closest entry in the 9 *df* row is 1.833, which is associated with a right tail area of 0.05. The power is close to 0.95. From a software package, the power is 0.947.

17.3. (Equal Variances) From Table 17.3, the sample sizes are small, only σ's are estimated, and the standard deviations are not significantly different (see Exercise 19.2); therefore, we choose a two-sample t test with equal standard deviations. $df = n_1 + n_2 - 2 = 14$. From Eq. (17.9),

$$s_d = \sqrt{\left(\frac{1}{n_1} + \frac{1}{n_2}\right)\left[\frac{(n_1 - 1)s_1^2 + (n_2 - 1)s_2^2}{n_1 + n_2 - 2}\right]}$$

$$= \sqrt{\left(\frac{1}{8} + \frac{1}{8}\right)\left[\frac{7 \times 0.88^2 + 7 \times 1.12^2}{8 + 8 - 2}\right]} = 0.5036.$$

From Eq. (17.8),

$$t = \frac{m_1 - m_2}{s_d} = \frac{96.41 - 97.86}{0.5036} = -2.88.$$

From Table II, the p-value for $t = 2.88$ standard deviations from the mean for 14 df is between 0.01 and 0.02. We have evidence to infer a difference in mean readings between ears. From statistical software, $p = 0.012$.

17.4. (Unequal Variances) By substituting in Eq. (17.10) and then Eq. (17.8), we find

$$s_d = \sqrt{\frac{s_1^2}{n_1} + \frac{s_2^2}{n_2}} = \sqrt{\frac{29451.98^2}{24} + \frac{50116.50^2}{22}} = 12260.06$$

and

$$t = \frac{21457.3 - 32174.5}{12260.06} = -0.874.$$

To find the critical value of t, we need df. Substituting in Eq. (17.9), we find

$$\text{approx. } (df) = \frac{\left(\frac{s_1^2}{n_1} + \frac{s_2^2}{n_2}\right)^2}{\frac{\left(\frac{s_1^2}{n_1}\right)^2}{n_1 - 1} + \frac{\left(\frac{s_2^2}{n_2}\right)^2}{n_2 - 1}} = \frac{\left(\frac{29451.98^2}{24} + \frac{50116.50^2}{22}\right)}{\frac{\left(\frac{29451.98^2}{24}\right)^2}{23} + \frac{\left(\frac{50116.50^2}{22}\right)^2}{21}}$$

$$= 33.35.$$

From Table II, the critical value of t for a two-tailed 5% α for 33 df is ± 2.03. The value -0.874 is not outside the ± 2.03 critical values, so H_0 is not rejected. From statistical software, $p = 0.388$.

Table 17.10

Analysis of Variance

Source	SS	df	MS	Calculated F	p
Mean	SSM $= 27234.20$	$k - 1 = 3$	MSM $= 27234.20/3$ $= 9078.07$	MSM/MSE $= 2.27$	0.120
Error	SSE $= 63953.60$	$n - k = 16$	MSE $= 63953.60/16$ $= 3997.10$		
Total	SST $= 91187.80$	$n - 1 = 19$	$(s^2 = 4799.36)$		

df, degrees of freedom; MS, mean square; MSE, mean square of error; MSM, mean square of means; SS, sum of squares; SSE, sum of squares for error; SSM, sum of squares for means; SST, sum of squares for the total.

17.5. From the formulas in Table 17.4, calculate SST $= (n - 1)s^2 = 19 \times 4799.358 = 91,187.805$; SSM $= 5[(291.4 - 314.9)^2 + (323.2 - 314.9)^2 + (274.8 - 314.9)^2 + (371.2 - 314.9)^2 = 27,234.20$; and SSE $=$ SST $-$ SSM $= 91,187.80 - 27,234.20 = 63,953.60$. Table 17.10 gives the ANOVA table. From Table V, the critical $F_{3,16df} = 3.24$. The calculated F $= 2.27$ is smaller, so H_0 is not rejected; mean parasite infestation is not shown to differ from batch to batch. (With a nonsignificant F, we are not concerned that a subordinate pair will be significant; thus, we do not pursue a multiple comparisons procedure.) From a statistical software package, the calculated $p = 0.120$.

Chapter 18

Multifactor Tests on Means of Continuous Data

18.1. CONCEPTS OF EXPERIMENTAL DESIGN

Purpose of Design

In too many medical studies, especially retrospective ones, we analyze as best we can the data that fate supplies. A great many uncontrolled factors influence the outcome we are measuring. The history of physics as a science has been one of controlling factors so that only the factor being studied influences the outcome. By contrast, the discipline of economics, which some would not call a science, is the victim of myriad uncontrolled influences. Biology, including medicine, has lain somewhere in between: Some factors can be controlled; others cannot. The subject matter of experimental design is that of gathering data in such a way as to *control statistically those factors that cannot be controlled physically*.

Factors Influencing Outcomes (as in Blood Loss in Surgery)

Suppose we want to know what influences blood loss in surgery. Using the methods of Chapter 17, we could compare the mean loss (dependent variable) according to a type-of-surgery factor (independent variable) containing two or three types of surgery. However, we could all name a number of other factors that influence blood loss, such as length of surgery, invasiveness of surgery, and the particular surgeon, among other factors. Suppose in our experiment to study influences, we are comparing two types of procedures to repair the same problem but also want to know if type of anesthetic (a second factor or independent variable) is an influence. The method of comparing mean level of a dependent variable for one independent variable and simultaneously comparing the mean level for another independent variable is called two-factor or two-way analysis of variance (ANOVA). It is possible to analyze simultaneously three or even more types of (categorical) independent variables, and even to combine an analysis of categorical and continuous independent variables.

ASSUMPTIONS UNDERLYING MULTIFACTOR DESIGNS

Recall that one-way ANOVA requires the assumptions that all observations have error components drawn randomly and independently from the same error distribution (as with most statistical tests, this assumption is so ubiquitous that it is seldom called out), that this distribution is normal, and that population variances for the several groups are equal. Multifactor ANOVA requires these same assumptions, but we note that "group" refers to each individual design cell defined by cross tabulation of the factors. This will become clearer with the development of the method. As with one-way ANOVA, we note that the methods are rather robust; that is, they are little influenced by moderate deviations from assumptions. If an extreme violation of an assumption is suspected, a test on that assumption may be made, as noted in Chapter 17.

BALANCE IN DESIGN

Definition

A *balanced* experimental design contains the same number of data readings in each cell of the categorical independent variables, that is, in each possible pairing of categories, so that every possible combination is represented an equal number of times. Such balanced designs provide the most efficient statistical control and assessment of the effect of influences on an experiment's outcome. Lack of balance implies a loss of efficiency.

Example

Suppose an orthopedic experiment consisted of comparing the mean rate of blood flow permitted by two modes of casting, each with two types of pinning, forming a two-factor ANOVA with interaction. (This could be designated a 2×2 design for the two categories in each factor. If there were three modes of casting, it would be a 3×2 design.) The first main effect results compare mean blood flow for the cast modes averaged over the pinnings; the second, for the pinning types averaged over the castings. The interaction effect contrasts the pinning results for each mode of cast to learn if the effect of pinning type is conditional on mode. If the orthopedist obtained six readings on each pinning type for one mode of cast, but only four on the other, the design would be unbalanced.

Balance Is Assumed in This Chapter

In the formulas given in this chapter, balance is assumed. Balanced designs are easier to understand and the formulas are simpler, which supports the goal of comprehension rather than manual calculation. The interpretation of ANOVA results for unbalanced designs will be similar to that from the result had balance been present. If the user cannot design a balanced experiment, in most cases, the statistical software adjusts the calculations for lack of balance and provides accurate degrees of freedom (*df*).

Missing Observations

If a very small number of observations in a balanced design are lost, perhaps one to two in a small sample or a few in a large sample, the missing observation(s) may be estimated and *df* reduced accordingly. Specifically, this estimation must follow one of a few technically defined methods, denoted *imputation*. Such a case might occur if an animal dies or a patient is lost to follow-up. Statistical software exists for imputation of missing observations. Lacking software, one approach (usually not the best) is just to substitute the average of the other observations in that cell. If many observations are missing, techniques for unbalanced designs exist, but are not simple; the help of a biostatistician should be sought.

18.2. TWO-FACTOR ANALYSIS OF VARIANCE

ORIENTATION BY EXAMPLE: HEART RATE EXAMINED BY TWO FACTORS

A 73-year-old male patient with elevated blood pressure complains of occasional faintness around mealtimes. Data were taken to evaluate this complaint. Blood pressure and heart rate (HR) were measured three times just before and 30 minutes after each meal (approximately 7:00 AM, noon, and 6:00 PM) in one day. Blood pressure provided no answers, but the HR results were interesting. Heart rate data are given in the following table. We ask the questions: (1) Does eating affect the mean HR? (2) Does time of day affect the mean HR? (3) Does the effect of eating interact with time of day to affect the mean HR?

Before or after meal	Heart rate (beats/min)		
	Meal 1	Meal 2	Meal 3
1	50	57	60
1	48	55	73
1	50	55	56
2	68	58	56
2	71	56	58
2	64	58	50

1, before meal; 2, after meal.

Answering Question (1) Using a *t* Test

If we had asked only question (1), we could have taken the nine observations before eating ($m_1 = 56.0000$; $s_1 = 7.4498$) and the nine readings after eating ($m_2 = 59.8889$; $s_2 = 6.5659$) and tested the difference using a two-sample *t* test. Calculating s_d from Eq. (17.9) and *t* from Eq. (17.8), we find $s_d = 3.3101$ and $t = -1.1749$. From Table II (see Tables of Probability Distributions), for $n_1 + n_2 - 2 = 16$ *df*, a |*t*| of 1.17 shows a two-tailed α exceeding 0.20. From statistical software, $p = 0.257$. There is no evidence that eating causes a difference in HR.

Examining the Effect of Eating on Mean Heart Rate Using One-Way Analysis of Variance

The t test is familiar to most users. In Section 17.5, the relationship $F_{1k\,df} = (t_{k\,df})^2$ is noted. To make comparisons of one- and two-way ANOVAs, we want question 1 answered by a one-way ANOVA. Let us use the formulas of Table 17.4. $k = 2$, $n = 18$, and $n_1 = n_2 = 9$. Sum of squares for means (SSM) = 68.0559. The sample variance for all $n = 18$ observations is $s^2 = 50.4085$ (which is also mean square of the total [MST]). Sum of squares for the total (SST) = $(n - 1)s^2 = 856.9445$. Sum of squares for error (SSE) = SST − sum of squares for means (SSM) = 788.8886. Mean square of means (MSM) = SSM, because $k - 1 = 1$. Mean square of error (MSE) = SSE/$(n - k)$ = 49.3055. Finally, $F_{1,16df}$ = MSM/MSE = 1.3803. From Table V (see Tables of Probability Distributions), the critical value for 5% α for 1,16 df is 4.49. Because $1.38 < 4.49$, we have no evidence that eating causes a difference in HR. From statistical software, $p = 0.257$, the same as using t. We note that $(-1.1749)^2 = 1.3804$, the same value to three decimal places.

Answering Question (2): One-Way Analysis of Variance

If we had asked only question (2), we could have taken the six readings from each mealtime and performed a one-way ANOVA. Required calculations are as follows:

Meal Number	M	s	s^2
1	58.5000	10.3102	106.3002
2	56.5000	1.3784	1.9000
3	58.8333	7.7050	59.3670
Total	57.9444	7.0999	50.4085

$k = 3$, $n = 18$, and $n_1 = n_2 = n_3 = 6$. Following the formulas of Table 17.4, we find SSM = 19.1108, SST = 856.9445, SSE = 837.8337, MSM = 9.5554, and MSE = 55.8556. F = MSM/MSE = 0.1711. From Table V, the critical value for 5% α for 2,15 df is 3.68, which is much greater than 0.1711. We have no evidence of a difference in HR caused by mealtime. From statistical software, $p = 0.844$.

How to Answer All Three Questions at Once: Two-Way Analysis of Variance

Neither of these analyses sheds light on question (3). Furthermore, the before-to-after eating means are averaged across mealtimes, so they include this influence in the denominator variability, the standard error of means, denoted MSE in ANOVA. Had we been able to remove this influence, we would have had a smaller MSE and the test on before-to-after would have been more sensitive. Similarly, the mealtime results are averaged across before-or-after and therefore include this influence in the denominator variability, the MSE. Had we been able to remove this influence, we would have had a smaller MSE and the test on mealtime would have been more sensitive. We would like to calculate a two-way ANOVA and compare results with the two one-way results, but first we need some formulas for the calculation. Let us examine the method and then return to results.

METHOD

The Basic Datum and Some Definitions

Let us designate the basic observation or reading as x. We have two factors and several readings for each pairing of factor indicators, so we need three subscripts to identify each reading, say, x_{ijk}. In the example, we have the factors eating (i) and time of day (j), with three readings (k) for eating-state/time-state pair. Let us denote the numbers of categories as r for the first and c for the second, and the number of repeatedly taken readings for each i, j pair, that is, *replications*, as w. We calculate the i, j mean using these w readings. These definitions can be summarized as follows:

$$i = 1, 2, \ldots, r$$
$$j = 1, 2, \ldots, c$$
$$k = 1, 2, \ldots, w$$
$$m_{ij} = \sum_{k=1}^{w} x_{ijk}/w$$

A Means Table

The first step is to set up the means we want to compare. We create a table with one of our two factors represented by rows and the other represented by columns. We can denote the row factor by R and the column factor by C (hence, r categories in R and c categories in C). As in Table 15.10, where we used a dot subscript to indicate the subscript summed over, let us use $m_{i\cdot}$ to indicate the mean over j, that is, $m_{i\cdot} = (m_{i1} + \cdots + m_{ic})/c$, and so on. The table appears as Table 18.1.

Table 18.1

Components of a Two-Factor Means

	C (columns)			
	1	$\ldots$	c	Combined
R (rows)				
1	m_{11}	$\ldots$	m_{1c}	$m_{1\cdot}$
$\vdots$	$\vdots$		$\vdots$	$\vdots$
r	m_{r1}	$\ldots$	m_{rc}	$m_{r\cdot}$
Combined	$m_{\cdot 1}$	$\ldots$	$m_{\cdot c}$	$m_{\cdot\cdot}$

An "Adjustment" Term

In Table 17.4, the formulas for sums of squares (SS) for mean (SSM) and total (SST) included the overall mean subtracted from each element in the sum. By using algebra, we could have carried this subtraction outside the parentheses and used a single "adjustment" or "correction" term for each SS. The concept is easier to understand as the table is given, but the calculations are easier with a single adjustment term. Because two- and higher-factor ANOVAs are complicated to calculate, the adjustment term, A, is used. It is just the total sample size multiplied by the overall mean.

Formulas for Calculation

The basic SS that appear in the ANOVA table are presented in Table 18.2.

Table 18.2
Formulas for Components in a Two-Factor Analysis of Variance[a]

$$A = rcw \times m_{..}^2$$
$$SST = \sum_i^r \sum_j^c \sum_k^w x_{ijk}^2 - A$$
$$SSR = cw \sum_i^r m_{i.}^2 - A$$
$$SSC = rw \sum_j^c m_{.j}^2 - A$$
$$SSI = w \sum_i^r \sum_j^c m_{ij}^2 - A - SSR - SSC$$
$$SSE = SST - SSR - SSC - SSI$$

[a]The symbolism $\sum_i^r$ implies "sum over i from 1 to r," and the equivalent is true for other indicator symbols.
SSC, sum of squares for columns; SSE, sum of squares for error (or residual); SSI, sum of squares for interaction (row-by-column); SSR, sum of squares for rows; SST, sum of squares for the total.

Analysis of Variance Table

The remaining calculations and results of the experiment are displayed in an ANOVA table (Table 18.3). Such a table provides the basis for interpreting the outcome of the experiment and is often presented in its entirety when the experiment is published.

Table 18.3
Two-Factor Analysis of Variance

Source	Sums of squares	*df*	Mean squares	F	*p*
Rows	SSR from Table 18.2	$r - 1$	MSR = SSR/*df*	MSR/MSE	
Columns	SSC from Table 18.2	$c - 1$	MSC = SSC/*df*	MSC/MSE	
Interaction (r × c)	SSI from Table 18.2	$(r-1)(c-1)$	MSI = SSI/*df*	MSI/MSE	
Error (residual)	SSE from Table 18.2	$rc(w-1)$	MSE = SSE/*df*		
Total	SST from Table 18.2	$rcw - 1$			

df, degrees of freedom; MSC, mean square of columns; MSE, mean square of error; MSI, mean squares for interaction (row-by-column); MSR, mean squares for rows; SSC, sum of squares for columns; SSE, sum of squares for error; SSI, sum of squares for interaction (row-by-column); SSR, sum of squares for rows; SST, sum of squares for the total.

Interpretation of the Analysis of Variance Table

The significance level α was chosen at the outset of the experiment. Most often, $\alpha = 0.05$ by convention. The critical values are found from the F table (see Table V). Numerator *df* is the *df* for the factor, that is, the source of variability, being tested. Denominator *df* is the *df* for error. For example, the critical value for the rows' factor is the Table V entry for $r - 1$ along the top and $rc(w - 1)$ along the side. The actual *p*-value can be found using statistical software. As with other tests of significance, a significant *p*-value implies evidence for differences among the means for that factor greater than would have occurred by chance. When a significant factor contains three or more means, we question which difference(s) between members of mean pairings account for the significance. This question can be answered using methods of multiple comparisons as addressed in Section 17.5.

TWO-FACTOR HEART RATE EXAMPLE COMPLETED

The first step is to generate the means table. The data are as follows:

	Breakfast	Lunch	Dinner	Across mealtimes
Before eating	49.3333	55.6667	63.0000	56.0000
After eating	67.6667	57.3333	54.6667	59.8889
Across eating	58.5000	56.5000	58.8333	57.9444

We note that $r = 2$, $c = 3$, and $w = 3$. Using Table 18.2, we find the following SS:

$A = 18(57.9444)^2 = 60435.9630$
$\mathrm{SST} = 50^2 + 48^2 + \cdots + 58^2 + 50^2 - 60435.963 = 856.9444$
$\mathrm{SSR} = 9(56^2 + 59.8889^2) - 60435.963 = 68.1601$
$\mathrm{SSC} = 6(58.5^2 + 56.5^2 + 58.8333^2) - 60435.963 = 19.1801$
$\mathrm{SSI} = 3(49.3333^2 + \cdots + 54.6667^2) - 60435.963 - 68.1601 - 19.1801 = 544.3777$
$\mathrm{SSE} = 856.9444 - 68.1601 - 19.1801 - 544.3777 = 225.2265$

Completing the ANOVA table, as in Table 18.3, we find the following data:

Source	Sums of squares	*df*	Mean squares	F	Critical F	*p*
Before vs. after eating	68.1601	1	68.1601	3.63	6.55	0.081
Mealtimes	19.1801	2	9.5901	0.51	5.10	0.613
Interaction	544.3777	2	272.1888	14.50	5.10	< 0.001
Error (residual)	225.2265	12	18.7689			
Total	856.9444	17				

Interpretation

A column has been added showing critical values of F from Table V. The first entry is $F_{0.95}$ for 1,12 *df*. The second and third entries are $F_{0.95}$ for 2,12 *df*. We can see that the calculated F is less than critical for the two main or primary factors, as we found in the one-way ANOVAs, but it is greater than critical for the interaction term. The significant interaction indicates that the eating factor has a pattern of mean differences that changes by mealtime. We can see from the means table that mean HR increases remarkably over breakfast but decreases over dinner, whereas it is little different at lunchtime. The increase at breakfast time and decrease at dinnertime obscured each other when pooled over time, so no significance appeared.

Note the Improvement over the *t* Test and the One-Way Analysis of Variance

Not only are we able to answer all three questions with one analysis, keeping our *p*-value from accumulating by repeated testing, but the tests of questions (1) and (2) are more sensitive. Using two-way ANOVA, we removed identifiable causes of variability from SSE present in the one-way ANOVAs, with resulting increases in F and decreases in *p*. Going from one- to two-way ANOVAs reduced *p* from 0.257 for the eating factor to 0.081 and from 0.884 for the mealtime factor to 0.613.

ADDITIONAL EXAMPLE: COOLING KIDNEYS BEFORE SURGERY

DB15 provides data on two methods of cooling kidneys to prevent necrosis during surgery. Six anesthetized pigs were opened and their kidneys cooled, one by infusing cold saline (treatment 1) and the other by packing in ice (treatment 2). Temperature (measured in degrees Celsius [°C]) was measured at baseline (time 1) and at 5 (time 2), 10 (time 3), and 15 (time 4) minutes. Because one kidney was cooled from the outside and the other from the inside, the depth measurement would be expected to be relevant. However, for the purpose of illustrating a two-factor ANOVA, temperatures at the two depths are taken as just further replications so that 2 depths on 6 pigs allows 12 replications (*w*) from which to calculate means for each treatment and time combination. This database is analyzed again in Section 18.5, which uses depth as a factor in three-factor ANOVA.

The first step is to generate the means table. The data are as follows:

	Baseline (0 minute)	5 minutes	10 minutes	15 minutes	Across time
Cold saline	34.4000	31.9417	28.7000	27.9583	30.7500
Ice	36.1167	16.9500	10.3833	7.5000	17.7375
Over treatments	35.2583	24.4458	19.5417	17.7292	24.2438

We note that $r = 2$, $c = 4$, and $w = 12$. Using Table 18.2, we find the following SS:

$A = 95(24.2438)^2 = 56424.6710$

$\text{SST} = 36.7^2 + 33.2^2 + \cdots + 16.6^2 + 9.6^2 - 56424.6710 = 11531.3562$

$\text{SSR} = 48(30.75^2 + 17.7375^2) - 56424.6710 = 4064.0365$

$\text{SSC} = 24(35.2583^2 + 24.4458^2 + 19.5417^2 + 17.7292^2) - 56424.6710 = 4462.0673$

$\text{SSI} = 12(34.4^2 + \cdots + 7.5^2) - 56424.6710 - 4064.0365 - 4462.0673 = 1826.4673$

$\text{SSE} = 11531.3562 - 4064.0365 - 4462.0673 - 1826.4673 = 1178.7851$

Completing the ANOVA table, as in Table 18.3, we find the following:

Source	Sums of squares	df	Mean squares	F	Critical F	p
Treatments (rows)	4064.0365	1	4064.0365	303.39	3.96	<0.001
Time (columns)	4462.0673	3	1487.3558	111.04	2.72	<0.001
Interaction	1826.4673	3	608.8224	45.45	2.72	<0.001
Error (residual)	1178.7851	88	13.3953			
Total	11531.3562	95				

All sources of mean differences are highly significant. (The F-values are huge.) We see that the means of kidney cooling are different for the two treatments, that the kidneys cool significantly over time regardless of the treatment, and that the pattern of kidney cooling over time is different for the two treatments. Looking at the means table, we see that the kidneys infused with saline lower in temperature, but not to the required 15°C, whereas those in ice do.

Exercise 18.1: Effectiveness of a Drug for Patients with Lower Right Quadrant Pain. Historically, results on effectiveness of a certain drug were mixed. An emergency department (ED) specialist[1] designed a two-way ANOVA to assess effectiveness for kidney stones versus other causes. Randomizing patients into drug versus placebo groups, the specialist collected visual analog pain scale (measure in millimeters) data on 14 patients with kidney stones and 14 patients with other causes of pain. Data, giving the change in pain scale readings at presentation minus 20 minutes after drug administration, were as follows[a]:

Patient no.	Drug	Kidney stones	Pain change (mm)
1	0	0	8
2	0	0	16
3	0	0	16
4	0	0	20
5	0	0	22
6	0	0	28
7	0	0	35
8	0	1	−5

(*Continued*)

Patient no.	Drug	Kidney stones	Pain change (mm)
9	0	1	−3
10	0	1	−3
11	0	1	−2
12	0	1	1
13	0	1	5
14	0	1	6
15	1	0	−9
16	1	0	−7
17	1	0	3
18	1	0	4
19	1	0	17
20	1	0	19
21	1	0	20
22	1	1	−3
23	1	1	−2
24	1	1	0
25	1	1	3
26	1	1	17
27	1	1	26
28	1	1	30

[a]Data are simplified to facilitate the example.
0, no; 1, yes.

Conduct the two-way ANOVA with interaction and interpret the results. Sums of squares were 24.14286 for drug, 531.57143 for kidney stones, 1032.14286 for drug–kidney stones interaction, and 4218.42857 for total.

18.3. REPEATED-MEASURES ANALYSIS OF VARIANCE

ORIENTATION BY EXAMPLE

One of two elderly patients (patient 1) experienced a mitral valve prolapse, and the other (patient 2) had high blood pressure. Heart rates were taken from each on arising, at midday, and in the evening (times 1, 2, and 3, respectively). Are their mean vital signs different? Do these signs vary over time? Do these signs vary over time for one patient differently from the other? (Simple data were chosen to exemplify the method; it is not suggested that they have research or clinical importance.)

At first, it might appear that a two-factor ANOVA would be appropriate. However, note that data taken through time are on the same patient. The through-time data contain influences from cardiac characteristics, whereas patient-to-patient data contain that influence and also person-to-person influences. If we use the same estimate of error variability (MSE) to divide by, we introduce a bias and change the sensitivity of the test. We need two estimates of random variability: one with and one without person-to-person differences.

Data and Means

The data appear as follows:

Patient no.	Replication no.	Time 1	Time 2	Time 3	Means
1	1	80	84	78	80.6667
1	2	77	74	78	76.3333
1	3	71	81	75	75.6667
2	1	50	57	60	55.6667
2	2	50	55	73	59.3333
2	3	48	55	56	53.0000

We need an additional set of means that does not appear in a two-way means table: those across the repeated measure (time) for each replication, which appear as the last column in the data table. These "among-times" means allow a within-times sum of squares (SSW) to be calculated. A means table of the sort seen for two-factor ANOVA follows:

Patient no.	Time 1	Time 2	Time 3	Across times
1	76.0000	79.6667	77.0000	77.5556
2	49.3333	55.6667	63.0000	56.0000
Across patients	62.6667	67.6667	70.0000	66.7778

The calculations appear after some needed formulas are presented.

METHOD

Goal

As with other multifactor ANOVAs, the goal is to obtain mean squares and then F-values for the main and interaction effects. However, we have one set of effects with variability caused by the treatment factor in addition to a case-to-case (e.g., patient-to-patient) factor and another set having variability without case-to-case differences, because it is repeated with each case. For example, suppose we are comparing mean pain levels immediately after surgery and then 4 hours after surgery reported by patients randomized to two types of anesthesia. Pain readings between anesthetics contain patient-to-patient differences, but pain readings between 0 and 4 hours postoperatively do not. Thus, to calculate F, we need one MSE that includes random variability influences of measures not repeated and one MSE that includes random variability influences of the repeated-measure-to-measure influences.

Two Sum of Squares for Error Terms

The SS for the main effects and interaction are the same as for a two-factor ANOVA (see Table 18.2). The difference, so far as calculation is concerned, is finding two SSE terms. One is a within-repeated-measures SSE, or SSE(W), that estimates error variability for causal terms involving the repeated factor. In the preceding example, this SSE(W) contains the influence of measures over the same patient. The other is a between-repeated-measures SEE, or SSE(B), that estimates error variability for causal terms not involving the repeated factor; that is, it contains the influence of the independent, or nonrepeated, factor.

Row and Column Designations

The references to rows and columns in Section 18.2 must be clarified. To be consistent with most software packages, we present the means table with repeated measures across the rows and independent measures down the columns. Thus, sum of squares for rows (SSR) will denote the SS for the repeated measure, and sum of squares for columns (SSC) will denote the SS for the independent measure.

Sum of Squares for Error Calculations

The first step is to find an interim "sum of squares across repeated measures," or SSAcross. This SS is the sum of squares for the means of each replication across the repeated measures, as shown in the rightmost column of the data table, $m_{i \cdot k}$ (Table 18.4). Table 18.2 gave formulas for calculating SST, SSR, SSC, and sum of squares for interaction (row-by-column) (SSI). Table 18.4 supplements Table 18.2 with the additional formulas required for repeated-measures ANOVA.

Table 18.4

Formulas Supplemental to Those in Table 18.2 for Components in a Repeated-Measures Analysis of Variance

$\text{SSAcross} = \sum_j^c m_{i \cdot k}^2 - A$
$\text{SSE(B)} = \text{SSAcross} - \text{SSC}$
$\text{SSE(W)} = \text{SST} - \text{SSAcross} - \text{SSR} - \text{SSI}$

SSAcross, sum of squares across repeated measures; SSC, sum of squares for columns; SSE(B), between-repeated-measures sum of squares for error; SSE(W), within-repeated-measures sum of squares for error; SSI, sum of squares for interaction (row-by-column); SSR, sum of squares for rows; SST, sum of squares for the total.

Analysis of Variance Table

Table 18.5 provides the repeated-measures (two-factor) ANOVA table.

Table 18.5

Repeated-Measures (Two-Factor) Analysis of Variance

Source	Sums of squares	df	Mean squares	F	p
Columns (groups)	SSC (Table 18.2)	$c-1$	MSC = SSC/df	MSC/MSE(B)	
Error between	SSE(B) (Table 18.4)	$c(w-1)$	MSE(B) = SSE(B)/df		
Rows (repeated)	SSR (Table 18.2)	$r-1$	MSR = SSR/df	MSR/MSE(W)	
Interaction	SSI (Table 18.2)	$(r-1)(c-1)$	MSI = SSI/df	MSI/MSE(W)	
Error within	SSE(W) (Table 18.4)	$c(r-1)(w-1)$	MSE(W) = SSE(W)/df		
Total	SST (Table 18.2)	$rcw-1$			

df, degrees of freedom; MSC, mean square of columns; MSE(B), between-repeated-measures mean square of error; MSE(W), within-repeated-measures mean square of error; MSI, mean squares for interaction (row-by-column); MSR, mean squares for rows; SSAcross, sum of squares across repeated measures; SSC, sum of squares for columns; SSE(B), between-repeated-measures sum of squares for error; SSE(W), within-repeated-measures sum of squares for error; SSI, sum of squares for interaction (row-by-column); SSR, sum of squares for rows; SST, sum of squares for the total.

Interpretation

Now that we have adjusted for the repeated measures, we can interpret the three F-values (and their *p*-values) in the same way as the ordinary two-factor ANOVA.

An Admonition

Recall that an assumption is the independence, that is, lack of correlation, among observations. Often, the repeated-measure observations are correlated. An example might be pain ratings arising from the same patient; one patient may be more sensitive and rate pain more highly across the board than another patient. In such a case, an adjustment to reduce the significance level of the repeated measure is necessary. Several statistical software packages contain such adjustments that can be designated when choosing the repeated-measures ANOVA option.

ADDITIONAL EXAMPLE

Experiment and Data

A widely used part of the treatment for rattlesnake bite is fasciotomy to relieve the pressure from edema. An ED specialist[83] questioned its benefit. He injected the hind legs of 12 anesthetized pigs with rattlesnake venom. The specialist treated six pigs, chosen randomly, with antivenin, and the other six without antivenin, performing fasciotomy on one leg but not the other. The first independent variable is antivenin treatment or not, which includes pig-to-pig influence, as well as influence due to the treatment and fasciotomy factors. The second independent variable is fasciotomy, which includes the treatment and fasciotomy influences, but not pig-to-pig influence, because fasciotomy versus no fasciotomy was performed on the same pig. As the outcome (dependent) variable, the specialist measured the percentage tissue necrosis at 8 hours. The data are as follows[a]:

| Pig no. | Antivenin treatment | 8-hour necrosis (%) | | Means across fasciotomy |
		With fasciotomy	Without fasciotomy	
1	No	48	0	24.0
2	No	57	52	54.5
3	No	6	23	14.5
4	No	9	15	12.0
5	No	25	8	16.5
6	No	13	4	8.5
7	Yes	5	30	17.5
8	Yes	10	0	5.0
9	Yes	16	0	8.0
10	Yes	24	0	12.0
11	Yes	40	15	27.5
12	Yes	12	10	11.0

[a]Data are simplified slightly to facilitate the example.

Means Table

The across-repeated-measures means, specifically across fasciotomy or not, are appended in the rightmost column of the data table. The means used to calculate the main- and interaction-effect SS appear in the following means table:

| | | Fasciotomy | | Across fasciotomy |
		Yes	No	
Antivenin treatment	No	26.3333	17.0000	21.6667
	Yes	17.8333	9.1667	13.5000
Across antivenin		22.0833	13.0833	17.5833

Calculations

Substitution in the formulas of Tables 18.2 and 18.4 yields the following SS:

$$A = 24 \times 17.5833^2 = 7420.1385$$
$$SST = 48^2 + \cdots + 10^2 - 7420.1385 = 6507.8361$$
$$SSC \text{ (treatment)} = 12 \times (21.6667^2 + 13.5000^2) - 7420.1385 = 400.1868$$

(Continued)

$$\text{SSR (fasciotomy)} = 12 \times (22.0833^2 + 13.0833^2) - 7420.1385 = 485.9746$$

$$\text{SSI (treatment} \times \text{fasciotomy)} = 6 \times (26.3333^2 + \cdots + 9.1667^2) - 7420.1385 -$$
$$400.1868 - 485.9746 = 0.6607$$

$$\text{SSAcross} = 2 \times (24^2 + 54.5^2 + \cdots + 11^2) - 7420.1385 = 3902.8615$$

$$\text{SSE(B)} = 3902.8361 - 400.1868 = 3502.6493$$

$$\text{SSE(W)} = 6507.8361 - 3902.8615 - 485.9746 - 0.6607 = 2118.3493$$

Analysis of Variance Table

Substitution in the formulas of Table 18.5 yields the *df*, mean squares, and F-values appearing in the ANOVA table.

Repeated-Measures Analysis of Variance for Rattlesnake Venom Experiment

Effect	Sums of squares	*df*	Mean squares	F	Critical F	*p*
Antivenin treatment	400.1868 (SSC)	1	400.1868	1.142	4.96	0.310
Between-pig error	3502.6493 [SSE(B)]	10	350.2650			
Fasciotomy	485.9746 (SSR)	1	485.9746	2.294	4.96	0.161
Treatment × Fasciotomy interaction	0.6607 (SSI)	1	0.6607	0.003	4.96	0.956
Within-pig error	2118.3647 [SSE(W)]	10	211.8349			
Total	6507.8361 (SST)	23				

Interpretation

The data show no significant effects. The antivenin treatment and the fasciotomy have not been shown to be effective. Certainly, the near-zero F for the interaction indicates that the effect of fasciotomy is not different with antivenin treatment than without it.

Exercise 18.2: Comparing Treatment of Severe Migraine Headaches. An ED specialist wanted to compare the pain-relieving effects of two drugs, diazepam and prochlorperazine, for patients suffering from acute migraine headaches.[84] The specialist randomly selected 19 of 38 patients for administration of one drug, whereas the remaining 19 patients received the other drug. The specialist measured pain levels between 0 and 100 by a visual analog scale at baseline, at 30 minutes, and at 60 minutes. The pain by drug contains patient-to-patient variability, whereas the pain by time is measured within each patient. Thus, a repeated-measures ANOVA is appropriate. The first few data are as follows:

Patient no.	Drug	Pain level (1–100)			Mean across
		Baseline (0 minute)	30 minutes	60 minutes	
1	Prochlorperazine	80	46	17	47.6667
2	Prochlorperazine	76	14	0	30.0000
⋮	⋮	⋮	⋮	⋮	⋮
37	Diazepam	43	6	1	16.6667
38	Diazepam	61	68	56	61.6667

A = 306697.7200, SST = 136171.0400, and SSAcross = 82177.4640. The means table is as follows:

Drug	Baseline (0 minute)	30 minutes	60 minutes	Across pain
		Pain level (1–100)		
Diazepam	69.8421	59.8421	55.3158	61.6667
Prochlorperazine	76.0526	32.7368	17.4210	42.0702
Across drug	72.9474	46.2895	36.3684	51.8684

Perform the repeated-measures ANOVA and interpret it.

18.4. ANALYSIS OF COVARIANCE

PURPOSE

It was stated earlier that the purpose of experimental design is to control statistically factors that cannot be controlled physically. The preceding sections examine how to control factors that fall into specific categories, perhaps modes of treatment. But what do we do if one factor is continuous rather than categorical, perhaps patient age? Such a factor has been called a covariate, and the method of analyzing the variability of mixed categorical and continuous factors is called analysis of covariance (ANCOVA).

ORIENTATION BY EXAMPLE: DOES SEX AND/OR AGE AFFECT THEOPHYLLINE LEVELS OF PATIENTS WITH EMPHYSEMA?

Data

In DB3, serum theophylline is measured for patients with emphysema before and during a course of azithromycin. The clinical expectation was that the antibiotic would increase theophylline levels. Let us denote the change in level as the outcome variable x = level at day 5 minus level at day 0, sex as g (for group), and age as the continuous independent variable u. Did the patient's sex and/or age affect y?[a] The resulting data set is as follows:

Patient no.	x (change)	g (sex)	u (age)
1	−11.8	1	61
2	−1.8	1	70
5	−0.2	1	64
8	−3.3	1	69
11	−1.7	1	65
13	−1.2	1	51

(Continued)

Patient no.	x (change)	g (sex)	u (age)
3	−2.3	2	65
4	0.4	2	65
6	1.6	2	76
7	4.2	2	72
9	0.0	2	66
10	−0.5	2	62
12	2.2	2	71
14	−2.9	2	71
16	0.6	2	50

[a]Data for a male patient (patient 15) in DB3 was deleted because the patient's age was unknown.

Let us denote subscript indicators and numbers of observations as follows: group (sex) denotes g_i, $i = 1$ or 2 ($1 = $ female; $2 = $ male); j denotes number within group, $j = 1, \ldots, n_i$, where $n_1 = 6$, $n_2 = 9$, and $n = n_1 + n_2 = 15$; change is x_{ij}; and age is u_{ij}.

Means

The mean, $m_{x.}$, of change in level x_{ij} overall is -1.11; $-3.33 = m_{x1}$ for female patients, and $0.37 = m_{x2}$ for male patients. The mean, $m_{u.}$, of age u_{ij} overall is 65.20; $63.33 = m_{u1}$ for female patients, and $66.44 = m_{u2}$ for male patients. Does the fact that male patients are older affect change in theophylline level? If age were categorized, perhaps as < 70 versus ≥ 70 years, we would have a 2×2 two-factor ANOVA. However, because age is continuous and not categorical, we cannot include it in an ANOVA.

What to Do About the Continuous Variable

If age is an influence and we do not adjust for it, it may bias the outcome. In addition to reducing potential bias, removing the variability of another influencing factor from the MSE will increase sensitivity and precision of the remaining analysis. With age continuous, we must use ANCOVA, an ANOVA with outcome adjusted for the continuous variable. Change x is potentially dependent on age u. We can investigate this dependency by the regression of x on u (see Chapter 8 or 24). Although any sort of regression model is possible to use, the explanation here is restricted to a straight-line regression. The slope of the regression line, b, is assumed to be the same for both sexes. (We could test this assumption if we are unsure.) However, the different means for male and female patients implies a regression line for each: $m_{x1} + b(u_{1j} - m_{u1})$ for male patients and $m_{x2} + b(u_{2j} - m_{u2})$ for female patients.

Each x_{ij} is adjusted by its respective regression for the deviation of the group means from the overall mean to become, for example, $x_{ij}^{(a)}$. We find that the adjusted female mean is $m_{x1}^{(a)} = -3.12$, and the adjusted male mean is $m_{x2}^{(a)} = 0.23$.

Analysis of Covariance Table

An SS for the categorical variable sex adjusted for age and an SS for the covariate age as a factor itself are calculated. (These are more involved than the clinical reader probably will care to follow.) The SSE, following the symbolism used in Chapter 8 with a "1" or "2" attached to the subscript to indicate sex, appears as

$$SSE = s_{x1}^2 + s_{x2}^2 - \frac{(s_{uy1} + s_{uy2})^2}{s_{u1}^2 + s_{u2}^2} = 127.16.$$

The SSE has $n_1 + n_2 - 3\ df$. The ANCOVA table, which looks similar to an ANOVA table, thus is generated as follows:

Analysis of Covariance for Sex and Age as Factors in Theophylline Levels

Source	Sum of squares	*df*	Mean square	F	Critical F	*p*
Sex	42.5901	1	42.5901	4.02	4.75	0.068
Age	2.2092	1	2.2092	0.21	4.75	0.656
Error	127.6541	12	10.5970			
Total	178.6573	14				

If we had tested sex unadjusted for age with a *t* test or a one-way ANOVA, we would have found $p = 0.044$, which is less than a significance cut point of 0.05. However, when adjusted for age, sex is greater than 0.05, losing significance. Thus, age is not a significant influence on theophylline level by itself ($p = 0.68$), nor is level significantly different between sexes (a test of mean ages for the two sexes yields $p = 0.434$), but age interacted with sex enough to alter theophylline level significance by sex.

ADDITIONAL EXAMPLE: COMFORT OF BANDAGE LENSES FOLLOWING EYE SURGERY

Part of the procedure for photorefractive keratometry (PRK) correction of vision is to remove the corneal membrane. To protect the eye and reduce pain, a soft contact lens is worn as a bandage. A study[14] compared comfort level between two types of lens for 100 post-PRK eyes wearing each type. A *t* test of mean comfort level between the lens types 2 days after surgery yielded $p = 0.011$, indicating a significant difference. However, the defect size, a continuous variable, could not be controlled, because the defect heals at different rates for different patients. Let us denote the lens types by 1 and 2. The comfort level means are 1.93 overall, 2.06 for lens 1, and 1.79 for lens 2. The day 2 defect sizes are 2.87 overall, 3.62 for lens 1, and 2.12 for lens 2. Does the defect size interact with lens type to alter the significance of comfort level difference by lens? An ANCOVA table is as follows:

Analysis of Covariance for Bandage Lens Type and Defect Size in Photorefractive Keratometry Comfort Level

Source	Sum of squares	*df*	Mean square	F	Critical F	*p*
Lens	4.1740	1	4.1740	7.50	3.89	0.007
Defect size	0.5794	1	0.5794	1.04	3.89	0.309
Error	109.5715	197	0.5562			
Total	113.8450	199				

When adjusting each lens' comfort level for defect size, the difference between lenses is slightly magnified; *p* reduces from 0.011 to 0.007. We are assured that the difference in defect size is not a major influence or biasing variable.

18.5. THREE- AND HIGHER-FACTOR ANALYSIS OF VARIANCE

ORIENTATION BY EXAMPLE: COOLING KIDNEYS EXTENDED TO ALL FACTORS

In the Additional Example of Section 18.2, the kidney-cooling comparison was addressed using a two-way ANOVA by treating readings on the factor of depth measurement within the kidney as replications. Because one treatment cooled from the outside and the other from the inside, depth is very relevant. Now we will reexamine the data using the depth measure in three-way ANOVA. Three-way ANOVA is conceptually a simple extension of two-way ANOVA. We have three main effect tests rather than two, three two-factor interactions rather than one, and the addition of a three-factor interaction. The main effect and two-factor interaction results are interpreted one by one in the same way as they are for two-way ANOVA. What is difficult about three-way ANOVA are means tabulation and the interpretation of the three-factor interaction. Let us provide the example and discuss these issues in context.

The SS, *df*, and mean squares for treatment, time, and their interaction are identical with those of a two-way analysis. We create a means table for treatment by depth, from which we calculate the SS for the depth main effect and the treatment-by-depth interaction following the patterns in Table 18.2. Number of *df* for depths is number of depths less one, or $2 - 1 = 1$; number of *df* for the treatment-by-depth interaction is *df* for treatment $\times$ *df* for depth, or $1 \times 1 = 1$. We create a means table for time by depth, from which we calculate the SS for the time-by-depth interaction following the two-factor interaction pattern in Table 18.2. $df = 1 \times 3 = 3$. We create a more extended means table that gives two factors along one axis and the third factor along the other axis, for example, each depth with each treatment pairing along the top and time down the side, from which we calculate the SS for the treatment-by-depth-by-time interaction as the product of number of elements in each main factor ($2 \times 2 \times 4$) times the raw SS of treatment $\times$ time $\times$ depth means $- A -$ SS for all main effects and second-order interactions. Degrees of freedom for the third-order interaction is the product of df for main effects, that is, $1 \times 1 \times 3 = 3$.

The reader is not carried through the actual computations, because no one is likely to perform three-factor ANOVA manually. The issue is to understand how the calculations found in a software-generated table came about. Such a software-generated table is as follows:

Source	Sums of squares	df	Mean squares	F	Critical F	p
Treatments	4063.8037	1	4063.8037	455.30	3.96	<0.001
Time	4461.8704	3	1487.2901	166.63	2.72	<0.001
Depth	145.0417	1	145.0417	16.25	3.96	<0.001
Treatment × time	1826.6405	3	608.8224	68.22	2.72	<0.001
Treatment × depth	201.8400	1	201.8400	22.61	3.96	<0.001
Time × depth	43.0092	3	14.3364	1.61	2.72	0.194
Treatment × time × depth	75.1142	3	25.0381	2.81	2.72	0.045
Error (residual)	714.0366	88	8.9255			
Total	11531.3562	95				

We note that MSE has reduced to two-thirds of its two-factor size because of removal of the variability due to the third factor. This, of course, increases all the F-values, making the tests on the original factors more sensitive. Time, treatment, and time-by-treatment interaction are all highly significant as before with the same interpretations. Depth is highly significant, as we anticipated. The treatment-by-depth interaction is also highly significant. By examining a means table, we could see that the depth effect is less in the cooled saline than in the ice treatment. The time-by-depth interaction is not significant, indicating that the pattern of reduction in mean temperature over time is only slightly different for the two depths. Finally, the third-order interaction is just barely significant. The interpretation of third- and higher-order interactions is usually not obvious and takes thought and understanding of the processes. In this case, it might best be explained by saying that the treatment-by-depth interaction changes over time.

18.6. MORE SPECIALIZED DESIGNS AND TECHNIQUES

Many specialized designs have been developed to meet unusual constraints on data availability, indeed more than could be discussed in any single book. The statistical literature abounds with them. This section describes a few major classes of specialized designs, with no pretension of being of direct use. If the reader recognizes a match to data to be analyzed, these are the names to seek in more specialized books or to discuss with a biostatistician.

FACTORIAL DESIGNS

Factorial designs are essentially multifactor designs that are balanced; therefore, all variable combinations are represented and have equal numbers. All the examples in the two- and three-way ANOVAs in this chapter are balanced in this way and are factorial designs. The kidney-cooling experiment, containing two treatments at two depths over four times was a $2 \times 2 \times 4$ three-way factorial ANOVA. There were $2 \times 2 \times 4 = 16$ readings on each pig; 6 pigs $\times$ 16 readings = 96 readings used.

SPARSE-DATA DESIGNS

In some situations, the investigator has considerable control over experimental variables, but data are very difficult or very expensive to obtain. In such a circumstance, we want to maximize the information to be obtained from every datum. Consider an experiment in which we want to investigate cancer patient survival based on combinations of radiation, chemotherapy, and surgery. We have three levels of radiation to examine, three dosages of chemotherapy, and three types of surgery. We might display a design of all combinations of the first two variables as follows:

		Chemotherapy		
		Dose 1	Dose 2	Dose 3
Radiation	Level 1			
	Level 2			
	Level 3			

It can be seen that each chemotherapy dose (C) is paired with each radiation level (R), forming nine combinations. If we were to pair each of the surgical options (S) with each of these pairings, we would have a $3 \times 3 \times 3$ (or 3^3) factorial ANOVA, requiring 27 patients (randomized into 27 treatment combinations) per replication. If, however, we overlay the two-way pattern of numbers with three letters, balanced by each occurring once and only once in each row and column, we then have a design in which the main effect results would be available with only nine patients. The design might look as follows:

| | | Chemotherapy | | |
		Dose 1	Dose 2	Dose 3
Radiation	Level 1	*A*	*B*	*C*
	Level 2	*C*	*A*	*B*
	Level 3	*B*	*C*	*A*

There are a number of different patterns satisfying the balance requirement, and the pattern can be chosen randomly. This design is named a *Latin square*, because of the (Latin) letters paired with the numerical combinations.

There are numerous other sparse-data designs. The *Graeco-Latin square*, in which a pattern of Greek letters, balanced with both the number pairings and the Latin letters, overlays a square, yields a design in which four main effects may be tested from the square. An *incomplete block* design is a design that lacks some of its balance. It usually arises from an inability to apply all treatments in one block. If three types of eye surgery are being compared on rabbits, which have only two eyes per rabbit, we might use an incomplete block design. A *split-plot* design is one in which data providing results for the most important effects occur in balance, but that for some lesser effects occur only within one or another primary effect. Some information is lost, but the lesser effects are not abandoned.

COMPONENTS OF ANALYSIS OF VARIANCE

If it is assumed that the data on which the ANOVA was conducted represent the population of such data, the ANOVA is called a fixed-effects model. Perhaps we select 10 patients for each of 3 treatments, all from the same hospital. If, however, it is assumed that the data for each component arises from its own subpopulation of the grand population, the ANOVA is called a components-of-variance model. The 10 patients for each treatment are selected from a different hospital, where the 3 hospitals are chosen randomly from a list of cooperating hospitals. The calculations and results are the same for the simpler analyses usually met in medical studies. The interpretation differs somewhat.

MULTIVARIATE ANALYSIS OF VARIANCE (MANOVA)

Up to this point in the chapter, we have addressed designs that had a single outcome as influenced by various factors. It may be that multiple outcomes can occur and need to be examined simultaneously. Suppose we have an experiment in which patients with breast cancer are treated with radiation, chemotherapy, and/or surgery. We want to know patient satisfaction levels for these treatments and their interactions, what is a pain/misery level resulting from treatment, and how this level interacts with survival. An ANOVA with more than one outcome measure is called a multivariate analysis of variance (MANOVA).

GENERAL LINEAR MODELS

General linear models (GLM) is an umbrella term for a mathematical program written with enough generality to include ANOVA, ANCOVA, MANOVA, and multiple regression methods in the same package. General linear models programs occur in some statistical software packages. The user must designate the parameters that specialize the package to the particular method desired and organize the data format accordingly.

ANSWERS TO EXERCISES

18.1. Means were as follows:

	Other causes	Kidney stones	Combined
Placebo	20.71	−0.14	10.29
Drug	6.71	10.14	8.43
Combined	13.71	5.00	9.36

The ANOVA table was as follows:

Two-Way Analysis of Variance with Interaction

Effect	Sums of squares	*df*	Mean squares	F	*p*
Drug	24.14286	1	24.14286	0.22	0.643
Kidney stones	531.57143	1	531.57143	4.85	0.038
Drug × kidney stones interaction	1032.14286	1	1032.14286	9.42	0.005
Residual (or error) component	2630.57143	24	109.60714		
Total	4218.42857	27			

We note that the drug is not shown to be effective ($p = 0.643$) overall. However, there is a significant difference in pain change between the patients with kidney stones and the other patients ($p = 0.038$), and the interaction is highly significant ($p = 0.005$), indicating that pain change has a different pattern between patients with kidney stones and other patients. Examining the means table, we see that, although combined groups no-drug-to-drug means are similar, mean pain is reduced by placebo (!) for the group with other causes of pain and is reduced by the drug for the group with kidney stones. This reversal has obscured the effect of the drug for the two groups and appears to account for the mixed result pattern seen in prior medical articles. We conclude that the drug should be given to patients with kidney stones but not to other patients.

18.2. The ANOVA table is as follows:

Repeated-Measures Analysis of Variance for Migraine Headache Trial

Effect	Sums of squares	df	Mean squares	F	Critical F	p
Drug treatment	10944.6404 (SSC)	1	10944.6404	5.531	4.12	0.024
Between-drug error	71232.8240 [SSE(B)]	36	1978.6896			
Time	27196.4737 (SSR)	2	13598.2368	58.439	3.13	< 0.001
Drug × time interaction	10043.4912 (SSI)	2	5021.7456	21.581	3.13	< 0.001
Within-time error	16753.6110[SSE(W)]	72	232.6890			
Total	136171.0400 (SST)	113				

A significant difference appears between drugs. From the means table, we can see that prochlorperazine reduces pain more than diazepam does. A highly significant difference appears in the repeated effects. Both drugs reduce the headache over time. However, their pattern through time is different. From the means table, we can see that prochlorperazine reduces pain through time more effectively than diazepam.

Chapter 19

Tests on Variances of Continuous Data

19.1. BASICS OF TESTS ON VARIABILITY

WHY SHOULD WE BE INTERESTED IN TESTING VARIABILITY?

The average does not tell the whole story. As a metaphor, consider two bowmen shooting at a target. Bowman A always hits the bull's-eye. Half of bowman B's arrows fall to the left of the bull's-eye and half to the right. Both bowmen have the same average, but bowman A is the better shot. As an example in medicine, small amounts of two orally administered drugs reach a remote organ. The mean level is the same for both, but drug B is more variable (has larger standard deviation) than drug A. In some cases, too little of drug B gets through to be effective, whereas in other cases, a dangerously high level gets through. Thus, the less variable drug is better in this example.

A TEST OF VARIABILITY SERVES TWO MAIN PURPOSES

A test of variability serves (1) to detect differences in variability per se, as just illustrated, and (2) to tests the assumption of equal variances used in means testing.

HOW ARE TWO VARIANCES COMPARED?

In dealing with variability, we usually use the *variance*, which is just the square of the standard deviation. The decisions made using it are the same, and the difficult mathematics associated with square roots are avoided in the derivation of probability functions. To compare the relative size of two variances, we take their ratio, adjusted for degrees of freedom (*df*). The tabulated critical value of the ratio for statistical significance also depends on the *df*.

A SAMPLE VARIANCE MAY BE COMPARED WITH A POPULATION VARIANCE OR ANOTHER SAMPLE VARIANCE

If we compare a sample variance s^2 against a population variance σ^2 (see Section 19.2), we find the significance level of chi-square $(\chi^2) = df \times s^2/\sigma^2$ in chi-square Tables III or IV (see Tables of Probability Distributions) for $df = n - 1$. If we compare two sample variances s_1^2 and s_2^2 (see Section 19.3), assigning subscript "1" to the larger, we find the significance level of F $= s_1^2/s_2^2$ in Table V (see Tables of Probability Distributions). This table involves df for both variances: $df_1 = n_1 - 1$ and $df_2 = n_2 - 1$.

A NONSIGNIFICANT TEST RESULT MUST BE INTERPRETED CAREFULLY

Tests of variance are often useful in assessing the validity of the equal-variance assumption required for normal (z) and t tests and the analysis of variance (ANOVA), but we must understand the limitation of this use. A significant difference between variances implies the conclusion that they are different, whereas, like all difference tests of hypotheses, a nonsignificant result implies only that no difference has been demonstrated, not that one does not exist. However, the usual approach is to use the method that makes the assumption of no difference in variances when we lack evidence of such a difference.

19.2. SINGLE SAMPLES

EXAMPLE POSED: PROSTATE-SPECIFIC ANTIGEN VARIABILITY

Is the prostate-specific antigen (PSA) variance of the first 10 urology patients (see Table DB1.1) different from that of the remaining 291 patients? Using the subscript "0" to denote the sample, $H_0: \sigma_0^2 = \sigma^2$. We have no reason to anticipate whether s^2 should be larger or smaller than σ^2, so we use $H_1: \sigma_0^2 \neq \sigma^2$; if α is chosen as 5%, we allow 2.5% for each tail. The reader might find it easier to think of the question in terms of standard deviations. Let us consider the 291 readings as a population with $\sigma = 17.19$ and the first 10 reading as a sample with $s = 1.61$. Could the difference between σ and s have occurred by chance? We use variances for our test to avoid the mathematics of square roots. $\sigma^2 = 295.50$ and $s^2 = 2.59$.

METHOD: TEST OF ONE SAMPLE VARIANCE

We ask whether the variance σ_0^2 of a population from which we draw a sample is the same as a theoretical variance σ^2. σ_0^2 is estimated by the sample variance s^2. We assume the sample data form a normal distribution. σ^2 may be known from some theory, or it may be estimated by a very large sample, because the sample variance converges on the population variance

---

as the sample size grows large. The null hypothesis is that s^2 is drawn from the population having variance σ^2, or $H_0: \sigma_0^2 = \sigma^2$, implying that the ratio s^2/σ^2 should be different from 1 by only a random influence. The statistic calculated is

$$\chi^2 = \frac{df \times s^2}{\sigma^2}. \tag{19.1}$$

This is a relatively simple test except for one aspect: The test statistic is distributed chi-square, which is not symmetric, so a different table must be used for each tail. The alternate hypothesis H_1 specifies the probability table(s) to be used. For a chosen α, the cases are as follows:

(a) $\sigma_0^2 > \sigma^2$: the chi-square critical value for α is found from Table III.
(b) $\sigma_0^2 < \sigma^2$: the chi-square critical value for α is found from Table IV.
(c) $\sigma_0^2 \neq \sigma^2$: split α; the left χ^2 critical value for $\alpha/2$ is found from Table IV, and the right χ^2 critical value for $\alpha/2$ is found from Table III.

The steps are as follows:

1. Verify that the sample frequency distribution is roughly normal.
2. Specify null and alternate hypotheses and choose α.
3. Identify the theoretical σ.
4. Look up the critical value(s) for the chosen α in the appropriate chi-square table(s).
5. Calculate the statistic from Eq. (19.1): $\chi^2 = df \times s^2/\sigma^2$.
6. Make the decision whether or not to reject the null hypothesis.

EXAMPLE COMPLETED: PROSTATE-SPECIFIC ANTIGEN VARIABILITY

Let us follow the steps given in the previous paragraph. We assume normality: We have seen that PSA data are roughly bell-shaped, although right skewed. We believe the deviations are not so great as to invalidate this rather robust test. The remaining 291 PSA readings form a sample large enough that its variance has converged closely to the population variance, so we may use the calculated variance as σ^2. We will test s^2/σ^2, which, when multiplied by df, is distributed chi-square. The left-tail critical value must be found from Table IV and the right-tail critical value from Table III, because the chi-square distribution is not symmetric like z or t distributions. $df = 10 - 1 = 9$. Table 19.1 gives a fragment of Table IV. From Table 19.1 (or Table IV), $\alpha = 0.025$ coupled with $df = 9$ yields $\chi^2 = 2.70$. If $df \times s^2/\sigma^2 < 2.70$, we reject H_0. Similarly, the same inputs into Table III (or Table 19.2) yield $\chi^2 = 19.02$; if $df \times s^2/\sigma^2 > 19.02$, we reject H_0. Now we can calculate our statistic. $\chi^2 = df \times s^2/\sigma^2 = 9 \times 2.59/295.50 = 0.0789$, which is far less than 2.70, the left critical value. We reject H_0 and conclude that the first 10 patients have smaller variability than the remaining 291 patients. Indeed, from Table 19.1, the χ^2 for $\alpha = 0.0005$ with 9 df is 0.97. Because $0.0789 < 0.97$, we have a less than 0.001 chance of error by rejecting H_0.

Table 19.1

Table 19.1

A Fragment of Chi-square Distribution, Left Tail[a]

α (area in left tail)	0.0005	0.001	0.005	0.01	0.025	0.05	0.10
$df = 9$	0.97	1.15	1.73	2.09	2.70	3.33	4.17

[a]Selected chi-square (χ^2) values (distances above zero) are given for 9 df for various α, the area under the curve in the left tail.
Table IV in Tables of Probability Distributions provides an expanded version of this table.

Table 19.2

A Portion of Chi-square Distribution, Right Tail[a]

α (area in right tail)	0.10	0.05	0.025	0.01	0.005	0.001	0.0005
$df = 2$	4.61	5.99	7.38	9.21	10.60	13.80	15.21
4	7.78	9.49	11.14	13.28	14.86	18.46	20.04
9	14.68	16.92	19.02	21.67	23.59	27.86	29.71
14	21.06	23.69	26.12	29.14	31.32	36.12	38.14

[a]Selected χ^2 values (distances above zero) are given for various df and for selected α, the area under the curve in the right tail.
Table III in Tables of Probability Distributions provides an expanded version of this table.

ADDITIONAL EXAMPLE: IS A TREATMENT FOR DYSPEPSIA IN THE EMERGENCY DEPARTMENT TOO VARIABLE?

In the additional example of Section 17.2, an emergency medicine physician tested the effectiveness, as measured on a 1–10 pain scale, of a "GI cocktail" (antacid plus viscous lidocaine) to treat emergency dyspeptic symptoms.[50] The physician concluded that the treatment was effective on average, but was it more variable? This question implies a one-tailed test. She will sample 15 patients; thus $df = 14$. Table 19.2 provides a portion of Table III. From Table 19.2 (or Table III), with 14 df, the critical value of χ^2 for $\alpha = 0.05$ is 23.69. The population standard deviation for the scoring difference of patients receiving treatment minus that for a large number of patients who did not receive treatment after a specified number of minutes was $\sigma = 1.73$. The specialist treated $n = 15$ patients, measuring the difference in pain before versus after treatment. Data were 6, 7, 2, 5, 3, 0, 3, 4, 5, 6, 1, 1, 1, 8, 6. The specialist calculated $m = 3.87$ and $s = 2.50$. She tested the variance of her sample ($s^2 = 6.27$) against the variance of untreated patients ($\sigma^2 = 2.99$) at the $\alpha = 0.05$ level of significance. By substituting in Eq. (19.1), she found

$$\chi^2 = \frac{df \times s^2}{\sigma^2} = \frac{14 \times 6.27}{2.99} = 29.36.$$

The calculated χ^2 is greater than the critical value of 23.69. Indeed, it is slightly greater than the 29.14 value associated with $\alpha = 0.01$. The specialist has evidence that her sample is more variable than is the population. From a statistical software package, $p = 0.009$.

Exercise 19.1: Is the Variability of Readings from a Tympanic Thermometer Too Large? We are investigating the reliability of a certain brand of tympanic thermometer (temperature measured by a sensor inserted into the patient's ear). Sixteen readings (measured in degrees Fahrenheit [°F]) were taken on a healthy patient at intervals of 1 minute[63]; data for left and right ears were pooled. (Data are given in Exercise 16.2.) In our clinical judgment, a reliable thermometer will be no more than 1°F from the correct temperature 95% of the time. This implies that 1°F is about 2σ, so that $\sigma = 0.5$°F. From the data, $s = 1.230$. At the $\alpha = 0.05$ level of significance, is the variability of the tympanic thermometer reading on a patient unacceptably large?

19.3. TWO SAMPLES

EXAMPLE POSED: DOES THE INITIAL UNREPRESENTATIVE VARIABILITY IN PROSTATE-SPECIFIC ANTIGEN LEVELS EXTEND TO LATER PATIENTS?

In the example of Section 19.2, we concluded that the PSA variances (and therefore standard deviations) for the first 10 and the remaining 291 patients were quite different. We want to know if the first 10 patients per se are unrepresentative or if the variability increased gradually or at a later point. Let us test the variance of the first 10 patients against each successive 10 patients and list the result. The normality-of-data assumption remains. By using "first" to denote the first sample of 10 and "later" to denote each later sample, $H_0: \sigma^2_{first} = \sigma^2_{later}$ and $H_1: \sigma^2_{first} < \sigma^2_{later}$.

METHOD: TEST OF TWO SAMPLE VARIANCES

Are two population variances, or standard deviations, the same? We assume that the data of both samples form a normal distribution. $H_0: \sigma^2_1 = \sigma^2_2$. Because we test the larger variance over the smaller, Table V may be used for any H_1. Choose α. Sample 1 has sample variance s^2_1 calculated from n_1 observations and sample 2 has s^2_2 from n_2, where the subscript "1" is assigned to the larger variance. The statistic is

$$F = \frac{s^2_1}{s^2_2}, \tag{19.2}$$

with $n_1 - 1$ and $n_2 - 1$ df. Look up the critical value of F in Table V or calculate the p-value with a statistical software package. The steps are as follows:

1. Verify that the data of each sample are roughly normal.
2. State the null and alternate hypotheses ($H_0: \sigma^2_1 = \sigma^2_2$, and $H_1: \sigma^2_1 > \sigma^2_2$); choose α.

Table 19.3

A Portion of F Distribution[a]

Denominator *df*		Numerator *df*		
		8	9	10
	9	3.23	3.18	3.14
	10	3.07	3.02	2.98
	11	2.95	2.90	2.85
	12	2.85	2.80	2.75

[a]Selected distances (F) are given for $\alpha = 5\%$, the area under the curve in the positive tail. Numerator *df* appears in column headings, denominator *df* appears in row headings, and F appears in the table body. Table V in Tables of Probability Distributions provides an expanded version of this table.

> 3. Look up the critical value for the chosen α in Table V.
> 4. Calculate the statistic $F = s_1^2/s_2^2$ (where s_1^2 designates the larger variance).
> 5. Decide whether or not to reject the null hypothesis.

EXAMPLE COMPLETED: DOES THE INITIAL UNREPRESENTATIVE VARIABILITY IN PROSTATE-SPECIFIC ANTIGEN LEVEL EXTEND TO LATER PATIENTS?

We choose $\alpha = 5\%$. Table 19.3 is a segment of Table V. From Table 19.3 (or Table V), we find that the critical value of F for 9,9 *df* is 3.18. The F ratio must exceed 3.18 to become significant.

Calculations

The second 10 readings (11–20) yielded a standard deviation of 2.11. $F = (2.11)^2/(1.61)^2 = 4.45/2.59 = 1.72$. Because $1.72 < 3.18$, we conclude that the variances (and, of course, the standard deviations) are not different for the first and second sets of 10 readings. Following is a list of comparisons of the first 10 readings with successive samples of 10, compiled using the same numerical procedures:

Sample numbers	F ratio
11–20	1.72
21–30	1.20
31–40	0.92
41–50	794.06
51–60	11.14
61–70	27.28
71–80	511.72
81–90	61.11
:	:

Interpretation

We note that the 41–50 set yielded a huge F. By examining the data, we find that PSA = 221 for patient 47! This turned out to be a useful way to discover *outliers*, those data so far different from the bulk of data that we suspect they arose from a unique population. However, even if we eliminated the 221 reading, the F ratio would equal 8.88 for the remaining data, which is still significant. It appears that there are patterns of nonhomogeneity scattered throughout the data.

ADDITIONAL EXAMPLE: COMPARE THE VARIABILITY OF TWO PAIN RELIEF DRUGS

In 1958, a study[50] compared a new postoperative pain relief drug (treatment 1) with the established Demerol (treatment 2). Data consisted of reduction in pain measured on a 1–12 rating scale. For patients who completed the protocol, data were as follows:

| Treatment 1 | 2 | 6 | 4 | 12 | 5 | 8 | 4 | 0 | 10 | 0 | | | |
| Treatment 2 | 2 | 0 | 3 | 3 | 0 | 0 | 7 | 1 | 4 | 2 | 2 | 1 | 3 |

To test the means, we need to know whether we can assume equal variances. We want to test $H_0: \sigma_1^2 = \sigma_2^2$ against $H_1: \sigma_1^2 \neq \sigma_2^2$ at $\alpha = 0.05$. $m_1 = 5.10$, $s_1 = 4.0125$, $s_1^2 = 16.1000$, $m_2 = 2.15$, $s_2 = 1.9513$, and $s_2^2 = 3.8077$. We substitute in Eq. (19.2) to find

$$F = \frac{s_1^2}{s_2^2} = \frac{16.10}{3.81} = 4.23.$$

From Table 19.3 (or Table V), the critical value of F for 9,12 *df* is 2.80. Because 4.23 is much larger, the variances (and, of course, the standard deviations) are significantly different. From a statistical software package, $p = 0.012$. A test of means must use either the unequal-variances form of the *t* test or the rank-sum test from Sections 6.6 or 16.3. From statistical software, $p = 0.012$.

Exercise 19.2: Is the Variability of the Tympanic Thermometer Readings the Same in Both Ears? We are investigating the reliability of a certain brand of tympanic thermometer (temperature measured by a sensor inserted into the patient's ear). Sixteen readings (measured in degrees Fahrenheit [°F]), eight per ear, were taken on a healthy patient at intervals of 1 minute, alternating ears.[63] (Data are given in Exercise 16.2.) We want to test the hypothesis that the mean temperatures are the same in the left (L) and right (R) ears (see Exercise 17.3), but we want evidence that the equal-variance *t* test is appropriate. $m_L = 96.41°F$, $s_L = 0.88°F$, $m_R = 97.86°F$, and $s_R = 1.12°F$. At the $\alpha = 0.05$ level of significance, are the variances (or standard deviations) of the two ears different?

19.4. THREE OR MORE SAMPLES

EXAMPLE POSED: CAN WE ASSUME EQUAL VARIANCES IN THE TEST OF CLASSIFYING PATIENTS BY RISK FOR PROSTATE CANCER?

Of our sample of 301 patients, 19 are known to have cancer already and therefore are not in the risk population; thus, we must delete their data to obtain an unbiased population, leaving 282 patients. Three groups remain, identified by a clinical decision algorithm: those whose risk for prostate cancer is (1) low, (2) moderate, and (3) high. We note that mean prostate-specific antigen density (PSAD) appears quite different for the three groups: 0.07, 0.21, and 0.35, respectively. We want to test this difference using a one-way ANOVA, but that requires the assumption of equal variances. Do we have evidence that we should not make that assumption?

METHOD: BARTLETT'S TEST OF HOMOGENEITY OF VARIANCE

We want to know whether three or more variances are the same. Bartlett's test compares the variances of k independent random samples, assumed to be distributed normal. It tests H_0: $\sigma_1^2 = \sigma_2^2 = \cdots = \sigma_k^2$ against the alternate H_1: not H_0. Choose α. Let i denote sample number, $i = 1, 2, \ldots, k$. There are n observations in total in the k samples with n_i observations x_{ij} in the ith sample. First, find the variance s_i^2 for each sample; this is done the same way the usual sample variance is found (where j is the index to sum over):

$$s_i^2 = \frac{\sum x_{ij}^2 - n_i m_i^2}{n_i - 1}.$$ (19.3)

Then pool the k sample variances to find the overall variance s^2 (now summing over i):

$$s^2 = \frac{\sum (n_i - 1)s_i^2}{n - k}.$$ (19.4)

Note that Eq. (19.4) is not the same as the overall variance of the n observations, because each is the variance about its own sample mean. The test statistic, Bartlett's M, is given by

$$M = \frac{(n - k)\ln(s^2) - \sum(n_i - 1)\ln(s_i^2)}{1 + \frac{1}{3(k - 1)}\left(\sum \frac{1}{n_i - 1} - \frac{k}{n - k}\right)}.$$ (19.5)

The α for M arises from the right tail of a chi-square distribution with $k - 1$ df. A critical value for M may be found in Table III. An M less than this critical value implies inadequate evidence to reject H_0; otherwise, reject H_0. The steps are as follows:

1. Note the null and alternate hypotheses (H_0: $\sigma_1^2 = \sigma_2^2 = \cdots = \sigma_k^2$, and H_1: not H_0).
2. Choose α.
3. Look up the critical value for the chosen α in Table III for $k - 1$ *df*.
4. Calculate the variances s_i^2 for each sample from Eq. (19.3) and the overall variance s^2. [Note that s^2 is not exactly the same as the variance for the entire n observations combined; calculate it from Eq. (19.4) using the several s_i^2].
5. Calculate Bartlett's M as in Eq. (19.5).
6. Make the decision to accept or reject the null hypothesis.

EXAMPLE COMPLETED: CAN WE ASSUME EQUAL VARIANCES IN THE TEST OF CLASSIFYING PATIENTS BY RISK FOR PROSTATE CANCER?

Is the normal assumption satisfied? We make plots of the PSAD frequency distributions for the three groups with normal fits superposed (Fig. 19.1). We are not satisfied with the one or two extremely high PSAD values in each group. It would be advisable to go to rank methods and perform a Kruskal–Wallis test

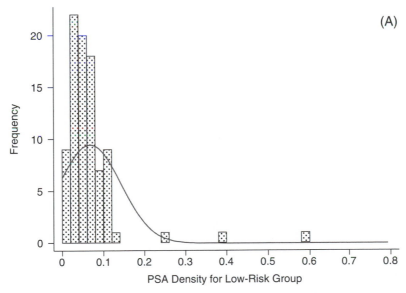

(A)

Figure 19.1 Plots of prostate-specific antigen density distributions of 282 patients for groups of low, moderate, and high risk for prostate cancer.   *Continued*

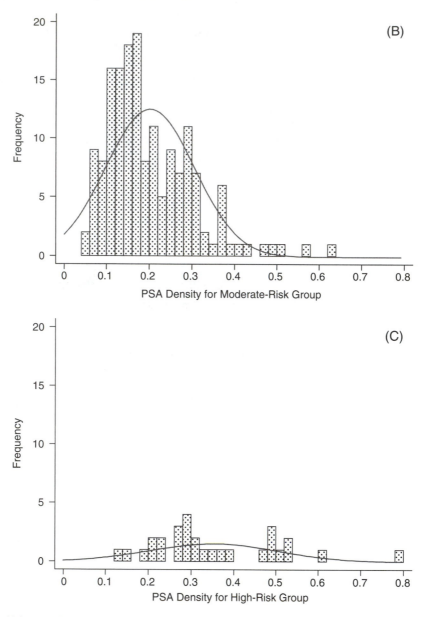

Figure 19.1—cont'd

from Section 16.4. However, we will continue with Bartlett's test for the sake of illustration.

Calculations

H_0: $\sigma_1^2 = \sigma_2^2 = \sigma_3^2$, and H_1: not H_0. We choose $\alpha = 0.05$. The 0.05 critical value from Table 19.2 (or Table III) for 2 df is 5.99; if $M > 5.99$, we reject H_0. We obtain the following values, carrying six significant digits in the calculations (so that we can take precise square roots). The s_i^2 values are just the ordinary variances (standard deviations squared). Those variances and the associated n's are as follows:

$n_1 = 89$	$s_1^2 = 0.00557546$
$n_2 = 164$	$s_2^2 = 0.0139188$
$n_3 = 29$	$s_3^2 = 0.0227769$

Substitution in Eq. (19.4) (summing over the index i) yields

$$s^2 = \frac{\sum (n_i - 1)s_i^2}{n - k} = \frac{88 \times 0.00557546 + 163 \times 0.0139188 + 28 \times 0.0227769}{282 - 3}$$

$$= 0.0121762.$$

Finally, substitution in Eq. (19.5) yields

$$M = \frac{(n - k)\ln(s^2) - \sum (n_i - 1)\ln(s_i^2)}{1 + \frac{1}{3(k - 1)}\left(\sum \frac{1}{n_i - 1} - \frac{k}{n - k}\right)}$$

$$= \frac{279\ln(0.0121762) - 88\ln(0.00557546) - \cdots - 28\ln(0.0227769)}{1 + \frac{1}{6}\left(\frac{1}{88} + \frac{1}{163} + \frac{1}{28} - \frac{3}{279}\right)}$$

$$= 29.1931.$$

The M of 31.04 is greater than the critical value of 5.99, so H_0 is rejected. In fact, M exceeds the critical value for the smallest chi-square in the table, 15.21 for $\alpha = 0.0005$. We may state that $p < 0.001$. We have strong evidence that the variances are quite different; thus an ANOVA would not be appropriate.

ADDITIONAL EXAMPLE: MEDICATION TO REDUCE EDEMA FOLLOWING RHINOPLASTY

Following rhinoplasty, swelling may cause deformity during healing. Steroids may decrease the swelling, but the required level of steroid was not known. Is the equal variance assumption satisfied for the one-way ANOVA of medication levels to reduce edema following rhinoplasty? A total of $n = 50$ rhinoplasty patients were randomized into $k = 5$ groups of increasing steroid level having $n_1 = \ldots = n_5 = 10$. Swelling reduction was measured by magnetic resonance images before and after administration of the steroid.[9] The means, variances, and first few data are shown in Table 19.4.

The requirement was to perform a one-way ANOVA to learn whether the means were all the same or whether there were differences among them (see Additional Example from Section 17.5). The ANOVA requires the assumption

Table 19.4

Data on Edema Following Rhinoplasty

	Swelling reduction (ml)				
	Level 1	Level 2	Level 3	Level 4	Level 5
Patients 1–5	1.6	4.4	5.5	4.0	8.0
Patients 6–10	2.3	5.8	6.4	8.3	6.2
Patients 11–15	2.3	6.4	6.9	5.8	3.1
⋮	⋮	⋮	⋮	⋮	⋮
Means m_i	3.77	5.00	5.67	6.79	6.35
Variances s_i^2	6.30	3.80	1.85	2.43	2.25

that the variances of the various groups are equal, which may be examined by Bartlett's M. We choose $\alpha = 0.05$. From Table 19.2 (or Table III), the critical value of M (which follows the χ^2 distribution) with $k - 1 = 4$ df is 9.49. By substituting in Eq. (19.4) and then Eq. (19.5), we find

$$s^2 = \frac{\sum (n_i - 1)s_i^2}{n - k} = \frac{9 \times 6.30 + 9 \times 3.80 + \cdots + 9 \times 2.25}{50 - 5} = 3.326$$

$$M = \frac{(n - k)\ln(s^2) - \sum (n_i - 1)\ln(s_i^2)}{1 + \dfrac{1}{3(k - 1)}\left(\sum \dfrac{1}{n_i - 1} - \dfrac{k}{n - k}\right)}$$

$$= \frac{45 \times \ln(3.326) - 9 \times \ln(6.30) - \cdots - 9 \times \ln(2.25)}{1 + \dfrac{1}{3 \times 4}\left(\dfrac{1}{9} + \cdots + \dfrac{1}{9} - \dfrac{5}{45}\right)} = 4.51.$$

Because $4.51 < 9.49$, we do not have enough evidence to reject H_0; the variances are not shown to be different. Because we have some evidence that the variances are not different, we will proceed with the one-way ANOVA. From a statistical software package, the calculated $p = 0.341$.

Exercise 19.3: Is There a Batch-to-Batch Variability of Laboratory Animals' Resistance to Parasitic Infestation? In a study[49] on the control of parasites, rats were injected with 500 larvae each of the parasitic worm *Nippostrongylus muris*. Ten days later, they were euthanized and the number of adult worms counted. In Exercise 17.5, we concluded that there was no batch-to-batch difference in *average* resistance to parasite infestation by groups of rats received from the supplier. However, is there a batch-to-batch difference in the *variability* of resistance? We answer this question using Bartlett's test on the homogeneity of variances. $k = 4$ batches of $n_i = 5$ rats each were tested. Data were given in Exercise 17.5. The four variances are 3248.30, 4495.70, 4620.70, and 3623.70. At the $\alpha = 0.05$ level of significance, test H_0: $\sigma_1^2 = \sigma_2^2 = \sigma_3^2 = \sigma_4^2$.

ANSWERS TO EXERCISES

19.1. H_0: $\sigma_0^2 = \sigma^2$, and H_1: $\sigma_0^2 > \sigma^2$. From Table III, the critical χ^2 value for $\alpha = 0.05$ with 15 df is 25.00. Substituting in Eq. (19.1), we find

$$\chi^2 = \frac{df \times s^2}{\sigma^2} = \frac{15 \times 1.230^2}{0.5^2} = 90.77.$$

Because the calculated χ^2 is much larger than the critical value, we reject H_0; the tympanic thermometer is too variable for clinical use. From a computer package, the p-value is zero to many decimal places; we would report $p < 0.001$.

19.2. H_0: $\sigma_L^2 = \sigma_R^2$, and H_1: $\sigma_L^2 \neq \sigma_R^2$. From Table V, the critical F-value for $\alpha = 0.05$ with 7,7 df is 3.79. By substituting in Eq. (19.2), we find F = $s_L^2/s_R^2 = 1.12^2/0.88^2 = 1.620$. Because $1.62 < 3.79$, we have no evidence to reject the null hypothesis; we may use the equal-variance t test. From a statistical software package, the actual $p = 0.272$.

19.3. From Eq. (19.4) and then Eq. (19.5), calculate

$$s^2 = \frac{\sum (n_i - 1)s_i^2}{n - k} = \frac{4 \times 3248.30 + \cdots + 4 \times 3623.70}{16} = 3997.10$$

and

$$M = \frac{(n - k)\ln(s^2) - \sum (n_i - 1)\ln(s_i^2)}{1 + \dfrac{1}{3(k - 1)} \left(\sum \dfrac{1}{n_i - 1} - \dfrac{k}{n - k} \right)}$$

$$= \frac{16 \times \ln(3997.10) - 4 \times \ln(3248.30) - \cdots - 4 \times \ln(3623.70)}{1 + \dfrac{1}{3 \times 3} \left(\dfrac{1}{4} + \cdots + \dfrac{1}{4} - \dfrac{4}{16} \right)}$$

$$= 0.1587.$$

From Table III, the critical $\chi_{3df}^2 = 7.81$. The calculated $M = 0.1587$ is much smaller, so we do not have evidence to reject H_0; variability of parasite infestation does not differ from batch to batch. From a statistical software package, the actual $p = 0.984$.

Chapter 20

Tests on the Distribution Shape of Continuous Data

20.1. OBJECTIVES OF TESTS ON DISTRIBUTIONS

WHAT DO WE USUALLY ASK ABOUT DISTRIBUTIONS?

Of the many questions that could be asked about distribution shape, two are most common: Is the distribution normal? and Do two distributions have the same shape?

TESTING NORMALITY

Normality of underlying data is assumed in many statistical tests, including the normal test for means, the t test, and analysis of variance. A user who is not sure that the assumption of normality is justified is on firmer ground by testing the sampling distribution(s) for normality before deciding on the appropriate test, although we must understand a limitation of this use. A significant outcome implies the conclusion that the distribution is not normal, whereas, like all difference tests of hypotheses, a nonsignificant outcome implies only that no deviation from normality has been demonstrated, not that it does not exist. An equivalence test would lead to a similar limitation; if we failed to reject the null hypothesis of difference, no equivalence has been demonstrated. At present, the usual procedure is to conduct the difference test and assume normality unless the null hypothesis of normality is rejected.

TESTING OTHER DISTRIBUTIONS

Note that the tests of normality are essentially various goodness-of-fit tests that may be used to test data fits to any known probability distribution. Other distributional forms will not be considered in this book, because the tests primarily are applied to the normal probability distribution. The methodology for other forms is similar.

TESTING EQUALITY

The user may want to know if two distributions have the same form, in some cases, to satisfy assumptions required for tests and, in other cases, to learn if the natural forces giving rise to these distributions are similar. The test does not address the issue of what the distribution shapes are, only whether or not they are the same.

20.2. TEST OF NORMALITY OF A DISTRIBUTION

TYPES OF NORMALITY TESTS AND THEIR COMPARISON

Keep in mind that the robustness of most tests (some more than others) allows the test to be used if the underlying data are *approximately* normal. Thus, barring quite unusual circumstances, any good test of normality will be adequate. As one might expect, the better tests are more complicated; also, as one might expect, the better tests are more recent. (The chi-square goodness-of-fit test was first published in 1900; the Kolmogorov–Smirnov, usually abbreviated KS, test was reported in 1933; and the Shapiro–Wilk test was published in 1965.) The choice of which test to use depends on both the sample size and whether the user would rather err on the side of being too conservative or the opposite. The Shapiro–Wilk test tends to reject the null hypothesis more readily than one would wish, whereas the KS and chi-square tests are too conservative, retaining the null hypothesis too often. Table 20.1 may provide assistance in making the test selection.

THE ROLE OF COMPUTER SOFTWARE

Statistical software packages usually are used in practice, but they are not required. Attention here is focused on what the KS and goodness-of-fit tests do and how they do it, not on commands required to conduct computer-based tests; the Shapiro–Wilk test is involved enough that it should be attempted only with a software package. Enough detail in the KS and chi-square goodness-of-fit methods is given to allow a user to apply the method even in the absence of software.

Table 20.1

Partial Guide to Selecting a Test of Normality of a Distribution

	Prefer less conservative test	Prefer more conservative test
Small sample (5–50)	Shapiro–Wilk test	Kolmogorov–Smirnov test (one-sample form)
Medium to large sample (>50)	Shapiro–Wilk test	Chi-square goodness-of-fit test

SHAPIRO–WILK TEST

Seek statistical computer software to use the Shapiro–Wilk test. One limitation of the Shapiro–Wilk test is that some software packages allow only the sample parameters (m and s, not μ and σ) to be used for the theoretical normal against which the data are being tested, whereas we would prefer to allow specification of the normal being tested by either theoretical or sample parameters.

KOLMOGOROV–SMIRNOV TEST (ONE-SAMPLE FORM)

The two-sample form of the KS test is used in Section 20.3 to test the equality of two distributions.

Example Posed: Do the Ages of the Patients in Table DB1.1 Form a Normal Distribution?

We might want to know if the sample of patients for whom a prostate biopsy was taken follows a normal distribution so that we could use the sample in a t test. H_0: The distribution from which the sample was drawn is normal. H_1: The distribution is not normal.

Method: One-Sample Kolmogorov–Smirnov Test

There are two questions we might ask: (1) Is the sample normal in shape without any *a priori* mean and standard deviation specified? and (2) Did the sample arise from a particular normal distribution with a postulated mean and standard deviation? The first question is the question we ask to satisfy the normality assumption in tests of hypothesis. H_0: The form of the distribution is normal. We use the sample's m and s in the method that follows. In the second question, we use the postulated normal distribution's mean μ and standard deviation σ in the method that follows. H_0: The distribution from which the sample was drawn is not different from *a specified* normal. Calculate the critical value from Eq. (20.1) for the desired α. (These critical values are approximations, but they are accurate to within 0.001.)

$$\frac{1.63}{\sqrt{n}} - \frac{1}{3.5n} \quad \text{for } \alpha = 0.01 \tag{20.1a}$$

$$\frac{1.36}{\sqrt{n}} - \frac{1}{4.5n} \quad \text{for } \alpha = 0.05 \tag{20.1b}$$

$$\frac{1.22}{\sqrt{n}} - \frac{1}{5.5n} \quad \text{for } \alpha = 0.10 \tag{20.1c}$$

1. Arrange the n sample values in ascending order.
2. Let x denote the sample value each time it changes. (If sample values are 1, 1, 2, 2, 2, 3, the first x is 2, the second 3, and so on.) Write these x's in a column in order.

---

3. Let k denote the number of sample members less than x. (When the sample values just mentioned changed from 1 to 2 and x became 2, there were $k = 2$ values less than x. At the next change, x takes on the value 3, and there are $k = 5$ values less than that x.) Write these k's next to their corresponding x's.

4. Let $F_n(x)$ denote k/n for each x. This is the sample cumulative frequency sum. Write down each $F_n(x)$ in the next column corresponding to the associated x.

5. Now we need the expected cumulative sum against which to test the sample. For each x, calculate $z = (x - m)/s$ for case (1) or $z = (x - \mu)/\sigma$ for case (2).

6. For each z, find an expected Fe(n) as the area under the normal distribution to the left of z. This area can be found using a statistical software package or from a very complete table of normal probabilities. If neither of these sources is available, interpolate from Table I (see Tables of Probability Distributions). Write down these $F_e(x)$ next to the corresponding $F_n(x)$.

7. Write down next to these the absolute value (value without any minus signs) of the difference between the F's, namely $|F_n(x) - F_e(x)|$.

8. The test statistic is the largest of these differences, say, L.

9. If L exceeds the critical value, reject H_0; otherwise, do not reject H_0.

Example Completed: Do the Ages of the 10 Patients Form a Normal Distribution?

$n = 10$. From Eq. (20.1b), choosing $\alpha = 0.05$, the critical value is found as

$$\frac{1.36}{\sqrt{n}} - \frac{1}{4.5n} = \frac{1.36}{\sqrt{10}} - \frac{1}{4.5 \times 10} = 0.408.$$

We set up Table 20.2 with headings representing the values that will be required in the calculation.

1. The first column consists of the ages in increasing order.
2. The second column (x) consists of the sample values at each change. The first change occurs as age goes from 54 to 61 years.
3. The third column (k) consists of the number of sample members less than x. For the first x, there is only one age less than 61, so $k = 1$.
4. The fourth column consists of the values k/n. For the first x, $k/n = 1/10 = 0.1$.
5. The fifth column consists of z, the standardized values of x. We note the mean $m = 65.1$ years and the standard deviation $s = 7.0$ years from Table DB1.1 and, for each x, we calculate $(x - m)/s$.
6. The sixth column consists of normal distribution probabilities, the area under the normal curve to the left of z. For the first x, we want to find

Table 20.2

Example Data (Ages of 10 Patients from Table DB1.1) with the Corresponding Calculations Required for the Kolmogorov–Smirnov Test of Normality

Age (years)	x	k	$F_n(x)$	z	$F_e(x)$	$\|F_n(x) - F_e(x)\|$
54						
61	61	1	0.1	$(61 - 65.1)/7 = -0.586$	0.279	0.179
61						
61						
62	62	4	0.4	$(62 - 65.1)/7 = -0.443$	0.329	0.071
62						
68	68	6	0.6	$(68 - 65.1)/7 = 0.414$	0.661	0.061
73	73	7	0.7	$(73 - 65.1)/7 = 1.129$	0.871	0.171[a]
74	74	8	0.8	$(74 - 65.1)/7 = 1.271$	0.898	0.098
75	75	9	0.9	$(75 - 65.1)/7 = 1.414$	0.921	0.021

[a]$L = 0.171$.

the area under a normal curve from the left tail up to the -0.586 standard deviation position (just over half a standard deviation to the left of the mean). A statistical package gives it as 0.279. If we do not have such access, we can interpolate from Table I. Because Table I gives only the right half of the curve, we use the curve's symmetry; the area to the left of -0.586 will be the same as the area to the right of $+0.586$. The value 0.586 lies 86% of the way from 0.50 to 0.60. We need the one-tailed area given by α for 0.50 plus 0.86 of the difference between the α's for 0.50 and 0.60, or $0.308 + 0.86 \times (0.274 - 0.308) = 0.27876$, which rounds to 0.279, as did the result from the statistical package.

7. The last column consists of the absolute difference (difference without regard to minus sign) between the entries in columns 4 and 6. For the first x, $|0.1 - 0.279| = |-0.179| = 0.179$.

8. The largest value in the rightmost column is $L = 0.179$.

9. Finally, we compare L with the critical value. Because $(L =) 0.179 \leq 0.408$ ($=$ critical value), we do not reject H_0: The distribution from which the sample was drawn has not been shown to be different from normal. From a statistical software package, $p = 0.472$.

Additional Example: A Test of Normality on a Potential Human Immunodeficiency Virus Vaccine

In a study on a potential vaccine for human immunodeficiency virus (HIV), the number of HIV per milliliter blood (denoted h) was measured on a baseline sample of $n = 45$ patients.[85] It was required that the data be approximately normal to conduct other statistical procedures. A frequency plot is clearly far from a normal distribution, which is superposed on the plot (Fig. 20.1A). Natural logarithms (ln) were taken in the hopes of transforming the data to normal. A plot of ln h with a superposed normal (based on the mean and standard deviation of the data) appears somewhat better (see Fig. 20.1B) but is deviant enough from bell-shaped to require a test of normality.

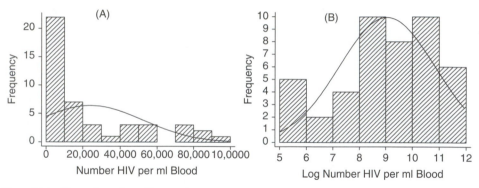

Figure 20.1 Plots of number of human immunodeficiency virus per milliliter blood (h) (A) and ln h (B) with superposed normal distributions based on m and s from the data.

Stepping Through the Test Procedure

α is chosen as 5%. The critical value of the test is found from Eq. (20.1b): substituting $n = 45$ yields 0.1978. Data arranged in ascending order appear in Table 20.3 in the format of Table 20.2. $m = 9.04$, and $s = 1.80$. Because there are no ties in ln h, x will be the same as the sample values. k is one less than the count number of each datum. As before, $F_n(x) = k/n$, $z = (x - m)/s$, and $F_e(x)$ is the area under the normal curve to the left of the z associated with that x. The last column is the absolute value of the difference between the two F's. The largest value in the last column is $L = 0.1030$. Because $L < 0.1978$, we do not reject the hypothesis of normality.

Exercise 20.1: Kolmogorov–Smirnov Test.   Hematocrit (Hct) readings need to be distributed normally to find a confidence interval. In the Additional Example of Section 16.3, we examined postoperative Hct on pyloromyotomized neonates[21] and concluded that Hct percentage was not different for laparoscopic versus open surgeries; thus, we pool the data. To put a confidence interval on mean Hct, we need to assume the distribution is normal. The $n = 16$ Hct readings in ascending order are 25.6%, 29.7%, 32.0%, 32.0%, 32.1%, 32.7%, 33.9%, 34.0%, 38.3%, 38.8%, 39.0%, 42.0%, 43.3%, 43.9%, 46.7%, and 52.0%. Test for normality using the KS test of normality.

LARGE-SAMPLE TEST OF NORMALITY OF A DISTRIBUTION: CHI-SQUARE GOODNESS-OF-FIT TEST

Characteristics of the Chi-square Goodness-of-Fit Test

The time-honored Pearson's goodness-of-fit test is a relatively easy concept, but it has limitations, including that it is an approximate test and that the calculated value of χ^2 depends on the user's choice of interval widths and starting points. The test is similar to a chi-square contingency table test but uses areas under the normal curve for expected values.

Table 20.3

Human Immunodeficiency Virus per Milliliter Blood Data in the Format of Table 20.2

ln h and x	k	$F_n(x)$	z	$F_e(x)$	$\|F_n - F_e\|$
5.2575					
5.3613	1	0.0222	−2.0437	0.0205	0.0017
5.6095	2	0.0444	−1.9058	0.0283	0.0161
5.8111	3	0.0667	−1.7938	0.0364	0.0302
5.8377	4	0.0889	−1.7790	0.0376	0.0513
6.2206	5	0.1111	−1.5663	0.0586	0.0525
6.4599	6	0.1333	−1.4334	0.0759	0.0575
7.4719	7	0.1556	−0.8711	0.1918	0.0363
7.7841	8	0.1778	−0.6977	0.2427	0.0649
7.8466	9	0.2000	−0.6301	0.2537	0.0537
7.9124	10	0.2222	−0.6264	0.2655	0.0433
8.0830	11	0.2444	−0.5316	0.2975	0.0530
8.1026	12	0.2667	−0.5208	0.3013	0.0346
8.1158	13	0.2889	−0.5134	0.3038	0.0149
8.4036	14	0.3111	−0.3536	0.3618	0.0507
8.4602	15	0.3333	−0.3221	0.3736	0.0404
8.5834	16	0.3556	−0.2537	0.3999	0.0443
8.6778	17	0.3778	−0.2012	0.4203	0.0425
8.8796	18	0.4000	−0.0891	0.4645	0.0645
8.9383	19	0.4222	−0.0565	0.4775	0.0552
8.9408	20	0.4444	−0.0551	0.4780	0.0336
9.1607	21	0.4667	0.6701	0.5267	0.0601
9.4340	22	0.4889	0.2189	0.5866	0.0977
9.5623	23	0.5111	0.2902	0.6141	0.1030[a]
9.5645	24	0.5333	0.2914	0.6146	0.0813
9.5988	25	0.5556	0.3104	0.6219	0.0663
9.7200	26	0.5778	0.3778	0.6472	0.0694
9.7600	27	0.6000	0.4000	0.6554	0.0554
9.7712	28	0.6222	0.4062	0.6577	0.0355
10.1036	29	0.6444	0.5909	0.7227	0.0783
10.1956	30	0.6667	0.6420	0.7396	0.0729
10.1963	31	0.6889	0.6424	0.7397	0.0508
10.4942	32	0.7111	0.8079	0.7904	0.0793
10.6105	33	0.7333	0.8725	0.8085	0.0752
10.6144	34	0.7556	0.8747	0.8091	0.0536
10.6794	35	0.7889	0.9108	0.8188	0.0410
10.8751	36	0.8000	1.0195	0.8460	0.0460
10.9183	37	0.8222	1.0435	0.8516	0.0294
10.9501	38	0.8444	1.0612	0.8557	0.0113
11.2159	39	0.8667	1.2089	0.8866	0.0200
11.2270	40	0.8889	1.2150	0.8878	0.0011
11.2511	41	0.9111	1.2284	0.8904	0.0208
11.3553	42	0.9333	1.2863	0.9008	0.0325
11.3634	43	0.9556	1.2907	0.9016	0.0540
11.4586	44	0.9778	1.3437	0.9105	0.0673

[a]$L = 0.1030$.

Example Posed: Is the Distribution of Ages of the 301 Urology Patients Normal?

We ask if the data are just normal, not drawn from a specific normal with theoretically given μ and σ; therefore, we use $m = 66.76$ and $s = 8.10$ from DB1. With a sample this large, we can use the chi-square goodness-of-fit test.

Method: Chi-square Goodness-of-Fit Test of Normality

The hypotheses are H_0: The distribution from which the sample was drawn is normal, or alternatively is normal with parameters μ and σ; and the two-tailed H_1: The distribution is different. Choose α. Look up the critical χ^2 in Table III (see Tables of Probability Distributions).

1. We define the data intervals, say, k in number, as we would were we to form a histogram of the data. We form a blank table in the format of Table 20.4.
2. To use a normal probability table, we standardize the ends of the intervals by subtracting the mean and dividing by the standard deviation. The "expected" normal is usually specified by the sample m and s, although it could be specified by a theoretical μ and σ.
3. To relate areas under the normal curve to the intervals, we find the area to the end of an interval from a table of normal probabilities, such as Table I, and subtract the area to the end of the preceding interval.
4. To find the frequencies expected from a normal fit (name them e_i), we multiply the normal probabilities for each interval by the total number of data n.
5. We tally the number of data falling into each interval and enter the tally numbers in the table. Name these numbers n_i.
6. Calculate a χ^2 value [the pattern is similar to Eq. (6.2)] using Eq. (20.2):

$$\chi^2 = \sum \frac{(n_i - e_i)^2}{e_i} = \sum \frac{n_i^2}{e_i} - n, \qquad (20.2)$$

where the first form is easier to understand conceptually, and the second form is easier to compute.
7. If calculated χ^2 is greater than critical χ^2, reject H_0; otherwise, accept H_0.

Table 20.4

Format for Table of Values Required to Compute the Chi-square Goodness-of-Fit Statistic

Interval	Standard normal z to end of interval	P	Expected frequencies (e_i)	Observed frequencies (n_i)
⋮	⋮	⋮	⋮	⋮
⋮	⋮	⋮	⋮	⋮

Example Completed: Is the Distribution of Ages Normal?

We choose $\alpha = 0.05$. The critical value of chi-square uses 9 df (number of intervals $-$ 1). From Table 20.5 (or Table III), chi-square (9 df) for $\alpha = 0.05$ is 16.92.

1. We define 10 intervals as <50, 50 up to but not including 55, ..., 85 up to but not including 90, and ≥90. We form a blank table in the format of Table 20.4 and complete it using the following steps, resulting in Table 20.6.
2. To use a normal probability table, we need to standardize the ends of the intervals by subtracting the mean and dividing by the standard deviation. We standardize the end of the first interval by $(50 - 66.76) / 8.1 = -2.0691$. The second interval is $(55 - 66.76) / 8.1 = -1.4519$.
3. We need to relate areas under the normal curve to the intervals, which we do by finding the area to the end of an interval and subtracting the area to the end of the preceding interval. Table I includes only positive z-values, so we use the normal symmetry: We change the sign of z and use α instead of $1 - \alpha$. For the first interval, $z = -2.0691$ interpolates as $0.023 + 0.691 \times (0.018 - 0.023) = 0.0195$. There is no prior interval yielding a probability to subtract, so 0.0195 is the probability for the

Table 20.5

An Excerpt of Chi-square Distribution, Right Tail[a]

α (area in right tail)	0.10	0.05	0.025	0.01	0.005	0.001
$df = 6$	10.64	12.59	14.45	16.81	18.54	22.46
8	13.36	15.51	17.53	20.09	21.95	26.10
9	14.68	16.92	19.02	21.67	23.59	27.86

[a] χ^2 values (distances to right of zero) are given for 6, 8, and 9 df for various α, the area under the curve in the right tail.
Table III in Tables of Probability Distributions provides an expanded version of this table.

Table 20.6

Table of Values Required to Compute the Chi-square Goodness-of-Fit Statistic

Interval	Standard normal z to end of interval	p	Expected frequencies (e_i)	Observed frequencies (n_i)
<50	−2.0691	0.0195	5.87	3
50–<55	−1.4519	0.0542	16.31	17
55–<60	−0.8346	0.1286	38.71	32
60–<65	−0.2173	0.2120	63.81	74
65–<70	0.4000	0.2407	72.45	62
70–<75	1.0173	0.1900	57.19	57
75–<80	1.6346	0.1035	31.15	37
80–<85	2.2519	0.0359	10.81	14
85–<90	2.8691	0.0145	4.36	4
≥90	∞	0.0019	0.57	1

first interval. For the second interval, $z = -1.4519$ gives $0.081 + 0.519 \times (0.067 - 0.081) = 0.0737$. Subtracting 0.0195 for the prior area results in $p = 0.0542$ for the second interval.

4. To obtain the expected frequencies (e_i), we multiply the probability associated with an interval by n. $e_1 = 0.0195 \times 301 = 5.87$, and so on.
5. To obtain the observed frequencies (n_i), we tally the data for each data interval and enter the frequencies $(3, 17, 32, \ldots)$ in the table.
6. The chi-square statistic is obtained as

$$\chi^2 = \sum \frac{n_i^2}{e_i} - n = \frac{3^2}{5.87} + \frac{17^2}{16.31} + \cdots - 301 = 7.895.$$

7. The calculated chi-square of 7.895 is less than the critical chi-square of 16.92, so the result is taken as not significant and we do not reject H_0; we treat the distribution from which the data arise as normal. From a statistical software package, $p = 0.555$.

Additional Example: Normality of Human Immunodeficiency Virus Data

Let us test the normality of the ln HIV data[85] introduced earlier using the chi-square goodness-of-fit test, even though the sample size is just borderline smaller than we would choose. We use seven intervals (see Fig. 20.1B) leading 6 df. We use $\alpha = 0.05$. From Table 20.5 (or Table III), the critical value of $\chi^2_{6df} = 12.59$. Recall that $m = 9.0408$ and $s = 1.8031$. Results in the format of Table 20.4 appear in Table 20.7. By substituting in Eq. (20.2), we find

$$\chi^2 = \sum \frac{n_i^2}{e_i} - n = \frac{5^2}{2.0655} + \frac{2^2}{3.7350} + \cdots + \frac{6^2}{3.9690} - 45 = 10.91.$$

Because $10.91 < 12.59$, we conclude that deviation from normality is not demonstrated. (The actual $p = 0.091$.)

Exercise 20.2: χ^2 Goodness-of-Fit Test. In the example discussed in Section 17.5, we assumed that the patient ages in the three groups formed a normal distribution. In particular, the distribution of $n = 164$ PSA readings in the equivocal 4–10 range was uncertain. Test this distribution for normality using the χ^2

Table 20.7

Human Immunodeficiency Virus per Milliliter Blood Data in the Format of Table 20.4

Interval	Standard normal z to end of interval	p	Expected frequencies (e_i)	Observed frequencies (n_i)
5–<6	−1.6864	0.0459	2.0655	5
6–<7	−1.1318	0.0830	3.7350	2
7–<8	−0.5772	0.1530	6.8850	4
8–<9	−0.0226	0.2091	9.4095	10
9–<10	0.5320	0.2117	9.5265	8
10–<11	1.0866	0.1588	7.1460	10
11–<12	1.6412	0.0882	3.9690	6

goodness-of-fit test at the $\alpha = 5\%$ level with the frequencies tallied for nine intervals from the following table. (The notation "45–<50" represents the interval including 45 and up to but not including 50, etc.). $m = 66.2988$, and $s = 7.8429$.

45–<50	50–<55	55–<60	60–<65	65–<70	70–<75	75–<80	80–<85	85–<90
2	9	19	44	30	36	15	6	3

20.3. TEST OF EQUALITY OF TWO DISTRIBUTIONS

THE TWO-SAMPLE KOLMOGOROV–SMIRNOV TEST

The two-sample KS test is used here to compare two data sets to decide whether they were sampled from population distributions of the same shape. (The one-sample form to test normality is discussed in Section 20.2.)

EXAMPLE POSED: ARE TWO PROSTATE-SPECIFIC ANTIGEN SAMPLES FROM DB1 DISTRIBUTED THE SAME?

We want to know if the distribution from which the first 10 PSA values were drawn is the same as that for the next 16 PSA values. We are suspicious that the first 10 are not representative of the remainder and may be subject to a sampling bias. The question of distribution shape accompanied by the tests for means and standard deviations will answer the question.

METHOD: THE TWO-SAMPLE KOLMOGOROV–SMIRNOV TEST

The hypotheses are H_0: The two samples arose from the same population distributions; and the two-tailed H_1: The samples arose from different population distributions. Choose α. The sample sizes are n_1 and n_2, where n_1 is the larger sample. Calculate the critical value for the chosen 1%, 5%, or 10% α from Eqs. (20.3a), (20.3b), or (20.3c), respectively. (These critical values are approximations. They are adequate for $n_2 > 10$. For smaller n's, they may be used if the calculated statistic is much greater or smaller than the critical value; if borderline, the user should arrange for calculation using a statistical software package.)

$$1.63\sqrt{\frac{n_1 + n_2}{n_1 n_2}} \quad \text{for } \alpha = 0.01 \tag{20.3a}$$

$$1.36\sqrt{\frac{n_1 + n_2}{n_1 n_2}} \quad \text{for } \alpha = 0.05 \tag{20.3b}$$

$$1.22\sqrt{\frac{n_1 + n_2}{n_1 n_2}} \quad \text{for } \alpha = 0.10 \tag{20.3c}$$

The test consists of calculating the cumulative data frequencies for the two samples and finding the probability of the greatest difference between these cumulative frequencies.

1. We form a blank table in the format of Table 20.8. Combine the two data sets, keeping track of which datum belongs to which sample, and enter in ascending order.
2. Going down the data list, each time a datum belonging to sample 1 is different from the datum above it, record an entry for k_1, the number of data in sample 1 preceding it. Repeat the process for sample 2 data, recording entries for k_2.
3. For every k_1, calculate $F_1 = k_1/n_1$ and enter it under the F_1 column. Wherever a blank appears (corresponding to sample 2 data), write down the F_1 from the line above. Repeat the process for sample 2 data, recording entries for F_2 and writing down the F_2 value from the preceding line to fill in blanks. The F's are the cumulative sums for the two samples.
4. Calculate and record $|F_1 - F_2|$, the absolute difference (difference without any minus signs) between the two cumulative sums, for every datum.
5. The test statistic is the largest of these differences; call it L.
6. If $L >$ critical value, reject H_0; otherwise, do not reject H_0.

Table 20.8

Format for Table of Data and Calculations Required for the Two-Sample Kolmogorov–Smirnov Test of Equality of Distributions

Ordered data	k_1	k_2	F_1	F_2	$|F_1 - F_2|$
:	:	:	:	:	:
:	:	:	:	:	:

The Two-Sample Kolmogorov–Smirnov on Large Samples

If the sample sizes are large, the number of differences to be computed may be reduced to a manageable number by collapsing the data into class intervals as one would do in making a histogram. The method, however, then acquires the faults of the goodness-of-fit test. It becomes an approximation and becomes dependent on the choice of interval position and spacing.

EXAMPLE COMPLETED: ARE TWO PROSTATE-SPECIFIC ANTIGEN SAMPLES FROM DB1 DISTRIBUTED THE SAME?

We choose $\alpha = 0.05$. The larger sample is designated number 1. Then, $n_1 = 16$ PSA readings, which are 5.3, 6.6, 7.6, 4.8, 5.7, 7.7, 4.6, 5.6, 8.9, 1.3, 8.5, 4.0, 5.8, 9.9, 7.0, and 6.9 ng/ml, and $n_2 = 10$, with the PSA values from Table DB1.1.

We use Eq. (20.3b) to find the critical value:

$$1.36\sqrt{\frac{n_1 + n_2}{n_1 n_2}} = 1.36\sqrt{\frac{26}{160}} = 0.5482.$$

1. Using the format (headings) as in Table 20.8, record in Table 20.9 the combined data sets in ascending order, keeping track of which datum belongs to which sample.
2. Going down the data list, each time a datum belonging to sample 1 is different from the datum above it, record an entry for k_1, the number of data in sample 1 preceding it. The 1.3 value has no data preceding it, so we enter $k_1 = 0$. The value 4.0 has one datum preceding it. The value 4.1 is not from sample 1; we skip to 4.6, which has two sample 1 data preceding it. Continue until reaching the PSA value 9.9, for which $k_1 = 15$. Repeat the process for sample 2 data, recording entries for k_2.
3. For every k_1, calculate $F_1 = k_1/n_1$ and enter it under the F_1 column. Wherever a blank appears (corresponding to sample 2 data), write down the F_1 from the line above. $F_1 = 0/16 = 0$ for the first PSA value. $F_1 = 1/16 = 0.0625$ for the second value. The third PSA value is from

Table 20.9

Example Data (Prostate-Specific Antigen Levels for 10 Patients from Table DB1.1 and the Next 16 Patients Presenting) Together with the Corresponding Calculations Required for the Two-Sample Kolmogorov–Smirnov Test of Equality of Distributions

Ordered data	k_1	k_2	F_1	F_2	$|F_1 - F_2|$
1.3	0		0	0	0
4.0	1		0.0625	0	0.0625
4.1		0	0.0625	0	0.0625
4.4		1	0.0625	0.1	0.0375
4.6	2		0.1250	0.1	0.0250
4.8	3		0.1875	0.1	0.0875
5.3	4		0.2500	0.1	0.1500
5.6	5		0.3125	0.1	0.2125
5.7	6		0.3725	0.1	0.2725
5.8	7		0.4375	0.1	0.3375[a]
5.9		2	0.4375	0.2	0.2375
6.1		3	0.4375	0.3	0.1375
6.6	8		0.5000	0.3	0.2000
6.8		4	0.5000	0.4	0.1000
6.9	9		0.5625	0.4	0.1625
7.0	10		0.6250	0.4	0.2250
7.6	11	5	0.6875	0.5	0.1875
7.7	12	6	0.7500	0.6	0.1500
7.9		7	0.7500	0.7	0.0500
8.0		8	0.7500	0.8	0.0500
8.5	13		0.8125	0.8	0.0125
8.9	14		0.8750	0.8	0.0750
9.0		9	0.8750	0.9	0.0250
9.9	15		0.9395	0.9	0.0395

[a]$L = 0.3375$.

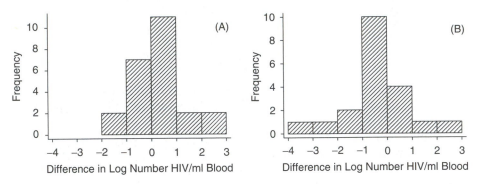

Figure 20.2 Plots of distributions of control (A) and vaccine-treated (B) differences in log number human immunodeficiency virus per milliliter blood before versus after treatment.

sample 2 and no k_1 was recorded, so we repeat 0.0625. We follow the same process for sample 2 data, recording entries for F_2 and writing down the F_2 value from the preceding line to fill in blanks. The F's are the cumulative sums for the two samples.

4. Calculate and record $|F_1 - F_2|$, the absolute difference (difference without any minus signs) between the two cumulative sums for every datum. For example, for the fourth PSA value, 4.4, $|F_1 - F_2| = |0.0625 - 0.1| = |-0.0375| = 0.0375$.

5. The test statistic L is the largest of these differences. We can see that it is 0.3375.

6. The L of 0.3375 is less than 0.5482, the critical value, so we do not reject H_0. (A statistical software package gives $p = 0.3995$.)

ADDITIONAL EXAMPLE: A POTENTIAL VACCINE FOR HUMAN IMMUNODEFICIENCY VIRUS

The study included a control sample of $n_1 = 24$ patients (placebo saline injection) and an experimental vaccine injection sample of $n_2 = 20$ patients, randomly allocated into the two groups.[85] The knowledge of whether or not the experimental and control groups follow the same distribution aided in the development of the physiologic theory. The number of HIV per milliliter blood (denoted h) was measured before treatment and at a fixed period after treatment. Its natural logarithm ($\ln h$) was calculated to reduce the extreme skewness of the frequency distribution. Then the difference $d(\ln h) = \ln h$ before $- \ln h$ after was taken. A frequency plot of the two samples appears in Fig. 20.2. The distributions are not obviously different, but enough so to warrant a test. We use the KS test of equality of distributions with a two-sided $\alpha = 0.05$.

The data and analysis operations appear in Table 20.10 in the format of Table 20.8. The values of $d(\ln h)$ are put in order. The number of data in the respective sample preceding the current observation is listed for control (k_1) and vaccine (k_2). $F_1 = k_1/24$, $F_2 = k_2/20$, and $|F_1 - F_2|$ are entered. Inspection of the last column shows that the largest value is $L = 0.4000$, which is indicated by

Table 20.10

Human Immunodeficiency Virus per Milliliter Blood Data in the Format of Table 20.8

$d(\ln h)$	k_1	k_2	F_1	F_2	$|F_1 - F_2|$
−3.4009		0	0	0	0
−2.4125		1	0	0.05	0.0500
−1.2169	0		0	0.05	0.0500
−1.1564	1		0.0417	0.05	0.0083
−1.1453		2	0.0417	0.10	0.0583
−1.0750		3	0.0417	0.15	0.1083
−0.7466		4	0.0417	0.20	0.1583
−0.6159	2		0.0833	0.20	0.1167
−0.5456	3		0.1250	0.20	0.0750
−0.5114		5	0.1250	0.25	0.1250
−0.4494		6	0.1250	0.30	0.1750
−0.4158		7	0.1250	0.35	0.2250
−0.4153		8	0.1250	0.40	0.2750
−0.4152	4		0.1667	0.40	0.2333
−0.3644	5		0.2083	0.40	0.1917
−0.3628		9	0.2083	0.45	0.2417
−0.2673	6		0.2500	0.45	0.2000
−0.2372		10	0.2500	0.50	0.2500
−0.2148		11	0.2500	0.55	0.3000
−0.1302		12	0.2500	0.60	0.3500
−0.1298		13	0.2500	0.65	0.4000[a]
−0.0851	7		0.2917	0.65	0.3583
−0.0030	8		0.3333	0.65	0.3167
0.6257	9		0.3750	0.65	0.2750
0.0989	10		0.4167	0.65	0.2333
0.1054	11		0.4583	0.65	0.1917
0.1756	12		0.5000	0.65	0.1500
0.2982	13		0.5417	0.65	0.1083
0.3122	14		0.5833	0.65	0.0667
0.4635	15		0.6250	0.65	0.0250
0.7133		14	0.6250	0.70	0.0750
0.7277	16		0.6667	0.70	0.0333
0.7371		15	0.6667	0.75	0.0833
0.7550	17		0.7083	0.75	0.0417
0.7907		16	0.7083	0.80	0.0917
0.8137	18		0.7500	0.80	0.0500
0.9115	19		0.7917	0.80	0.0083
0.9980		17	0.7917	0.85	0.0583
1.1957	20		0.8333	0.85	0.0167
1.4076	21		0.8750	0.85	0.0250
1.6083		18	0.8750	0.90	0.0250
2.1146	22		0.9167	0.90	0.0167
2.3530	23		0.9583	0.90	0.0583
2.7834		19	0.9583	0.95	0.0083

[a]$L = 0.4000$.

an asterisk. The critical value for the test is found from Eq. (20.3b):

$$1.36\sqrt{\frac{n_1 + n_2}{n_1 n_2}} = 1.36\sqrt{\frac{24 + 20}{24 \times 20}} = 0.4118.$$

Because $0.4000 < 0.4118$, we do not reject H_0 and proceed on the premise that the distributions are not different.

Exercise 20.3: Are the Distributions of Functionality the Same for Two Types of Ankle Repair? An orthopedist installs hardware in the repair of broken ankles in $n = 19$ patients.[28] He randomly selects two groups: the hardware is left in place after adequate healing for one group ($n_1 = 10$), and it is removed for the other group ($n_2 = 9$). The orthopedist judges the posthealing percentage functionality of the ankle joint. In assessing the relative success of the two groups, he observes that their distributions are far from normal in shape. However, are the distributions both the same, whatever they may be? His data are as follows:

Leave hardware (%)	95	98	95	45	90	93	85	100	72	99
Remove hardware (%)	75	65	100	90	35	63	78	70	80	

Perform a KS test of equality of distributions using $\alpha = 0.05$.

ANSWERS TO EXERCISES

20.1. For $n = 16$, our 5% critical value is $1.36/4 - 1/(4.5 \times 16) = 0.326$. The data are given in ascending order. We prepare Table 20.11 in the format of Table 20.2. x is Hct each time the value is different from the preceding value. k is the number of values preceding x. From the Hct readings, we calculate $m = 37.25$ and $s = 6.9937$. We calculate $z = (x - m) / s$. F_e is the area under the normal curve to the left of z. And finally, the last column is the absolute value of the difference between the F-values. From inspection of the last column, the largest is $L = 0.1164$. Because $0.1164 < 0.326$, we do not reject the hypothesis of normality; we will apply the data as if it were normal.

20.2. With nine intervals, $df = 8$. From Table 20.5 (or Table III), the critical value of χ^2 is 15.51. The data and calculations are set up in Table 20.12 in the format of Table 20.4. By using Eq. (20.2), we find

$$\chi^2 = \sum \frac{n_i^2}{e_i} - n = \frac{2^2}{3.0832} + \frac{9^2}{9.1840} + \cdots - 164 = 9.2112,$$

which is less than the critical value. (From statistical software, $p = 0.325$.) We conclude that the distribution is not significantly different from normal and that our initial judgment is correct.

Table 20.11

Hematocrit Data in the Format of Table 20.2

x	k	$F_n(x)$	z	$F_e(x)$	$|F_n - F_e|$
25.6					
29.7	1	0.0625	−1.0795	0.1402	0.0777
32.0	2	0.1250	−0.7507	0.2264	0.1014
32.0					
32.1	4	0.2500	−0.7364	0.2308	0.0192
32.7	5	0.3125	−0.6506	0.2577	0.0548
33.9	6	0.3750	−0.4790	0.3160	0.0590
34.0	7	0.4375	−0.4647	0.3211	0.1164[a]
38.3	8	0.5000	0.1501	0.5597	0.0597
38.8	9	0.5625	0.2216	0.5877	0.0252
39.0	10	0.6250	0.2502	0.5988	0.0262
42.0	11	0.6875	0.6792	0.7515	0.0640
43.3	12	0.7500	0.8651	0.8065	0.0565
43.9	13	0.8125	0.9501	0.8292	0.0167
46.7	14	0.8750	1.3512	0.9117	0.0367
52.0	15	0.9375	2.1090	0.9825	0.0450

[a]$L = 0.1164$.

Table 20.12

Prostate-Specific Antigen Data in the 4–10 Range in the Format of Table 20.4

Interval	Standard normal z to end of interval	p	Expected frequencies (e_i)	Observed frequencies (n_i)
45–<50	−2.0782	0.0188	3.0832	2
50–<55	−1.4406	0.0560	9.1840	9
55–<60	−0.8031	0.1361	22.3204	19
60–<65	−0.1656	0.2233	36.6212	44
65–<70	0.4719	0.2473	40.5572	30
70–<75	1.1094	0.1849	30.3236	36
75–<80	1.7470	0.0933	15.3012	15
80–<85	2.3845	0.0318	5.2152	6
85–<90	3.0220	0.0073	1.1972	3

20.3. The 5% critical value from Eq. (20.3b) is

$$1.36\sqrt{\frac{n_1 + n_2}{n_1 n_2}} = 1.36\sqrt{\frac{9 + 10}{9 \times 10}} = 0.6249.$$

The orthopedist rank orders the values of percentage functionality in Table 20.13 in the format of Table 20.8. The number in the respective sample preceding each current observation is listed for leave in place (k_1) and remove (k_2). $F_1 = k_1/10$ and $F_2 = k_2/9$ are entered, and finally, $|F_1 - F_2|$. Inspection of $|F_1 - F_2|$ shows the largest value as $L = 0.5778$, which is denoted by an asterisk in Table 20.13. L is a little smaller than critical; there is not enough evidence to conclude that the distributions are different.

Table 20.13

Functionality of Repaired Ankles Data in the Format of Table 20.8

Functionality (%)	k_1	k_2	F_1	F_2	$|F_1 - F_2|$
35		0	0	0	0
45	0		0	0	0
63		1	0	0.1111	0.1111
65		2	0	0.2222	0.2222
70		3	0	0.3333	0.3333
72	1		0.1	0.3333	0.2333
75		4	0.1	0.4444	0.3444
78		5	0.1	0.5556	0.4556
80		6	0.1	0.6667	0.5667
85	2		0.2	0.6667	0.4667
90		7	0.2	0.7778	0.5778^a
90	3		0.3	0.7778	0.4778
93	4		0.4	0.7778	0.3778
95	5		0.5	0.7778	0.2778
95	6		0.6	0.7778	0.1778
98	7		0.7	0.7778	0.0778
99	8		0.8	0.7778	0.0222
100		8	0.8	0.8889	0.0889
100	9		0.9	0.8889	0.0111

[a]$L = 0.5778$.

Chapter 21

Equivalence Testing

21.1. CONCEPTS AND TERMS

CONCEPT

In most of the testing methods examined in previous chapters of this book, namely *difference tests*, we have posed a statistical null hypothesis of no difference, say, between outcome means of a treatment and a placebo, and use a test to see if there is sufficient evidence to reject it and assert that the treatment is superior. In contrast, it is becoming ever more common to ask if a new treatment is as good, on average, as an established one, perhaps because it is less invasive or less costly. In this case, the null hypothesis might state: The effectiveness of a new treatment is outside a magnitude judged to represent clinical similarity. The alternate hypothesis states: The mean difference is within the clinically relevant magnitude, and thus the treatments are deemed similar enough to be used interchangeably in clinical practice. Rejecting the null hypothesis implies evidence that the treatments are not different. The relatively recent development in statistical methodology of *testing for no difference* has been termed *equivalence testing*.

EQUIVALENCE TESTS ARE SIMILARITY TESTS

Equivalence tests might better be termed *similarity* tests, because they test whether or not means are approximately the same, that is, similar enough to be not clinically different, rather than exactly the same.

EQUIVALENCE VERSUS NONSUPERIORITY (OR NONINFERIORITY)

We could ask if one treatment is no better than another *(nonsuperiority)*, leading to a *one-sided test*, or if neither treatment is better than the other *(equivalence)*, leading to a *two-sided test*. The one-sided tests also may be referred to as *noninferiority* tests, which address the question of one treatment being no worse than another. The distinction lies in which mean is subtracted from the other.

The two names refer to the same test. The name nonsuperiority is used primarily in this book.

EQUIVALENCE OF MEANS, PROPORTIONS, AND OTHER PARAMETERS

This chapter is devoted to tests of one or two means or proportions. (Indeed, the conceptual paragraphs focus on means for simplicity.) Because most equivalence tests are conducted on means and proportions, these are the tests for which the concepts have been most thoroughly developed. Equivalence concepts apply in a number of other methods that involve testing—analysis of variance, regression, rank-order tests, and others. Because equivalence testing is relatively recent in the development of statistics, methods for many of these cases have not been detailed in the literature. However, the logic is similar. In some cases, the enterprising reader may be able to extend the ideas sufficiently to perform appropriate equivalence testing.

21.2. BASICS UNDERLYING EQUIVALENCE TESTING

THE NULL HYPOTHESIS IS KEY

In both difference and equivalence tests, we are assessing a difference between means; the distinction lies in the null hypotheses. In a difference test, we hypothesize no difference. In an equivalence test, we hypothesize a difference of at least some clinically relevant amount, say, Δ. In some studies, a value for Δ will be obvious from clinical considerations; in other studies, the size of Δ will not be clear at all and must be almost arbitrary. In the latter case, an arbitrary 20% of baseline is often chosen. Δ is the value we test against. The test statistic consists of the sample difference, say, d, plus or minus the probable number of standard errors designated by α. If we are testing paired data, for example, the change from before to after treatment, d can be taken as the mean of this change, m_d. If we are testing a single sample mean m versus a theoretical mean μ, $d = m - \mu$, as in the numerator of Eq. (17.1) or (17.2). If we are testing the mean of one sample, m_1, against that of another, m_2, $d = m_1 - m_2$, as in the numerator of Eq. (17.5) or (17.8).

DIFFERENCE SYMBOLS

The word *difference* has been used so much that it might be confusing. Let us be clear in distinguishing the three difference terms. (1) There exists a true but unknown *theoretical* (or population) difference between two outcomes; we denote this difference as δ. This would be the measured difference if we could measure the entire population. (2) From our data, we can calculate an estimate of δ, the *sample* difference; we denoted this difference as d. This is the measured difference that we actually use. (3) The null hypothesis states that the superiority (or inequality) of the one outcome compared with the other is at least a certain difference, Δ, our *clinically relevant* difference. This is the size of difference we want to rule out.

WHAT THE NULL HYPOTHESIS STATEMENT REALLY MEANS

The null hypothesis should be stated as an equality rather than an inequality, because we attach a test of a specific probability parameter to it. In a nonsuperiority test, for example, we want to test the null hypothesis that the theoretical difference δ is *at least* equal (i.e., equal to or greater than) to the clinically relevant difference Δ. However, we state a null hypothesis of this sort: $H_0: \delta = \Delta$. Rejecting H_0 implies that δ is not as large as Δ and therefore cannot be larger.

ONE-SIDED VERSUS TWO-SIDED TESTS

A one-sided (nonsuperiority) test is rather straightforward. We postulate Δ, look up a critical value to demark difference from not difference, and compare d with Δ to conclude a relationship between δ and Δ. A two-sided test (equivalence) is trickier. Whereas in difference testing, the null hypothesis (the hypothesis we test) lies in a single region of the axis and the alternate lies in two regions, in equivalence testing, the reverse is true: it is the null hypothesis that lies in two regions. We hypothesize that the theoretical difference δ either is Δ or more, or alternatively is $-\Delta$ or less. The logically consistent approach to this problem has evolved to be the use of two tests, one for each side. This approach has been termed *two one-sided tests*, or *TOST* (sometimes Schuirmann's TOST). TOST is further discussed in Section 21.4 after more groundwork has been laid.

USING A CONFIDENCE INTERVAL INSTEAD OF A TEST STATISTIC

Other chapters have stated that confidence intervals use much the same calculations and provide much the same information as hypothesis tests, but the purpose is different; confidence intervals estimate and tests test. That being said, many investigators find it easier to use confidence intervals than hypothesis tests in equivalence testing. We can place a confidence interval on δ using d, σ_d (or s_d), and the appropriate statistic, such as $z_{1-\alpha}$ (or $t_{1-\alpha}$). If this confidence interval fails to enclose Δ, we have evidence to reject H_0 and to conclude equivalence.

21.3. METHODS FOR NONSUPERIORITY TESTING

EXAMPLE POSED: EFFECT OF SILICONE IMPLANTS ON PLASMA SILICON LEVELS

DB5 gives data on the effect of silicone implants on plasma silicon levels. If postoperative plasma silicon level is not greater than the preoperative level, the implants had no deleterious effect on this measure, so we want a nonsuperiority test on the mean of postoperative minus preoperative readings. We theorize that silicone implants cannot decrease plasma silicon levels, and therefore we have a one-sided test; if postoperative plasma silicon levels should show a decrease in our sample, it would be a random, not caused, effect. We judge that anything less than a 20% increase in postoperative mean is not clinically relevant. The preoperative mean is 0.2257 µg/g (dry weight). Thus, $0.20 \times 0.2257 = 0.0451$ µg/g increase is our Δ. The null hypothesis is H_0: $\delta = 0.0451$. The alternate hypothesis is H_1: $\delta < 0.0451$, which equates in medical practice to no clinically relevant difference.

METHOD

Hypotheses

We want a sample difference d, estimating the population difference δ. For a single sample mean m versus a theoretical mean μ, $d = m - \mu$. For a change time-to-time, treatment-to-treatment, and so on, d = mean change. For a single-sample proportion p versus a theoretical proportion π, $d = p - \pi$. For two-sample means m_1 versus m_2, $d = m_1 - m_2$. For two-sample proportions p_1 versus p_2, $d = p_1 - p_2$. Changing the sign of d will toggle between nonsuperiority and noninferiority, so we will concentrate on the nonsuperiority case. To show evidence that the difference δ is less than a clinically important amount Δ, we want to reject a null hypothesis that claims δ is Δ or more. However, we have to calculate a probability based on an equality; we cannot calculate the probability of an unspecified "or more." Thus, the hypothesis is posed statistically as H_0: $\delta = \Delta$, implying that if we reject δ being as large as Δ, we certainly reject it being larger. The alternate hypothesis is H_1: $\delta < \Delta$. We choose α, the risk for error if we reject the null hypothesis when it is true.

Test Statistics

We want to test if d is more than a critical value distance from Δ in standardized units. We standardize by dividing $\Delta - d$ by the standard error of d, s_d. The test statistics are

$$z = \frac{\Delta - d}{\sigma_d} \tag{21.1}$$

or

$$t = \frac{\Delta - d}{s_d}, \tag{21.2}$$

according to standard deviation known or estimated, respectively. Formulas for the standard error s_d and the names of the probability table to be used to find the critical value $z_{1-\alpha}$ or $t_{1-\alpha}$ may be found Table 21.1. For one-sample tests, rows 1 and 2 correspond to Eq. (21.1) and row 3, to Eq. (21.2). For two-sample means or proportions tests, rows 4, 5, and 6 correspond to Eq. (21.1) and row 7, to Eq. (21.2). If $z \geq z_{1-\alpha}$ or $t \geq t_{1-\alpha}$, reject H_0. We have evidence of nonsuperiority.

Table 21.1

Test Cases and Standard Deviations with Corresponding Standard Errors and Probability Tables from Which to Find Critical Values[a,b]

Case	Standard deviation(s)	s_d (standard error)	Probability table[c]
$d = m - \mu$	σ	$\dfrac{\sigma}{\sqrt{n}}$	z table (Table I)
$d = p - \pi$	Assumed[d]	$\sqrt{\dfrac{\pi(1-\pi)}{n}}$	z table (Table I)
$d = m - \mu$	s $df = n - 1$	$\dfrac{s}{\sqrt{n}}$	t table (Table II)
$d = m_1 - m_2$	σ_1, σ_2 $\sigma_1 = \sigma_2 = \sigma$	$\sigma\sqrt{\dfrac{1}{n_1} + \dfrac{1}{n_2}}$	z table (Table I)
$d = m_1 - m_2$	σ_1, σ_2 $\sigma_1 \neq \sigma_2$	$\sqrt{\dfrac{\sigma_1^2}{n_1} + \dfrac{\sigma_2^2}{n_2}}$	z table (Table I)
$d = p_1 - p_2$	Assumed[d]	$\sqrt{\left(\dfrac{1}{n_1} + \dfrac{1}{n_2}\right)\left(\dfrac{n_1 p_1 + n_2 p_2}{n_1 + n_2}\right)\left(1 - \dfrac{n_1 p_1 + n_2 p_2}{n_1 + n_2}\right)}$	z table (Table I)
$d = m_1 - m_2$	s_1, s_2 $df = n_1 + n_2 - 2$	$\sqrt{\left(\dfrac{1}{n_1} + \dfrac{1}{n_2}\right)\left[\dfrac{(n_1 - 1)s_1^2 + (n_2 - 1)s_2^2}{n_1 + n_2 - 2}\right]}$	t table (Table II)

[a]Subscripts correspond to sample number.
[b]Expressions for proportions assume moderate to large sample sizes.
[c]See Tables of Probability Distributions.
[d]Standard deviations for proportions are assumed from binomial approximation.

STEPS IN NONSUPERIORITY TESTING

Following are the steps in nonsuperiority testing:

(1) Pose a clinically important difference Δ.

(2) State null hypothesis H_0: $\delta = \Delta$; the theoretical difference is Δ. (By implication, the same conclusion will be reached if $\delta > \Delta$.) State an alternate hypothesis H_1: $\delta < \Delta$; the theoretical difference is less than clinically important.

(3) Choose α.

(4) Look up the critical values $z_{1-\alpha}$ or $t_{1-\alpha}$ in the appropriate table for the chosen α.

(5) Calculate the appropriate statistic from Eq. (21.1) or (21.2) and Table 21.1.

(6) If the test statistic is as large as the critical value, reject H_0, which implies accepting nonsuperiority.

USING A CONFIDENCE INTERVAL

For known σ, H_0 is rejected if $(\Delta - d)/\sigma_d \geq z_{1-\alpha}$. Simple algebra can put the statement into the form: H_0 is rejected if $\Delta \geq d + z_{1-\alpha}\sigma_d$, the interior of a one-sided confidence interval probability statement. For s estimated from the sample, a similar statement occurs: H_0 is rejected if $\Delta \geq d + t_{1-\alpha}s_d$.

EXAMPLE COMPLETED: SILICONE IMPLANTS AND PLASMA SILICON LEVELS

We calculate postoperative minus preoperative differences from the data, plot frequencies, and see that the distribution is not far from normal, and then calculate the mean difference as $d = -0.0073$. $s = 0.1222$ and $n = 30$; therefore, $s_d = 0.0223$. From Table II (see Tables of Probability Distributions), one-tailed $t_{0.95}$ for 29 df is 1.699. Using Eq. (21.2), we find that $(\Delta - d)/s_d = [0.0451 - (-0.0073)]/0.0223 = 2.3498$, which is larger than 1.699. We have evidence to reject H_0. The plasma silicon level after implantation has been shown to be clinically equivalent to the preoperative levels. Computer software shows the actual $p = 0.013$.

ADDITIONAL EXAMPLE: WHEN DOES EXERCISE-INDUCED BRONCHOSPASM CAUSE A CHANGE IN EXHALED NITRIC OXIDE?

DB14 gives data on the effect of exercise-induced bronchospasm (EIB) as measured by exhaled nitric oxide (eNO). Exhaled nitric oxide is not different between patients with and without EIB before exercise. Earlier it was shown that the eNO difference became significant by 20 minutes of exercise. Do the data show that the eNO decrease at 5 minutes of exercise in $n_1 = 6$ patients with EIB has not yet become greater than that in $n_2 = 32$ patients without EIB? Let us follow the steps in nonsuperiority testing. (1) We deem a change of 20% from preexercise level, averaging 29.26 parts per billion (ppb) to be clinically important. $\Delta = 0.20 \times 29.26 = 5.85$. (2) H_0: $\delta = 5.85$, for example, the true-but-unknown difference is at least 5.85. H_1: $\delta < 5.85$. (3) $\alpha = 0.05$. (4) The sample difference is the difference between the two means of eNO change, or $d = m_1 - m_2$, and the standard deviations are calculated from the sample, so we use the last row of Table 21.1. From the t table (see Table II) for $n_1 + n_2 - 2 = 36\ df$, $t_{0.95} = 1.689$ (by interpolation). (5) The mean and standard deviation are 0.70 and 4.64 for patients with EIB and 4.82 and 7.41 for patients without EIB, respectively.

Substitution in the Table 21.1 entry yields

$$s_d = \sqrt{\left(\frac{1}{6} + \frac{1}{32}\right)\left(\frac{5 \times 21.53 + 31 \times 54.91}{6 + 32 - 2}\right)} = 3.15.$$

The observed difference between means is $d = 0.70 - 4.64 = -3.94$. From Eq. (21.2), the test statistic $t = (\Delta - d)/s_d = [5.85 - (-3.94)]/3.15 = 3.11$. Because $3.11 > 1.689$, H_0 is rejected; we have evidence that the eNO between patients with and without EIB has not increased by 5 minutes.

Exercise 21.1. Data on recovery after orthopedic surgery is given in DB10. We ask if triple hop time on the operated leg is not larger than (nonsuperior to) that on the nonoperated leg. We judge that recovery, that is, Δ, is defined as operated leg time no greater than 10% of nonoperated leg time. Do we have evidence of adequate recovery?

21.4. METHODS FOR EQUIVALENCE TESTING

This section addresses two-sided equivalence testing.

EXAMPLE POSED: CARDIAC INDEX BY BIOIMPEDANCE VERSUS THERMODILUTION

A method of measuring cardiac index (CI; cardiac output normalized for body surface area) is thermodilution (TD), in which a catheter is placed in the heart. A proposed noninvasive method is bioimpedance (BI), in which an instrument attached to the patient's skin by electric leads indicates CI. For BI to be clinically useful, we need to provide evidence that it is equivalent to TD.

METHOD

Hypotheses

A two-sided equivalence test asks the clinical question: Is m_1 similar to m_2? In contrast to the difference test, in which H_0 lies in one region of the axis and H_1 lies in two separated regions, the equivalence test requires H_0 to lie in two regions, whereas H_1 lies in one. The question is posed statistically as three hypotheses: two null, subscripted "L" for left and "R" for right, and the alternate. They are $H_{0L}: \delta = -\Delta$; $H_{0R}: \delta = \Delta$; and $H_1: -\Delta < \delta < \Delta$. (Recall that the equal signs are used for exact calculation of a probability and are interpreted as "δ at most $-\Delta$" and "δ at least Δ," respectively.) Δ can occur on only one side, and we want α probability of error on that side, in contrast to a two-sided difference test in which α is split into two sides. However, we do not know in advance on which side of d we will find Δ, which gives rise to the concept of TOST, with error risk α for each. Two one-sided tests do not increase the work of testing; we simply go through the motions twice, once for each side. If we use the confidence interval approach, however, we have to form a $1 - 2\alpha$ level of confidence to have error risk α in each tail. Figure 21.1 is a diagram of the δ-axis showing illustrative locations of Δ, d, and the confidence interval for the case of the t distribution. Because the confidence interval does not include Δ, there is evidence to reject $H_{0B}: \delta > \Delta$ and conclude equivalence.

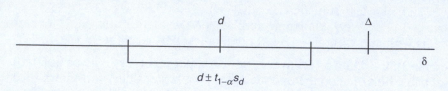

Figure 21.1 Diagram of the δ-axis illustrating d, its confidence interval when using a critical *t*-value, and a possible location of Δ.

Test Statistics

The test statistics are given in Eqs. (21.3.) and (21.4), with standard error and error probability table as in Table 21.1:

$$z_L = \frac{d - (-\Delta)}{\sigma_d}, \quad z_R = \frac{\Delta - d}{\sigma_d} \tag{21.3}$$

or

$$t_L = \frac{d - (-\Delta)}{s_d}, \quad t_R = \frac{\Delta - d}{s_d}. \tag{21.4}$$

For Eq. (21.3), if both $z_L \geq z_{1-\alpha}$ and $z_R \geq z_{1-\alpha}$, reject H$_0$. For Eq. (21.4), if both $t_L \geq t_{1-\alpha}$ and $t_R \geq t_{1-\alpha}$, reject H$_0$.

STEPS IN EQUIVALENCE TESTING

(1) Pose a clinically important difference Δ.

(2) State a pair of null hypotheses H$_{0L}$: $\delta = -\Delta$, the theoretical difference is $-\Delta$, and H$_{0R}$: $\delta = \Delta$, the theoretical difference is Δ. (By implication, the same conclusion will be reached if $\delta < -\Delta$ or if $\delta > \Delta$.) State an alternate hypothesis H$_1$: $-\Delta < \delta < \Delta$, the theoretical difference is within a clinically unimportant interval.

(3) Choose α.

(4) Look up the critical values $z_{1-\alpha}$ or $t_{1-\alpha}$ in the appropriate table for the chosen α.

(5) Calculate the appropriate left and right statistics from Eq. (21.3) or (21.4) and Table 21.1.

(6) If both test statistics exceed the critical value, reject H$_0$, which implies accepting equivalence.

USING A CONFIDENCE INTERVAL

For known σ, H$_0$ is rejected if Δ falls outside the confidence interval $d - z_{1-\alpha}\sigma_d < \delta < d + z_{1-\alpha}\sigma_d$. For s estimated from the sample, a similar statement occurs: H$_0$ is rejected if Δ falls outside the confidence interval $d - t_{1-\alpha}s_d < \delta < d + t_{1-\alpha}s_d$.

EXAMPLE COMPLETED: CARDIAC INDEX BY BIOIMPEDANCE VERSUS THERMODILUTION

In keeping with what is becoming a convention in equivalence testing in the absence of a Δ chosen for clinical importance, we will judge BI to be equivalent if its mean is within 20% of the TD mean, which is known from repeated large samples to be $\mu_T = 2.75$ (L/min/m^2). Because μ_T is known, we will be testing a sample mean against a population mean. Because we want BI to be neither greater nor less than TD, we need a two-sided test. We use TOST with $\alpha = 0.05$ in each tail. Twenty percent of 2.75 is $0.55 = \Delta$, so H_{0L}: $\delta = (\mu_B - \mu_T) = -0.55$; H_{0R}: $\delta = 0.55$; and H_1: $-0.55 < \Delta < 0.55$. We sample $n = 96$ patients[70] and find BI has mean $m_B = 2.68$ (L/min/m^2). $d = 2.75 - 2.68 = 0.07$ with standard deviation $s = 0.26$, so $s_d = s/\sqrt{n} = 0.0265$. From Table II, $t_{0.95}$ for 95 $df = 1.661$. Substituting in Eq. (21.4), we find $t_L = [0.07 - (-0.55)]/0.0265 = 23.40$, which exceeds 1.661, and $t_R = (0.55 - 0.07)/0.0265 = 18.1132$, which also exceeds 1.661. Both H_{0L} and H_{0R} are rejected. We have evidence of equivalence, that is, that BI readings are clinically similar to TD readings. The result could have been reached somewhat more easily using a confidence interval. Substituting in $d - t_{1-\alpha}s_d < \delta < d + t_{1-\alpha}s_d$ yields $0.026 < \delta < 0.114$. $\Delta = 0.55$ is outside the confidence interval, so H_0 is rejected. There is evidence for equivalence.

ADDITIONAL EXAMPLE: EFFECT OF PRIOR SURGERY ON RATE OF INTUBATION

DB12 gives data on prior surgery and intubation for patients undergoing carinal resection. We ask if the rate of intubation is the same for 34 patients with and 100 without prior surgery. We will follow the steps in equivalence testing. (1) We deem that a change of 5% in the intubation rate because of prior surgery is clinically significant. Then, $\Delta = 0.05$. (2) H_{0L}: $\delta = -0.05$; H_{0R}: $\delta = 0.05$; and H_1: $-0.05 < \delta < 0.05$. (3) $\alpha = 0.05$. (4) Row 6 of Table 21.1 indicates Table I (see Tables of Probability Distributions) to find the critical value. From Table I, in the "one-tailed $1 - \alpha$" column, $z_{1-\alpha} = z_{0.95} = 1.645$, used twice, once for each H_0. (5) We calculate the intubation rate as 17.65% for patients with prior surgery and 13.00% for those without prior surgery. The difference in rates is $d = 0.0465$. From the formula in row 6 of Table 21.1, we find σ_d as follows:

$$\sigma_d = \sqrt{\left(\frac{1}{34} + \frac{1}{100}\right)\left(\frac{34 \times 0.1765 + 100 \times 0.1300}{34 + 100}\right)\left(1 - \frac{34 \times 0.1765 + 100 \times 0.1300}{34 + 100}\right)}$$

$$= 0.0693.$$

From Eq. (21.3), we find

$$z_L = \frac{d - (-\Delta)}{\sigma_d} = \frac{0.0465 + 0.05}{0.0693} = 1.39,$$

and

$$z_R = \frac{\Delta - d}{\sigma_d} = \frac{0.05 - 0.0465}{0.0693} = 0.05.$$

Because neither 1.39 nor 0.05 exceeds 1.645, we have no evidence for equivalence. The data fail to demonstrate similarity.

Exercise 21.2.   DB3 gives baseline serum theophylline levels for patients with emphysema. Perform an equivalence test to learn if the data are free from an initial sex bias, that is, if mean baseline level is equivalent for $n_1 = 6$ women and $n_2 = 10$ men. We consider a difference in means of at least 2 to indicate a bias. Sample calculations are $m_1 = 12.67$, $s_1 = 3.46$, $m_2 = 9.68$, and $s_2 = 3.65$.

ANSWERS TO EXERCISES

21.1. Mean nonoperated time is 2.54 seconds, so we take $\Delta = 0.25$ second. The mean of differences between leg times is $d = 0.1600$, and the standard deviations of differences is $s = 0.2868$. From the second row of Table 21.1, $s_d = s/\sqrt{n} = 0.2868/\sqrt{8} = 0.1014$. The standard deviation is estimated from the sample, so we use the t distribution. For 7 df, $t_{0.95} = 1.895$. If the test statistic of Eq. (21.2) ≥ 1.895, there is evidence to reject H_0. The calculated $t = (\Delta - d)/s_d = (0.25 - 0.16)/0.1014 = 0.8876$ does not exceed 1.895; therefore, we have no evidence to reject the null hypothesis of difference. We have not shown nonsuperiority.

21.2. Follow the steps in equivalence testing. (1) $\Delta = 2$. (2) H_{0L}: $\delta = -2$; H_{0R}: $\delta = 2$; and H_1: $-2 < \delta < 2$. (3) $\alpha = 0.05$. (4) Row 6 of Table 21.1 indicates Table II to find the critical value. From Table II, in the column "0.95" for one-tailed $1 - \alpha$, for $n_1 + n_2 - 2 = 14$ df, the critical value of 1.761 is to be used twice, once for each H_0. (5) We calculate $d = m_1 - m_2 = 2.99$, and $s_d = 1.8684$. The test statistics are $t_L = (2.99 + 2)/1.8684 = 2.6707$ and $t_R = (2 - 2.99)/1.8684 = 0.5299$. The data fail to establish that baseline theophylline level is unbiased by sex, because both t-values do not exceed the critical value. To perform the test using a confidence interval, we put the calculated values in the form $d - t_{1-\alpha}s_d < \delta < d + t_{1-\alpha}s_d$, specifically $2.99 - 1.761 \times 1.8684 < \delta < 2.99 + 1.761 \times 1.8684$, or $-0.30 < \delta < 6.28$. $\Delta = 2$ falls within the interval. The data do not reject the null hypotheses.

Chapter 22

Sample Size Required in a Study

22.1. OVERVIEW

Chapter 7 examines the basic ideas of estimating the minimum sample size required in a study. More exactly, it is the size required in a test or confidence interval, because sample size may be estimated for multiple tests in a study. The conservative strategy is to choose for the study the largest of those sample sizes. It was pointed out that statistical needs may compete with limitations of time, money, support facilities, and ethics of patient use, but speaking solely from a statistical viewpoint, *the larger the sample, the better,* because the character of the sample approaches that of the population as the sample grows larger. Larger samples provide better estimates, more confidence, and smaller test errors. Ideally, we would obtain all the data our time, money, support facilities, and ethics will permit. The purely statistical purpose of minimum sample size estimation is to *verify that we will have enough data* to make the study worthwhile.

NUMBER NEEDED TO TREAT

A related concept is the number needed to treat, as in the number of mammograms required to detect one breast cancer that would otherwise have been missed. The methods for this question are rather different and are examined in Chapter 23.

SUMMARY OF MINIMUM SAMPLE SIZE CONCEPT

Chapter 7 introduces the concept of minimum sample size as estimating that sample size n that has $1 - \beta$ probability of detecting a clinically chosen difference d (or δ) when it is present and $1 - \alpha$ probability of rejecting such a difference when it is absent. The calculation of the probabilities depends on the standard error of the mean (SEM), σ_m. Figure 7.1 illustrates the effect of increasing the sample size from n to $4n$, which halved σ_m and reduced the error rates α and β.

The difference d was termed *clinical relevance*. The method of estimating minimum required sample size is often called *power analysis*, because $1 - \beta$, the power of the test, is a component in the estimation process.

TEST SIDEDNESS

Sample size estimation may be based on a one- or two-sided alternate hypothesis. For one-sided cases, the more commonly used two-sided form of the normal tail area $z_{1-\alpha/2}$ is replaced by $z_{1-\alpha}$ wherever it appears. For example, replace the two-tailed 5% α's $z = 1.96$ by the one-tailed 5% α's $z = 1.645$.

TEST PARAMETERS

Although $\alpha = 5\%$ and power $= 80\%$ ($\beta = 20\%$) have been the most commonly selected error sizes in the medical literature, a 20% β is larger than appropriate in many cases. Furthermore, the false-negative/false-positive rate (β/α) ratio of 4:1 may affect the care of patients in treatments based on the outcomes of the study, because sometimes the false-positive result is worse for the patient than the false-negative result. The β/α ratio should be chosen with thought and care for each study.

CLINICAL RELEVANCE

The size of the difference between treatment outcomes (often denoted d or δ) is the value, or cut point, that delineates some change in clinical practice. It is the statistical parameter that most influences the sample size. It may also affect patient care. Too often it is chosen rather casually. Because a larger difference will require a smaller n, the temptation exists to maximize the difference to allow for a small enrollment in the study. The choice should be based on clinical rather than statistical grounds.

SEQUENTIAL ANALYSIS IN RELATION TO SAMPLE SIZE ESTIMATION

The concept of sequential analysis is to test the hypothesis successively with each new datum. The result of the test falls into the following classes: accept hypothesis, reject hypothesis, or continue sampling. At first, because of a very small sample size, the test result falls into the continue-sampling class. Sampling is continued until it falls into one of the other classes, at which point it is stopped. The purpose is to minimize the sample size required in a study. Although it does serve this purpose, it does not give the investigator any advance idea of required sample size. On the contrary, it prevents a planned sample size. Thus, sequential analysis is categorized as a time-dependent analysis and is addressed in Chapter 25.

22.2. RELATION OF SAMPLE SIZE CALCULATED TO SAMPLE SIZE NEEDED

THE CALCULATED SAMPLE SIZE IS JUST A (RATHER POOR) ESTIMATE

Remember that the calculation of the minimum required sample size estimate is based on judgmental inputs and data different from that that will be used in the ensuing study. This estimate, therefore, is one of low confidence. Inasmuch as the SEM used in the sample size calculation arose from a sample other than that of the study, there is a *wrong data* source of possible error in the calculation. Also, sample size is estimated on data that are subject to randomness; therefore, there is also a *randomness* source of possible error in the estimate. Because of these errors, the calculated sample size is likely to be different from that actually needed. In case the calculated size is smaller than truly needed, the calculated size should be increased by a "safety factor" to allow for these two sources of possible error. The size of this safety factor is an educated guess.

22.3. SAMPLE SIZE FOR TESTS ON MEANS

WHAT IS NEEDED FOR THE CALCULATION

Chapters 6 and 17 review methods for testing means of continuous data. Values used in the calculation of minimum sample size for such tests are the error risks (α and β), the standard deviation of the data (σ or s), and the difference between the two means being tested (d, or sometimes δ). This difference should be chosen as the difference clinically important to detect. The minimum sample size depends more on this difference than on the other inputs. The sample size grows large quickly as this difference grows small.

WHERE THE FORMULAS COME FROM

The explanation will treat a test of a sample mean against an established mean using the normal distribution; concepts for other tests are similar. A population exists with mean μ. We sample from some population that may or may not be the same. Our sample distribution has a population mean μ_s (subscript s denotes "sample"). We want to learn whether μ_s is the same as μ, or, for example, it is larger, implying hypotheses H$_0$: $\mu_s = \mu$ and H$_1$: $\mu_s > \mu$. However, we do not know the value of μ_s, so we estimate it with the sample mean m. Thus, to decide if $\mu_s = \mu$, we test m against μ. Figure 22.1 shows the two distributions involved: the null distribution, with its mean $\mu = 0$ (standard normal) indicated by a vertical line, and a possible alternate distribution, with its mean μ_s, estimated by m, indicated by a vertical line at about 3. σ is the standard deviation of the population data, so σ_m, the SEM (the standard deviation of distribution shown in the figure), is $\sigma/\sqrt{n}$, where n is the sample size we seek. We use the form of a test for a significant difference between means, $m - \mu$, with $\alpha = 5\%$ and $\beta = 20\%$ (power $= 1 - \beta = 80\%$). The critical value (here, $\mu + 1.645\sigma_m$) is the

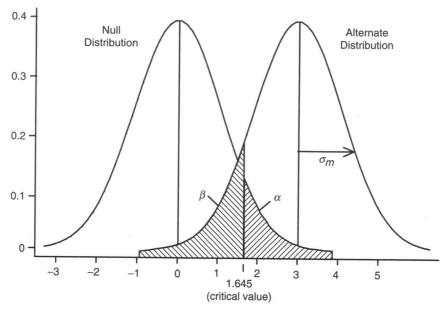

Figure 22.1 Distributions of null and alternate hypotheses showing Types I and II errors separated by the critical value, chosen at 1.645 (standard errors) to provide $\alpha = 5\%$. We specify the difference in standard errors between the means (in this example, about 3) that we want the test to detect. The error probabilities are dictated by the overlap of distributions, which, in turn, is dictated by the standard error $\sigma_m = \frac{\sigma}{\sqrt{n}}$. As we increase n, σ_m shrinks; therefore, the areas under the overlap giving the error sizes shrink until the specified error probabilities are reached. The resulting n is the minimum sample size required to detect the specified difference in means.

position separating the two types of error; this is shown in Fig. 22.1 as the number of standard errors to the right of μ that yields a 5% α (area under a tail of the null distribution; 1.645 is the critical value from the normal table for $\alpha = 5\%$). Similarly, β is the area under a tail of the alternate distribution, specified by the number of standard errors to the left of m that the critical value lies, or $m - 0.84\sigma_m$ (where 0.84 is the value from the normal table for 20% in the tail area). Because these two expressions both equal the critical value, we set them equal to each other, or $\mu + 1.645\sigma/\sqrt{n} = m - 0.84\sigma/\sqrt{n}$. Solution of the equation for n yields $n = (1.645 + 0.84)\sigma^2/(m - \mu)^2$. Other formulas for minimum required sample size follow equivalent logic.

SUPPOSE WE DO NOT KNOW σ

Note that the methods given here primarily use normal distribution theory and the population or large-sample σ, rather than a small-sample s. If only data for s are available, we just use s in place of σ. The reason for this is twofold. First, by not knowing n, we have no degrees of freedom (df) to use in finding the appropriate t-values to use in the calculation. Second, the nicety of using t would be lost in the grossness of the approximation, because the process depends on pilot data or results from other studies and not the data to be used in the actual analysis.

CASE 1: ONE MEAN, NORMAL DISTRIBUTION

Example Posed: Patients with Benign Prostatic Hypertrophy Treated with Hormonal Therapy

We want to determine whether or not patients with benign prostatic hypertrophy (BPH) treated with an experimental hormonal therapy have larger prostates than those of patients without BPH. The prostate volume (measured in milliliters) of our 296 patients without BPH (see DB1) has mean $\mu = 35.46$ and standard deviation $\sigma = 18.04$. What is the smallest sample size that can detect a sample mean m (of the experimental group) at least 10 ml larger than μ (implying a one-sided alternate hypothesis)? We choose $\alpha = 0.05$ and $\beta = 0.20$, which are values commonly chosen in medicine.

Method

What size of sample do we need to decide if m is different from μ? (More exactly, we are trying to determine whether or not the mean of the sample's population as estimated by m is different from μ.) Choose the smallest distance d between m and μ, that is, $d = m - \mu$, that you want to detect with statistical significance. Choose the risk you will accept of concluding there is a difference when there is not (α) and of concluding there is no difference when a difference does exist (β, or $1 -$ power). Look up the z-values in Table I (see Tables of Probability Distributions) that are associated with these two risks, $z_{1-\alpha/2}$ and $z_{1-\beta}$. Substitute the z-values and d, together with the standard deviation σ in Eq. (22.1), to find n, the minimum sample size required.

$$n = \frac{\left(z_{1-\alpha/2} + z_{1-\beta}\right)^2 \sigma^2}{d^2} \tag{22.1}$$

Example Completed: Patients with Benign Prostatic Hypertrophy

The selection of $\alpha = 0.05$ and $\beta = 0.20$ yields $z_{1-\alpha} = 1.645$ and $z_{1-\beta} = 0.84$. Substitution in Eq. (22.1) yields $n = (1.645 + 0.84)^2 \times 18.04^2/10^2 = 6.175 \times 325.44 \div 100 = 20.1$; we must round up to 21. In line with Section 22.2, we should choose a slightly larger n.

Additional Example: Treatment of Emergency Dyspepsia

An emergency medicine physician wants to test the effectiveness of a "GI cocktail" (antacid plus viscous lidocaine) to treat emergency dyspeptic symptoms as measured on a 1–10 pain scale.[50] The standard deviation without treatment has been scored for a large number of patients as $\sigma = 1.73$. The physician considers a reduction in pain rating of $d = 1.5$ points as clinically meaningful. He believes that the treatment cannot increase pain; therefore, a one-sided test will be used. By taking $\alpha = 0.05$ and $\beta = 0.20$ (power $= 80\%$), $z_{1-\alpha} = 1.645$ and $z_{1-\beta} = 0.84$.

By substituting in Eq. (22.1), the physician finds

$$n = \frac{(z_{1-\alpha/2} + z_{1-\beta})^2 \sigma^2}{d^2} = \frac{(1.645 + 0.84)^2 \times (1.73)^2}{1.5^2} = 8.21.$$

He sees that only nine patients are needed to be able to just detect significance. However, the physician anticipates that the posttreatment standard deviation might be larger and wants a safety factor in line with Section 22.2; therefore, he decides to sample 14 patients.

Exercise 22.1: One Mean. An emergency medicine physician wants to know if mean heart rate after a particular type of trauma differs from the healthy population rate of 72 beats/min.[63] He considers a mean difference of 6 beats/min clinically meaningful. He takes σ to be the 9.1 beats/min reported in a large study. How many patients will be needed? Use $\alpha = 0.05$ and power $= 0.80$.

CASE 2: TWO MEANS, NORMAL DISTRIBUTIONS

The following case also is detailed in Section 7.4, together with an example in which mean ranges of motion between two types of artificial knee are compared.

Example Posed: Testing Age as Related to Survival

As another example, let us find the minimum required sample size to detect a 2-year mean difference in age at carinal resection surgery (DB12) between patients who survive and those who die. We estimate standard deviations, which we take as σ-values, from DB12 data. For the 117 surviving patients, $s_1 = 16.01$; for the 17 patients who died, $s_2 = 15.78$. We might propose older patients to be at greater risk, but unanticipated factors that might create a risk in younger patients cannot be ruled out; therefore, a two-tailed test is appropriate. α is chosen as 5%, and $1 - \beta$ (the power) is chosen as 80%.

Method

We ask if μ_1 (estimated by m_1) is different from μ_2 (estimated by m_2). Choose the smallest distance d between μ_1 and μ_2, that is, $d = \mu_1 - \mu_2$, that you want to detect with statistical significance. This is usually a clinical choice: the size of the difference that is clinically important. Also find σ_1^2 and σ_2^2, as estimated by s_1^2 and s_2^2. Choose the risk you are willing to accept of concluding there is a difference when there is not (α) and of concluding there is no difference when there is (β, or $1 - $ power). Look up the z-values in Table I for these two risks, $z_{1-\alpha/2}$ and $z_{1-\beta}$. Substitute the z-values and d, together with the variances in Eq. (22.2), to find $n_1 (= n_2)$, the minimum sample size required in *each* sample. Replace $z_{1-\alpha/2}$ with $z_{1-\alpha}$ for a one-sided test.

$$n_1 = n_2 = \frac{(z_{1-\alpha/2} + z_{1-\beta})^2 (\sigma_1^2 + \sigma_2^2)}{d^2} \qquad (22.2)$$

Example Completed: Testing Age as Related to Survival

Variances (squares of the standard deviations) are $\sigma_1^2 = 256.32$ and $\sigma_2^2 = 209.09$. From Table I, $z_{1-\alpha/2} = 1.96$ and $z_{1-\beta} = 0.84$. Substitution in Eq. (22.2) yields

$$n_1 = n_2 = \frac{\left(z_{1-\alpha} + z_{1-\beta}\right)^2 \left(\sigma_1^2 + \sigma_2^2\right)}{d^2} = \frac{2.80^2 \times 465.41}{2^2} = 912.20.$$

The required minimum sample size is 913. For the reasons discussed in Section 22.2, including a few more data would be advisable.

Additional Example: Testing Two Treatments of Dyspepsia

An emergency medicine physician wants to test the relative effectiveness of two treatments, a "GI cocktail" (antacid plus viscous lidocaine) (treatment 1) versus intravenous ranitidine hydrochloride (treatment 2) to treat emergency dyspeptic symptoms as measured on a 1–10 pain scale.[50] Not having data on the pain ratings of the treatments, he estimates s_1 and s_2 as the standard deviation without treatment $\sigma = 1.73$. The physician considers a reduction in pain rating of $d = 1.5$ points as clinically meaningful. Either treatment can either be more effective; therefore, a two-sided test will be used. Taking $\alpha = 0.05$ and $\beta = 0.20$ (power = 80%), $z_{1-\alpha/2} = 1.96$ and $z_{1-\beta} = 0.84$. By substituting in Eq. (22.2), the physician finds

$$n_1 = n_2 = \frac{\left(z_{1-\alpha/2} + z_{1-\beta}\right)^2 \left(\sigma_1^2 + \sigma_2^2\right)}{d^2} = \frac{(1.96 + 0.84)^2 (2 \times 1.73^2)}{1.5^2} = 20.86.$$

He requires a minimum of 21 patients per sample to show significance. However, the physician suspects that sample standard deviations will be larger than the untreated σ and plans a larger sample size.

Exercise 22.2: Two Means. An emergency medicine physician wants to know if mean heart rate after two particular types of trauma is different.[63] He considers a mean difference of 6 beats/min as clinically meaningful. From pilot data, the physician finds $s_1 = 6.13$ beats/min and $s_2 = 6.34$ beats/min, which he uses to estimate σ_1 and σ_2. How many patients are needed in each group? Use $\alpha = 0.05$ and power = 0.80.

CASE 3: NONNORMAL (POORLY BEHAVED OR UNKNOWN) DISTRIBUTIONS

One relationship that is sometimes helpful in sizing samples needed to detect a difference between m and μ arises in an inequality named for the Russian mathematician P. L. Chebychev (other spellings include Tchebysheff). The relationship was actually discovered by I. J. Bienaymé (1835); Chebychev discovered it independently a bit later. The inequality is known as the Law of Large Numbers. Of interest to us is a form of the inequality relating sample size n to the deviation of the sample mean from the population mean. *This relationship is useful when an underlying data distribution is very poorly behaved or when nothing is known about the distribution. It is also useful when there are no data at all to estimate the variability.* Because this inequality is solved without any information as to the distribution of the statistic involved, it is a rather gross overestimate of the sample size that would be required were the distribution known and well behaved. The sample size given by Chebychev's inequality will certainly be enough in any case and is therefore a conservative sample size, but *it is more than required for most applications.*

Example Posed: Testing Drug Effect on Intraocular Pressure

Suppose we want to know the required minimum sample size to find a difference between mean intraocular pressure (IOP in mm Hg) between patients who have been treated with a new drug and those who have not.[6] The standard deviation is 4 mm Hg. We decide as a clinical judgment that we want to detect a 2 mm Hg decrease in IOP. We choose $\alpha = 0.05$; that is, we are willing to take a 5% risk of being wrong in our choice.

Method

Choose k, the difference you want to detect between the sample mean and population mean (a clinical choice), expressed as the number of standard deviations of distance between them. The form of Chebychev's inequality that is useful in this case is

$$P[-k \leq m - \mu \leq k] \geq 1 - \frac{\sigma^2}{k^2 n}, \tag{22.3}$$

where m denotes the sample mean, μ represents the hypothesized population mean, and σ represents the population standard deviation. Choose α, the risk for error that you are willing to accept. The left side of Eq. (22.3) will be $1 - \alpha$, so $1 - \alpha = 1 - \sigma^2/k^2 n$, which reduces to

$$n = \frac{\sigma^2}{\alpha k^2}. \tag{22.4}$$

Substitution for k in terms of σ allows the σ_2 values to cancel each other out and provides a number for n. The value of σ is not even required if the user can express the difference k in standard deviation units.

Example Completed: Drugs and Intraocular Pressure

The units of k must be standard deviations; that is, a standard deviation is one unit. In this case, we want to detect a difference of 2 mm Hg and $\sigma = 4$ mm Hg, so $k = 0.5\sigma$. $\alpha = 0.05$. Using these quantities, we calculate Eq. (22.4) as

$$n = \frac{\sigma^2}{\alpha k^2} = \frac{\sigma^2}{0.05 \times 0.25\sigma^2} = 80.$$

CASE 4: NO OBJECTIVE PRIOR DATA

If we have neither data nor experience with a phenomenon being studied, we have no way to guess a required sample size. However, if we have some experience with the quantitative outcome of the variable to be measured but no objective data at all, we have recourse to a rough idea of needed sample size as a starting point.

Example Posed: Effectiveness of an Herbal Remedy in Treating Colds

The wife of an internist has been treating her common colds with an herbal remedy for 3 years and claims it reduces the number of days to disappearance of symptoms. Her best time was 8 days, and the worst was 15 days. The internist decides to conduct a prospective, randomized, double-masked study to evaluate the remedy's efficacy.[63] How many data should he take? He decides that a reduction of $d = 1$ day would be clinically meaningful. He chooses $\alpha = 5\%$ and power $= 80\%$.

> **Method**
>
> From experience, guess the smallest and largest values of the variable you have noticed. Take the difference between them as a guess of the interval: mean $\pm$ 2 standard deviations. This is equivalent to assuming that the data are roughly normal and your experience covers about 95% of the possible range. Then, σ is estimated as 0.25(largest $-$ smallest). Use this value of σ in the method of case 1. Clearly, this "desperation" estimate is of extremely low confidence, but it may be better than picking a sample size arbitrarily.

Example Completed: Herbal Remedy

The assumption that the remedy cannot increase the time to symptom disappearance implies a one-sided test. The chosen α and β yield one-tailed z-values of 1.645 and 0.84, respectively. In the absence of objectively recorded data, the investigator guesses σ from the observed range: $\sigma \approx 0.25$ (largest $-$ smallest) $= 0.25(15-8) = 1.75$. (The symbol "$\approx$" is often used to indicate "approximately equal to.") Substitution of these values in Eq. (22.1) yields

$$ n = \frac{\left(z_{1-\alpha} + z_{1-\beta}\right)^2 \sigma^2}{d^2} = \frac{(1.645 + 0.84)^2 \times (1.75)^2}{1^2} = 18.91. $$

Nineteen data in each group (the experimental and the placebo-treated groups) is indicated as a minimum. He chooses a large safety factor (see Section 22.2) because of the poor estimation of σ and plans to include 25 data in each group.

22.4. SAMPLE SIZE FOR CONFIDENCE INTERVALS ON MEANS

EXAMPLE POSED: EXTENT OF CARINAL RESECTION

In DB12, the extent of carinal resection (measured in centimeters) form an approximately normal distribution, and the sample size is large enough to use the estimated standard deviation 1.24 as σ. Suppose that in a new study we require the sample mean m of the extent of resection to be no more than 0.5 cm from the population mean μ, that is, $d = |m - \mu| = 0.5$. How large a sample is needed to obtain 95% confidence on this interval?

METHOD

In Chapter 4, which discusses confidence intervals, the end values of a $1 - \alpha$ confidence interval on the mean μ of a normal distribution with known σ was given by Eq. (4.5) as $\mu = m \pm z_{1-\alpha/2}\sigma_m$. As before, the form using σ rather than s is used regardless of sample size. We subtract m from μ to generate $d = |\mu - m|$, the difference between the observed and theoretical means. We square throughout and substitute $\sigma^2/n = \sigma_m^2$ to obtain $d^2 = (z_{1-\alpha/2}\sigma)^2/n$, or

$$n = \frac{(z_{1-\alpha/2})^2\sigma^2}{d^2}. \qquad (22.5)$$

The n calculated in Eq. (22.5) is a very minimum; it would be wise to take a slightly larger sample for the reasons discussed in Section 22.2. For 95% confidence, we need only replace $z_{1-\alpha/2}$ by 1.96.

β IS ABSENT FROM CONFIDENCE INTERVAL SAMPLE SIZE ESTIMATION

Confidence intervals are estimation tools. There is no alternate hypothesis and therefore no Type II error, the risk of which is β. Although sample size estimation for a test involves limiting the two error risks jointly, sample size estimation for a confidence interval addresses only lack of confidence in the estimate at issue.

EXAMPLE COMPLETED: EXTENT OF CARINAL RESECTION

The need for 95% confidence implies that $z_{1-\alpha/2} = 1.96$ (the frequently used entry from Table I). By substituting that value and the values of σ and d in Eq. (22.5), we find

$$n = \frac{z_{1-\alpha/2}^2\sigma^2}{d^2} = \frac{1.96^2 \times 1.24^2}{0.5^2} = 24.$$

Therefore, a sample of at least 24 patients is needed.

ADDITIONAL EXAMPLE

In the example from DB12 just completed, suppose we required greater confidence, say, 99%, and greater accuracy, say, 0.25 cm. σ was taken as 1.24, derived from the data already at hand. $d = |m - \mu| = 0.25$. In Table I, the 0.990 entry in the column under two-tailed $1 - \alpha$ lies in the row for $z_{1-\alpha/2} = 2.576$. By substituting in Eq. (22.5), we find

$$n = \frac{z_{1-\alpha/2}^2 \sigma^2}{d^2} = \frac{2.576^2 \times 1.24^2}{0.25^2} = 163.25.$$

Therefore, a minimum of 164 patients is required.

Exercise 22.3. In DB7, the distribution of bone density is not far from normal. $m = 154$, and $s = 24$. Estimate n for a 95% confidence interval on a deviation of the sample mean from the theoretical mean of no more than $d = 10$.

22.5. SAMPLE SIZE FOR TESTS ON RATES (PROPORTIONS)

Chapters 6 and 15 examine methods for testing categorical data. From categorical data, proportions can always be obtained; minimum sample size is calculated for such proportions. Values used in such calculation are the error risks α and β, the proportions involved, and the difference between the proportions. This difference, denoted d in this textbook, is the value to be tested. It should be chosen as the difference clinically important to detect. The minimum sample size depends more on this difference than on the other inputs. The sample size grows large quickly as this difference grows small.

CONTINGENCY TABLES

Methods for the estimation of minimum required sample size are not well developed for contingency tests, but we can use the method for tests of proportion. The cell entries over their marginal totals (the totals for each category) provide proportions. If the contingency table is bigger than 2×2, an *ad hoc* approach is to calculate the required sample size for the various pairs by using the method for two proportions repeatedly and then accept as the estimate the largest sample size that emerges.

CASE 1: TEST OF ONE PROPORTION

One proportion (from a sample) is tested against a theoretical or previously established proportion. A proportion follows the binomial or Poisson distributions, depending on whether the proportion is central (is *not* near 0 or 1) or extreme (*is* near 0 or 1), respectively. A proportion from an at least moderately sized sample has an approximately normal distribution for either the binomial or Poisson. The proportion behaves like a mean for either, leaving the only difference in the method as the mode of estimating the standard deviation.

Formulas for the binomial form, proportion π known or estimated by p, respectively, are

$$\sigma = \sqrt{\frac{\pi(1-\pi)}{n}}, \qquad s = \sqrt{\frac{p(1-p)}{n}}, \tag{22.6}$$

and for the Poisson form, the formulas are

$$\sigma = \sqrt{\frac{\pi}{n}}, \qquad s = \sqrt{\frac{p}{n}}. \tag{22.7}$$

Because of the normal approximation, the required sample size estimation follows the concept and logic depicted in Fig. 22.1.

Example Posed: Rate of Positive Prostate Cancer Biopsy Results

Section 6.2 proposes testing an observed 30% rate of positive prostate cancer biopsy results against a smaller theoretical proportion $\pi = 0.25$. How many patients do we need for a p proportion $= 0.30$ of positive test results to be significant with a one-tailed $\alpha = 0.05$ and power $= 0.80$? (Note that we have a one-sided alternate hypothesis.)

Method

We assume normal theory, with mean difference $p - \pi$ divided by the appropriate standard deviation from Eq. (22.6) or (22.7), which gives rise to the sample size Eqs. (22.8) and (22.9). We know the theoretical proportion π, and we calculate the sample proportion p from our data. If π is central (not near 0 or 1), we use a binomial form, Eq. (22.8). If π is extreme (near 0 or 1), we use a Poisson form, Eq. (22.9). We choose the risk required of a false-positive (α) and a false-negative (β, or $1 -$ power) result. We look up the z-values in Table I associated with these two risks (areas in the tails of the normal curve), $z_{1-\alpha/2}$ and $z_{1-\beta}$. We then substitute the z-values, π, and p in Eq. (22.8) or (22.9) to find n, the minimum sample size required.

$$n = \left[\frac{z_{1-\alpha/2}\sqrt{\pi(1-\pi)} + z_{1-\beta}\sqrt{p(1-p)}}{p - \pi} \right]^2 \tag{22.8}$$

$$n = \left[\frac{z_{1-\alpha/2}\sqrt{\pi} + z_{1-\beta}\sqrt{p}}{p - \pi} \right]^2 \tag{22.9}$$

It is important to note that $p - \pi$ is a major determinant in the sample size estimation; therefore, p should be chosen carefully in line with its clinical implications. For a one-sided test, replace $z_{1-\alpha/2}$ by $z_{1-\alpha}$.

EXAMPLE COMPLETED: RATE OF POSITIVE BIOPSY FOR PATIENTS WITHOUT CANCER

From Table I, the z-values are 1.645 and 0.84, respectively. Because π is not near 0 or 1, we use the binomial form, Eq. (22.8). By substituting the z-values, π, and p, we obtain $n = [1.645 \times \sqrt{0.25 \times 0.75} + 0.84 \times \sqrt{0.30 \times 0.70}]^2/(0.30 - 0.25)^2 = 481.6$. Therefore, at least 482 biopsies are required.

Additional Example: Rate of Schistosomiasis

A navy specialist in internal medicine who has been sent to an African nation must decide if female residents of a particular rural region lying along an often stagnant river have a prevalence of schistosomiasis greater than the national average, implying the hypotheses H_0: $\pi_s = \pi$ (subscript s for "sample") and H_1: $\pi_s > \pi$ (a one-sided alternative).[67] The specialist plans an informal study to compare the local mean with the national mean and must estimate how many patients need to be sampled. Clinical judgment leads him to believe that a difference of 2% needs to be detected. The specialist finds an article in the literature that quotes a prevalence of 24% over 1600 patients examined. He chooses one-tailed $\alpha = 5\%$ and power $= 80\%$. The z-values from Table I are 1.645 and 0.84. By substitution in Eq. (22.8), he finds

$$n = \left[\frac{z_{1-\alpha}\sqrt{\pi(1-\pi)} + z_{1-\beta}\sqrt{p(1-p)}}{p - \pi} \right]^2$$

$$= \left[\frac{1.645\sqrt{(0.24)(0.76)} + 0.84\sqrt{(0.26)(0.74)}}{0.02} \right]^2 = 2867.63.$$

The specialist will require nearly 2900 patients to be able to detect a 2% difference.

Exercise 22.4 (One Proportion). At one time, radial keratotomy was performed by residents in ophthalmology in a certain hospital. A review of records showed that 18% of residents required enhancements in more than one fifth of their cases.[5] This frequency of enhancements was thought to arise from the surgery learning curve. Training in the surgery by a computer simulation may shorten the learning curve. How many residents would have to be monitored to detect a decrease of 6% (i.e., a decrease from 18% to 12%) in the number of residents requiring enhancements in more than one fifth of their cases? Use $\alpha = 5\%$ and power $= 80\%$.

CASE 2: TEST OF TWO PROPORTIONS

A test of two proportions (see also Section 15.6) can be used in place of a contingency test (see Sections 6.3 and 15.2) but is usually less convenient because it does not appear in many software packages. We use it here reformatted for sample size estimation.

Example Posed: Rate of Personality Disorder in Criminals

A psychiatrist wants to know if the proportion of people having a personality disorder is the same for those committing violent crimes (p_1) and those committing nonviolent crimes (p_2).[50] Theoretically, either p could be the larger; therefore, she chooses a two-sided alternate hypothesis. She examines a few of her past records to serve as a pilot survey and estimates p_1 as 0.06 and p_2 as 0.02. How many patients are needed to detect a difference $p_1 - p_2 = 0.04$, significant at two-tailed $\alpha = 0.05$ and power $= 0.80$?

Method

We assume normal theory. We calculate the sample proportions p_1 and p_2 from our data and then p_m (m for mean) as the average of p_1 and p_2:

$$p_m = \frac{p_1 + p_2}{2}$$

If p_m is central (not near 0 or 1), we use Eq. (22.10), which is derived from the binomial form:

$$n_1 = n_2 = \left[\frac{z_{1-\alpha/2}\sqrt{2p_m(1 - p_m)} + z_{1-\beta}\sqrt{p_1(1 - p_1) + p_2(1 - p_2)}}{p_1 - p_2}\right]^2 \tag{22.10}$$

If p_m is extreme (near 0 or 1), we use Eq. (22.11), which is derived from the Poisson form:

$$n_1 = n_2 = \left[\frac{(z_{1-\alpha/2} + z_{1-\beta})\sqrt{p_1 + p_2}}{p_1 - p_2}\right]^2 \tag{22.11}$$

We choose α, the risk for a wrong rejection of H_0, and β ($1 -$ power), the risk for a wrong acceptance of H_0. We look up the z-values in Table I associated with these two risks (areas in the tails of the normal curve), $z_{1-\alpha/2}$ and $z_{1-\beta}$. We substitute the z-values and the p-values in Eq. (22.10) or (22.11) to find n, the minimum sample size required in each group. If it is a one-sided test, replace $z_{1-\alpha/2}$ by $z_{1-\alpha}$.

Estimates on the Borderline Between Binomial and Poisson

If π is close to the borderline for using the Poisson approximation, a correction of the normal approximation is appropriate, especially if n_1 is small. The corrected n for each group will be n_{corr} from

$$n_{corr} = \frac{n_1}{4}\left[1 + \sqrt{1 + \frac{4}{n_1 \times |p_1 - p_2|}}\right]^2. \tag{22.12}$$

Example Completed: Rate of Personality Disorder in Criminals

Because p_m is near 0, the Poisson form, Eq. (22.11), is appropriate. From Table I, the z-values are 1.96 and 0.84, respectively. p_m, the mean p, is $(p_1 + p_2)/2 = 0.04$.

$$n_1 = n_2 = \left[\frac{(z_{1-\alpha/2} + z_{1-\beta})\sqrt{p_1 + p_2}}{p_1 - p_2} \right]^2 = \left[\frac{(1.96 + 0.84)\sqrt{0.08}}{0.04} \right]^2 = 392$$

The psychiatrist will need a minimum of 392 patients in each group.

Additional Example: Sex Difference in Fever Reporting

An emergency medicine specialist finds that 15% of patients report having a fever.[50] She wants to know if there is a difference between reporting rates of men and women. The specialist does not know which group would have the greater rate, which implies a two-tailed test. How many patients would she need to monitor to find a difference of 6% (e.g., 12% for one group and 18% for the other) with $\alpha = 0.05$ and power $= 0.80$? $z_{1-\alpha/2} = 1.96$, and $z_{1-\beta} = 0.84$. $p_m = 0.15$, which is not near 0 or 1; therefore, the binomial form, Eq. (22.10), is used.

$$n_1 = n_2 = \left[\frac{z_{1-\alpha/2}\sqrt{2p_m(1 - p_m)} + z_{1-\beta}\sqrt{p_1(1 - p_1) + p_2(1 - p_2)}}{p_1 - p_2} \right]^2$$

$$= \left[\frac{1.96\sqrt{2 \times 0.15 \times 0.85} + 0.84\sqrt{0.12 \times 0.88 + 0.18 \times 0.82}}{0.12 - 0.18} \right]^2 = 554.16$$

The minimum sample size per group is 555 patients.

Exercise 22.5 (Two Proportions).   At one time, an eye surgeon was performing radial keratotomies using hand surgery and considered using a laser device.[4] His hand surgery record showed that 23% of eyes required surgical enhancement. The surgeon planned a prospective study in which patients would be randomized into hand and laser groups. On how many eyes per method would he have had to operate to detect an improvement of 10% (i.e., a reduction in enhancement rate from 23% to 13%)? The surgeon believes the laser devise can only improve the enhancement rate; therefore, a one-sided test is used. Use $\alpha = 5\%$ and power $= 80\%$.

22.6. SAMPLE SIZE FOR A CONFIDENCE INTERVAL ON A RATE (PROPORTION)

EXAMPLE POSED: RATE OF PATIENTS SATISFIED WITH ANESTHESIA ADMINISTERED DURING ORAL SURGERY

In the additional example of Section 14.5, patients undergoing oral surgery were anesthetized by a combination of propofol and alfentanil, and 89.1% of patients rated the anesthetic as highly satisfactory.[43] We placed a 95% confidence interval on π as 83% to 95%. The width of the half interval $w = |p_s - \pi|$ is 6%. How many patients are needed to reach $w = 5\%$?

METHOD

Conceptually, we may think of estimating the sample size needed for a confidence interval as specifying the interval we require and back-solving the equation for n. Note that power, that is, $1 - \beta$, is not involved (see the comment on this issue in Section 22.3). However, there are also other considerations. The width of the interval is $w = |p - \pi|$, where we do not have p yet and usually will never know π exactly; thus, this interval is rather arbitrary. Because we need p to estimate σ but will not have it until data have been gathered, we must use some rough indicator of p, such as the proportion found in a pilot study or a similar study found in the literature. As a result of these uncertainties, the n obtained is not an accurate sample size but rather just a rough idea of the size. We would be better assured of reaching our target confidence if we take a slightly larger n, although there is no way to know just how much larger. Solving one side of the confidence interval Eqs. (14.9) (ignoring the $1/2n$ continuity correction) and (14.10) for n for 95% confidence, we find

$$n = \frac{1.96^2 p(1 - p)}{w^2} \tag{22.13}$$

for the case of central π (not near 0 or 1), and

$$n = \frac{1.96^2 p}{w^2} \tag{22.14}$$

for extreme π (near 0 or 1). Confidence levels other than for 95% can be found by replacing the 1.96 with the appropriate probability from Table I. [Notably, because the maximum value of $p(1 - p) = 0.25$, the numerator of Eq. (22.13) cannot exceed 0.96 and the numerator of Eq. (22.14) cannot exceed 3.842.]

EXAMPLE COMPLETED: SATISFACTION WITH ORAL SURGERY ANESTHESIA

π is not near 0 or 1, so we use Eq. (22.13).

$$n = \frac{1.96^2 p_s(1 - p_s)}{w^2} = \frac{1.96^2 \times 0.891 \times 0.109}{0.05^2} = 149.2$$

A minimum of 150 patients are needed, and we advise adding a few more patients to the sample size. Suppose we wanted to be 98% confident, that is, to risk 1% chance for error on each tail. From Table I, the 0.98 two-tailed $1 - \alpha$ yields a corresponding z of 2.326. Replacing the 1.96 in the preceding formula by 2.326, we obtain

$$n = \frac{2.326^2 p_s(1-p_s)}{w^2} = 210.2.$$

ADDITIONAL EXAMPLE: EFFICACY OF A DERMATOLOGIC TREATMENT

A dermatologist is studying the efficacy of tretinoin in treating $n = 15$ women for postpartum abdominal stretch marks.[77] Tretinoin was used on a randomly chosen side of the abdomen, and a placebo was used on the other side. Neither patient nor investigator knew which side was medicated. The dermatologist rated one side or the other as better where he could make a distinction (in 13 of the 15 abdomens); afterward, the code was broken and the treated side identified. The treated side was chosen in 9 of 13 abdomens for an observed proportion of 0.69. If the treatment were of no value, the theoretical proportion π would be 0.5. How many patients would be needed to have 95% confidence on the $w = |p - \pi| = 0.19$? π is not near 0 or 1; therefore, the dermatologist uses Eq. (22.13). By substituting, he finds

$$n = \frac{1.96^2 p(1-p)}{w^2} = \frac{3.8416 \times 0.69(1-0.69)}{(0.69-0.5)^2} = 22.8.$$

He requires a minimum of 23 patients.

Exercise 22.6.   A pediatric surgeon is studying indicators of patient condition following pyloromyotomy (correction of stenotic pylorus) in neonates.[21] The surgeon finds 14 of 20 infants, that is, a proportion of 0.7, experienced emesis after surgery. How large a sample is required to be 95% sure the surgeon is within 0.1 of the true proportion?

22.7. SAMPLE SIZE FOR SIGNIFICANCE OF A CORRELATION COEFFICIENT

EXAMPLE POSED: REPAIRED ANKLE PLANTAR FLEXION CORRELATED WITH AGE

In a study on broken ankle repair,[28] an orthopedist found the correlation coefficient between age and plantar flexion of a repaired ankle to be 0.1945. How large a sample would have been required to show evidence of flexion depending on age?

METHOD

When a correlation coefficient between two variables is small enough to have occurred by chance alone, we cannot infer evidence of a relationship between these variables. What sample size will make the correlation coefficient r larger than would occur by chance? More precisely, how large a sample is required for the 95% confidence interval to exclude $\rho = 0$? When $\rho = 0$ is true and the correlated variables are approximately normal, a transformation to the t distribution can be made, namely

$$t = \frac{r\sqrt{n-2}}{\sqrt{1-r^2}}. \tag{22.15}$$

Simple algebra will solve this equation for n and provide Table 22.1. The last lines of Table 22.1 carry the sample size smaller (about <12) than would be wise to use in practice. It is given to show the pattern of reduction in n. What is done if r is negative? A symmetry property of the t distribution allows the same n to emerge as if the r were positive, so just drop the minus sign for sample size purposes. Note that power, that is, $1 - \beta$, is not involved (see the comment on this issue in Section 22.3).

Table 22.1

Samples Sizes Required to Infer a ρ Larger Than Would Occur by Chance at 95% Confidence for Various Values of the Sample Correlation Coefficient r

r	n	r	n	r	n
0.02	—	0.21	88	0.40	25
0.03	4269	0.22	80	0.41	23
0.04	2401	0.23	74	0.42	22
0.05	1538	0.24	68	0.43	21
0.06	1068	0.25	63	0.44	20
0.07	785	0.26	58	0.45	19
0.08	601	0.27	54	0.46	19
0.09	475	0.28	50	0.47	18
0.10	385	0.29	47	0.48	17
0.11	319	0.30	44	0.49	17
0.12	268	0.31	41	0.50	16
0.13	228	0.32	39	0.55	13
0.14	197	0.33	37	0.60	11
0.15	172	0.34	34	0.65	9
0.16	149	0.35	32	0.70	8
0.17	134	0.36	30	0.75	7
0.18	120	0.37	29	0.80	6
0.19	107	0.38	27	0.85	5
0.20	97	0.39	26	0.90	4

STATISTICAL SIGNIFICANCE AND CLINICAL MEANING

The first few rows of Table 22.1 show that an investigator can find a correlation coefficient to be "larger than chance" just by taking a large enough sample. The fact of being larger than chance statistically may not imply that the correlation is clinically useful. For example, a 0.10 correlation coefficient between a disease and a marker may be significant, but it does not allow inference of the disease from the marker. Increasing sample size is useful up to the value of the coefficient that is clinically meaningful; beyond that, further increases are just a statistical exercise that should be avoided because the result could be misleading.

EXAMPLE CONCLUDED: PLANTAR FLEXION AND AGE

From Table 22.1, $r = 0.1945$ corresponds to an n between 97 and 107; the surgeon would have needed at least a few more than 100 patients.

ADDITIONAL EXAMPLE: CORRELATION BETWEEN SURGICAL DIFFICULTY AND SURGICAL DURATION

An anesthesiologist wanting to predict the requirement for deep sedation during surgery found a correlation coefficient between surgical difficulty and surgical duration of 0.17 on a small sample of patients.[50] If this sort of coefficient holds, how many patients would she need to conclude that surgical difficulty is a factor in predicting surgical duration? From Table 22.1, $r = 0.17$ corresponds to $n = 134$.

Exercise 22.7. An ophthalmologist suspects that the effect of a beta-blocker (timolol) on IOP diminishes with age.[6] He takes a pilot sample of $n = 30$ patients, recording their ages and their reduction in IOP 4 weeks after initiation of treatment. He calculates the sample correlation coefficient as $r = -0.23$. Based on this pilot result, how large a sample is needed to conclude that the population ρ is truly less than 0?

22.8. SAMPLE SIZE FOR TESTS ON RANKED DATA

Chapters 6 and 16 examine methods for testing ranked data. Not much theoretical development has occurred for estimating minimum required sample sizes for rank-order (nonparametric) data, because the largest application is for cases in which the distributions are unusual or "poorly behaved." Sample size methods depend on these probability distributions that are generally unknown. If the distributions are not extremely deviant from those of established methods, the established methods may be used, but increasing the sample size somewhat to be conservative. After all, sample size estimation is based on judgmental inputs combined with data other than that from the actual sample to be obtained, and are therefore very approximate. For extremely deviant cases of this sort, the method applied to case 3 (see Section 22.3) may be used, although this method is conservative and tends to overestimate the minimum required sample size.

22.9. SAMPLE SIZE FOR TESTS ON VARIANCES, ANALYSIS OF VARIANCE, AND REGRESSION

Estimation of sample size n for tests on these methods is not easy. The chi-square (χ^2) and F probabilities used to estimate n depend on n. A few such estimates exist as tables, but they are based only on the Type I error and thus are not dependable. It is possible to develop estimates based on both types of error, but this entails using the difficult mathematical distributions known as noncentral χ^2 and noncentral F, which are far beyond the level of this textbook. Recently, software in some statistical packages has appeared that treats one or another of these problems. The use of this software depends on "effect size." The concept of effect size is the outcome result contrast, say, between two arms of the experiment, in ratio to the variability of the sampling distributions of these arms. This ratio controls the sizes of the risks α and β. Unfortunately, in much of the software, effect size is poorly defined or even undefined, or it requires a spate of judgmental inputs over and above the α, β, and d already required. The best advice for an investigator who requires such sample sizes is to seek the assistance of an accomplished statistician.

ANSWERS TO EXERCISES

22.1. The posttrauma heart rate can be either greater than or less than the healthy rate; thus, a two-sided test is used. $z_{1-\alpha/2} = 1.96$, and $z_{1-\beta} = 0.84$. From Eq. (22.2),

$$n_1 = n_2 = (z_{1-\alpha/2} + z_{1-\beta})^2 \sigma^2/d^2 = (1.96 + 0.84)^2(9.1)^2/6^2 = 18.0.$$

The physician needs at least 18 patients. However, for reasons set forth in Section 22.2, he should choose a few more than 18.

22.2. Either trauma type could evoke a greater heart rate; thus, a two-tailed test is used. $z_{1-\alpha/2} = 1.96$, and $z_{1-\beta} = 0.84$. From Eq. (22.2),

$$n_1 = n_2 = (z_{1-\alpha/2} + z_{1-\beta})^2(6.13^2 + 6.34^2)/6^2 = 16.9.$$

The physician needs at least 17 patients in each group. However, for reasons set forth in Section 22.2, we advise including a few more than 17 patients.

22.3. By substituting in Eq. (22.1), we find

$$n = \frac{(z_{1-\alpha/2})^2 \sigma^2}{d^2} = \frac{1.96^2 \times 24^2}{10^2} = 22.1.$$

A sample of at least 23 patients is needed; however, per the explanation in Section 22.2, a few more patients should be included.

22.4. $\pi = 0.18$. Because π is not near 0 or 1, the binomial form, Eq. (22.8), will be used. A one-sided test is used (the training will not increase the number of enhancements); therefore, $z_{1-\alpha} = 1.645$.

$$n = \left[1.645\sqrt{0.18 \times 0.82} + 0.84\sqrt{0.12 \times 0.88}\right]^2/(0.12 - 0.18)^2 = 227.5$$

At least 228 residents would be required, which is too large a number to evaluate the training device in one medical center within a reasonable period; the plan for this study must be either dropped or arranged as a multicenter study.

22.5. The one-sided test uses $z_{1-\alpha} = 1.645$. $p_m = (0.23 + 0.13)/2 = 0.18$, which is not near 0 or 1; therefore, the binomial form, Eq. (22.10), is used. Substitution yields

$$n_1 = n_2 = \left[1.645\sqrt{2 \times 0.18 \times 0.82} + 0.84\sqrt{0.23 \times 0.77 + 0.13 \times 0.87}\right]^2 /0.10^2$$

$$= 181.25.$$

The surgeon would need a minimum of 182 eyes per method.

22.6. π is central; therefore, the surgeon uses Eq. (22.13). $p = 0.7$, $w = 0.1$, and, for 95% confidence, he retains the 1.96 given in Eq. (22.13). Substitution yields

$$n = \frac{1.96^2 p(1-p)}{w^2} = \frac{3.8416 \times 0.7(1 - 0.7)}{0.1^2} = 80.7.$$

22.7. The predicted required sample size will be the same as if the r were positive, because of the symmetry property. In Table 22.1, $r = 0.23$ corresponds to $n = 74$.

Chapter 23

Modeling and Clinical Decisions

23.1. OVERVIEW OF MODELING

DEFINITIONS AND GOALS

In the context of this book, a *model* is *a quantitative or geometric representation of a relationship or process*. Modeling was introduced in Section 8.1. The example given is an increase of leukocyte count (white blood cell [WBC] count) with the level of infection (count of a cultured bacterium on a microscope slide grid). This was modeled (1) as a straight line and (2) only for certain values of the two variables (e.g., for WBC count, from the upper limit of normal to a large but not astronomical value). The model may not represent the true relationship. Usually, biological processes are so complicated that a model can represent only certain aspects of the process and in only certain ranges of the variables. White blood cell count depending on infection might start rather flat and curve upward with infection count in an exponential fashion. Alternatively, it might start that way but then flatten again, approximating a plateau, as in a biological growth curve. The flexibility of model shape is not so much weakness in modeling as strength; the model can be changed with increasing knowledge and data and become ever closer to approximating the true event. Through such evolving models, we pass through the sequence of scientific learning from describing a process to testing its causal factors to being able to predict it. When we understand the causes of a process or an event and can predict it reliably, we have considerable knowledge that we can apply to clinical decision making.

TYPES OF MODELING AND CLINICAL DECISION MAKING

Four types of models are exemplified in this book: algebraic models, recursive partitioning, screening-type models (including number needed to treat [NNT]), and outcomes analysis.

Algebraic Models

In algebraic models, the dependent variable or outcome factor is equal to, that is, represented by, a mathematical combination of independent variables, the causal factors. This mathematical expression is termed a *function* of the variables. The functional form of the model may often be visualized by a geometric relation among the variables, as the WBC count was modeled as a straight-line fit to infection count. There are various methods of fit, the most common of which and the only one addressed in this book being regression, a method that is examined in some detail in Chapter 24. An algebraic model is not restricted to a straight line; it may be curved. Section 23.3 examines curved models. An algebraic model is not restricted to one dependent and one independent variable; it may have more than one of either. Models having one variable dependent on several independent variables are termed *multivariate*; multivariate models are examined in Section 23.5. The name *multivariable* tends to refer to multiple dependent variables, which is a topic too advanced for this book.

Recursive Partitioning

In their daily work, clinicians pass through a logical sequence of questions (decision points) in ruling out potential causes of illness and closing in on a diagnosis. The decision points consist of partitioning information into indicators of one disease but not another. Done repeatedly, or recursively, the partitioning leads to diagnosis or optimal therapy. Whereas the use of this method is second nature in common cases and usually is not even recognized, its formal use sometimes is of assistance in more complicated cases or in developing diagnostic or therapeutic procedures for newly discovered diseases. Section 23.6 examines clinical decision models using recursive partitioning, although they are in rather simple forms so that the user may concentrate on the method rather than the subject matter.

Screening-Type Models

Some models arise from the methods of mathematical logic (often combined with probability theory). The simplest forms are the "If A then B, but if not A then C" sort of logical analysis. Screening results and the NNT are often of this type; this book limits discussion of logical models to screening and NNT (see Section 23.7 for further details on this topic).

Outcomes Analysis

Clinical decisions based on the eventual overall outcome of treatment rather than on a single physiologic measure are part of this class. Section 8.6 introduces outcomes analysis, and Section 23.8 provides a more detailed discussion of the topic. An overall outcome seldom is based on a single measure but rather usually involves a combination of indicators. Finding the nature of relationship among these indicators and combining them into a single measure of effectiveness (MOE) is a form of modeling.

23.2. STRAIGHT-LINE MODELS

Review of Section 8.2

Let us think of a graph with x as the horizontal axis and y as the vertical axis. A straight line is determined by two pieces of information. A useful form of expressing a straight-line relationship between medical variables is a slope β_1 (inclination of the line from horizontal) and a point. Most commonly, the point is the y intercept $(0, \beta_0$; slope-intercept form). However, actual data often place the intercept at an awkward distance; therefore, a form using the x and y means, (m_x, m_y), is often preferable. Equation (8.1) is an example of this form. The algebraic expressions of these two forms are:

$$\text{slope-intercept form: } y = \beta_0 + \beta_1 x; \tag{23.1a}$$

and

$$\text{slope-means form: } y - m_y = \beta_1(x - m_x). \tag{23.1b}$$

Straight lines are termed *first degree*, because no x in the expression has an exponent greater than 1. Other common algebraic forms for determining straight lines are a slope and a point anywhere on the line and two points on the line. In mathematics and physics, some variables are exactly related by a straight line and the fit is just a mathematical exercise. In biology and medicine, however, multiple causal factors are usual, so a straight-line relationship is only approximate and the quality of a fit must be expressed in probability. Indeed, a straight-line relationship may not exist for all values of x and y; it may be useful only in certain ranges. We are familiar with the estimates of m_x and m_y and how they are calculated. The best fit for the slope may be calculated by different criteria of "best," but the most common is named *least squares*, that is, finding the smallest possible sum of squares of vertical distances from the data points to the line. A straight-line fit to data subject to variability is a form of *regression*. Section 8.3 introduces the topic of regression, and Chapter 24 provides further examination.

23.3. CURVED MODELS

TYPES OF CURVED MODELS

Relationships between x and y may have almost any shape. However, this chapter presents only the most common shapes. Because fitting functional forms to data is examined in Chapter 24, this section is limited to displaying common shapes in data relationships and noting the algebraic form that may express that relationship. Graphs of medical data are given, and the user is asked to recognize the form.

ADDING A SQUARED TERM

Let us start with the slope-intercept form of a straight line, $y = \beta_0 + \beta_1 x$, and add a squared term to obtain $y = \beta_0 + \beta_1 x + \beta_2 x^2$. We now have a curved line, part of a parabola. A parabola is shown in Fig. 23.1A for negative β_2; the

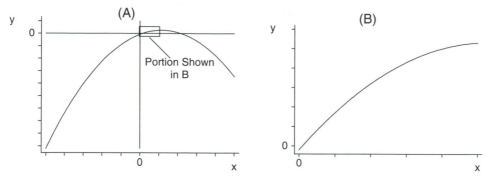

Figure 23.1 (A) A parabola, concave downward as a result of a negative β_2. (B) Portion of the parabola that is useful in the example in Fig. 23.2.

parabola would be concave upward for a positive β_2. All or any portion of it could be used to model a curved relationship. An example is shown in the box, which is enlarged in Fig. 23.1B. When 2 is the highest power, the expression is termed *second degree*.

EXAMPLE: AZITHROMYCIN FOR EMPHYSEMA

In DB3, the effect of azithromycin on serum theophylline levels of patients with emphysema was recorded. Figure 23.2 shows day 10 levels as predicted by baseline levels. What is the shape of the relationship? A case could be made for

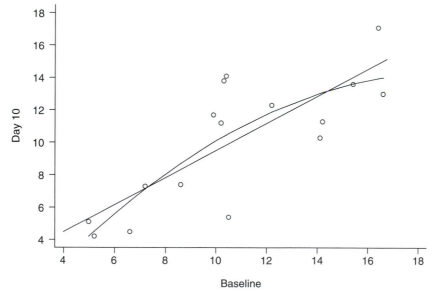

Figure 23.2 Serum theophylline levels at day 10 as predicted by baseline are shown, with straight line and parabolic fits superposed. Note that the curved fit is similar to the portion of the parabola shown in Fig. 23.1B.

either a first- or second-degree fit. Both fits were made using the least squares method and are shown superposed on the data. Chapter 24 addresses evaluation of these fits.

METHOD

The shape of a model is best derived from knowing the physiologic forces that generate the relationship. This shape may be verified or "fine-tuned" using curve-fitting methods. Positing a relationship directly from inspection of a data plot is reserved for data exploration at the very beginning of knowledge acquisition. It is helpful in the interpretation of relationships to recognize general shapes from a data plot. After the general shape is noted and portions of a theoretical curve that are irrelevant to the relationship are deleted, the fit may be adjusted by sliding the curve horizontally or vertically on the axes or by stretching or shrinking the curve horizontally or vertically on the axes. These mechanisms are examined in the next section.

COMMON SHAPES OF RELATIONSHIPS

Aside from a straight line (first degree) and a downward-opening parabola (second degree), which are seen in the preceding example, common curves are illustrated in Fig. 23.3. These include the upward-opening parabola (perhaps weight depending on height; see Fig. 23.3A), a third-degree curve (adding an x^3 term to a parabolic model; see Fig. 23.3B), a logarithmic curve (see Fig. 23.3C), an exponential curve (see Fig. 23.3D), a biological growth curve (see Fig. 23.3E), and a sine wave (see Fig. 23.3F). Sign changes will flip the third-degree curve top for bottom in shape. The logarithmic curve looks similar to a portion of the downward-opening parabola, but it never reaches a maximum as does the parabola, increasing, however, more and more slowly with increasing x. The exponential curve looks similar to a portion of the upward-opening parabola, but it increases more rapidly. Growth curves fit many growth patterns, for example, animal (and human) weight over time or the volume of a cancer. Periodic curves, of which the sine wave is a simple case, are frequently seen in cardiopulmonary physiology.

Note that Figs. 23.3A and 23.3B are called *linear models* even though they are not straight lines, because they are composed of simple algebraic terms added together. However, Fig. 23.3C–F are *nonlinear models*, because their terms are not simple algebraic expressions but rather contain logarithmic, exponential, and trigonometric expressions.

ADDITIONAL EXAMPLE: SURVIVAL OF MALARIAL RATS

In part of the experiment giving rise to DB11, rats infected with malaria (100 in each sample) were treated with healthy red blood cells (RBCs) and with a placebo (hetastarch). Number (same as percentage) in each sample surviving

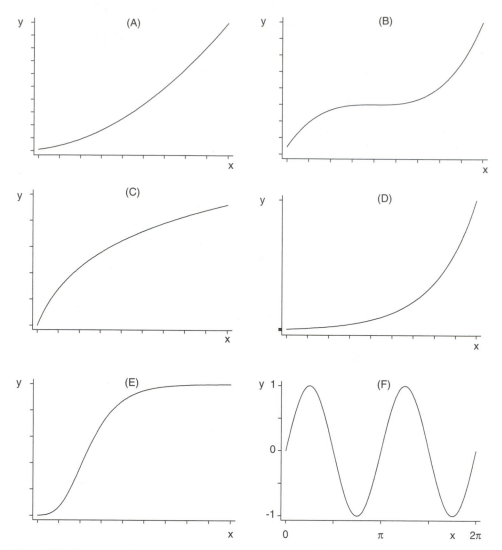

Figure 23.3 Representative portions of the following classes of curve are illustrated: (A) second-degree curve (parabola) opening upward; (B) third-degree curve; (C) logarithmic curve; (D) exponential curve; (E) biological growth curve; and (F) sine wave.

by day for 10 days is plotted in Figs. 23.4A and 23.4B. The RBC curve appears to follow a sequence of plateaus but is not clearly a common shape other than a straight line. The placebo survival diminishes more rapidly and appears to be approaching a minimum survival percentage toward the end of the period, looking more like a segment of an upward-opening parabola (second-degree expression). These two curves were fit to the data and superposed to form the combined graph in Fig. 23.4C. Under the proper conditions, these fits will follow certain probability distributions, and statistical tests may be made of their shape and their similarity; Chapter 24 addresses such methods.

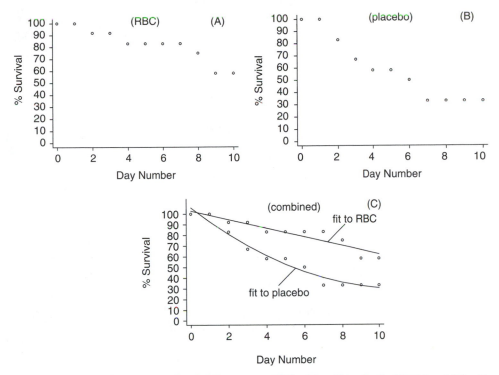

Figure 23.4 Survival percentage of malarial rats treated (A) with red blood cells (RBCs) and (B) with a placebo. The combined graph (C) shows a straight-line fit to the RBC data and a parabolic fit to the placebo data.

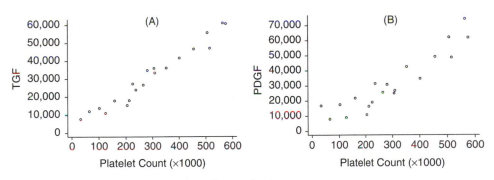

Figure 23.5 Growth factor levels as depending on platelet counts.

Exercise 23.1. Using data from DB9, Fig. 23.5A shows transforming growth factor (TGF) plotted as depending on platelet count, and Fig. 23.5B shows a similar plot for platelet-derived growth factor (PDGF). Recognize the shape and choose the curve (from those presented earlier in this section) that seems by eye to provide the best fit for each.

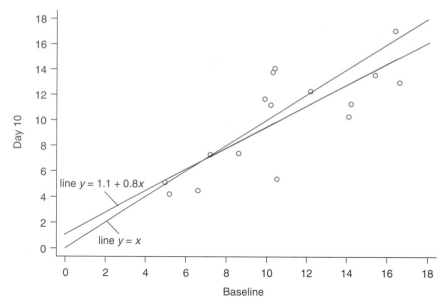

Figure 23.6 Ten-day serum theophylline level as related to baseline level. The line $y = 0 + 1x$ is shown to illustrate the change in equation parameters required for the best fit $y = 1.1 + 0.8x$.

23.4. CONSTANTS OF FIT FOR ANY MODEL

EXAMPLE POSED: FIT 10 DAYS AFTER START OF TREATMENT TO BASELINE

In DB3, serum theophylline level 10 days after the beginning of antibiotic treatment would be expected to be related to baseline level, although not exactly on a straight line. If we started with the straight line $y = x$, which is a line with slope 1 passing through an intercept of 0, what manipulations would provide a better fit? Figure 23.6 shows that data with the line $y = x$ and a better fitting line $y = 1.1 + 0.8x$.

METHOD

When we are faced with a relationship between factors in a physiologic process shown on a plot of data representing this process, we can imagine the relationship as a geometric curve typifying the essence of the dependence of one factor on the other. Often, we can see the family of curves by inspection, as we did in Fig. 23.2. But how do we pass from a family of curves to the particular member of that family? In most cases, the family member can be specified by sliding, stretching, or shrinking the general family model on one or both of the axes. This is done by adding or multiplying by one or more of four possible parameters, where the values of these parameters at this stage are estimated roughly by eye.

The Purpose of Such Manipulations

Of course, we do not attempt to fit models to statistical data by manipulating parameters with judgmental values; we use optimized mathematical methods. The use of examining the parameters is twofold. First, understanding the role of the parameters helps us relate the mathematical expression of models to the physiologic processes generating the data. Second, by understanding the roles of the parameters, we can look at a process or at data and often anticipate the functional form of the model. This anticipation also prevents gross errors that sometimes arise by using statistical software without careful thought.

The Four Parameters Are Scale and Position Constants for x and y

Suppose we have a relationship in which the way y depends on x is expressed by $y = f(x)$. This f expresses the form of the relationship and usually is called the *function*. White blood cell count is a straight-line function of time. However, the nature of the form does not tell us necessary details. We need to select the scales and the starting positions for x and y that will make the relationship realistic. The four constants are termed *parameters* because they remain constant only for the one member of a family of functions but vary from member to member. They will specify the horizontal and vertical scales and the horizontal and vertical positions. These parameters are $a, b, c,$ and d, for example. Let us attach them to the function so that we have a multiplier and an additive for each of x and y, giving rise to the form

$$y = d + cf(ax + b). \tag{23.2}$$

Conceptually, the multipliers "stretch" or "shrink" the curve in the x or y direction to adjust the scale, and the additives "slide" the curve along the respective scale to adjust the position. Table 23.1 gives the effect of each of the four parameters used in fitting a function.

Table 23.1

Effects of Fit Constants (Parameters) That Change the Function $y = f(x)$ to $y = d + cf(ax + b)$ and Effects of the Values They May Take On

Fit constant	Along which axis	General effect	Detailed effect
a	x	Stretches–shrinks curve	$a < 1$: stretches $a > 1$: shrinks
b	x	Slides curve	$b > 0$: to left $b < 0$: to right
c	y	Stretches–shrinks curve	$c > 1$: stretches $c < 1$: shrinks
d	y	Slides curve	$d > 0$: up $d < 0$: down

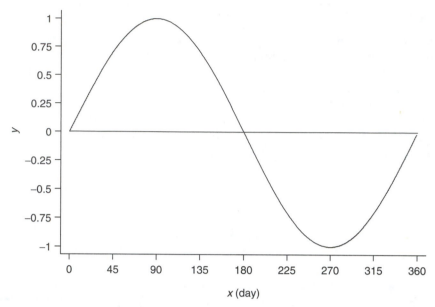

Figure 23.7   The first cycle of a sine wave.

EXAMPLE COMPLETED: 10-DAY TO BASELINE FIT

Starting with the form $y = x$, we can see that the slope needs to be less steep. This is done by stretching the x-axis, which, from Table 23.1, can be done by multiplying x by a constant less than 1. Because the slope needs to be only a bit less steep, we can estimate by eye that a multiplier of about 0.8 or 0.9 is reasonable; let us guess 0.85. We can further see that the resulting line needs to slide up the y-axis about 1 unit, which, from Table 23.1, can be done by adding a constant of about 1 to the right side. We then have the form $y = 1 + 0.85x$, which is rather close to the best fit $y = 1.1 + 0.8x$ (see Fig. 23.6). The best fit is actually the least squares fit.

ADDITIONAL EXAMPLE: ASPERGILLOSIS OCCURRENCE BY SEASON

An allergist suspects from his experiences with patients that aspergillosis occurs more often in the winter than in the summer, although this has not been shown previously.[73] He collects a number of cases over 4 years and averages the incidence by season. If his intuition is correct, average incidence should follow the form of a sine wave, $y = \sin(x)$ (see Fig. 23.3F for an example of the shape). More exactly it would follow a portion of a sine wave. A seasonal pattern might fit the first cycle of a sine wave, where x is given as days of a 360-day year (to coincide with the 360° of a circle; Fig. 23.7).

Definition of a Sine Wave

The reader probably learned about sine waves in high school. Think of a clock having numbers around a circle with a hand extending from the center to the perimeter. x- and y-axes are superposed on the clock, with the x-axis

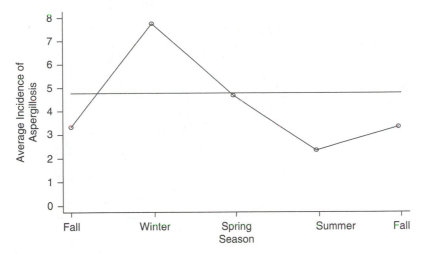

Figure 23.8   Average incidence of aspergillosis by season.

extending from 9 o'clock to 3 o'clock and the y-axis extending from 6 o'clock to 12 o'clock. As the hand travels from the rightmost x-axis (3 o'clock) around the 360° of the circle, there is a vertical distance from the x-axis to the end of the hand. The sine is defined as the ratio of this length to the length of the hand.

Aspergillosis Data and Fitting

The allergist finds mean number of cases he treats by season to be the following: fall, 3.33; winter, 7.75; spring, 4.67; and summer, 2.33. The overall average is 4.77 (Fig. 23.8).

Relating Observed Data to a Family of Curves

Observing the similarity of form in Figs. 23.7 and 23.8 will show that the data start in the fall near the mean, increase to a maximum at 90 days (winter), decline through 180 days (spring) to a minimum at 270 days (summer), and finally move back toward the average at 360 days (return to fall). The form of the model, following Eq. (23.2), will be

$$y = d + c\sin(ax + b). \qquad (23.3)$$

If the seasonal designations are chosen as fall at day 0, winter at day 90, and so on, the data follow the sine shape in the horizontal direction; the allergist does not need to adjust the scale or slide the curve on the x-axis to fit; therefore, $a = 1$ and $b = 0$. For the y-axis, he wants the incidence average to lie at $y = 0$ (on the x-axis) so that the seasonal effect deviates from average. He slides the sine curve up to this level by selecting $d = 4.77$, the average. This leaves the model $y = 4.77 + c\sin(x)$, with c to be estimated from the data.

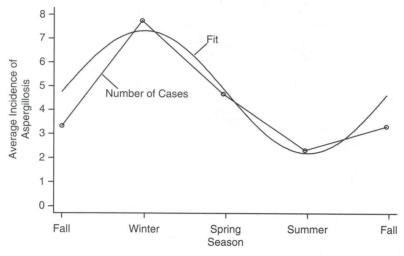

Figure 23.9 Average incidence of aspergillosis by season with the least-squares sine wave fit superposed.

Estimating c

The allergist wants to find a c-value that will minimize the $\sum(\text{obs} - \text{fit})^2/\text{fit}$. Eventually, this will be done mathematically and data substituted in the result. At this stage, the purposes are understanding the process and verifying that the correct family of models has been chosen. In a few iterations on a calculator, taking "fit" as $4.77 + (\text{evolving } c) \times \sin(x)$, the allergist finds that $c = 2.56$; therefore, the model becomes $y = 4.77 + 2.56\sin(x)$. Interestingly, the sum of squares that he minimized is the chi-square goodness-of-fit statistic discussed in Chapter 20 (see Eq. 20.2). The 2.56 value of c yields a chi-square of 0.9021 and a p-value of 0.924, indicating that the fit is quite acceptable. Figure 23.9 shows the fitted sine wave superposed on the aspergillosis data.

Exercise 23.2. Using data from DB10, Fig. 23.10 shows a plot of the time to hop as related to the distance hopped. Visualize a straight line passing through the points and choose the parameters a, b, c, and d required to express its approximate equation.

23.5. MULTIPLE-VARIABLE MODELS

CONCEPT

The previous sections have reviewed models with only two variables, which can be represented by the x- and y-axes; y is thought of as depending on x. We know that y may well depend on more than one variable; for example, systolic blood pressure (SBP) depends on physical condition, age, recency of

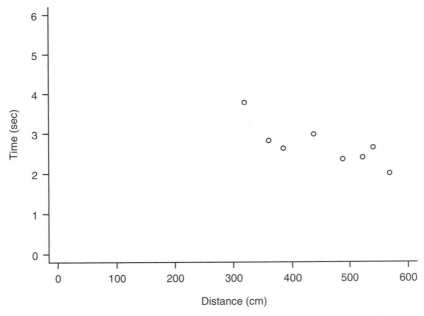

Figure 23.10 Time (in seconds) to perform a triple hop on the postoperative leg related to the distance (in centimeters) hopped after surgery on hamstrings or quadriceps.

exercise (in minutes), and other factors. Such a relationship can be expressed algebraically.

VISUALIZING THREE DIMENSIONS

The case of y depending on two x's can be visualized geometrically, as with the expression $y = \beta_0 + \beta_1 x_1 + \beta_2 x_2$. If y is SBP, x_1 might be age in years and x_2 might be minutes of vigorous exercise just before the measurement. The β's are the parameters of fit, conceptually similar to those considered in Section 23.4. To visualize the relationship, think of y as the intersection of two walls in the corner of a room and the two x's as the intersections of these walls with the floor. An observation with values (y, x_1, x_2) is represented as a position in the space above and in front of the corner. x_1 and x_2 are readings on the independent variables (e.g., age and minutes of exercise), and y is the reading on the dependent variable (e.g., SBP). Just as a line may be fit to a sample of points in two dimensions, a plane may be fit to a sample of points in three dimensions. Figure 23.11 represents a three-dimensional space showing a single point and a plane that might have been fit to a set of such points. For this figure, β_0 is the y-intercept (where the plane cuts the y-axis), β_1 is the slope of the line formed by the plane cutting the x_1,y-plane, and β_2 is the slope of the line formed by the plane cutting the x_2, y-plane.

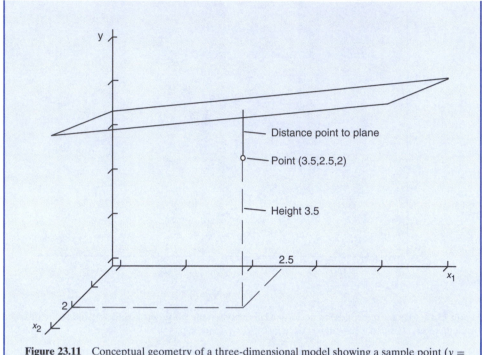

Figure 23.11 Conceptual geometry of a three-dimensional model showing a sample point ($y = 3.5$, $x_1 = 2.5$, and $x_2 = 2$) and a plane of fit to a set of points in the space.

A CURVED SURFACE IN THREE DIMENSIONS

Just as the straight line $y = \beta_0 + \beta_1 x$ may be generalized to represent a curve by adding a squared term, as $y = \beta_0 + \beta_1 x + \beta_2 x^2$, the plane $y = \beta_0 + \beta_1 x_1 + \beta_2 x_2$ may be generalized to become a curved surface by adding one or two squared terms, for example, perhaps $y = \beta_0 + \beta_1 x_1 + \beta_2 x_2 + \beta_3 x_2^2$. A curved surface rather than a flat surface in three dimensions still may be easily visualized.

MORE THAN THREE DIMENSIONS

Suppose the dependent variable y depends on more than two dimensions, as if we added a measure of general cardiovascular health x_3. Algebraically, we have no problem. We can add as many variables in whatever powers we want and the model may be readily used in analysis. Visualization of a relationship in four or more dimensions is difficult, if possible at all. Four-dimensional displays have been attempted using time (moving images), colors or shades, and other mechanisms, but with limited success, because depiction of the fourth dimension is necessarily different in nature from that of the first three dimensions. There is no problem using algebraic methods for modeling in more than three dimensions.

An algebraic expression of a four-dimensional plane would $y = \beta_0 + \beta_1 x_1 + \beta_2 x_2 + \beta_3 x_3$. Only visualization is limited to three dimensions.

THE TERM *LINEAR MODEL*

The term *linear model* or *general linear model* (see Section 18.6) is often used in analyses and software packages. A linear model is a model in which the terms are added, such as has been used so far in this section, rather than multiplied or divided. A linear model is not restricted to a straight line or its analogue in higher dimensionality. Sometimes we see a reference to a curved model that is still called linear, which may seem oxymoronic but arises from particular mathematical connotations. The curved shape, $y = \beta_0 + \beta_1 x + \beta_2 x^2$, is a linear model because terms are added and only one variable appears per term. The model $y = \beta_0 x_1 / x_2$ is not linear.

MULTIPLE DEPENDENT VARIABLES

Suppose we are concerned with both systolic and diastolic blood pressure as predicted by the several independent variables; we have y_1 and y_2 depending on the several x's. Such models usually are denoted multivariable (not multivariate; see Section 23.1). Methods exist to analyze such models, but they are beyond the scope of this book.

Exercise 23.3. We would like to predict the postoperative hop distance on anticipated hamstring or quadriceps surgery. DB10 provides data on the hop time and distance on both the repaired and the uninjured leg for eight patients. What might be the dependent and independent variables? What is the nature of the geometric picture?

23.6. CLINICAL DECISION BASED ON RECURSIVE PARTITIONING

EXAMPLE POSED: DIFFERENTIATING MYELOMAS

A set of 20 patients is known clinically to suffer from a myeloma, some from multiple myeloma (designated 0) and some from a syndrome known as MGUS (designated 1).[50] Associated measures are immunoglobulin (Ig) type (IgM: 0; IgG: 1; and IgA: 2), sex (female: 0; male: 1), and protlev (protein level of patient divided by the midrange of the healthy protein level). Part of the way through the partitioning process, it was noticed that protlev seems to differentiate between preceding partitions at about a level of 2; thus, protlev was then categorized as protein level category (protlev < 2: 0; protlev ≥ 2: 1). The resulting data appear as Table 23.2. We want to follow the logic of recursive partitioning and form a chart that can be used in future differentiation.

Table 23.2

Ordering of Myeloma Categories as Related to Immunoglobulin, Sex, Protein Level Multiple, and Protein Level Category

Myeloma	Ig	Sex	Protlev	Protcat
0	1	0	6.5	1
0	1	0	6.0	1
0	1	0	5.3	1
0	1	1	4.4	1
0	1	1	2.6	1
0	1	1	1.9	0
0	2	0	5.4	1
1	0	0	1.1	0
1	0	0	1.0	0
1	0	0	0.9	0
1	0	1	1.7	0
1	1	0	2.1	1
1	1	0	1.9	0
1	1	0	1.9	0
1	1	0	1.9	0
1	1	0	1.6	0
1	1	0	1.6	0
1	1	0	1.5	0
1	1	0	1.1	0
1	2	0	0.7	0

Ig, immunoglobulin; Protcat, protein level category; Protlev, protein level multiple.

METHOD

Often, a set of indicators will diagnose a particular malady or predict the outcome of a particular treatment. A patient having symptoms A, B, and C has disease 1. However, if the list of indicators is long, the number of possible combinations becomes unwieldy. It works better to pass through a sequence of branching for each indicator. In mathematics, the selection of the particular branch of an indicator is termed *partitioning;* therefore, the sequential process is termed *recursive partitioning.* Given a set of data on a list of indicator signs and symptoms and the illnesses that may be associated, recursive partitioning may identify which indicators are decisive in diagnosis and which are not. It may be able to distinguish the logical sequence through which the indicators should be considered. Finally, it can compare the logical sequence of partitions used successfully in a new set of data with a traditionally used sequence to either reinforce or replace the established mechanism.

Steps to Develop the Decision Process

This section addresses small indicator groups and not more than three categories per indicator. Mathematical sorting routines for large sample sizes and more complicated categorizations exist, but they are beyond the scope of this book. Sort the possible outcomes (diagnoses or result of treatment), listing those alike in adjoining rows. Identify the related indicator (symptom/sign or condition) most consistently related to the outcome, and then the next, and so on. Draw a decision box with arrows extending out, depending on initial partitioning. Draw the next decision boxes with appropriate arrows extending to them, and more arrows extending out, depending on the next round of partitions. Continue until reaching the final outcomes. The boxes should be ordered so that no looping back can occur and no box is repeated. There is usually some trial and error in establishing the best sequence. The best way to perceive the process is to follow one of the examples.

EXAMPLE COMPLETED: LOGIC AND CHARTING IN DIFFERENTIATING MYELOMAS

Inspection of Table 23.2 shows that each outcome appears in each diagnosis, except for Ig type: Ig type 0 (IgM) appears only with diagnosis 1 (MGUS). Thus, we can start Fig. 23.12 with a decision box on Ig type = 0 with an arrow leading to outcome box MGUS. We note that Ig 2 occurs in both outcomes but only one sex, so sex will not distinguish between outcomes. However, diagnosis 0 (multiple myeloma) is associated with protein level category (protcat) 1, and diagnosis 1 (MGUS) with protcat 0, so protcat will partition Ig 2 into the diagnoses. We draw another decision box for protcat with arrows passing from the Ig box's exit 2 to the protcat box and from there to the decision boxes. We note that, for Ig 1, all male patients are associated with diagnosis 0; therefore, we draw a

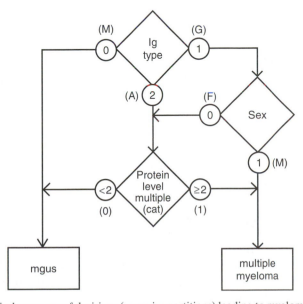

Figure 23.12   Logical sequence of decisions (recursive partitions) leading to myeloma differentiation.

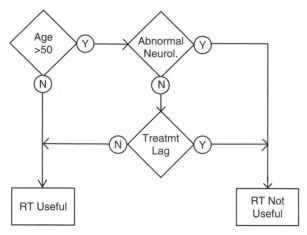

Figure 23.13 Logical sequence of decisions (recursive partitions) leading to decision to treat with radiotherapy.

Table 23.3

Ordering of Usefulness of Radiotherapy as Related to Malignant Glioma Categories of Age, Karnofsky Performance Status, Condition of Neurology, and Period Between First Symptoms and Initiation of Treatment

RT useful	Age > 50 years	High KPS	Abnormal neurology	Treatment lag > 3 months
Yes				
Yes			X	X
Yes		X		
Yes	X			
Yes	X	X		
No	X	X		X
No	X		X	

KPS, Karnofsky performance status; RT, radiotherapy.

sex box and appropriate arrows. Finally, we note that, for Ig 1 and for female subjects, protcat will distinguish the diagnosis. We connect the final arrows to complete Fig. 23.12.

ADDITIONAL EXAMPLE: USEFULNESS OF RADIOTHERAPY IN PATIENTS WITH GLIOMA

In a sample of 1500 patients with malignant glioma, radiotherapy (RT) was found to be useful to some patients but not to others. (Scott[74] inspired this study, but our sample is simplified and one datum has been fabricated to complete the example.) The patients were grouped by possible indicators of RT usefulness (Table 23.3). We note that all the younger patients are in the useful class, which comprises the first branching box in Fig. 23.13.

Now we concentrate on the four groups older than 50 years. One class with high Karnofsky performance status (KPS) and one with low KPS lies in each of RT useful and RT not useful; therefore, KPS is not a helpful discriminant. In contrast, we note that the RT not useful groups have either abnormal neurology or treatment lag, but not both, whereas the RT useful groups have neither. Thus, the logic path passes through branching boxes for these discriminants; if either is "yes," the arrow goes to RT not useful; otherwise, it goes to RT useful. We now have a logical recursive partitioning path to use in deciding whether or not to use RT on a patient.

Exercise 23.4. The disease of patients whose ocular pressure has tested high must be classified as glaucoma, which is damaging to the optic nerve, or ocular hypertension, which is rather benign.[62] The pressure, the angle between the iris and peripheral cornea, the field of vision, and the disc (optic nerve head) are examined. High pressure coupled with a narrow angle signals glaucoma. If the angle is wide, either a bad field or a bad disc in conjunction with the high pressure indicates glaucoma. Otherwise, ocular hypertension is inferred. Draw a decision chart representing the logical partitioning.

23.7. NUMBER NEEDED TO TREAT OR TO BENEFIT

NUMBER NEEDED TO TREAT: SCREENING FOR DISEASE

Number Needed to Treat If No Cases Are Detected by Other Means

The simplest case of NNT is the number of people screened to detect one case to treat. For example, how many mammograms must be run on randomly chosen women of a certain age to find one case of breast cancer? How many vaccinations for hepatitis C are required to prevent one case?

Example Posed: Screening for Lung Cancer in Baltimore

A population at risk, or catchment in epidemiologic terminology, was composed of male smokers in Baltimore in 1980.[86] A total of 10,387 people were screened, and 47 cases of advanced lung cancer were found. How many members of the catchment had to be screened to find one case?

Method

Let us denote by n_s the number of subjects in the population or catchment screened and by n_d the number of disease cases discovered by screening. The ratio $p_d = n_d/n_s$ is the proportion of disease cases discovered by screening. If all disease cases are found by screening, p_d also estimates the disease prevalence. The NNT is the reciprocal of the proportion discovered, or

$$\text{NNT} = \frac{1}{p_d} = \frac{n_s}{n_d}. \tag{23.4}$$

437

Example Completed: Screening for Lung Cancer in Baltimore

The prevalence is $p_d = 47/10387 = 0.0045$. NNT $= 1/0.0045 = 222$. The authorities must screen 221 subjects without advanced lung disease for each one they discover.

Additional Example

Oral leukoplakia, which frequently progresses to overt squamous cell carcinoma, was treated with 13-*cis*-retinoic acid in $n_s = 44$ patients.[30] $n_d = 24$ patients responded histologically. NNT $= n_s/n_d = 1.83$ patients treated per responder.

Number Needed to Treat for Additional Detections

In most screening programs, some of the disease cases would be discovered through ordinary medical practice. In the Baltimore screening, some of the 47 cases found by screening would have been discovered through patient visits to their primary care providers. Let us denote by p_m the proportion of cases that would be found through common medical practice. Then, $p_d - p_m$ is the proportion of additional detections due to the screening and

$$\text{NNT} = \frac{1}{p_d - p_m}. \tag{23.5}$$

Exercise 23.5. In a study[17] on radiographic screening in a correctional facility in New York City, the rate of tuberculosis among entering inmates was 0.00767. A total of 4172 entering inmates were screened. How many cases were detected by the radiographic screening? Of those detected, 25 had entered with a prior diagnosis of tuberculosis. What is the NNT for those newly diagnosed by the screening?

Cost of Number Needed to Treat

An essential ancillary issue is the cost of the resources (time, facilities, personnel, and money) expended for one detection and trading off this cost against the gain of treating that patient. The gain is usually intangible and the cost-effectiveness becomes a matter of comparing events from two different value bases. This might be likened to comparing the value of coins from two different money systems; we have to find or develop a medium of exchange, that is, an MOE (see Sections 8.6 and 23.8), that will compare the cost of the NNT with the gain to the patient.

Examining only monetary costs, let us denote by c_s the cost for the screening program, that is, the cost to screen n_s people. Then the cost per subject screened is c_s/n_s, and the cost per detection c_d is

$$c_d = \frac{c_s}{n_s} \times \text{NNT}. \qquad (23.6)$$

If NNT is the number needed per additional detection, then c_d is the cost per additional detection.

Detection Compared with Other Measures of Effectiveness, Such as Mortality

Is detection the final criterion of screening efficacy? It certainly is not the only one. Already we have, in addition to NNT, the number detected by the screening program (n_d) and the cost per detection (c_d). However, even a statistically significant portion of the population being detected by the screening is not evidence that the screening is beneficial. Suppose more diseased patients are detected, but the detection does not help them. In a German study,[89] 41,532 men born between 1907 and 1932 were screened for lung cancer with chest fluorography every 6 months for 10 years and compared with age-matched men screened similarly every 18 months. No significant reduction in lung cancer mortality or in overall mortality was found. When mortality rather than detection rate was taken as the MOE, tripling the screening rate provided no improvement. An investigator, and certainly a reader of medical articles, should carefully consider the MOE used. The next section further considers MOEs.

NUMBER NEEDED TO BENEFIT: ASSESSING THE BENEFIT OF A NEW TREATMENT

A change in treatment or mode of diagnosis often requires new instrumentation, laboratory tests, and training, among other resources. Is the benefit worth the changeover effort and cost? One measure of the benefit is number needed to benefit (NNB), that is, the number of patients needed to be treated or diagnosed with the new procedure to obtain one additional successful outcome. A confidence interval on NNB would be an additional help in making an administrative decision about a changeover in treatment or diagnosis.

Example Posed: Number Needed to Benefit from Switching Tattoo Ink

The majority of people with occasional tattoos get their tattoos as young adults and want removal in middle age. The commonly used titanium ink is difficult to remove. How many patients must be switched to nontitanium ink to facilitate one additional successful removal (NNB), and what would the 95% confidence interval be on this NNB? From DB6, 35 removals of titanium ink tattoos were attempted, with 5 successful outcomes. Of nontitanium ink tattoos, 15 removals were attempted, with 8 successful outcomes.

Method

Let us denote by p_1 the proportion of successful outcomes using the established treatment or mode of diagnosis and by p_2 that of the new treatment or mode. Then, $d = p_2 - p_1$ is the difference in success rates.

$$\text{NNB} = \frac{1}{d} \tag{23.7}$$

The confidence interval requires two steps. A confidence interval is a probability statement that an interval calculated from a sample bounds the theoretical parameter being estimated. Let us denote by δ the theoretical difference in success rates being estimated by d and by NNβ the theoretical number needed to benefit being estimated by NNB. A confidence interval on the theoretical difference, δ, estimated by d, is found, and then the components are inverted [as in Eq. (23.7)] to provide confidence on the theoretical NNβ, estimated by NNB. The standard error of the difference (SED) is as follows:

$$\text{SED} = \sqrt{\frac{p_1(1 - p_1)}{n_1} + \frac{p_2(1 - p_2)}{n_2}} \tag{23.8}$$

Using SED, the $1 - \alpha$ confidence interval on δ is given by Eq. (23.9):

$$P[d - z_{1-\alpha/2}\text{SED} < \delta < d + z_{1-\alpha/2}\text{SED}] = 1 - \alpha \tag{23.9}$$

To convert Eq. (23.9) to confidence on NNB, the components inside the brackets are inverted. However, recall from algebra that such inversion reverses the inequality signs. The confidence interval on NNβ becomes as follows:

$$P\left[\frac{1}{d + z_{1-\alpha/2}\text{SED}} < \text{NN}\beta < \frac{1}{d - z_{1-\alpha/2}\text{SED}}\right] = 1 - \alpha \tag{23.10}$$

Example Completed: Number Needed to Benefit from Switching Tattoo Ink

$p_1 = 5/35 = 0.14$. $p_2 = 8/15 = 0.53$. The difference between the p-values is $d = 0.39$. SED $= \sqrt{0.14 \times 0.86/35 + 0.53 \times 0.47/15} = 0.14$. The 95% confidence interval is $P[0.39 - 1.96 \times 0.14 < \delta < 0.39 + 1.96 \times 0.14] = P[0.116 < \delta < 0.664] = 0.95$. Taking reciprocals within the brackets, changing the inequality signs, and then reversing the order of components to return the inequalities to their usual position yields the confidence on a theoretical NNB as $P[1.51 < \text{NN}\beta < 8.7]$.

Our estimate of the true NNB is 3, and we are 95% sure it lies between 1 and 9.

23.8. CLINICAL DECISION BASED ON MEASURES OF EFFECTIVENESS: OUTCOMES ANALYSIS

REVIEW OF SECTION 8.6

Section 8.6 introduces the MOE concept. It was noted that clinical decisions are usually made on the basis of an MOE, or often more than one; there is no unique MOE. Several commonly used MOEs are listed, including the probability of cure; the probability of cure relative to other treatments; the risk for debilitating side effects; the cost of treatment relative to other treatments; and, for a surgeon, time in surgery, time to heal, and patient survival rate. Patient satisfaction with the treating physician might be informative. What might be best under one MOE may not be under another. For example, when comparing two drugs, one of which is effective but expensive and the other is somewhat effective but not costly, one drug is better under an efficacy MOE and the other is better under a cost MOE. Some investigators claim that the ultimate MOE is satisfaction with state of health.

OUTCOMES ANALYSIS

Simple measures are often precise and unequivocal, but the content of information may be shallow. It has been noted that MOEs related to the eventual "outcome" of the treatment rather than an interim indicator of progress have come to be termed *outcomes analysis*. Certainly, outcomes are often more difficult to quantify and often take much longer to obtain, but they are without doubt better indicators of the patient's general health and satisfaction. The reader is encouraged to review the example of Section 8.6 about an elderly stroke patient.

EXAMPLE POSED: CARPAL TUNNEL SYNDROME

Carpal tunnel syndrome involves hand paresthesia, pain, weakness, and sometimes sensory loss and wasting as a result of pressure on the median nerve from edema arising from one or more of a variety of causes. In primary or idiopathic carpal tunnel syndrome, a common surgical treatment is to relieve the pressure by cutting the transverse carpal ligament. A nonsurgical approach is to induce passive traction for some minutes on alternate days for a month by a controlled pneumatic device (e.g., the CTD-Mark I; Para Tech Industries, Dayton, OH). It is claimed that this treatment slightly extends the wrist area, reducing the pressure, which allows the edema to be resorbed by the lymphatic system. In a randomized controlled trial to compare these two treatments,[63] patients were randomized to treatment; blinding was not possible. The issue of this example is the MOE used for comparison. What might it be?

METHOD

The First Step

Measures of effectiveness might be better thought of as being "built" than being discovered. There is no formula to build MOEs. Building an MOE is as much art as technology. Like so many aspects of medicine, an MOE is a matter of thoughtful judgment. The first step is to ask the question: What are we really trying to accomplish? For example, is our goal a perfect surgery? Patient survival rate? Minimum pain? Minimum cost? We might think of combining several measures. As with a medical study, we start with clearly and unequivocally defining our goal.

The Second Step

We must identify the components of the MOE that will satisfy the goal. These must be quantifiable variables. We must specify how they will be quantified, in what units, and with what precision.

The Third Step

We must identify the relationship among the components, at first verbally, but eventually quantitatively. We must remember that they are mostly in different units of measurement. We must weight each component to adjust for units. We must note that the precision of the entire MOE is no greater than that of the least precise of its variables.

The Fourth Step

The components of the MOE (the variables) represent different levels of relative importance. A serious question is how categories of relative importance are weighted. A major issue is weighing the cost of treatment against the chance of it being effective. Some weights occur rather naturally, but with others the best an investigator can do is set forth good judgment and allow the readers to accept or reject it for themselves.

Subsequent Steps

Of course, a study having a clearly defined and planned MOE must follow the usual steps of any good scientific study, which has been addressed in various sections throughout this book.

Different Measures of Effectiveness

Recall that there is no unique MOE for a problem. The MOE is just a window through which we seek the truth. We may see the truth through several different windows, and one is not by definition better than another. This is not to say, however, that all MOEs are acceptable. Each must be evaluated with reasoned judgment and rejected if it is likely to yield spurious information.

EXAMPLE COMPLETED: CARPEL TUNNEL SYNDROME

What are some components of the MOE that will satisfy the goal? Not only the improvement but also its duration is important; the MOE will be taken at 1 month, 1 year, and 2 years postoperatively. Individual measures posed are median nerve conduction amplitude (measured in microvolts) and latency (measured in milliseconds), grip and pinch strength (kilograms), pain (visual analog scale of 0–100), and percentage normality of function (0–100). A single MOE is developed as a weighted sum of nerve conductions (x_1, x_2) and strengths (x_3, x_4) (four scores each with a maximum of 100) as percentage of normal, pain (x_5; one score with maximum of 100), and normality of function (x_6), which is defined as percent ability for the hand to carry out its normal functions (one score with maximum of 100). x_2 and x_5 are negative, because smaller values indicate better condition. By denoting weights as respective b's, the MOE becomes $b_1x_1 - b_2x_2 + \cdots + b_6x_6$. What are b-values? At this point, the development of the MOE becomes somewhat arbitrary, because judgment of the relative importance of the component measures will vary with the investigator. One set of evaluations follows. Nerve conductions and strengths are taken as equal: $b_1 = b_2 = b_3 = b_4 = 1$. Pain is taken as important as these four combined: $b_5 = 4$. Normality of function is taken as equal to pain level: $b_6 = 4$. This last might engender disagreement, but note that the two are related: pain will impair normality of function. The final MOE is

$$MOE = x_1 - x_2 + x_3 + x_4 - 4x_5 + 4x_6.$$

The MOE is measured for the two treatment groups and the resulting data compared using appropriate statistical methods. Larger values of the MOE signal better treatment.

ADDITIONAL EXAMPLE

In photorefractive keratectomy (PRK), the surface of the cornea is reshaped by laser to alter the angle of light refraction entering the eye. Photorefractive keratectomy has been shown to be an effective surgical correction for myopia, as measured by postoperative visual acuity. However, some patients report glare, especially at night.[71] An outcomes analysis would assess success in terms of the overall patient satisfaction with the procedure, rather than using only the single success measure of visual acuity. One possible MOE might be a preoperative to postoperative difference of measures of visual acuity, v (diopters of correction for distance correction); glare, g (patient rating from no glare [0] to severe impairment caused by glare [4]); and "Would you do it again under the same circumstances?" satisfaction, w (patient rating from definitely not [0] to without doubt [0]). The measures could be made roughly equivalent to be additive by multiplying visual acuity by 2 (2 diopters of correction should represent moderate dissatisfaction) and willingness-to-repeat by 2 (twice as important as glare), after reversing its range in direction so that 0 is desirable by subtracting it from 4. Then, the MOE would be $2v + g + 2(4 - w)$. Smaller MOE values are associated with improvement. The preoperative minus postoperative MOE is a difference, say, d, that would be zero if there were no change because of the surgery; thus, H_o: $d = 0$ can be tested by a paired t test if d is distributed approximately normal or by the signed-rank test otherwise.

Exercise 23.6. An older ophthalmologic surgical procedure to treat myopia is radial keratotomy (RK), in which corneal incisions radiating around a clear central zone allow a controlled amount of collapse of the cornea, altering the angle of light refraction entering the eye. Radial keratotomy provides almost 20/20 uncorrected visual acuity immediately after surgery, but the acuity tended to drift afterward, becoming hyperopic over some years, and many patients reported impaired night vision. What might be a useful MOE to compare the overall outcome of RK with PRK?

ANSWERS TO EXERCISES

23.1. The TGF curve appears to be a straight-line fit. The PDGF curve appears to be well fit by a section of a parabola, opening upward. The figures are repeated here with those respective fits.

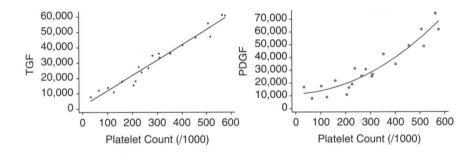

23.2. Although we could use the means of the two variables for the b and d constants, the plot is drawn such that we can see the intercept and need to use only d. By drawing a line along a ruler with the edge passing through the points as a rough eye fit, we see that it passes through a value around 5 on the y-axis. Because the functional form of a straight line $f(x)$ can be taken as just x, either a or c may be used; let us, for example, use a. a is the amount of change in y per unit increase in x and is negative, because the line decreases from left to right. The line drawn with the ruler shows a drop of about 1 second per 200-cm increase in distance, or a is approximately 0.005. Thus, the roughly fitted equation is: seconds $= 5 - 0.005 \times$ centimeters.

23.3. The dependent variable y is the postoperative hop distance. There are two independent variables, x_1 and x_2, which represent hop time and distance on the uninjured leg. The geometric picture would be a plane in three dimensions, similar to Fig. 23.11.

23.4.

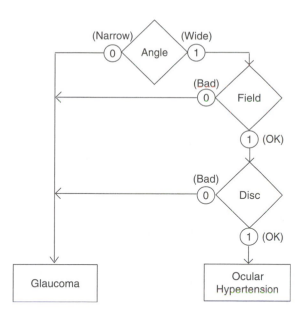

23.5. $p_d = 0.00767$. $n_s = 4172$. $n_d = p_d \times n_s = 32$. $p_m = 25/4172 = 0.00595$. The additional proportion detected by screening is $p_d - p_m = 0.00767 - 0.00599 = 0.00168$. NNT $= 1/0.00168 = 595$ entering inmates screened for every new tuberculosis case discovered that would not have been found otherwise.

23.6. One possible MOE might be a 10-year postoperative weighted sum of measures of visual correction, v; glare, g; night vision, n [patient rating from no interference (0) to nonfunctional vision at night (4)]; and "Would you do it again under the same circumstances?" satisfaction, w. The measures could be made roughly equivalent to be additive by multiplying visual correction by 2 (2 diopters of correction should represent moderate dissatisfaction) and glare and night vision by 1. Willingness to repeat must have its range reversed in direction so that 0 is desirable by subtracting it from 4, and then might have a 2 multiplier (twice as important as glare or night vision). Then, the MOE is $2v + g + n + 2(4 - w)$. The smaller the MOE is, the better the patient's condition is. The mean MOE for RK and PRK patient groups could be contrasted using the two-sample t test if the data distributions are approximately normal in shape with similar standard deviations or using the rank-sum test if these assumptions are not satisfied. The user may not agree with these weightings; they were chosen for illustration.

Chapter 24

Regression and Correlation Methods

24.1. REGRESSION CONCEPTS AND ASSUMPTIONS

WHAT IS REGRESSION?

Regression (see introduction to topic in Section 8.3) is a statistical tool that describes and assesses the relationship between two or more variables, with at least one being an independent variable, that is, putative causal factor, that predicts the outcome, and at least one being a dependent variable, the predicted outcome factor. The description is based on an assumed model of the relationship. This model may describe, or fit, the relationship well or poorly; the method tells us how well the model fits. If the model is inadequate, we may seek a better fitting model. Chapter 23 discusses methods of modeling. The model may be a straight line (simple regression; see Section 24.3), a curved line of any functional form (curvilinear regression; see Section 24.7), or a single dependent variable predicted by two or more independent variables (multiple regression; see Section 24.8). (The possibility of two or more dependent variables, that is, simultaneous outcomes, called multivariable regression, is not addressed in this book.) Prior to Section 24.9, the outcome, or predicted, variable in regression is taken as a continuous variable. Section 24.9 lists types of regression in which the predicted variable is of another type, and Section 24.10 examines one of these types.

PREDICTIVE VERSUS EXPLORATORY USES OF REGRESSION

Regression is most useful when the model is dictated by an understanding of the underlying physical or physiologic factors causing the relationship. In this case, regression allows us to predict the value of the outcome that will arise from a particular clinical reading on the independent variable and to measure the quality of this prediction. However, in the absence of this causal understanding between the variables, regression can be used as an exploratory tool suggesting which factors are related and/or the nature of such relationship. It can serve the sequence of steps in acquiring scientific knowledge: detecting the existence of a relationship between variables, describing this relationship, testing the quality of this description, and, finally, predicting outcome values based on causal values.

FIVE ASSUMPTIONS UNDERLYING REGRESSION

Five assumptions underlying regression are as follows:

(1) There is the usual assumption in statistical methods that *the errors in data values* (i.e., the deviations from average) *are independent from one another.* Four other assumptions remain that are important to note, because they often are not recognized or assessed in day-to-day clinical research.

(2) Obviously, *regression depends on the appropriateness of the model used in the fit.* Figure 24.1 shows two fits on a data set. The first-degree model, $y = b_0 + b_1 x$, provides a horizontal regression line ($b_1 = 0$), indicating no relationship between x and y. A parabolic (second-degree) model, $y = b_0 + b_1 x + b_2 x^2$, shows an almost perfect fit.

(3) *The independent (x) readings are measured as exactly known values* (measured without error). This arises because the least-squares method minimizes the sum of squared errors vertically, that is, on the y-axis but not on the x-axis. This assumption would be reasonable if we were relating yearly number of illnesses to patient age, because age is recorded rather exactly. However, if we were relating yearly number of illnesses to patient self-reported obesity (body mass index calculated from patient-reported height and weight), this assumption would be violated, because this measure of obesity is rather inaccurate. What do we do in the face of a violation? We cannot eliminate the inaccuracies, and

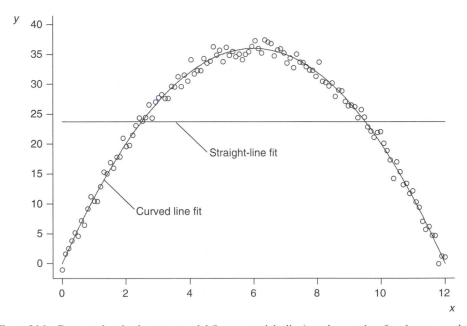

Figure 24.1 Data set showing how one model (here, a straight-line) can be a useless fit, whereas another model (here, a curved model) can be an almost perfect fit.

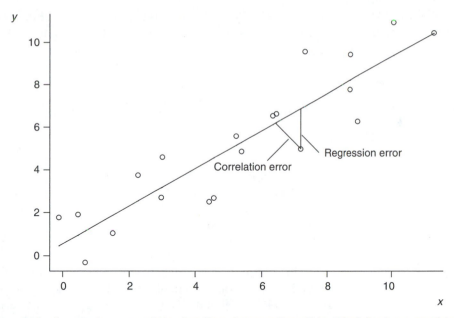

Figure 24.2 A regression or correlation line (data chosen so line will be either) fitted to a set of data. The deviations, or "errors," between a point and the line as assumed in regression and in correlation are labeled. Note that the regression error is vertical, indicating no *x* variability, whereas the correlation error allows variability in both *x* and *y*.

refusal to calculate the regression would deny us useful information; we can only proceed and make careful note that the quality of, and therefore confidence in, the regression is weakened. Figure 24.2 shows a regression line fitted to data with a vertical error shown. (Note that this assumption is not required in correlation.)

(4) *The variance of y is the same for all values of x.* As we move from left to right along the regression line, the standard deviation of *y* about this line is constant.

(5) *The distribution of y is approximately normal for all values of x.* As we move from left to right along the regression line, distribution of data remains normal for each *x* with mean at the regression line. Figure 24.3 illustrates

Figure 24.3 A regression line shown on an *x, y* plane laid flat with the assumed probability distribution rising vertically for two representative points on the line. Note that the distribution is normal for both and shows the same variability for both.

449

assumptions (4) and (5) simultaneously. Regression is reasonably robust to these last two assumptions; one need not worry too much unless the violations are severe.

24.2. CORRELATION CONCEPTS AND ASSUMPTIONS

WHAT IS CORRELATION?

Correlation as a general term implies simply a relationship among events. In statistics, it refers to a quantitative expression of the interrelationship, namely a *coefficient of correlation*. The population coefficient, which may be hypothesized but is generally unknown, is denoted by ρ and its sample estimate by r. In the context of statistics, the term *correlation* usually refers to a correlation coefficient; this chapter uses the term in this context. The correlation coefficient for continuous variables is introduced in Section 3.2, where its calculation was given as the covariance of x and y divided by both standard deviations. This division has the effect of standardizing the coefficient so that it always represents points scattered about a 45° line. In correlation, the level of association is not measured by the slope of the line as in regression but rather by how tightly or loosely the x, y observations cluster about the line. Because this coefficient is standardized by dividing the standard deviations, it lies in the range -1 to $+1$, with 0 representing no relationship at all and ± 1 representing perfect predictability. A positive coefficient indicates that both variables tend to increase or decrease together, whereas with a negative coefficient, one tends to increase as the other decreases. We would agree that a correlation coefficient of 0.10 implies little, if any, relationship between the two variables and that one of 0.90 indicates a strong relationship. Figure 24.4 illustrates the relationship between the coefficient and the scatter pattern of data. We might ask the question: What is the value of a correlation coefficient lying at the intersection of the probably coincidental (due to chance) and the probably associative (the measures vary together)? Such questions are addressed in Section 24.6.

ASSUMPTIONS UNDERLYING CORRELATION

Let us compare assumptions about continuous-variable correlation with the five regression assumptions.

(1) Correlation requires the same assumption: *The errors in data values are independent from one another.*

(2) *Correlation always requires the assumption of a straight-line relationship.* A large correlation coefficient implies that there is a large linear component of relationship but not that other components do not exist. In contrast, a zero correlation coefficient only implies that there is not a linear component; there may be curved relationships (see Fig. 24.1).

(3) The assumption of exact readings on one axis is *not* required of correlation; *both* x *and* y *may be measured with random variability* (see Fig. 24.2).

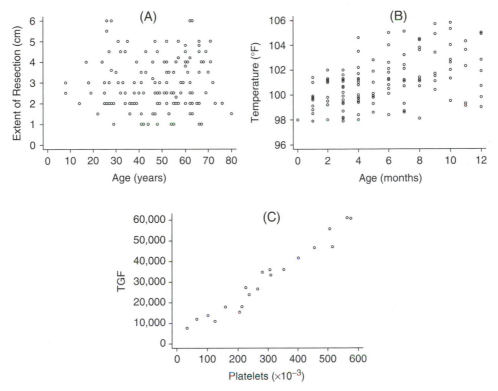

Figure 24.4 Low-, moderate-, and high-correlation coefficients as related to the scatter pattern of data from three sources. (A) $r = 0.02$ for the extent of carinal resection as related to patient age (from DB12). (B) $r = 0.48$ for temperature as related to age for infants with pulmonary complications in their first year of life (one extreme datum omitted to facilitate illustration). (C) $r = 0.98$ for transforming growth factor count as related to platelet count (from DB9).

(4) and (5) These assumptions take on a different form, because x and y vary jointly; thus, what is assumed for y relative to x is also assumed for x relative to y. x *and* y *are assumed to follow a bivariate normal distribution,* that is, sort of a hill in three dimensions, where x and y are the width and length and the height is the probability (could be read relative frequency) of any joint value of x and y. The peak of the hill lies over the point specified by the two means, and the height of the hill diminishes in a normal shape in any direction radiating out from the means point. If the bivariate normal assumption is badly violated, it is possible to calculate a correlation coefficient using rank methods.

Rank Correlation

Sometimes, the data to be analyzed consist of ranks rather than continuous measurements. At other times, the assumption of a bivariate normal distribution is violated, requiring that the ranks of the data replace the continuous measurements. In these cases, a measure of correlation based on ranks is required. If the formula for the ordinary correlation coefficient is used with ranks rather than continuous-type observations, a coefficient emerges that may be interpreted the same way.

NAMES FOR CORRELATION

The correlation coefficient between two continuous variables was originated by Francis Galton. British statistician Karl Pearson (who credits Galton, incidentally), as well as with Francis Edgeworth and others, did a great deal of the work in developing this form of correlation coefficient, so sometimes it is referred to as Pearson's correlation. Another name for this coefficient sometimes used is product–moment correlation. Moments are mathematical entities related to the descriptors we use, such as the mean (first moment) and the variance (second moment about the first moment). Moments derived using a product of x and y are called product moments. The covariance used in calculating the correlation coefficient is a form of a product moment. These names are not really necessary except in the context of distinguishing correlation coefficients based on continuous variables from those based on ranks. The rank correlation coefficient was first written about by C. E. Spearman, who also simplified the formula. The subscript s (for Spearman) is attached to the population ρ or sample r to signify this form. (In the early part of the twentieth century, before the use of Greek for population and Roman for sample convention became standard, ρ sometimes was used for the Spearman coefficient; if you encounter it, consider it a historical leftover.)

24.3. SIMPLE REGRESSION

REVIEW OF EXAMPLES FROM CHAPTER 8

A regression example of white blood cell (WBC) count as depending on level of infection is given in Section 8.3. Mean bacterial culture count m_x was 16.09 ($\times 10^{-9}$), mean WBC count m_y was 40.96 ($\times 10^{-9}$), and the slope b_1 was 0.1585, giving rise to the regression equation $y - 40.96 = 0.1585(x - 16.09)$. The graph was shown in Fig. 8.3. WBC count seems to increase with level of infection. A second example, number of days of hospital treatment for psychiatric patients depending on intelligence quotient (IQ), is given in Section 8.4. Mean number of days m_x was 10.5, mean IQ m_y was 100, and the slope b_1 was 0.3943, giving rise to the regression equation $y - 10.5 = 0.3943(x - 100)$. The graph was shown in Fig. 8.4. Time spent in treatment at a psychiatric hospital seems to increase with IQ.

EXAMPLE POSED: DAY 10 THEOPHYLLINE LEVEL PREDICTED BY BASELINE LEVEL

DB3 contains data on levels of serum theophylline (e.g., a vasodilator to treat asthma, emphysema) just before administering an antibiotic (baseline) and 10 days later. The relationship between the two levels was modeled as a straight-line fit in Section 23.4 (see Fig. 23.6). The inputs to the equation for this regression line are calculated as $m_x = 10.8000$, $m_y = 10.1438$, $s_{xy} = 11.8253$, and $s_x^2 = 14.1787$. What is the equation for the regression line?

METHOD

The Regression Equation

A straight line, including simple regression, is determined by two pieces of information (see Section 8.2). A useful form of a regression line, using the slope b_1 of the line and the mean point (m_x, m_y) in the x, y plane, is given by Eq. (8.2) as

$$y - m_y = b_1(x - m_x). \qquad (24.1)$$

Recall that theoretical or population coefficients in models are denoted by β's, whereas sample-based estimates of the β's are denoted by b's. The estimate of β_1, b_1, is the covariance of x and y divided by the variance of x, or s_{xy}/s_x^2. (The formulas for the m's and s's may be found in the Chapter Summaries [see section for Chapter 3].) This is the best fit by least squares (and also other mathematical criteria) expressing the relationship between x and y.

Predicting y from x

The most likely value of y, that is, its prediction, for a chosen value of x is just the value of y, say, $y|x$ (read "y given x"), arising from substituting the x value in Eq. (24.1) and solving for y. If many predicted values are to be calculated, an easier form for substitution is the slope-intercept form, $y = b_0 + b_1 x$, in which $b_0 = m_y - b_1 m_x$:

$$y|x = b_0 + b_1 x \qquad (24.2)$$

Notably, a prediction is valid only over the range of existing data. (In the example in Section 24.7, extending the days after malarial infection far enough will appear to bring dead rats back to life.)

EXAMPLE COMPLETED: DAY 10 THEOPHYLLINE LEVEL PREDICTED BY BASELINE LEVEL

Substitution in Eq. (24.1) yields the line $y - 10.1438 = 0.8340(x - 10.8000)$. Solution for y yields a simpler form for prediction: $y = 1.1363 + 0.8340x$. The line drawn in Fig. 23.6 was simplified by rounding to $y = 1.1 + 0.8x$. Suppose a patient has a baseline serum theophylline level of 10 and we wish to predict the postantibiotic level, $y|x$, where $x = 10$. We substitute $x = 10$ to obtain $y|10 = 9.4763$, or about 9.5. (Note that substitution of $x = 10$ in the rounded formula yields 9.1, which illustrates the advisability of carrying several decimal places of precision during calculation and rounding afterward.)

ADDITIONAL EXAMPLE: PREDICTING TEMPERATURE OF LUNG-CONGESTED INFANTS

Lung congestion often occurs in illness among infants, but it is not easily verified without radiography. Are there indicators that could predict whether or not lung opacity will appear on a radiograph? A study[82] of 234 infants included age (measured in months), respiration rate (RR; in breaths/min), heart rate (HR; in beats/min), temperature (in degrees Fahrenheit), pulse oximetry (as percentage), clinical appearance of illness on physical examination, and lungs sounding congested on physical examination. Lung radiographs were taken to be "truth" and were recorded as clear or opaque. However, the use of a binary outcome (yes or no, + or −, 0 or 1, etc.) as a dependent variable requires logistic regression and is addressed in Section 24.10. Until that section, relationships among continuous measurements are used in this chapter.

Predicting Temperature by Age

The following question arises: Is temperature associated solely with illness, or is it influenced by age? Let us fit a regression line to temperature as predicted by age. If the line is horizontal, knowledge of the age gives us no ability to predict temperature. However, if the line has a marked slope (Section 24.4 discusses testing the slope statistically), age provides some ability to predict temperature and would seem to be one factor influencing lung congestion. Regression line inputs from age (taken as x) and temperature (y) were calculated as $m_x = 10.9402$, $m_y = 101.5385$, $s_{xy} = 5.6727$, and $s_x^2 = 45.5329$. $b_1 = s_{xy}/s_x^2 = 0.1246$. The slope-means form of the regression line, from Eq. (24.1), is $y - 101.5385 = 0.1246(x - 10.9402)$. $b_0 = 100.1754$, leading to the slope-intercept form as $y = 100.1754 + 0.1246x$. If an infant is 20 months old, the most likely prediction of temperature would be $y|20 = 100.1754 + 0.1246 \times 20 = 102.6674$, or about 102.7°F. The data and the regression line are shown in Fig. 24.5.

Exercise 24.1. As part of a study on circulatory responses to intubation during laryngoscopy,[75] the following question arose: Could changes in systolic blood pressure (SBP; y-axis) caused by laryngoscopy be predicted by skin vasomotor reflex amplitude (SVmR; x-axis)? Readings (these readings from a graph may be off from the original by some decimal units) were taken on $n = 26$ patients with the following results: $m_x = 0.1771$, $m_y = 25.6538$, $s_{xy} = 2.2746$, and $s_x^2 = 0.0241$. Calculate b_1 and b_0. Write the slope-means and slope-intercept forms for the regression line. Sketch the SVmR and SBP axes and the regression line. If the SVmR amplitude is measured as 0.3, what is the most likely prediction of change in SBP?

24.4. CORRELATION COEFFICIENTS

EXAMPLE POSED: THEOPHYLLINE LEVEL EXAMPLE CONTINUED

In the example from Section 24.3, DB3 data were to predict serum theophylline level 10 days after antibiotic treatment from the baseline level. The covariance s_{xy} was given as 11.8253, and the regression line as

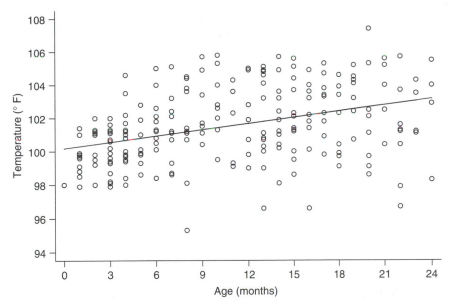

Figure 24.5 Temperature depending on age for 234 infants.

$y = 1.1363 + 0.8340x$. Additional statistics are the standard deviations $s_x = 3.7655$, calculated from the variance, and $s_y = 3.9816$. What is the correlation coefficient r? What is the rank correlation coefficient r_s? Are they similar?

METHOD

When to Use Correlation as Opposed to Regression

Correlation is used when the question of interest is how closely x and y are associated (in a straight-line relationship). They are not thought of as dependent and independent variables. There is no intent to predict one from the other as in regression, where x is a factor (partially) causing y, or at least related to such causal factors.

Calculating Correlation

The calculation of the correlation coefficient is given in Section 8.3 as

$$r = \frac{s_{xy}}{s_x s_y} \left(= b_1 \frac{s_x}{s_y} \right). \tag{24.3}$$

The first expression, the covariance divided by the standard deviations, usually is used. However, if a regression has already been calculated, the second expression, the regression slope multiplied by the ratio of standard deviations, may be more convenient.

Calculating the Rank Correlation

For each of n pairs of x and y ranks in the sample, find the difference between x-rank minus y-rank; these are denoted d_i, the difference in x and y ranks for the ith patient. Square and add together these d_i values. Spearman's formula (simplified from the formula for the continuous coefficient) is

$$r_s = 1 - \frac{6 \sum d_i^2}{n(n^2 - 1)}. \qquad (24.4)$$

EXAMPLE COMPLETED: THEOPHYLLINE LEVEL EXAMPLE

By using the relationship of Eq. (24.3), $r = 11.8253/(3.7655 \times 3.9816) = 0.7887$. Note that the second part of Eq. (24.3) can also be used, yielding $r = 0.8340 \times 3.7655/3.9816 = 0.7877$. This correlation coefficient of about 0.79 is rather high, indicating a close straight-line association. Were the assumptions underlying r justified? The frequency distribution of baseline is not far enough from normal to cause much worry, but the day 10 readings are. It is preferable to use r_s. Table 24.1 shows the DB3 data with ranks and d_i^2. $n = 16$, and $\sum d_i^2 = 192$. Substituting in Eq. (24.4), we find $r_s = 0.7176$, about 0.72, which is not far different from $r = 0.79$.

Table 24.1

DB3 Data (Serum Theophylline Levels) for Baseline and 10 Days After Beginning Antibiotic, with Associated Ranks and Squared Difference Between Ranks

Patient no.	Base	Rank	Day 10 level	Rank	d_i^2
1	14.1	12	10.3	7	25
2	7.2	4	7.3	5	1
3	14.2	13	11.3	9	16
4	10.3	8	13.8	14	36
5	15.4	14	13.6	13	1
6	5.2	2	4.2	1	1
7	10.4	9	14.1	15	36
8	10.5	10	5.4	4	36
9	5.0	1	5.1	3	4
10	8.6	5	7.4	6	1
11	16.6	16	13.0	12	16
12	16.4	15	17.1	16	1
13	12.2	11	12.3	11	0
14	6.6	3	4.5	2	1
15	9.9	6	11.7	10	16
16	10.2	7	11.2	8	1

d_i^2, squared differences between ranks.

ADDITIONAL EXAMPLE: INFANT LUNG CONGESTION CONTINUED

In the additional example from Section 24.3 on infant lung opacity,[82] we posed the question: Is temperature associated with age? The covariance is $s_{xy} = 5.6727$, and the standard deviations are $s_x = 6.7478$ and $s_y = 2.2557$. $r = 5.6727/(6.7478 \times 2.2557) = 0.3727$. The correlation coefficient is not high, but it also is not negligible. Is it statistically significant, that is, greater than what is likely to occur by chance? This question is answered in Section 24.6. If we had been investigating rather than illustrating, at the beginning of the exercise we would have asked if the assumptions underlying correlation were satisfied. If we plot quick frequency distributions of temperature and age, we find temperature is approximately normal in distribution but age is bimodal, although the two modes are rather weak, and is not far from a uniform distribution (nearly the same frequency for all age intervals). The assumptions for age are not satisfied; we should use r_s rather than r. On ranking the data, then finding, squaring, and adding together the rank differences, and substituting in Eq. (24.4), we find $r_s = 0.3997$. This is rather close to the r value of 0.3727; therefore, the violation of an assumption did not cause a major error in r.

Exercise 24.2. As part of a study on circulatory responses to laryngoscopy and intubation (see Exercise 24.1),[75] the relationship between change in SBP caused by laryngoscopy and SVmR is of interest. Readings were taken on $n = 26$ patients with the following results: $s_{xy} = 2.2746$, $s_x = 0.1552$, and $s_y = 20.0757$. Calculate r. What does this r indicate about the association of SBP change and SVmR? If $\sum d_i^2 = 863$, what is r_s? How does r_s agree with r?

24.5. TESTS AND CONFIDENCE INTERVALS ON REGRESSION PARAMETERS

EXAMPLE POSED: TESTING THE THEOPHYLLINE PREDICTION

In the example from Section 24.3 using DB3, serum theophylline level at day 10 was predicted from baseline level by the regression line $y = 1.1363 + 0.8340x(b_1 = 0.8340)$. The calculated components needed are $n = 16$, $m_x = 10.8000$, $s_{xy} = 11.8253$, $s_x = 3.7655$, and $s_y = 3.9816$. Is the prediction of y by x significant? How strong is this prediction? What is the confidence in the prediction of average y? Of individual y? Let us also predict the 5-day serum level by baseline. Are the two regression lines different?

METHOD

Assessing Regression

Five questions might be asked of the regression and are considered in turn: (1) Does the regression show y is significantly predicted by x? (2) How strong is this prediction? (3) By using regression methods for prediction, what will be confidence limits on average y for a given value of x? (4) What will be confidence limits on the best predicted *y* for an individual patient's *x*? (5) Given two samples, are the regression slopes different?

Standard Error Notation

In Part I, A Study Course of Fundamentals, the standard error of the mean was abbreviated as SEM, which is fairly common in medicine. However, in regression, we shall have to find the standard errors of other statistics and use them in formulas, so a more succinct and flexible notation is needed. Because a standard error of a statistic is just its standard deviation, it will be convenient to use the standard deviation symbol *s* with the respective statistic indicated as a subscript. Thus, the SEM becomes s_m. We will need four more standard errors: the standard error of the residual (residuals are the observations' deviations from the regression line), usually denoted s_e in statistics (*e* for "error"); the standard error of the estimate of the regression slope b_1, s_b; the standard error of the estimate of the mean values of *y* for each *x*, $s_{m|x}$; and the standard error of the estimate of the individual predictions of *y* for each *x*, $s_{y|x}$. Because the residual standard error, s_e, is used in the others, it is given first. Starting with the sum of squares of deviations from the regression line, we define

$$s_e^2 = \sum (y - b_0 - b_1 x)^2 / (n - 2),$$

where $n - 2$ will be degrees of freedom (*df*), because we lost 2 *df* by calculating the two *b*'s. Some algebra will yield the following forms. Use whichever form is easier computationally.

$$s_e = \sqrt{\frac{n-1}{n-2}\left(s_y^2 - b_1^2 s_x^2\right)} = s_y \sqrt{\frac{n-1}{n-2}\left(1 - R^2\right)} \qquad (24.5)$$

(R^2, the coefficient of determination (see Section 8.4) is discussed in greater detail in the section How Strong Is the Predictive Ability? which discusses the strength of the *x,y* relationship.)

---

(1) Does the Regression Show That *x* Is a Significant Predictor of *y*?

If the regression line is horizontal, the predicted *y* is the same for every *x*; *x* has no ability to predict *y*. If the line is sloped, each *x* yields a different *y*; there is predictive ability. Is this slope significantly different from horizontal, or could an apparent slope be due to chance only? In a test of this question, $H_0: \beta = 0$ versus $H_1: \beta \neq 0$ (or <0 or >0 if one tail is impossible or clinically irrelevant). In the hypothesis test of a mean, we asked if the estimate of the mean was significantly larger than the standard error of that mean, that is, if the ratio of statistic to its standard error

---

was larger than a critical *t*-value. The pattern here is the same. We find the standard error of the slope as

$$s_b = \frac{s_e}{\sqrt{n-1}s_x}. \tag{24.6}$$

To test H_0, *t* with $n-2$ *df* is simply

$$t_{(n-2)df} = \frac{b_1}{s_b}, \tag{24.7}$$

and we reject H_0 if b_1/s_b is farther out in the tails of the *t* distribution than the critical *t*. If the calculated *t* is not rejected, then *either* the relationship is improbable *or* the relationship does not follow a straight line.

Testing Against a Theoretical Slope

Suppose the slope is to be compared not with zero slope but with a theoretical slope, say, β_1. Situations in which such a β_1 might arise are, for example, when a relationship has been posed on the basis of established physiology or when comparing with previously published results. The standard error s_b remains the same, as does the $n-2$ *df*. The hypotheses are $H_0: \beta = \beta_1$ versus $H_1: \beta \neq \beta_1$. The *t* statistic becomes

$$t_{(n-2)df} = \frac{b_1 - \beta_1}{s_b}, \tag{24.8}$$

which is tested and interpreted the same as for zero slope.

(2) How Strong Is the Predictive Ability?

The *coefficient of determination* (see introduction of this topic in Section 8.4) indicates how well the model predicts the dependent variable. It might be defined as the proportion of total predictive capability represented by this model. Generally designated R^2, it is sometimes denoted r^2 in the case of a single *x* predicting *y* by a straight line, but we shall use R^2 in all cases to be consistent. In the single *x* straight-line case, R^2 is just the square of *r*, the correlation coefficient. In the case of a curved line or multiple *x*, it is more complicated. Note well that R^2 does not evaluate how well *x* predicts *y* but rather how well *this model of the* x,y *relationship* predicts *y*. Sometimes, a straight line is a poor predictor, whereas a curved line such as a parabola is a good predictor (see Fig. 24.1). In this case, $y = b_0 + b_1 x$ yields a small R^2 (0.002 in Fig. 24.1), but $y = b_0 + b_1 x b_2 x^2$ yields a large R^2 (0.970 in Fig. 24.1). We know that an *r*, and therefore R^2, of 0 indicates no relationship and that an *r* of -1 or $+1$, and therefore an R^2 of 1, indicates perfect predictability. At what value of R^2 does the relationship become greater than chance? It turns out that this value is the same as the value at which b_1 becomes significantly greater than zero, as tested by Eq. (24.7). Algebra permits this form to be rewritten as a test of R^2 (for straight-line single x regression only!). The null hypothesis H_0: $\rho^2 = 0$ is tested by

$$t_{(n-2)df} = \sqrt{\frac{(n-2)R^2}{1-R^2}}. \qquad (24.9)$$

Because Eq. (24.9) tells us nothing new compared with Eq. (24.7), R^2 usually is used in interpreting rather than testing the x, y relationship.

Confidence Interval on the Slope

Sometimes in describing the slope of the regression relationship, a confidence interval on the slope will be informative. Section 4.4 presents the general pattern of confidence intervals. To follow that pattern, we need the estimate of the statistic (sample slope b_1), the critical value of the probability distribution ($t_{1-\alpha/2}$), and the standard error for that statistic (s_b), all of which are given earlier. The confidence interval on the population slope β_1 is

$$P[b_1 - t_{1-\alpha/2}s_b < \beta_1 < b_1 + t_{1-\alpha/2}s_b] = 1 - \alpha, \qquad (24.10)$$

where the critical t has $n - 2$ df. A tight confidence interval indicates a strong relationship.

(3) What Is the Confidence Interval on the Regression Prediction of the Average y for a Given Value of x?

Because there is only one regression line, there is only one prediction of y from a given x. The average y for a given x is the same as the most likely y for a given x. However, the confidence interval for a mean y is not the same as that for an individual patient's y, so these confidence intervals are given separately in this and the next paragraph. The confidence interval follows the same pattern as before: the estimated value on the y-axis plus or minus a critical t multiplied by the standard error. The estimate and the critical t are the same as in Eq. (24.10); only the standard error differs. We symbolize the sample mean value of y for a given x as $m|x$, which estimates the population mean value μ for a given x, $\mu|x$. Its standard error is

$$s_{m|x} = s_e\sqrt{\frac{1}{n} + \frac{(x - m_x)^2}{(n-1)s_x^2}}. \qquad (24.11)$$

Thus, the $1 - \alpha$ confidence interval, where the critical t's have $n - 2$ df, is

$$P[m|x - t_{1-\alpha/2}s_{m|x} < \mu|x < m|x + t_{1-\alpha/2}s_{m|x}] = 1 - \alpha. \qquad (24.12)$$

(4) What Is the Confidence Interval on the Regression Prediction of y for a Given Patient's Individual x?

The average regression more often is of interest in research. In clinical practice, the predicted value for an individual patient often is of interest. The prediction will be the same; the average is the most likely value and our best guess. However, the standard error is slightly different, leading to an altered confidence interval. We symbolize the most likely value of y for a given x as $y|x$, estimating the population expected value of y for a given x, $E(y)|x$. Its standard error is

$$s_{y|x} = s_e \sqrt{1 + \frac{1}{n} + \frac{(x - m_x)^2}{(n-1)s_x^2}}. \tag{24.13}$$

The $1 - \alpha$ confidence interval, where the critical t's have $n - 2$ df, is

$$P\left[y|x - t_{1-\alpha/2}s_{y|x} < E(y)|x < y|x + t_{1-\alpha/2}s_{y|x}\right] = 1 - \alpha. \tag{24.14}$$

(5) Given Two Samples, Are the Regression Slopes Different?

To distinguish the statistics calculated from the two samples, let us denote n as n_1 and n_2 for samples 1 and 2, respectively; b_1 as $b_{1:1}$ and $b_{1:2}$; s_x^2 as s_1^2 and s_2^2; and s_e^2 as $s_{e:1}^2$ and $s_{e:2}^2$. The test of the null hypothesis H_0: $b_{1:1} = b_{1:2}$ is a t test with the usual format, having $n_1 + n_2 - 4$ df. The standard error of difference in slopes is

$$s_{b:1-2} = \sqrt{\left(\frac{(n_1 - 2)s_{e:1}^2 + (n_2 - 2)s_{e:2}^2}{n_1 + n_2 - 4}\right)\left(\frac{1}{(n_1 - 1)s_1^2} + \frac{1}{(n_2 - 1)s_2^2}\right)}, \tag{24.15}$$

which yields a t as

$$t = \frac{b_{1:1} - b_{1:2}}{s_{b:1-2}}. \tag{24.16}$$

If the calculated t is larger than a two-tailed critical t for $n_1 + n_2 - 4$ df, H_0 is rejected.

Example Completed: Testing the Theophylline Prediction

(1) Is x a Significant Predictor of y?

From Eq. (24.5), $s_e = 2.5335$. To test H_0: $\beta_1 = 0$, we need s_b. Substitution in Eq. (24.6) yields $s_b = 0.1737$. We use Table II (see Tables of Probability Distributions) to find a 5% two-tailed critical value of t for $n - 2 = 14$ df

as 2.145. From Eq. (24.7), the calculated $t = 4.80$, which is much larger than critical. We conclude that the slope of the line, and therefore the prediction of y by x, is significant.

(2) How Strong Is the Predictive Ability?

To evaluate the strength of the predictive ability, we need to ask: Is baseline theophylline level a major predictor of serum theophylline level at day 10? Use of Eq. (24.3) yields $r = 0.7887$ (see Section 24.4); thus, $R^2 = 0.6220$. We can say that about 62% of the possible capability to predict serum theophylline level at day 10 from baseline level is provided by this straight-line model; 38% remains for all other factors combined plus randomness. Although baseline level does not predict the level at day 10 exactly, we can conclude that baseline level is the major predictor. To put a confidence interval on the slope, we substitute b_1, s_b, and critical t in Eq. (24.10) to find the confidence interval:

$$P[0.8340 - 2.145 \times 0.1737 < \beta_1 < 0.8340 + 2.145 \times 0.1737]$$
$$= P[0.46 < \beta_1 < 1.21] = 0.95$$

We can state with 95% confidence that the predicted slope is not more than about three-eighths of a unit off in y per unit of x. Although not perfect, the prediction is rather good.

(3) What Is the Prediction's Confidence Interval of Average y for a Given x?

Suppose we have a group of patients with a mean baseline serum level of 10 mg/dl. The predicted $m|x$ is, of course, the value of the regression obtained by substitution of that x, in this case, 9.476. We want a confidence interval on the mean of this group. Use of Eq. (24.11) yields the standard error of this prediction statistic as $s_{m|x} = 0.6484$. We found the 5% two-tailed critical value of t for 14 df to be 2.145. By substituting in Eq. (24.12), we find the confidence interval on mean level at day 10 for a baseline level of 10 mg/dl to be

$$P[9.476 - 2.145 \times 0.6484 < \mu|10 < 9.476 + 2.145 \times 0.6484]$$
$$= P[8.085 < \mu|10 < 10.867] = 0.95$$

(4) What Is the Prediction's Confidence Interval of y for an Individual Patient's x?

Suppose we have an individual patient with baseline serum level of 10 mg/dl. The predicted $y|x$ is still 9.476. We want a confidence interval on this individual's prediction. Use of Eq. (24.13) yields the standard error of this prediction statistic as $s_{y|x} = 3.270$. The critical value of t remains 2.145. By substituting in Eq. (24.14), we find the confidence interval on the patient's level at day 10 for a baseline level of 10 mg/dl to be

$$P[9.476 - 2.145 \times 2.615 < E(y)|10 < 9.476 + 2.145 \times 2.615]$$
$$= P[3.867 < E(y)|10 < 15.085] = 0.95$$

The confidence interval on the individual's predicted 10-day level is much wider than that on the mean.

(5) Are Two Regression Slopes Different?

By using the symbols to distinguish the two samples as in the Method section, we found the regression of the 10-day serum level to have slope $b_{1:1} = 0.8340$, $s_1^2 = 14.1787$, and $s_{e:1}^2 = 6.4186$. Let us also consider the regression of the 5-day serum level on baseline. In returning to the data of DB3, we find the slope to be $b_{1:2} = 0.8210$, $s_2^2 = s_1^2$, and $s_{e:2}^2 = 12.5230$. $n_1 = n_2 = 16$. Substitution in Eq. (24.15) yields $s_{b:1-2} = 0.3089$, and substitution in Eq. (24.16) yields $t = 0.0421$. The critical two-tailed 5% α-value of t from Table II with $n_1 + n_2 - 4 = 28$ df is 2.048, which is much larger than the calculated t. (The actual p-value is calculated as 0.967.) There is no evidence that the slopes are different.

ADDITIONAL EXAMPLE: INFANT LUNG CONGESTION CONTINUED

In the lung congestion study[82] introduced in the additional example from Section 24.3, we ask if temperature can be predicted by infant age. The regression line was calculated to be $y = 100.1752 + 0.1246x (b_1 = 0.1246)$. The regression prediction for a 20-month infant is $102.6674°F$. The interim statistics we need are $n = 234$, $m_x = 10.9402$, $s_{xy} = 5.6727$, $s_x = 6.7478$, and $s_y = 2.2557$. From Eq. (24.5), $s_e = 2.0977$.

(1) Is x a Significant Predictor of y?

To test H$_0$: $\beta_1 = 0$, we need s_b. Substitution in Eq. (24.6) yields $s_b = 0.0204$. We interpolate in Table II to find a 5% two-tailed critical value of t for $n - 2 = 232$ df, which is about 1.97. From Eq. (24.7), the calculated $t = 6.1078$, which is much larger than the critical t. Indeed, because t is larger than the interpolated Table II $t = 3.33$ for two-sided $\alpha = 0.001$, we know that $p < 0.001$. We conclude that the slope of the line, and therefore prediction of y by x, is significant. This tells us that age is a real causal factor in temperature but not whether it is a major or a minor cause.

(2) How Strong Is the Predictive Ability?

Use of Eq. (24.3) yields $r = 0.3727$ (see Section 24.4); thus, $R^2 = 0.1389$. We can say that about 14% of possible capability to predict temperature from age is provided by this straight-line model, the remaining 86% arise from other causal factors and randomness. This indicates that temperature is a minor causal factor. To put a confidence interval on the slope, we substitute b_1, s_b, and critical t in Eq. (24.10) to find the confidence interval:

$$P[0.1246 - 1.97 \times 0.0204 < \beta_1 < 0.1246 + 1.97 \times 0.0204]$$

$$= P[0.08 < \beta_1 < 0.16] = 0.95$$

(3) What Is the Prediction's Confidence Interval of Average y for a Given x?

Let us consider a group of 20-month-old infants. The predicted $m|x$ is, of course, the value of the regression obtained by substitution in that x, in this case, 102.6672. We want a confidence interval on the mean of this group.

Use of Eq. (24.11) yields the standard error of this prediction statistic as $s_{m|x} = 0.2299$. We found the 5% two-tailed critical value of t for 232 df to be 1.97. By substituting in Eq. (24.12), we find the confidence interval on mean temperature of 20-month-old infants to be

$$P[102.6674 - 1.97 \times 0.2299 < \mu|20 < 102.6674 + 1.97 \times 0.2299]$$
$$= P[102.2 < \mu|20 < 103.1] = 0.95.$$

(4) What Is the Prediction's Confidence Interval of y for an Individual Patient's x?

Suppose we have an individual 20-month-old infant. The predicted $y|x$ is still 102.6674. We want a confidence interval on this individual's prediction. Use of Eq. (24.13) yields the standard error of this prediction statistic as $s_{y|x} = 2.1103$. The critical value of t remains 1.97. By substituting in Eq. (24.14), we find the confidence interval on the 20-month-old infant to be

$$P[102.6674 - 1.97 \times 2.1104 < E(y)|20 < 102.6674 + 1.97 \times 2.1104]$$
$$= P[98.5 < E(y)|20 < 106.8] = 0.95.$$

Although the confidence interval on the mean might be useful in research, the confidence interval on temperature as predicted by age for an individual infant is useless, encompassing almost the entire range of possible readings. If we want to predict temperature for clinical use, we will have to incorporate more important causal factors in our model, probably going to a multiple regression.

Exercise 24.3. Exercise 24.1[75] poses the question: Could change in SBP (y) caused by laryngoscopy be predicted by SVmR (x)? Required interim statistics are $n = 26$, $m_x = 0.1771$, $s_{xy} = 2.2746$, $s_x = 0.1553$, and $s_y = 20.0757$. Find s_e and s_b. Look up the 95% critical t. Test $H_0: \beta_1 = 0$. Is the slope significantly greater than 0? Find and test R^2. Find the 95% confidence interval on β_1. Is SVmR a major predictor? What percentage of the predictive capability remains for other predictors and randomness? For SVmR $= 0.3$, calculate $s_{m|x}$ and $s_{m|y}$ and find the 95% confidence intervals on $m|x$ and $y|x$.

24.6. TESTS AND CONFIDENCE INTERVALS ON CORRELATION COEFFICIENTS

EXAMPLE POSED: ARE THEOPHYLLINE LEVEL CORRELATIONS DIFFERENT BY DAY?

In the emphysema example, the correlation coefficient of baseline serum theophylline level with that 10 days after beginning an antibiotic was found to be $r_{0,10} = 0.7887$ (see Section 24.4). Baseline standard deviation was found to be 3.7655. In Section 24.5, the slope of 5-day level predicted by baseline was given as $b_{0,5} = 0.8210$. With the additional information that 5-day level standard deviation is 4.6093, the correlation coefficient between 5-day and baseline levels can be found from Eq. (24.3) as $r_{0.5} = 0.8210 \times 3.7655 \div 4.6093 = 0.6707$. Are the two correlation coefficients significantly different from 0? Is there probabilistic evidence that they differ from one another, or is the observed difference due only to random fluctuations? What is a confidence interval on the 10-day versus baseline coefficient?

METHOD

A Significance Test for $\rho = 0$

The sample correlation coefficient, r, estimates the population correlation coefficient, ρ. It indicates how closely a scattergram of x, y points cluster about a 45° straight line. A tight cluster (see Fig. 24.4) implies a high degree of association. The coefficient of determination, R^2 (introduced in Section 24.5), indicates the proportion of ability to predict y that can be attributed to the model using the independent (predictor) variables. In the case of a single predictor x in a straight-line relationship with y, R^2 is just the square of r. It has been noted that Eq. (24.9) provides a test of the hypothesis that R^2, and therefore r, is zero, that is, that x and y are independent of each other. Rewritten using r, the test of H_0: $\rho = 0$ is

$$t_{(n-2)df} = \sqrt{\frac{(n-2)r^2}{1-r^2}}. \tag{24.17}$$

If the calculated t is greater than a critical t from Table II, H_0 is rejected.

A Test for ρ Other Than Zero

Suppose the correlation coefficient between two blood test measures for repeated samples of healthy people has proven to be some ρ_0, a theoretical correlation coefficient other than zero, perhaps 0.6, for example. We obtain a sample of ill patients and would like to know if the correlation coefficient between the blood tests is different for ill patients versus well subjects. The (unknown) population coefficient from the ill patients is ρ. The null hypothesis becomes H_0: $\rho = \rho_0$. It has been shown mathematically that the expression

$$m = \frac{1}{2} \ln \left(\frac{1+r}{1-r} \right) \tag{24.18}$$

is distributed approximately normal (for larger samples, $n > 50$) with mean

$$\mu = \frac{1}{2} \ln \left(\frac{1+\rho_0}{1-\rho_0} \right) \tag{24.19}$$

and standard deviation

$$\sigma = \frac{1}{\sqrt{n-3}}. \tag{24.20}$$

The test is just the usual z test on the standardized normal,

$$z = \frac{m - \mu}{\sigma}. \tag{24.21}$$

If the calculated z from Eq. (24.21) is larger than a critical z found from Table I (see Tables of Probability Distributions), H_0 is rejected.

A Test of Two Correlation Coefficients

Suppose we have large-sample correlation coefficients between blood test measures for type 1 and type 2 diseases; we want to compare two sample correlation coefficients, r_1 and r_2. We test the null hypothesis that the two ρ's, the correlation coefficients for the populations of all patients with these diseases, are equal, that is, H_0: $\rho_1 = \rho_2$. We calculate m_1 and σ_1 for sample 1 and m_2 and σ_2 for sample 2 in the forms of Eqs. (24.18) and (24.20). The test is a z test conducted as with Eq. (24.21), where

$$z = \frac{m_1 - m_2}{\sqrt{\sigma_1^2 + \sigma_2^2}}. \tag{24.22}$$

No good tests have been developed for these cases for small samples.

A Confidence Interval on ρ

Given a sample r, we can find a confidence interval for the population ρ from which r was drawn (see Section 14.6). However, the expression and calculation are somewhat bothersome. It is preferable to use a confidence interval on the regression β_1 if appropriate; however, if this is not possible, a few minutes with a capable calculator will provide the confidence interval using Eq. (24.23):

$$P\left[\frac{1 + r - (1 - r)e^{\frac{2z_{1-\alpha/2}}{\sqrt{n-3}}}}{1 + r + (1 - r)e^{\frac{2z_{1-\alpha/2}}{\sqrt{n-3}}}} < \rho < \frac{1 + r - (1 - r)e^{\frac{2z_{1-\alpha/2}}{\sqrt{n-3}}}}{1 + r + (1 - r)e^{\frac{2z_{1-\alpha/2}}{\sqrt{n-3}}}} \right] = 1 - \alpha. \tag{24.23}$$

Example Completed: Are Theophylline Level Correlations Different by Day?

Significance Tests of the Correlation Coefficients

In Section 24.5, addressing the prediction of the 10-day level by baseline level, we test H_0: $\beta_1 = 0$ using Eq. (24.7), concluding that the t of 4.80 showed the prediction to be significant. We note that Eq. (24.7) gives the same result as Eq. (24.9) or (24.17). Substitution of $r = 0.7887$ in Eq. (24.17) yields $t = 4.80$, which is indeed the same result. This sample t is much larger than 2.145, the two-tailed critical $t_{0.95}$ for 14 df, yielding $p < 0.001$ and indicating a significant association. Similarly, substitution of the 5-day versus baseline correlation of 0.6707 in Eq. (24.17) yields $t = 3.38$, which is also larger than the critical 2.145, with $p = 0.002$. There is strong evidence that both $\rho_{0,5}$ and $\rho_{0,10}$ are greater than zero.

Test of Two Correlation Coefficients

First, we must note that we have only $n = 16$ readings in our sample, which is too small for a proper approximation; we shall carry out the calculations only for illustration, not for a legitimate medical conclusion. Substitution of 0.7887 and 0.6707 in turn in Eq. (24.18) yields $m_1 = 1.0680$ and $m_2 = 0.8122$. From Eq. (24.20), both variances are $1/13 = 0.0769$, yielding a pooled standard deviation of 0.3923. By substituting in Eq. (24.22), we find $z = 0.6520$. A critical z, using two-tailed $\alpha = 0.05$, is the familiar 1.96. The calculated z is far less than the critical z; we cannot reject the null hypothesis of no difference. We have no evidence that the two correlation coefficients are different.

Confidence Interval on the Correlation Coefficient

Using the estimated correlation coefficient $r = 0.7887$, what are 95% confidence limits on the population ρ? Substitution of r, $z_{1-\alpha/2} = 1.96$, and $n = 16$ in Eq. (24.23) yields $P[0.48 < \rho < 0.92] = 0.95$.

Additional Example: Infant Lung Congestion Continued

A Significance Test of the Correlation Coefficient

In the example of lung opacity in $n = 234$ infants,[82] the correlation coefficient between temperature and age was found to be 0.3727. Testing H_0: $\rho = 0$ by substitution in Eq. (24.17) yields a t-value of 6.12, which is much larger than the critical $t_{0.95}$ for 232 df of 1.97; $p < 0.001$. The x, y association is highly significant.

Test of Two Correlation Coefficients

Of the 234 infants, 78 proved to have lung opacity on radiography (sample 1) and 156 did not (sample 2). The correlation coefficients between temperature and age for these groups were $r_1 = 0.4585$ and $r_2 = 0.3180$. By substituting

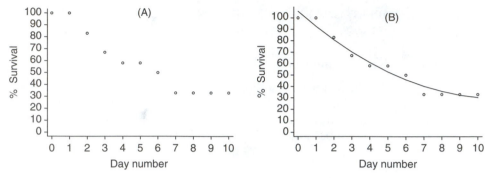

Figure 24.6   Percentage of survival by day of rats infected with malaria. (A) Data. (B) Parabolic regression curve fitted to data.

these values in turn in Eq. (24.18), we find $m_1 = 0.4954$ and $m_2 = 0.3294$. Similarly, by using Eq. (24.20), $\sigma_1^2 = 0.0133$ and $\sigma_2^2 = 0.0065$. Substitution of these values in Eq. (24.22) yields $z = 1.18$, which is smaller than the 5% two-tailed critical z of 1.96. There is no evidence of a difference between correlation coefficients.

Confidence Interval on the Correlation Coefficient

What is the 95% confidence interval on the population ρ? The sample $r = 0.3727$, $z_{1-\alpha/2} = 1.96$, and $n - 3 = 231$. Substitution of these values in Eq. (24.23) yields $P[0.26 < \rho < 0.48] = 0.95$.

Exercise 24.4.   Let us continue with the SVmR example,[75] in which the correlation between SVmR and SBP is 0.7295 for 26 patients. Exercise 24.3 tests the hypothesis $H_0: \beta_1 = 0$. Repeat the test in the form of Eq. (24.17) to test $H_0:$ $\rho = 0$. Suppose physiologic theory posed a ρ of 0.5. Does this sample agree with or differ from that theory?

24.7. CURVED REGRESSION

EXAMPLE POSED: A BETTER FIT TO THE SURVIVAL OF MALARIAL RATS

DB11 contains data on percentage of survival of malaria-infected rats (placebo group) as dependent on day number. The data plot (see Fig. 23.4A) is redisplayed as Fig. 24.6A. We want to fit a regression curve expressing the general pattern of survival over time.

Choosing the Model

Survival drops over time, but the drop rate appears to diminish. The data suggest a second-degree model, opening upward. We select the model $y = \beta_0 + \beta_1 x + \beta_2 x^2$, where y is percentage of survival and x is number of days. How is this model fitted?

METHOD

Concept

The concept of curved (more exactly, curvilinear) regression is the same as simple regression throughout, except that the form of the model is not restricted to a straight line. We now can generally refer to the regression curve, which includes the straight line as a subordinate case. Section 23.3 considers model forms at length. The most frequently used curves are the parabola, which is like a simple regression with an x^2 term added, and the logarithmic and exponential curves, which are like a simple regression with the x term replaced by a $\log x$ or e^x term. However, any mathematical function may be appropriate.

Choosing the Model

The method of choosing the model varies with goals of the study. If the study is being used to assess a theoretical relationship (e.g., physiology), the form of the model will arise from the theory and the regression significance will be used to validate the theory. If the study is being used to develop the predictor of an established form, that form dictates the model and the regression is used to identify the parameters (constants) used in the prediction. If the study is used to explore relationships, the form of the model will be suggested by shape and pattern in the data plot. The model need be appropriate only within the range of existing data (see Admonition section at the end of the completed example).

Data Input

The model that is chosen dictates the inputs. Basically, we have a y depending on some function of x. We just substitute x- and y-values in the forms given by the model. Where y appears, y-values are put in; where x^2 appears, x-values are squared and those squares are put in; where $\ln(x)$ appears, logarithms of x-values are found and put in; and so forth.

Solution

The regression curve calculated is the best-fit curve according to various mathematical criteria of "best," as appropriate to the form and assumptions. In simple regression, the criterion was least squares used for linear models (see Section 23.2). Other criteria are maximum likelihood (used more for nonlinear models), unbiasedness, minimum variance, and so on, or combinations of these characteristics. For simple regression, the least-squares solutions gave rise to relatively simple formulas (see earlier discussion). However, for more complicated models, sets of linear equations derived using calculus must be solved by methods of matrix algebra, which is not done easily manually. These now have come to be solved using computer programs contained in statistical software packages. It is reasonable to assume that any investigator using curvilinear regression will have access to such a package. Sections 24.7, 24.8, and 24.10 do not attempt to present solutions but rather concentrate on choosing the model, entering the data, and interpreting the results.

Results to Select and Their Interpretation

Statistical software packages usually display a number of results, many of which are needed only occasionally. The user must select what is needed. The following types of results are the most likely to be of use. (1) In validating a model or exploring data to identify relationships, the *p*-value of a test of the model (usually an F test) indicates whether the relationship between x and y is probably real or probably just caused by sampling fluctuations. For relationships that test significant, the coefficient of determination R^2 (no longer the same as r^2) indicates whether or not the predictive capability is clinically useful. The value $1 - R^2$ tells us the proportion of predictive capability attributable either to causal factors not contained in the model or to the x factor in a different model form and to random effects. Thus, in a computer display of regression results, we seek the *p*-value of a test of the model and R^2. (2) In evaluating the contribution of model components, *p*-values of *t* tests on model components indicate whether or not to retain them in the model. (3) In developing a prediction equation, coefficients for the model equation can be identified.

EXAMPLE COMPLETED: A BETTER FIT TO THE SURVIVAL OF MALARIAL RATS

Data Input

We select regression with its appropriate model in whatever statistical software package we choose. We enter percentage of survival into the y position and number of days into x. We square the number of days and enter these values into x^2.

Results Output

Different packages use different results display formats, but most are labeled well enough to select the values we need. We can perform a check on the correctness of our selection by anticipating a gross approximation to the values we desire. In this example, we find the sample estimates of the β's to be: $b_0 = 105.77$, $b_1 = -13.66$, and $b_2 = 0.61$. Substitution in the model yields the equation representing the prediction of percentage survival by number of days as $y = 105.77 - 13.66x + 0.61x^2$. The curve given by this equation is shown superposed on the data in Fig. 23.6B. By looking further at the results, we find p for the F test of the model to be less than 0.001 and R^2 to be about 0.97.

Interpretation

The significant *p*-value indicates that a real (curved) predictive relationship of percentage of survival by number of days does exist. The very large R^2 indicates that survival percentage can be quite well predicted by this model, with only 3% of the survival variability left for randomness and other causal factors.

An Admonition

This prediction is valid up to the modeled 10 days only; if we extend the model to periods outside the experiment, the model will claim that the rats come back to life and that the percentage of survival begins to increase. This nonsensical result illustrates the necessity of inferring conclusions only within the limits modeled.

ADDITIONAL EXAMPLE: INFANT LUNG CONGESTION CONTINUED

Let us again consider the dependence of temperature on age in infants with pulmonary complications[82] (example is continued from the additional example of Section 24.3). At that time, our simple regression fit was $y = 100.18 + 0.12x$. Furthermore, in Section 24.5, we found that the p-value for a t test on the regression slope was less than 0.001 and $R^2 = 0.1389$.

Choosing the Model

In Fig. 24.5, the line appears to be centered on the data in the right part of the plot but lies a bit below center in the 6- to 15-month region and above center on the left. That causes us to conjecture that temperature, which we showed by simple regression to increase with age, increases rapidly with age in newborns and slows its increase as the infant grows older. This suggests a logarithmic fit. We would like to see if we obtain a better fit by using the model $y = \beta_0 + \beta_1 \ln(x)$.

Data Input

We enter our model into the software package. We calculate the (natural) log of x for each x and enter them and the y-values. We instruct the computer to calculate the regression of temperature on log age. The calculations are carried out just as in simple (straight-line) regression but use log age rather than age.

Results Output

The p-value for the F test on the model and coefficient of determination R^2 for the two models are shown in Table 24.2, and the log fit on the data is shown in Fig. 24.7.

Table 24.2

p-Value and R^2 for Straight-Line and Logarithmic Fits to Infant Pulmonary Data

Model	p-value for F test	R^2
Straight line	<0.001	0.1389
Logarithmic	<0.001	0.1651

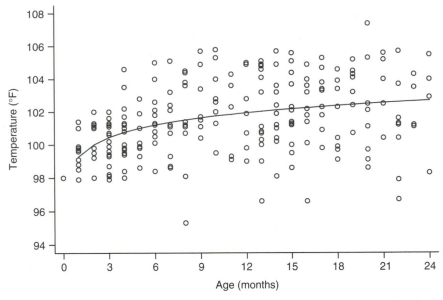

Figure 24.7 Temperature depending on age for 234 infants with pulmonary complications, with logarithmic regression fit shown.

Interpretation

We can see from the relatively low R^2 in both cases that, whereas age is a significant predictor, it is not very important clinically. However, R^2 has increased from about 14% to about 17%; the log fit describes the physiologic process slightly better.

Exercise 24.5. Let us continue the question of circulatory response to intubation during layrngoscopy.[75] We found that a simple regression of SBP on SVmR gave the fit $y = 8.94 + 94.38x$, that the p-value of a significance test on the model was less than 0.001, and that $R^2 = 0.5322$.

> *Choosing the model:* Figure 24.8 shows the data. Although the simple regression fit seems to be useful, a case can certainly be made for curvature, opening downward. Write the equation for a parabolic fit, using SBP and SVmR rather than y and x.

Data input: What data would be entered into the computer?

> *Results and interpretation:* The outputs are read as $b_0 = -2.92$, $b_1 = 289.00$, $b_2 = -414.41$, $p < 0.001$, and $R^2 = 0.6822$. Write the final predictive equation. Is the relationship between SBP and SVmR significant in both models? Is the modeled SVmR a major predictor in both models? Which model provides a better fit? Calculate an approximate y-value for the left, center, and right regions of SVmR and roughly sketch the curve.

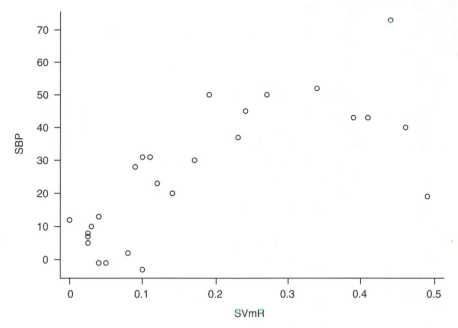

Figure 24.8 Change in systolic blood pressure as depending on skin vasomotor reflex amplitude in intubated patients.

24.8. MULTIPLE REGRESSION

EXAMPLE POSED: PREDICTING LENGTH OF HOSPITAL STAY

A psychologist would like to be able to predict length of hospital stay (measured in days) for inpatients at time of admission.[22] Information available before examining the patient includes IQ (mean, 100; standard deviation, 10), age (in years), and sex (0,1). Data are available for $n = 50$ inpatients. How can length of hospital stay be predicted by three variables simultaneously, with each contributing as it can?

METHOD

Concept

Multiple regression is the term applied to the prediction of a dependent variable by several (rather than one) independent variables. For example, in an investigation of cardiovascular syncope, no one sign would be sufficient to predict its occurrence. The investigator would begin with, at least, SBP, diastolic BP, and HR, and then would extend the list of predictors to include perhaps the presence or absence of murmurs, clicks, vascular bruits, and so on. The concept of multiple regression is similar to that of simpler regression, in that the dependent variable is related to the independent variables by a best fit. However, the geometry is extended from a line in x, y dimensions to a plane in x_1, x_2, y dimensions, or to what is called a hyperplane $x_1, x_2, x_3, \ldots, y$ dimensions, which is just a plane extended to more than three dimensions. A hyperplane can be treated similarly to a plane mathematically, although it cannot be visualized. A multiple regression model with two independent variables is $y = \beta_0 + \beta_1 x_1 + \beta_2 x_2$. Models of this sort are considered in Section 23.5 and visualized in Fig. 23.11. More generally, we are not even restricted to a plane, that is, to first-degree terms. The model can contain second-degree or other terms of curvature that lead to a curved surface or, in several dimensions, a curved hypersurface. An example of such a model might be $y = \beta_0 + \beta_1 x_1 + \beta_2 x_2 + \beta_3 x_2^2$. The foregoing conceptualizations may be more confusing than enlightening to some readers. If so, it is sufficient to remember just that several independent variables combine to predict one dependent variable.

Choosing the Model

We can consider each predictor x one at a time and enter its relationship to the dependent variable y as if it were alone. If y is related to x_1 in a straight line, we add $\beta_1 x_1$ to the model. If y is related to x_1 in a second-degree curve, we add $\beta_1 x_1 + \beta_2 x_1^2$ to the model. Then we proceed with x_2, and so on. (Components combining variables in the same term are possible but form nonlinear models, which are unusual and outside the realm of this book.)

Setting Up Nominal Variables

Independent variables used to predict the dependent variable may be used just as recorded if they are continuous measurements or they can be put in order (e.g., small to large). The only type of variable that must be altered for use in multiple regression is nominal data that cannot be ordered (e.g., ethnic group or disease type). A dichotomy, perhaps malignant versus benign, would not require special treatment because of symmetry. However, a categorization with three or more classes cannot be used as is. A racial heritage categorization classified as European, African, or Asian, for example, if assigned recording codes 1, 2, and 3, respectively, would predict a disease state differently than if assigned 1, 3, and 2. In this case, a strategy to use such classes as predictors is to reduce them to a set of dichotomies, that is, to create "dummy variables."

Variables are created for each class, denoting that class as, say, 1 and not as 2. Thus, we would have a "European variable" with Europeans denoted 1 and non-Europeans 2, an "African variable" with Africans denoted 1 and non-Africans 2, and so on. For k classes in the categorization, we need only $k - 1$ variables to replace the original variable. The last category's variable is redundant, because all the information needed is contained in the first $k - 1$ variables.

Data Input

Data are entered into the computer software model just as in simpler regressions. Where x_1 appears in the model, we enter the x_1 data; where x_1^2 appears, we square each datum and enter the squares; and so forth.

Solution

To estimate the β's that provide a best fit, we have to solve a set of linear equations simultaneously, as in curved regression. This process is better done by statistical software on a computer. More detail on the reason for this may be found in the curvilinear regression section (see Section 24.7). After calculating the best fit, the significance of the model is tested by F and the associated p-value given. For a model with k terms (i.e., the variable-containing components), F has k (numerator) and $n - k - 1$ (denominator) df. A regression with model $y = \beta_0 + \beta_1 x_1 + \beta_2 x_2$ for 50 patients would have 2 and $50 - 2 - 1 = 47$ df. Also, for each predictor, that part of its contribution that is not also provided by another predictor is tested by t and the associated p-value given. These results are conceptually not very different from the equivalent results in simpler regressions. The coefficient of determination R^2 is also given. R^2 is $1 - $ (residual variance/total variance), where total variance is s_y^2, and residual variance is the variance of differences of data from the model.

Results to Select and Their Interpretation

As with other regression forms, the software packages often provide a variety of results; the user must select those that answer the questions being asked of the data. As with curvilinear regression, the following types of results are the most likely to be of use. (1) In validating a model or exploring data to identify relationships, the p-value of a test of the model (usually an F test) indicates whether the relationship between y and the model is probably real or probably just caused by sampling fluctuations. (2) For relationships that test significant, the coefficient of determination R^2 indicates whether or not the predictive capability is clinically useful. The value $1 - R^2$ indicates the proportion of predictive capability attributable to causal factors not contained in the model, to a different model form, and to random effects. (3) In evaluating the contribution of model components, p-values of t tests on model components indicate how to rank the clinical usefulness of the components in the model. (4) R^2 also helps identify the clinically useful predictors. For a model with k components, note its R^2. Remove a component you suspect may not be useful (perhaps that with the smallest t test p-value) and recalculate the regression. Note R^2 for the reduced model. The reduction in R^2 evoked by removing the variable indicates how useful that variable is as a predictor. (5) In developing a prediction equation, coefficients for the model equation can be identified so that its predictive equation can be written down.

An Admonition

The b's estimating the β's used in the model depend on the units chosen. Temperatures measured in degrees Fahrenheit will yield different coefficients from temperatures measured in degrees Celsius. However, the statistical results, F, t, and R^2, will not change.

Stepwise Multiple Regression

If we were to start with one predictor variable and add more variables one at a time, we would be following a form of forward stepwise regression. We would monitor the R^2 to see how much predictive capability each additional variable added and, given we added in the order of predictive strength ascertained from single-predictor analyses, would stop when additional variables added only an unimportant level of predictive capability. More frequently, the entire set of predictors under consideration is used at the start, the strategy being to eliminate the least contributive variables one by one until elimination excessively reduces the predictive capability. This form is known as backward stepwise regression. Statistical software is capable of performing this task, with the added benefit that at each elimination all previously eliminated variables are retried. This process corrects the error of correlation between two variables, leading to removal of one that would not have been removed had the other one not been present at the outset. However, the addition–removal decisions are made on the basis of mathematics, and the investigator loses the ability to inject physiologic and medical information into the decision. For example, if two correlated variables contribute only slightly different predictive ability and one should be removed, the software may remove one that occurs in every patient's chart while leaving one that is difficult or costly to measure.

EXAMPLE COMPLETED: PREDICTING LENGTH OF HOSPITAL STAY

Choosing the Model

The dependent variable days in hospital is y. The IQ, age, and sex are x_1, x_2, and x_3, respectively. The model becomes $y = \beta_0 + \beta_1 x_1 + \beta_2 x_2 + \beta_3 x_3$. With three variable-containing terms in the model, the F for the model contains $k = 3$ and $n - k - 1 = 50 - 3 - 1 = 46 \, df$.

Data Input

In the software package, enter the data for days in hospital into the y position, the data for IQ into the x_1 position, age into the x_2 position, and sex into the x_3 position.

Results to Select and Their Interpretation

(1) The computer result gives the p-value of the F test to be 0.030. We conclude that the predictive ability of the model is real, not due to chance. (2) $R^2 = 0.175$. This indicates that, even though the model is significant as a predictor, it represents less than 18% of the predictive capability; 82% remains for other predictors and the influence of randomness. (3) Interpolating from Table II,

the critical value of $t_{0.95}$ for 46 *df* is 2.01. The values of t calculated for the independent portion of each x are as follows: IQ, 2.258; age, 1.624; and sex, 0.244. Only IQ is greater than the critical value, but the issue is to identify the predictors that contribute to the prediction. Sex produced the smallest t and appears not to be a useful predictor. Let us delete sex and recalculate the multiple regression, using the new model $y = \beta_0 + \beta_1 x_1 + \beta_2 x_2$.

(1) The p for the F test is 0.011, which is actually somewhat improved. (2) $R^2 = 0.174$. We lost nothing by dropping sex; it is not clinically useful. (3) The values of t calculated for the independent portion of each x are as follows: IQ, 2.312; and age, 1.688. Intelligence quotient is still significant ($p = 0.025$), whereas age is not ($p = 0.098$). We ask how much age contributes to the prediction. We delete age and recalculate the regression, using the reduced model $y = \beta_0 + \beta_1 x_1$. The significance of the model is changed only slightly ($p = 0.012$), but $R^2 = 0.124$, which is a notable drop in predictive ability. We conclude that we should retain age for prediction purposes, even though it is not of significant use. We return to the model $y = \beta_0 + \beta_1 x_1 + \beta_2 x_2$. (4) The computer output lists the estimates of the β's as $b_0 = -32.5938$ (constant), $b_1 = 0.3576$ (coefficient of IQ), and $b_2 = 0.1686$ (coefficient of age). The resulting predictive model is days in hospital $= -32.5938 + 0.3576(\text{IQ}) + 0.1686(\text{age})$. Patient 10 remained in the hospital for 4 days. His IQ was 90.4, and he was 24 years old. His predicted days in hospital was $-32.5938 + 0.3576 \times 90.4 + 0.1686 \times 24 = 3.8$.

ADDITIONAL EXAMPLE: INFANT LUNG CONGESTION CONTINUED

Let us again consider the prediction of temperature in infants with pulmonary complications[82] (see the additional example from Section 24.3). Age was found to be a significant predictor. Data on HR and pulse oximetry also had been recorded. Are they useful predictors?

Choosing the Model

The model for age (x_1) as a predictor of temperature (y) was $y = \beta_0 + \beta_1 x_1$. Adding terms for HR ($x_2$) and pulse oximetry ($x_3$) gives the model $y = \beta_0 + \beta_1 x_1 + \beta_2 x_2 + \beta_3 x_3$. There are $n = 234$ patients. The number of variable-containing terms is $k = 3$. The model has $n - k - 1 = 234 - 3 - 1 = 230$ *df*.

Data Input

In the software package, the sample's temperature values are selected to enter in the y position, age value in the x_1 position, HR value in the x_2 position, and so on.

Results to Select and Their Interpretation

(1) The computer result gives the p-value of the F test to be less than 0.001. We conclude that the predictive ability of the model is real, not due to chance. (2) $R^2 = 0.289$. This indicates that the model represents only about 29% of the predictive capability (71% remains for other predictors and the influence of randomness), which may be clinically useful, although not definitive. (3) The values of t calculated for the independent portion of each x are as follows: age, 7.117;

HR: 6.895; and pulse oximetry, 0.111. The first two values are large compared with the critical t of 1.97, but pulse oximetry appears not to be a useful predictor. Let us delete pulse oximetry and recalculate the multiple regression, using the new model $y = \beta_0 + \beta_1 x_1 + \beta_2 x_2$.

(1) The p for the F test is still less than 0.001. (2) $R^2 = 0.284$. We lost only a negligible amount of the predictive ability by dropping pulse oximetry; it is not clinically useful. (3) The values of t calculated for the independent portion of age and HR were significantly large, so we conclude that both are clinically useful predictors. (4) The computer output lists the estimates of the β's as $b_0 = 94.4413$ (constant), $b_1 = 0.1333$ (coefficient of age), and $b_2 = 0.0354$ (coefficient of HR). The resulting predictive model is temperature $= 94.4413 + 0.1333(\text{age}) + 0.0354(\text{HR})$. Infant patient 1 had an observed temperature of 101.4°F. She was 1 month old, and her HR was 180 beats/min. Her predicted temperature was $94.4413 + 0.1333 \times 1 + 0.0354 \times 180 = 100.9°\text{F}$, which is within half a degree.

Exercise 24.6. By using the data of DB10, we may be able to predict the strength of hamstrings or quadriceps muscles or tendons following surgery by using the strength and control of the nonoperated leg.

Choosing the model: Strength following surgery is measured by the distance covered in a triple hop on the operated leg. Possible predictors are the equivalent on the nonoperated leg (x_1) and the time to perform the hop (x_2). Write down the equation of the model.

Data input: What are the data to be entered into a software package?

Results selected and their interpretation: F's $p < 0.001$. $R^2 = 0.974$. t-values are 11.617 for x_1 and 1.355 for x_2. By omitting x_2, the F test's $p < 0.001$, $R^2 = 0.965$, and $t = 12.832$. For the simple regression of y on x_1, $b_0 = -300.868$ and $b_1 = 1.466$. Is the predictive ability of the model using both x's real? Is it clinically useful? Judging from the t values of each x, should we consider dropping x_2? If we do, is the reduced model significant? Clinically useful? What is the predictive equation for the reduced model? If the triple-hop distance for a particular patient is 504 cm, what is the predicted distance for the operated leg? How does this agree with an observed distance of 436 cm from DB10 data?

24.9. TYPES OF REGRESSION

CLASSIFYING TYPES OF REGRESSION MODEL

Previously in this chapter, treatment of regression has been restricted to prediction of continuous variables as outcomes. However, investigators often require other types of outcome. Several classes of regression models available to address the various types of outcome are enumerated in Table 24.3. Note that these classes refer only to the predicted variable, not the predictors. Each type may have first-degree, higher-degree, or nonlinear predictor variables. Each type may have a single or multiple predictor variables.

Table 24.3

Classes of Regression Model by Type of Outcome Variable They Predict

Nature of outcome (dependent) variable	Name of regression type
Continuous	(Ordinary) regression
Ranks	Ordered regression
Categorical: two categories	Logistic regression
Categorical: several categories	Multinomial regression
Counts (number of occurrences)	Poisson regression
Survival (independent variable: time)	Cox (proportional hazards)

EXAMPLES OF THE SEVERAL TYPES

An example of an outcome variable in clinical research associated with each regression type follows.

Ordinary regression: Prostate-specific antigen (PSA) levels are predicted by body mass index. The outcome is a continuous variable.

Ordered regression: Insomniacs are asked to rank the effectiveness of four sleeping aids. The outcome is a rank-order variable.

Logistic regression: Breast cancer in patients who have received radiation therapy (RT) may or may not have recurred, as predicted by RT dose. The outcome is a binary variable.

Multinomial regression: One year after treatment for actinic keratoses, patients may be classified as cured, recurred keratoses, or progressed to squamous cell carcinoma. The outcome is a three-category variable.

Poisson regression: Diet plus supplements, antibiotic use, and exercise levels are used to predict the number of infectious illnesses over a 3-year period. The outcome variable is composed of number of occurrences (counts) as outcome.

Cox proportional hazards regression: Survival depending on time as predicted by certain modes of treatment. The outcome of survival versus death is a binary, time-dependent variable.

REGRESSION MODELS IN STATISTICAL SOFTWARE

Even modestly capable software packages can analyze ordinary regression, including multiple regression. Curvilinear and simpler nonlinear regression models are included, because the analysis is the same as multiple regression, except that higher-degree or nonlinear terms are substituted for additional variables. For example, a second-degree model with one predictor is just the first-degree model with two predictors, $y = \beta_0 + \beta_1 x_1 + \beta_2 x_2$, with the squared term replacing x_2 to form $y = \beta_0 + \beta_1 x_1 + \beta_2 x_1^2$. The other forms listed in Table 24.3 appear in various software packages. The more complete packages (e.g., SAS, S-Plus, SPSS, Stata, Statistica) provide the capability for all or most of the types. Often, identification of the required data format and just what commands to use is somewhat difficult.

WHERE REGRESSION TYPES ARE FOUND IN THIS BOOK

The goal of this book is to present the most frequently needed forms of analysis. This chapter addresses (ordinary) regression with straight-line (see Section 24.3), curved (see Section 24.7), and multivariable (see Section 24.8) prediction models, and logistic regression (see Section 24.10). The regression aspect of survival data in Chapter 25 touches on Cox proportional hazards regression. Other types of regression are not met or are not required frequently in current medical research. That said, statistical sophistication is increasing every year. In another decade, it is likely that the medical investigator frequently will require analyses not presented in this edition.

24.10. LOGISTIC REGRESSION

EXAMPLE POSED: PREDICTING SURVIVAL FOR PATIENTS WITH TRACHEAL CARINAL RESECTION

The data of DB12 arose from a study on resection of the tracheal carina. A patient who dies is recorded with a data code of 1, and a patient who survives the operation is recorded with a data code of 0. Independent variables on which survival may depend are age at surgery (years), whether patient had prior tracheal surgery (1) or not (0), extent of resection (measured in centimeters), and whether intubation was required at end of surgery (1) or not (0). Can we predict survival from the independent variables?

METHOD

Concept

In medicine, a large number of situations arise in which we wish to predict a binary outcome (e.g., survival or not, a patient heals faster than usual or not, a disease recurs or not), which can be coded 1 or 0 for numerical treatment. If we set the binary outcome y equal to a regression line, $\beta_0 + \beta_1 x_1$, we get nonsensical results, usually anything except 0 or 1.

On returning to basics, we note that, for a continuous y, the ordinary regression prediction yields the expected y (most likely value of y). For a binary y, the expected value is the probability that a code of 1 (e.g., survival) occurs, estimated by the proportion of 1's in the sample. Appropriately, this proportion is the mean of the sample, that is, the sum of the 0 and 1 codes divided by the number in the sample. Let us affix the subscript m for mean so that the population probability is denoted π_m and the sample proportion is denoted p_m. Note that $p_m/(1 - p_m)$ gives the odds that a code of 1 will occur in the sample on any one opportunity. If we take the logarithm of this expression, we have the *log odds ratio* (see Section 15.5). Because the probability distribution of the log odds ratio is known, it can be used in the statistical tests needed for regression. Thus, rather than the simple regression model $y = \beta_0 + \beta_1 x_1$, when we face binary outcomes we will use the model

$$\ln\left(\frac{p_m}{1 - p_m}\right) = \beta_0 + \beta_1 x. \tag{24.24}$$

Logistic regression is just a transformation of the dependent variable to the log odds ratio, after which the usual regression procedures are followed. Curvilinear and multiple logistic regression are used just the same as in ordinary regression, with the dependent variable transformed. The right side of Eq. (24.24) may be extended to include whatever terms are required.

Choosing the Model

We choose the right side of the model just as in simple, curvilinear, or multiple regression. If unorderable nominal variables, for example, three or more ethnic groups or disease categories, are to be used, see the explanation for setting up the predictor variables in Section 24.8.

Data Input

Independent variable data are entered into the computer software model just as in ordinary regressions. The 0 or 1 binary codes are entered in the same way as the y-values in ordinary regression; we are ready for the computer package to make the transformation to the log odds ratio.

Solution

Because of the transformation to the log odds ratio, as well as solving simultaneous linear equations in cases more complicated than a single first-degree x, the computation process is better done by statistical software on a computer. This software provides a best fit (where the appropriate mathematical criterion of "best" is maximum likelihood for logistic regression). The fitting process estimates the β's, the parameters of fit. The estimates are denoted as b's. The right side of Eq. (24.24) becomes $b_0 + b_1 x$. To provide a regression prediction, we require one step more than that for continuous regression. The prediction desired is the estimate of the probability of survival, patient improvement, or treatment success, that is, of the code of 1. We have denoted this estimate as p_m. When the observed x-values for a particular patient or case are substituted in the right side of the model, the log odds ratio results must be solved to provide p_m. If we take anti-logs (exponentials) of both sides of Eq. (24.24) and solve for p_m, we obtain

$$p_m = \frac{e^{b_0 + b_1 x}}{1 + e^{b_0 + b_1 x}}. \tag{24.25}$$

For any model other than $b_0 + b_1 x$, the $b_0 + b_1 x$ of Eq. (24.25) is just replaced by the right side of that model.

Results to Select and Their Interpretation

As with other regression forms, the software packages often provide a variety of results; the user must select those that answer the questions being asked of the data. The following types of results are the most likely to be of use. (1) In validating a model or exploring data to identify relationships, the p-value of a test of the model (in logistic regression, a chi-square [χ^2] rather than an F) indicates whether the relationship between y and the model is probably real or probably just caused by sampling fluctuations. (2) For relationships that test significant, the coefficient of determination R^2 helps indicate if the predictive capability is clinically useful. However, because of the transformation, the usual R^2 cannot be calculated the same way. Most software packages provide an approximate equivalent to R^2, or to an R that can be squared, which can be interpreted in much the same fashion. We can use the value $1 - R^2$ to indicate the proportion of predictive capability attributable to causal factors not contained in the model, to a different model form, and/or to random effects. (3) In evaluating the contribution of model components, p-values of tests on model components, in this case, normal (z) tests, indicate how to rank the clinical usefulness of the components in the model. (4) The R^2 equivalent also helps identify the clinically useful predictors by noting its change caused by changing the model. If a legitimate R^2 is absent, the value of the χ^2 statistic used to test the model can be used in much the same way. (5) In developing a prediction equation, coefficients for the model equation can be identified. A prediction may then be made using Eq. (24.25) (or its equivalent with a more sophisticated model substituted for the $b_0 + b_1 x$).

EXAMPLE COMPLETED: PREDICTING SURVIVAL FOR PATIENTS WITH TRACHEAL CARINAL RESECTION

Choosing the Model

The left side of the model is the log odds ratio form as in Eq. (24.24). The right side of the model is a four-variable multiple regression, $\beta_0 + \beta_1 x_1 + \beta_2 x_2 + \beta_3 x_3 + \beta_4 x_4$, where x_1 is age, x_2 is prior surgery, x_3 is extent of resection, and x_4 is intubation.

Data Input

The survival or dying (0 or 1) codes are entered into the statistical software package in the position for the left side of the model. The recordings for age, prior surgery, extent, and intubation are entered into the corresponding positions for the right side of the model.

Results Selected and Their Interpretation

(1) From the χ^2 test of the model, $p < 0.001$, which indicates that the relationship between y and the model is probably real and not caused by sampling fluctuations. (2) The coefficient of determination, R^2, is about 0.34, which indicates that these predictors account for about a third of the predictive ability; this set of predictors is important and likely useful, but by no means completely decisive. (3) The p-values of the variables, ranked by size, are as follows: intubation, <0.001; extent, 0.031; prior surgery, 0.244; and age, 0.947. Clearly, intubation

and extent are significant, and age contributes almost nothing. Prior surgery is far from significant and likely contributes little. Let us assess its contribution by recalculating the regression with prior surgery and age removed.

(1) The new logistic regression based on intubation and extent retains the p-value of less than 0.001 of the model's χ^2 test. These variables are real rather than chance predictors. (2) The new R^2 is about 33%, still indicating useful but not perfect prediction. (3) The p-values for tests on the predictors are less than 0.001 for intubation and 0.044 for extent. Both are still significant. (4) The R^2 of 33% shows that the regression has lost only about 1% of predictive ability by omitting prior surgery and age. The χ^2 statistic, 34.5 for the four variables, is now 33.2, which is a negligible reduction. We conclude that prior surgery and age are not predictive and that we can do as well without them. (5) The coefficients for the model equation are identified as follows: constant, -4.6171; intubation, 2.8714; and extent, 0.5418. The right side of the final model is $b_0 + b_1 x_1 + b_2 x_2 = -4.6171 + 2.8714(\text{intubation}) + 0.5418(\text{extent})$. A prediction may be made for any one patient by substituting this expression in Eq. (24.25). For example, the expression for a patient who was intubated (substitute 1 for x_1) after a 2-cm resection (substitute 2 for x_2) is -0.6621. Substituting this for the exponent of e in Eq. (24.25) yields

$$p_m = e^{-0.6621}/(1 + e^{-0.6621}) = 0.3403.$$

The chance of death for this patient is predicted by this model to be about 34%.

ADDITIONAL EXAMPLE: PREDICTION OF LUNG CONGESTION IN INFANTS

It would be useful to predict the presence or absence of lung congestion in infants with respiratory problems[82] without resorting to obtaining x-ray films. In a study of 234 infants, variables recorded were age (measured in months), RR, HR, temperature, pulse oximetry (as a percentage), clinical appearance of illness on physical examination (physex; 1 = appeared ill; 0 = did not appear ill), and sound of congestion in lungs on physical examination (lungex; 1 = sounded congested; 0 = did not sound congested). Lung radiographs were taken to be "truth" and were recorded as clear (0) or opaque (1).

Choosing the Model

The left side of the model is the log odds ratio form as in Eq. (24.24). The right side of the model is a seven-variable multiple regression: $\beta_0 + \beta_1 x_1 + \beta_2 x_2 + \beta_3 x_3 + \beta_4 x_4 + \beta_5 x_5 + \beta_6 x_6 + \beta_7 x_7$, where x_1 is age, x_2 is RR, x_3 is HR, x_4 is temperature, x_5 is pulse oximetry, x_6 is physex, and x_7 is lungex.

Data Input

The 0 or 1 codes denoting clear or opaque radiographs are entered in the position for the left side of the model; the recordings for age, RR, HR, temperature, pulse oximetry, physex, and lungex are entered in the positions for the right side of the model.

Results to Select and Their Interpretation

(1) From the χ^2 test of the model, $p = 0.001$, which indicates that the relationship between y and the model is probably real and not caused by sampling fluctuations. (2) The coefficient of determination, R^2, is about 0.09, which indicates that these variables, although significantly related to lung opacity, are rather poor predictors and unlikely to be of great help clinically. (3) The p-values of the variables, ranked by size, are as follows: RR, 0.005; lungex, 0.022; age, 0.118; pulse oximetry, 0.186; physex, 0.228; temperature, 0.273; and HR, 0.855. Recall that these p-values are calculated for that portion of each variable that does not overlap in prediction capability with another variable, that is, the independent portions of the variables. (For example, earlier sections show that temperature is related to age. We would expect that the predictive ability of the two combined would be less than the sum of that for each alone.) We decide to retain only the first three variables, because of the larger p-values of the others. We retain age despite it not being significant, because we suspect its independent portion might become significant when the correlated temperature is removed. Now we have the right side of the model $\beta_0 + \beta_1 x_1 + \beta_2 x_2 + \beta_3 x_3$, where x_1 is RR, x_2 is lungex, and x_3 is age.

(1) The p-value for the χ^2 test of the model is less than 0.001. These are significant predictors. (2) R^2 is a slightly more than 7%. The prediction will not be very useful clinically. (3) The p-values for the tests on the individual variables are 0.002, 0.018, and 0.008, respectively. The age did indeed become significant. (4) R^2 dropped from its previous 9% to a slightly more than 7%, and the χ^2 statistic for the model test dropped from 25.1 to 20.7. Although these are noticeable reductions, the bulk of the predictive ability remains. (5) The coefficients to be used in a prediction are $b_0 = -3.3591$, $b_1 = 0.0410$, $b_2 = 0.7041$, and $b_3 = 0.0607$, leading to the right side of the model as $-3.3591 + 0.0410(\text{RR}) + 0.7041(\text{lungex}) + 0.0607(\text{age})$. Suppose an 8-month-old infant with respiratory problems has an RR of 30 breaths/min and congested-sounding lungs. The model's right side becomes $-3.3591 + 0.0410 \times 30 + 0.7041 \times 1 + 0.0607 \times 8 = -0.9394$. Substitution of this value for the exponent of e in Eq. (24.25) yields $p_m = e^{-0.9394}/(1 + e^{-0.9394}) = 0.2810$. The chance that this infant's lungs will appear opaque on a radiograph is predicted by this model to be about 28%, or just over one chance in four. However, we have little faith in this prediction, because about 93% of the ability to predict lung opacity lies with other factors and with randomness.

Exercise 24.7. DB1 includes several variables related to prostate cancer (CaP). The "truth" indicator is positive or negative biopsy result (coded 1 or 0, respectively). Try to predict biopsy result on the basis of PSA, volume, transurethral ultrasound (TRU; $1 = $ indicated CaP; $0 = $ CaP not indicated), digital rectal examination (DRE; $1 = $ indicated CaP; $0 = $ CaP not indicated), and age. What is the right side of the model? (1) The p-value of the test of the model is less than 0.001. Is the model a significant predictor? (2) $R^2 = 0.1513$. Is the prediction clinically useful? (3) The individual p-values are as follows: PSA, <0.001; volume, 0.002; TRU, 0.014; DRE, 0.336; and age, 0.794. What should be dropped from the model? (Reduced model 1 to 3) The reduced model retains $p < 0.001$ and has $R^2 = 0.1488$. Is the reduced model satisfactory? (4) Was much predictive ability lost by the reduction? (5) The coefficients to be used in a prediction are $b_0 = -0.8424$, $b_1 = 0.1056$ (PSA), $b_2 = -0.0315$ (volume), and $b_3 = 0.7167$ (TRU). One of the patients had a PSA level of 9.6 ng/ml, a volume of 62 ml, and a positive TRU. What does this model predict the probability of CaP to be?

ANSWERS TO EXERCISES

24.1. $b_1 = 94.3817$, and $b_0 = 8.9388$. The slope-mean form is $y - 25.6538 = 94.3817(x - 0.1771)$, and the slope-intercept form is $y = 8.9388 + 94.3817x$. If SVmR $= 0.3$, the change in SBP is predicted to be 37.2533. The axes and regression are shown on the accompanying graph, together with the data.

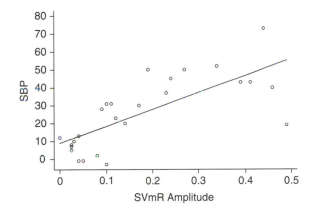

24.2. $r = 2.2746/(0.1552 \times 20.0757) = 0.7300$. A correlation coefficient of 0.73 is rather high; there is clearly an association between the variables. $r_s = 0.7050$, which is close to r.

24.3. $s_e = 14.00$, and $s_b = 18.04$. The two-tailed 95% critical t for 24 df is 2.064. Calculated t to test the slope is 5.23, which is significantly greater than the critical t; the slope of the regression line is greater than chance. $R^2 = 0.5329$. SVmR is a major predictor; 47% of the predictive capability remains for other predictors and randomness. t for testing $R^2 = 5.23$, which is far greater than 2.064; R^2 is highly significant. $P[57 < \beta_1 < 132] = 0.95$. $s_{m|0.3} = 3.53$, and $s_{y|0.3} = 14.45$. $P[30 < \mu|0.3 < 45] = 0.95$. $P[7.4 < E(y) | 0.3 < 67.1] = 0.95$.

24.4. $r^2 = 0.5329$. The two-tailed 95% critical t for 24 df is 2.064. The calculated t is $\sqrt{24 \times 0.5329/0.4671} = 5.2327$, which is much larger than the critical t. The population correlation coefficient is in probability greater than zero. (Slight differences appearing in prior calculations based on these data are due to rounding.) Denote ρ_s as the correlation coefficient of the population from which the sample was drawn. To test the null hypothesis H_0: $\rho_s = 0.5$, substitute the appropriate values in Eqs. (24.18) through (24.21). $m = 0.9287$, $\mu = 0.5493$, $\sigma = 0.2085$, and $z = 1.82$. The critical $z = 1.96$. The calculated z is less; therefore, we must conclude that there is inadequate evidence to reject H_0. The correlation has not been shown to be different from the theoretical value.

24.5. *Model:* $SBP = b_0 + b_1(SVmR) + b_2(SVmR)^2$. *Data:* The square of SVmR would be calculated. The values for SBP, SVmR, and $SVmR^2$ would be entered. *Interpretation:* The final predictive equation is $SBP = -2.92 + 289.00(SVmR) - 414.42(SVmR)^2$. The fit is shown in the figure. The *p*-values for tests of both models are less than 0.001, which is significant. The modeled prediction accounts for the majority of variability in both models; SVmR is a major predictor in either model. Moving from the simple model to the parabolic model increases the R^2 from about 53% to about 68%, indicating that the curved model is the better fit.

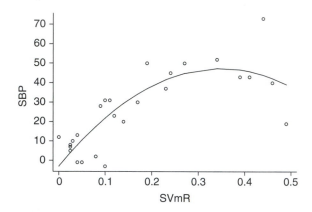

24.6. *Model:* $y = \beta_0 + \beta_1 x_1 + \beta_2 x_2$. *Data:* Enter data for the triple-hop distance using the operated leg into the *y* position, enter data using the nonoperated leg into the x_1 position, and enter the time to hop using the nonoperated leg into the x_2 position. *Interpretation:* The predictive ability using both *x*'s is real and clinically useful. Dropping x_2 should be considered. The reduced model also is significant and almost as clinically useful. The predictive equation is $y = -300.868 + 1.466x_1$. For a nonoperated leg hop distance of 504 cm, the predicted distance on the operated leg is 438 cm, which agrees well with the observed 436 cm.

24.7. $\beta_0 + \beta_1 x_1 + \beta_2 x_2 + \beta_3 x_3 + \beta_4 x_4 + \beta_5 x_5$. (1) Yes. (2) Not very. (3) Drop DRE and age. Reduced model: (1 to 3) The model is still a significant predictor and is still not a very useful one. (4) The resultant R^2 is still about 15%. No predictive capability was lost by the reduction. (5) The right side of the model becomes $-0.8424 + 0.1056 \times 9.6 - 0.0315 \times 62 + 0.7167 \times 1 = -1.0649$. Substitution for the exponent of *e* in Eq. (24.25) yields $p_m = 0.2564$. The probability is about 0.25, one chance in four, that this patient's biopsy result is positive. (Of possible interest: it was not positive.)

Chapter 25

Survival and Time-Series Analysis

25.1. TIME-DEPENDENT DATA

Time-dependent data are common in medicine. Many monitors in surgery, including a variety of cardiac and pulmonary measures, are time-dependent data. Cardiac stress measures are time-dependent data. Electroencephalograms are time-dependent data. Many epidemiologic measures are time dependent, including survival.

TIME SERIES

A large number of observations recorded through time on the same variable usually is called a time series in statistics. Although taken literally, time-dependent data would include before and after observations and posttreatment follow-up; data sets such as these are not considered time series. To be susceptible to time-series analysis, a sequence of observations should number in the order of multiples of 10. Survival data sets often have fewer observations than a true time series. However, both share a focus on a process through time rather than on a specific event and thus are included together in this chapter.

25.2. SURVIVAL CURVES: ESTIMATION

SURVIVAL DATA

Basic survival data are times to death of members of a cohort (demographic group). Note that "survival" is not restricted to remaining alive but may represent other terminal events (e.g., surviving without the recurrence of cancer); we use the remaining alive connotation in this chapter for convenience. The proportion of a cohort's survival through time can be calculated from the raw data. Section 9.7 introduces the historic life table. This table gives the proportion of a cohort surviving to the end of each time interval. If no patients are lost to follow-up, the proportion surviving is simply surviving number divided by initial number.

CENSORED DATA

Most survival data sets include some patients who are lost to follow-up before death. Their data are designated as *censored*. The method given in Section 9.7 includes the treatment of censored data: using them while they are known to be alive and removing them from the database when they are lost.

DATA AND CALCULATIONS REQUIRED FOR A LIFE TABLE

A sample life table is given as Table 25.1.[42] As described in Section 9.7, basic data for a life table on *n* patients are the following: time intervals; *begin*, the number at the beginning of each time interval; *died*, the number dying in each time interval; and *lost*, the number lost to follow-up in each time interval. The rest of the table is calculated from these basic data. It was pointed out that the method used here is the simplest in that it assumes the time of death or loss to occur at the end of the time interval. Other methods giving sophisticated adjustments exist. For convenience, a second line is entered for a time interval having censored data. Calculation for a column 5 entry is *end* = *begin* − *died* − *lost*. Calculation for a column 6 entry is *S* (the proportion surviving) = *S* for last period × (*end* for this period ÷ *end* for last period). The reason for this comes from a basic law of probability: the probability that two independent events occur together is the product of their probabilities. If a coin is tossed twice, the chance of a head appearing on both tosses is the chance of a head appearing on the first multiplied by the chance of a head appearing on the second, or 1/2 × 1/2 = 1/4. Similarly, the chance of surviving to the end of the current interval is the chance of surviving to the beginning of the interval multiplied by the chance of surviving during the interval.

LIFE TABLE FOR MEN WITH DIABETES MELLITUS

Table 25.1 provides basic and calculated data for a life table on the survival of 319 men. (The equivalent table for women is presented as Table 9.3 in Exercise 9.7.) In the first period (>0–2 years), *died* = 16 and *lost* = 0; thus,

Table 25.1

Survival Data of 319 Men in Rochester, Minnesota, with Adult-Onset Diabetes Mellitus Who Were Older Than 45 Years at Onset During 1980–1990

Interval (years)	Begin	Died	Lost	End	S (survived)
0 (outset)	319	0	0	319	1.0000
>0–2	319	16	0	303	0.9498
>2–4	303	19	0	284	0.8902
>4–6	284	19	0	265	0.8306
>6–8	265	8	0	257	0.8055
	257	0	2	255	
>8–10	255	6	0	249	0.7865

end $= 319 - 16 - 0 = 303$. $S = 1.00 \times (303/319) = 0.9498$. In the second period, $S = 0.9498 \times (284/303) = 0.8902$. In the third period, $S = 0.8902 \times (265/284) = 0.8306$. In the fourth period, eight patients died and two were lost to follow-up, which were separated on two lines (the group of those who died is first). S for that period is $0.8306 \times (257/265) = 0.8055$. At the end of that period, we subtracted the 2 *lost* patients, leaving 255. Because we assumed that they remained alive to the end of the period, they did not reduce the survival but they are removed for calculating survival in the next period. At the end of 10 years, about 79% of the men remain alive. This may also be interpreted as follows: The probability that a man randomly chosen at the outset will remain disease free longer than 10 years is estimated to be 0.79.

SURVIVAL CURVES

Section 9.7 introduces the graphical display of survival information. That section mentions that a method of estimating survival functions developed by E. L. Kaplan and P. Meier in 1958 is more accurate than a life table and should be used when statistical software is available. Lacking that, or for the purpose of understanding the concepts, a simple mode of display is just to graph the survival data from the life table against the time intervals. A survival datum stays the same for the period of an interval, dropping at the end, which produces a stepped pattern. The survival curve for Table 25.1 is shown as Fig. 25.1. Note that the number lost to follow-up (censored) is shown as a small integer over the line for the period in which they were lost, distinguishing those lost from those dying.

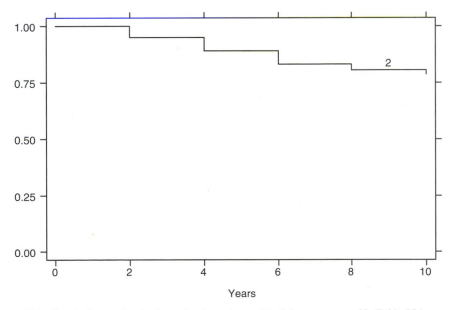

Figure 25.1 Survival curve for the data of male patients with diabetes presented in Table 25.1.

CONFIDENCE INTERVALS

The survival curve shows estimates of a population's survival pattern based on the data from a sample. How confident of that estimate are we? As with many other estimates, we can find confidence intervals. In this case, there will be a confidence interval on each survival proportion, which leads to confidence curves enclosing the survival curve. The confidence curves have a similar stepped appearance. Different methods exist to calculate the confidence intervals on survival proportions. The most general are evolved from a method originated by M. Greenwood in 1926, but these are difficult to calculate. They are good to use if statistical computer software offers the capability. Otherwise, a much simpler method by Peto and colleagues[59] may be used, as long as the user remembers that they are rougher approximations. The pattern of a confidence expression is the same for confidence intervals on other types of estimates, namely that given as Eq. (4.2). (The probability that a population statistic from a distribution of estimates of that statistic is contained in a specified interval is given by the area of the distribution over that interval.) Section 13.6 states that a proportion is distributed approximately normal for moderate to large samples. If S_i denotes the estimate of the proportion survival at the end of interval i and SEE denotes the standard error of the estimate, the probability is $1 - \alpha$ that the true proportion survival is bracketed by $S_i \pm z_{1-\alpha/2} \times$ SEE. For 95% confidence, $z_{1-\alpha/2} = 1.96$. SEE, as for proportions discussed in Section 13.6, depends only on S_i and the number at the beginning of each time period, for example, n_{i-1}. (n_{i-1} replaces *begin* to keep track of the interval involved.)

$$\text{SEE} = S_i \sqrt{\frac{1 - S_i}{n_{i-1}}} \tag{25.1}$$

The 95% confidence interval on the true survival proportion for each time period becomes

$$S_i \pm 1.96 \times \text{SEE} = S_i \pm 1.96 \times S_i \sqrt{\frac{1 - S_i}{n_{i-1}}}. \tag{25.2}$$

Because these confidence intervals are not exact, we must use common sense to avoid letting them exceed 1 or fall below 0 in reporting or graphing them.

Example

Table 25.2 shows confidence intervals calculated using Eq. (25.2) for the 319 diabetic men listed in Table 25.1. Figure 25.2 shows the Kaplan–Meier survival curve of Fig. 25.1 with the confidence intervals as calculated in Table 25.2.

Table 25.2

Survival Data of Table 25.1 with Confidence Intervals Calculated

Period (years)	Begin (n_{i-1})	S_i	SSE	Confidence interval
0 (outset)	319	1.0000		
>0–2	319	0.9498	0.0119	0.9265, 0.9731
>2–4	303	0.8902	0.0169	0.8571, 0.9233
>4–6	284	0.8306	0.0203	0.7908, 0.8704
>6–8	265	0.8055	0.0218	0.7628, 0.8482
>8–10	255	0.7865	0.0228	0.7418, 0.8312

SSE, sum of squares for error.

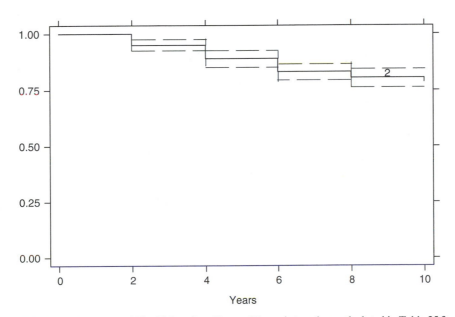

Figure 25.2 Survival curve of Fig. 25.1 enclosed by confidence intervals as calculated in Table 25.2.

The user may note that the confidence limits associated with a particular period in Table 25.2 are drawn about the succeeding period in Fig. 25.2. The reason is that the proportion survival is related to the end of the period and is maintained until the next death figure, at the end of the following period. If we had data giving the exact time of death rather than a period during which it occurred, this pictorial lag would not occur.

Exercise 25.1. Life table data for survival of diabetic women[42] from 1970 to 1980 are shown in Table 25.3. Complete the life table. Graph a survival curve. Calculate and graph the 95% confidence intervals on this survival curve.

Table 25.3

Survival Data of 274 Women in Rochester, Minnesota, with Adult-Onset Diabetes Mellitus Who Were Older Than 45 Years at Onset During 1970–1980

Interval (years)	Begin (n_{i-1})	Died	Lost	End (n_i)	S_i (survived)	Confidence interval
0 (outset)	274	0	0	274	1.0000	
>0–2	274	14	0			
>2–4	260	13	0			
>4–6	247	14	0			
>6–8	233	18	0			
	215	0	1			
>8–10	214	19	0			

25.3. SURVIVAL CURVES: TESTING

EXAMPLE POSED: SURVIVAL OF MEN VERSUS WOMEN WITH DIABETES

Figure 25.3 superposes the survival curves for 319 men (see Fig. 25.1) and 370 women (see figure in answer to Exercise 9.8) with diabetes mellitus during the 1980–1990 decade.[42] We see a difference by inspection, but is this difference significant?

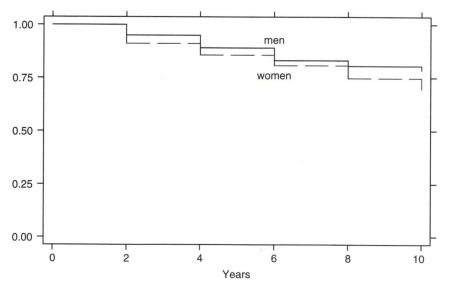

Figure 25.3 Survival curves for 370 women and 319 men in Rochester, MN, with adult-onset diabetes mellitus who were older than 45 years at onset during the 1980–1990 decade.

Method

Are Two Survival Curves Different?

One statistical procedure that answers the question of whether or not two survival curves are different is the *log-rank test*. This test uses a chi-square (χ^2) statistic based on the difference between the observed survival and the survival that would be expected if the curves were not different, in the same way that a χ^2 goodness-of-fit test [see Eq. (19.2)] uses the sum of squares of weighted differences between the observed and expected curves. However, the χ^2 of the log-rank test is more complicated to calculate. It uses matrix algebra, multiplying vectors of differences for the time periods and the matrix of variances and covariances. A statistical software package should be used for this calculation. The result of the calculation is a χ^2 statistic, which may be compared with a χ^2 critical value from Table III (see Tables of Probability Distributions) for degrees of freedom (df) = number of survival curves − 1. When the two curves of Fig. 25.3 are tested, $df = 1$. If the calculated statistic is greater than the critical value, the curves are significantly different.

The Log-Rank Test Compared with Other Tests of Survival Curves

Two alternative tests of survival curves that might be considered for use are the Mantel–Haenszel test and the Cox proportional hazards test. The Mantel–Haenszel test is almost the same as the log-rank test. Indeed, both Mantel and Haenszel contributed to the theory of the log-rank test. However, the Mantel–Haenszel test is restricted to two curves, whereas the log-rank test may use more than two curves. Therefore, the log-rank test is recommended. The Cox proportional hazards test allows the risk for death to vary within the model, whereas the log-rank test assumes it to be the same throughout. The Cox assumption leads to rather complicated mathematics and the methods must be used carefully and exactly to avoid a number of potential criticisms. The user is advised to seek guidance from a biostatistician if the death rates vary within the data set.

Example Completed: Survival of Men Versus Women with Diabetes

We use the log-rank test to compare survival for men and women with diabetes. From Table III, giving χ^2 values, the critical χ^2 for 1 df is 3.84. The log-rank test yields $\chi^2 = 7.2$, which is greater than 3.84. We conclude that men have a significantly better survival outcome than women. (The actual log-rank $p = 0.007$.)

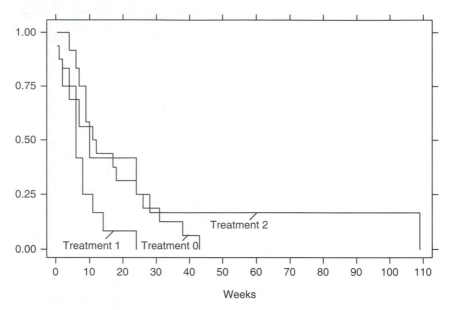

Figure 25.4 Kaplan–Meier survival curves of 44 patients with advanced cancer. Treatment 0 implies no treatment; treatments 1 and 2 are experimental treatments of unknown efficacy.

ADDITIONAL EXAMPLE: TESTING THREE CANCER SURVIVAL CURVES

Figure 25.4 shows survival curves for patients with advanced cancer simulated for classroom use by a radiation oncologist.[34] In this example, treatment 0 implies no treatment was given to the patient and treatments 1 and 2 refer to experimental treatments of unknown efficacy. Because three curves are being compared in the log-rank test, $df = 2$. The critical χ^2 for 2 df from Table III is 5.99. A software package yields $\chi^2 = 6.35$. Because the calculated χ^2 is larger than the critical χ^2, the null hypothesis of no difference is rejected. (The actual $p = 0.042$.) It appears that treatment 1 is worse for survival than no treatment.

Exercise 25.2. The survival data[42] for 370 diabetic women during the 1980–1990 decade are given as Table 9.3 (in the answer to Exercise 9.7). The equivalent data on 274 women for the 1970–1980 decade appear in Table 25.7. A log-rank test of the difference between the two curves yields $\chi^2 = 0.63$. Find the critical value, compare the χ^2 values, and interpret the resulting decision.

25.4. SEQUENTIAL ANALYSIS

Suppose you were advising Ignaz Semmelweis in Vienna in 1847 on an experiment to show the effect on patient mortality of hand washing between patients.[65] You plan to randomize 1000 patients into two arms—one treated by physicians who wash between patients and the other treated by physicians who do not—and then plan to compare the mortality rates. After data for 100 patients in each arm have been analyzed, you have found only 1 death from sepsis in the arm with physicians who practiced hand washing but 18 deaths in the arm with physicians who did not wash between examining patients. The probability that such a difference would occur by chance is less than 1 in 10,000. If you allow the experiment to continue at these rates, 68 additional patients in your study will die who can be saved if you immediately institute hand washing between patients for all physicians. The ethical challenge is obvious. In addition to mortality, carrying studies past clear decision points may cause unnecessary morbidity, inconvenience to patients, increased study costs, unnecessary demand on facilities, and delay in reporting useful information. In sequential analysis, a decision is made after acquiring each datum regarding whether to (1) accept the null hypothesis, (2) accept the alternate hypothesis, or (3) continue sampling. This method, however, has been available at least since Wald's pioneering 1947 book[88] and has hardly proved to be a panacea. Limitations exist, and they are examined here together with the method.

EXAMPLE POSED: EFFECT OF HAND WASHING ON DEATH RATE

Sequential analysis might have helped Semmelweis had he done the study just described. As in the introductory paragraph, we suppose that 1 death resulted in the first 100 patients of the experimental group. For illustration, we let this death occur early, as the fourth patient in the sequence. The null hypothesis states that the probability of a patient dying is not different from the old (nonwashing) procedure, or H_0: population death rate = 0.18. The alternate hypothesis poses the new (washing) rate, or H_1: population death rate = 0.01. At what point in the sampling process can we accept one of the hypotheses?

METHOD

α, β, and n in Traditional Hypothesis Tests

A hypothesis test compares a decision statistic, for example, a sample mean hematocrit (Hct) or proportion of surgical patients having complications, with a pair of hypotheses about the population value of this statistic and selects the appropriate hypothesis based on the size of the risk of being wrong. The relationship among risks and probability distributions in making this selection is shown in Fig. 25.2. A critical value divides possible outcomes of a decision statistic into acceptance and rejection regions. The outcome of the test is controlled by three quantities: α, the risk of selecting H_1 when H_0 is true; β, the risk of selecting H_0 when H_1 is true; and n, the sample size. In traditional hypothesis testing, β cannot be calculated because the alternate value of the population parameter being tested (a mean or proportion, for example) is unknown; therefore, α and n are selected, and β falls where it may. In the rare case in which a specific alternate value of the decision statistic is known, both α and β

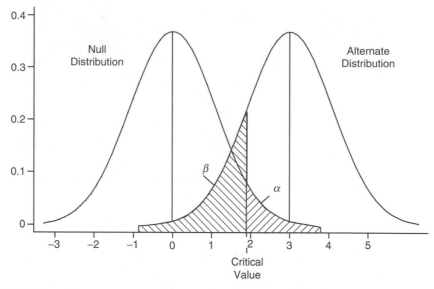

Figure 25.5 Depiction of null and alternate distributions along with the error probabilities arising from a statistical decision. A critical value of a decision statistic divides the horizontal axis into two regions. If the sample value of the decision statistic falls to the right of the critical value, H_0 is rejected; if the sample value falls to the left, H_0 is not rejected.

are selected and the minimum value of n may be calculated. (This is, in fact, the basis of minimum sample size estimation, the so-called power analysis, in which a clinically relevant alternate hypothesis value of the decision statistic is conjectured.)

Concept of Sequential Analysis

Sequential analysis requires the alternate hypothesis value to be known. In this case, α and β can be specified and the sample size depends on them. Two critical values are chosen and the test's outcome is assigned to one of three regions, that is, one region for each for the three decisions: accept H_0, accept H_1, or continue sampling. The relation among risks and probability regions are illustrated in Fig. 25.6, which is Fig. 25.5 altered to allow the third decision. Because now the sample size is not an input but depends on α and β, it may be treated as a variable. Let us designate this increasing sample size as k to distinguish it from a fixed n. Initially, a decision is made for a small value of k. If the decision is to continue sampling, k is increased and the decision is made again. This continues until the decision statistic falls in either the acceptance or rejection region (see Fig. 25.7). The data points show possible outcomes of the decision statistic for increasing k. When a value of the statistic falls outside the continue-sampling corridor, the final decision is made and sampling is stopped.

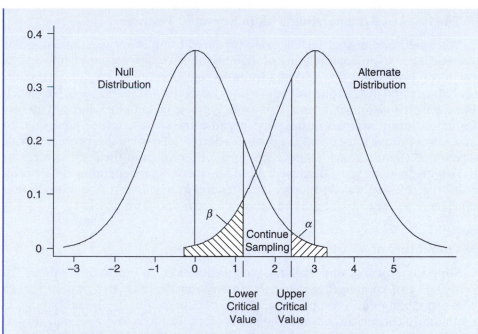

Figure 25.6 Diagram of probability distributions associated with the null and alternate hypotheses with α and β error risks shown. As long as the decision statistic (calculated from the sample of increasing size) falls between the shaded areas, sampling continues.

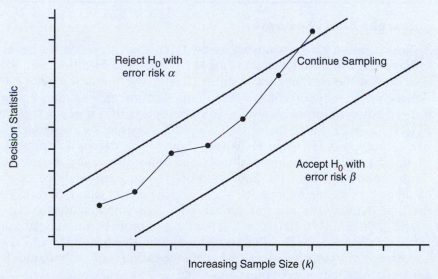

Figure 25.7 Graph of the decision statistic plotted against k, the increasing sample size. As long as the statistic falls between the lines, sampling continues. If the statistic falls above the upper line, H_0 is rejected with risk α of making the wrong decision. If the statistic falls below the lower line, H_0 is accepted with risk β of making the wrong decision.

The Issue of Accumulating Risk in Repeated Testing

We noted earlier that testing using repeated samples on the same decision statistic leads to an accumulating probability of error. For example, if five tests, each with probability $\alpha = 0.05$ of a spurious rejection of H_0, are conducted, the probability that at least one will be spurious becomes $1 - (0.95)^5 = 0.226$, nearly 1 in 4, which is no longer an acceptable risk. Sequential sampling does not involve testing different samples in this way, because the same sample augmented by new observations is used, but the problem persists. To solve this problem, Wald developed the test using a theoretical basis different from the least squares we have seen on many tests. Because k could be expressed as depending on the error risks, the solution was developed for a general k, whatever value it might take on. Thus, the sequential test is based on a fixed α and β for any k.

Probability of a Particular Sequence Occurring

Suppose a form of surgery led to complications (C) a fourth of the time. The probability of complications, say, P, on any randomly chosen patient is 0.25. The probability of no complication (N) is, of course, $1 - P$. The probability of a randomly chosen surgery resulting in no complications and a second surgery resulting in complications, that is, the sequence NC, is $(1 - P) \times P = 0.75 \times 0.25 = 0.1875$. For four patients, the sequence $NCNN$ has probability $(1 - P) \times P \times (1 - P) \times (1 - P) = P \times (1 - P)^3 = 0.25 \times 0.75^3 = 0.1055$. In this fashion, we can calculate the probability for any particular sequence. Note well that the probability is for this particular sequence, not for any other ordering of outcomes with 3 N's and 1 C.

Test of a Binary Sequence

Now suppose we are studying a new type of surgery that we believe to have the lower complication rate 0.10. We pose H_0: the complication rate of the new surgery is still 0.25, and H_1: the new rate is 0.10. We record 1 as the outcome when a complication appears and 0 when it does not. The probability of a complication is 0.25 if H_0 is true, or $P(1|H_0) = 0.25$. Similarly, $P(1|H_1) = 0.10$. As usual, we assign α, the risk of accepting H_1 when H_0 is true, and β, the risk of accepting H_0 when H_1 is true. The sequential test's decision statistic, say, D, is based on the following ratio: the likelihood of H_1 being true divided by the likelihood of H_0 being true for a given sequence of observations. For the first patient, the ratio of probabilities that complications occur is $P(1|H_1)/P(1|H_0) = 0.10/0.25 = 0.4$; the ratio of probabilities that complications do not occur is $[1 - P(1|H_1)]/[1 - P(1|H_0)] = 0.90/0.75 = 1.2$. For a sequence of k patients, we can calculate the ratio D_k of probabilities for any particular sequence of outcomes using

$D_k = D_{k-1} \times P(1|H_1)/P(1|H_0)$ if complications occur, *or*

$D_k = D_{k-1} \times [1 - P(1|H_1)]/[1 - P(1|H_0)]$ if complications do not occur.

(25.3)

Thus, the increasing sequences N, NC, NCN, $NCNN$ have decision statistics $D_1 = 1.2$; $D_2 = D_1 \times 0.4 = 0.48$; $D_3 = D_2 \times 1.2 = 0.576$; and $D_4 = D_3 \times 1.2 = 0.6912$. The critical values used to decide which decision region the D_k is in have been shown theoretically to be approximately the following, where the subscript a denotes acceptance and r, rejection:

Accept H_0 when $\qquad D_a \le \dfrac{\beta}{1 - \alpha}$

Reject H_0 when $\qquad D_r \ge \dfrac{1 - \beta}{\alpha}$ $\qquad$ (25.4)

Continue sampling otherwise.

One of these three choices is selected after each observation.

EXAMPLE COMPLETED: EFFECT OF HAND WASHING ON DEATH RATE

We posed H_0: death rate $= 0.18$, and H_1: death rate $= 0.01$. Of course, we do not know the actual sequence of outcomes, but our supposition poses it as in Table 25.4. (The data are consistent with historical record.) We select α, the risk of rejecting H_0 when it is true, as 0.05, and β, the risk of accepting H_0 when it is false, as 0.10. The two critical values are calculated from Eq. (25.4) as $D_a = \beta/(1 - \alpha) = 0.10/0.95 = 0.1053$ and $D_r = (1 - \beta)/\alpha = 0.90/0.05 = 18$. We designate a patient surviving as 0 and dying as 1. The first patient survived, so $D_1 = P(0|H_1)/P(0|H_0) = [1 - P(1|H_1)]/[1 - P(1|H_0)] = 1.2073$. The multiplying probability ratios used to find the succeeding D_k will be $P(1|H_1)/P(1|H_0) = 0.01/0.18 = 0.0556$ when the patient died and $P(0|H_1)/P(0|H_0) = 0.99/0.82 = 1.2073$ when the patient survived. We enter D_k in Table 25.4 using calculations from Eq. (25.3). By the rule of Eq. (25.4), we will accept H_0 the first time the decision statistic D_k becomes less than 0.1053, or we will reject H_0 the first time D_k becomes greater than 18. In Table 25.4, D_1 was 1.2073. After that, D_k starts to increase, drops dramatically with the death of patient 4, and then starts to build again. With patient 32, D_{32} has exceeded 18 and we reject H_0; hand washing has proved beneficial. We note not only that the number of patients required is far less than initially planned but also that much of this small sample requirement is due to the dramatic difference in the old and new death rates.

Table 25.4

Possible Outcomes from Semmelweis' Experiment with the Associated Decision Statistic[a]

k (observation no.)	1	2	3	4	5	6–29	30	31	32
Outcome	0	0	0	1	0	All 0	0	0	0
D_k		1.2073	1.4576	1.7597	0.0978	0.1179	...	13.1040	15.8204	19.1000

[a]Observation outcome is 0 for survival and 1 for death of the patient.

Test of a More General Sequence

For binary data (success–failure, occur–not, die–survive, etc.), the calculations of the basic probabilities usually are as simple as counting the number of occurrences, because there are only two outcomes. In the case of a test on continuous data, such as means or standard deviations, the probability ratio used to multiply D_{k-1} to find D_k varies with each sampling and must be recalculated for each additional observation. For a test on means, we know that the distribution of the means is approximately normal (see discussion of Central Limit Theorem in Section 2.8), so we use the normal distribution in calculations. For a test on standard deviations (actually on variances), we would use the χ^2 or F distributions. More detail is required than is given in the tables of this book. We need electronic tables or very complete printed tables.

Steps to Conduct a More General Sequential Test

The sequential test procedure, extended to include continuous statistics, may be easier to follow step by step. Let us denote by x_k the observation on the kth patient. This will be the reading on whatever is being studied, such as Hct, PSA, died: 1 or survived: 0, and so forth. Denote by $P(x_k|H_0)$ the probability of x_k occurring if H_0 is true and denote by $P(x_k|H_1)$ the probability of x_k occurring if H_1 is true.

(1) Pose the null and alternate hypotheses.
(2) Assign error risk values to α and β and calculate the critical values.
(3) Find the probability ratio $D_1 = P(x_1|H_1)/P(x_1|H_0)$.
(4) Fill in Table 25.5, calculating

$$D_k = D_{k-1} \times P(x_k|H_1)/P(x_k|H_0). \tag{25.5}$$

(5) After each patient, make a decision using the criteria of Eq. (25.4). If H_0 is accepted or rejected, stop; if not, sample another patient and repeat steps (4) and (5).

Table 25.5

Sequence of Observations with Associated Decision Statistic Values

k (observation no.)	1	2	3	4	5	6	...
Observation outcome							
D_k							

Step (4) Specified for a Test of Proportions

If the observation is binary, x_k can be expressed as either 0 or 1. The calculation of D_k will be as in Eq. (25.3).

Step (4) Specified for a Test of Means

Denote the mean under the null hypothesis as μ_0 and the mean under the alternate as μ_1; denote the variance as σ^2. Simplifying the ratio of two normal probabilities yields the multiplication increment in Eq. (25.5) as

$$\exp\left\{-\frac{1}{2\sigma^2}\left[(x_k - \mu_1)^2 - (x_k - \mu_0)^2\right]\right\} \tag{25.6}$$

(where "exp{...}" implies "$e^{\{...\}}$").

A Graphical Solution

If the user wishes, it is possible to solve the expressions in Eq. (25.4) for a general k number of observations, which results in two straight-line equations, and set up a graph as in Fig. 25.7. It is more work for binary data than is calculation, but often less for continuous data, because equations can be derived for drawing the critical value lines.

The Average Sample Size Will Be Smaller in the Long Term

We want to know what the average sample size using sequential analysis would be and how it would compare with fixed sample size methods. However, the average sample size depends on the population's value of the decision statistic, and we have only an estimate that is changing with each new sample element. It is possible to find the average sample size based on a conjecture of what the population value would be, but the resulting size would then be a conjecture itself. In general, in the long term, sample sizes from sequential sampling techniques are smaller than those from fixed sample size techniques.

Disadvantages of Sequential Analysis

If the sample size generally will be smaller with sequential analysis, why is it not used more frequently? The primary drawback is the requirement to specify the alternate hypothesis value of the decision statistic, which is seldom known. Why would Semmelweis have believed that hand washing would have reduced the mortality rate to 1% as opposed to 5% or 10%? In addition, the method is demanding computationally and has not found its way into many statistical software packages.

ADDITIONAL EXAMPLE: SILICON LEVEL IN BREAST IMPLANTS

In DB5, the change in plasma silicon levels of 30 women was measured after they received silicone breast implants. Mean difference was 0.0073, and standard deviation was 0.1222. A quick plot of the data shows approximate normality. A one-sample t test on the difference with a null hypothesis of no difference yields $p = 0.745$. The mean difference is far from significant, and most investigators would conclude that implantation did not change the plasma silicon level. However, did we need $n = 30$? Let us conduct a sequential test following the five steps. (1) Formulate the hypotheses. μ is the population mean difference; H_0: $\mu = 0$. The alternate hypothesis is a problem, because we have no idea what the population mean difference is if it is not 0. Let us conjecture H_1: $\mu = 0.1$. We can make a test and come to a conclusion, but it is a test of only *this* alternate, not any other. Furthermore, we do not know σ^2. Suppose we find some published data from prior studies indicating that the standard deviation is about 0.2. (2) Let us choose $\alpha = 0.05$, which is commonly used in medicine. The implication of a Type I error is to say that women receive no increase in plasma silicon level from implantation when, in fact, they do; this would encourage implantation. A Type II error would be to say that women do receive an increase in plasma silicon level from implantation when they do not; this would discourage implantation. Because the decision about breast implants is more often a cosmetic than a medical decision, a quantification of the difference in loss from the two types of error is not clear-cut; let us choose β to be the same as α. From Eq. (25.4), the two critical values are $\beta/(1-\alpha) = 0.05/0.95 = 0.0526$ and $(1-\beta)/\alpha = 0.95/0.05 = 19$. When the accumulating D_k falls below 0.0526 or exceeds 19, the test is concluded. (3) By substituting μ_0, μ_1, and σ in Eq. (25.6) and simplifying, we find that expression to be $\exp\{-12.5(0.01 - 0.2x_k)\}$. From the data of DB5, the first difference is -0.06. $D_1 = 0.7595$. The second difference is -0.11. The multiplying probability ratio is 0.6703. $D_2 = 0.7595 \times 0.6703 = 0.5092$. The progress through the sequence can be seen in Table 25.6, varying up and down until decreasing to less than 0.0526 in D_{18}. On the 18th patient, the test results in accepting H_0: $\mu = 0$; the plasma silicon level is not changed by the implantation.

Table 25.6

DB5 Data Showing Sequence of Plasma Silicon Differences Before Minus After Implantation, Multiplying Probability Ratio, and Decision Statistic D_k[a]

k (patient no.)	Plasma silicon difference	Multiplying probability ratio	D_k (decision statistic)
1	−0.06		0.7596
2	−0.11	0.6703	0.5092
3	0.29	1.8221	0.9277
4	0.08	1.0779	1.0000
5	0.11	1.1618	1.1618
6	0.17	1.3499	1.5683
7	0.02	0.9277	1.4549
8	−0.03	0.8187	1.1912
9	−0.11	0.6703	0.7984
10	0.04	0.9753	0.7787
11	−0.12	0.6538	0.5091
12	−0.13	0.6376	0.3246
13	−0.23	0.4966	0.1612
14	−0.26	0.4607	0.0743
15	0.04	0.9753	0.0724
16	0.06	1.0253	0.0743
17	−0.07	0.7408	0.0550
18	−0.04	0.7985	0.0439

[a] H_0 is accepted when D_k drops below 0.0526 or is rejected when D_k exceeds 19. H_0 is accepted with patient 18.

Exercise 25.3. From the data of DB12, 5.2% of nonintubated patients with carinal resection die. Through experience, the surgeon believes that intubated patients have a much higher death rate, appearing to have a 50:50 chance. You want to test this belief. Take intubated patients from the data set one by one and decide from a sequential test whether their death rate is the same as for nonintubated patients or is 50%.

25.5. TIME SERIES: DETECTING PATTERNS

"Long-series" data appear most often as a sequence of data through time, so most analytic methods for long-series data are found under the *time series* heading. Types of long sequences other than time-dependent sequences are also important and are illustrated here by examples, but time as the independent variable is used for convenience. Some examples of time-dependent data from recent literature are incidence of nosocomial infection in a hospital, wave-form analysis of neurologic potentials in infants, the course of Ménière's disease through time, the course of opioid use during bone marrow transplantation, seasonal variation in hospital admissions for specific diseases, critical care monitoring, analysis of respiratory cycles, velocity of eye movements during locomotion, trends in juvenile rheumatoid arthritis, movement effects in magnetic resonance imaging, and assessing the level of anesthesia. Indeed, time series are encountered in almost every field of medicine.

QUESTIONS ADDRESSED BY TIME SERIES

The questions most commonly asked of time series are as follows: (1) Is there a trend in the event through time? Identifying the best-fit regression curve will give a view of trend through time (see Chapters 8 and 24). (2) Is the event cyclic through time? Autocorrelation (see introduction of topic in Section 9.7) may detect cycles and describe their nature. (3) Is the event correlated with other events in time? Cross-correlation (which also is introduced in Section 9.7) may detect other events that follow related patterns through time. (4) Is there a point in time at which the event changes its pattern? Change-point estimation is addressed in Section 25.6.

THE TIME-SERIES METHODS INTRODUCED HERE ARE BASIC

As with much of statistics, there exist more sophisticated methods than are introduced in this chapter. For example, a term sometimes encountered in contemporary medical articles is *autoregressive moving average* (ARMA), a weighted moving mean adjusted for the influence of autoregression. This book will have accomplished its purpose if the user understands the basic ideas of time series and can appreciate what can be done.

THE NEED FOR SMOOTHING PROCESSES

Often, a potential pattern of a sequential event is obscured by variability and is better discerned if the variability about the pattern is reduced, or "smoothed." Smoothing is unnecessary for very small samples, because patterns are not perceptible anyway, whatever the method. Smoothing becomes useful when used on sequences with dozens, or better hundreds, of data.

EXAMPLES POSED: PROSTATE-SPECIFIC ANTIGEN LEVEL, PROSTATE VOLUME, AND AGE

Sequential data from DB1 are shown in Figs. 25.8A and 25.8B. In addition to the major ideas, these examples illustrate long-series data that are not time series.

Relation of Prostate Volume to Prostate-Specific Antigen Level

Figure 25.8A shows the prostate volumes of 301 men with urologic problems as related to the rank order of PSA levels in increasing sequence. Any pattern of change in volume with PSA level is rather well obscured by the variability.

Relation of Prostate-Specific Antigen Level to Age

In Fig. 25.8B, PSA levels from the same men are shown in the sequence of the men's ages. Is there a pattern of change in PSA level with increasing age?

METHOD

Moving Samples

Smoothing is based on the concept of a moving sample. Consider a sequence composed of many data, perhaps 100, and subsets of n data, say, 10. We take the first 10 data and calculate some statistic, perhaps an average. Then we drop off the 1st datum, add the 11th datum, and recalculate the average. We continue, dropping the 2nd and adding the 12th, dropping the 3rd and adding the 13th, and so on. We have a sample moving through the sequence, always with size n, providing a statistic moving through the sequence. We may name this sequence of samples a *moving sample* and also apply "moving" to the statistic, as a *moving average*. The moving average mutes the variability that obscures patterns.

Moving Averages

There are two relevant moving averages: the moving mean and the moving median. The moving mean subdues the variability, bringing extremes closer to the overall path through time but allowing them some influence. In a moving mean of 20 in a sequence of 1000 visual-evoked potential readings, a "hump"

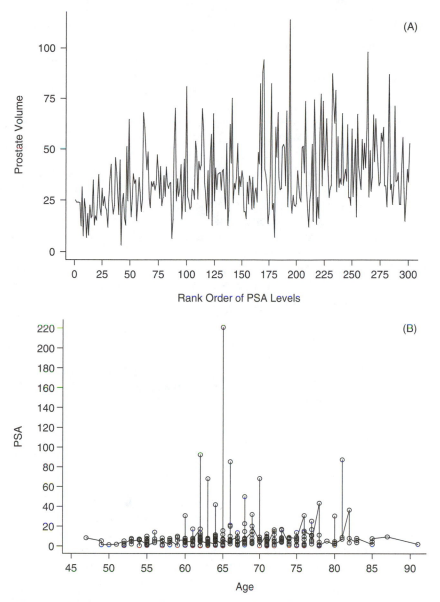

Figure 25.8 (A) Prostate volume of 301 men as depending on increasing levels of prostate-specific antigen. (B) Prostate-specific antigen levels of 301 men as depending on increasing age.

of 50 beats due to a stimulus image being presented will still appear, although it will be somewhat muted. However, a single extreme reading caused by a blink will also appear as a slightly enlarged mean. The moving median, even one as small as three readings, completely eliminates single outliers, because the greatest reading of the three will always be replaced by the middle one. Thus, we would think of a moving mean to smooth but not eliminate causal influences and of a moving median to remove single outliers. Other more subtle benefits will appear to the user with practice.

Calculation

A moving mean or median is just the ordinary mean or median of the moving sample. However, because the sample drops the leftmost reading and adds one to the right end, most of the last calculated average will be the same; it needs only to be adjusted for the two readings changed. If a moving sample of readings x is of width n with rightmost element $k-1$ (it is composed of readings $x_{k-n}, x_{k-n+1}, \ldots, x_{k-1}$), and we denote its mean by m_{k-1}, m_k will be $m_{k-1} + (x_k - x_{k-n})/n$. If we had readings 2, 4, 3, 5, 1, 6, and wanted a moving mean of 3, m_3 would be $(2+4+3)/3 = 3$ and m_4 would be $3 + (5-2)/3 = 4$. The values of a moving median of 3 would be 3, 4, 3, 5. There are missing elements at the beginning unless they are supplied by specially defined values. For example, in the moving mean of 3 from the sequence just illustrated, the m_2 might be twice the first value plus the second value divided by 3, or $(2 \times 2 + 4)/3 = 2.67$. Some statistical software packages and even Microsoft Excel provide moving mean capability. Usually, these packages place the moving average at the center of the moving sample. The scheme used here will aid in testing (see the next section). To plot this scheme with means at the center of the moving average, you just plot m_k at position $k - (n-1)/2$.

Serial Correlation

Other types of patterns we might wish to discern may be detectable using serial correlation. A relationship between two variables through time may be identified by cross-correlation, and the repetition of a pattern, or periodicity, by autocorrelation. Chapter 9 introduces these concepts in connection with epidemiology, but they may also be useful in clinical medicine.

Cross-correlation

If the matching data sets are taken sequentially through time, the correlation between them is termed *cross-correlation*. It tells us how closely related the two variables are through time. Blood pressure and cardiac output through a sequence of exercise and rest would be expected to be cross-correlated. The correlation is based on the relationship between them, not on their individual behavior through time. Thus, if they both rise and fall together, the correlation is high whatever the time-dependent pattern may be. The calculation may be easily understood if we think of lining the two data sets in adjacent columns of a table and calculate the ordinary correlation coefficient between the columns. Also, cross-correlation can be calculated with one of the sets lagged behind the other. For example, the appearance of symptoms of a disease having a 2-week incubation period can be correlated with exposure to the disease, where exposure observations are paired with the symptom observations that occurred 2 weeks later. By varying the lag, we may be able to find the incubation period. For example, bacterial vaginosis has been found to facilitate acquired immunodeficiency syndrome. We would expect to see a cross-correlation between incidences of the two diseases, although not a high one because each occurs alone. In this case, the issue is not the size of the coefficient but the lag at which it is maximum, which might provide insight into the nature and timing of exposures.

Again, the calculation may be thought of as finding the ordinary correlation coefficient between variables listed in two columns of a table, but now one column is slid down the table a number of rows equal to the desired lag.

Autocorrelation

Observations through time may be correlated with a lagged version of themselves. If the autocorrelation coefficient retreats from 1.00 as the lag increases but then returns to almost 1.00, we know that we have a periodically recurring disease, as in symptoms of malaria. If the observed lag is 12 months, the disease is seasonal. In the additional example of Section 23.4, we looked at aspergillosis as a seasonal phenomenon and found that mean incidence by season fitted a sine wave with a period of 1 year. We ask if older patients are more susceptible in winter. If we calculate the autocorrelation of mean age of infected patients per season with mean age 1 year later, we find the coefficient to be -0.22, which verifies what we expected: age of infected patient is not seasonal. We also ask if the percentage infected by sex is seasonal, which we do not expect. The autocorrelation for percentage male with a 1-year lag yields an autocorrelation coefficient of -0.14, indicating that sex is not seasonal. Interestingly, the cross-correlation coefficient of age and percentage male is 0.58, indicating that older patients tend to be male.

EXAMPLES COMPLETED: PROSTATE-SPECIFIC ANTIGEN LEVEL, PROSTATE VOLUME, AND AGE

Relation of Prostate Volume to Prostate-Specific Antigen Level

We need to smooth the data to perceive a pattern. We create a moving average (more precisely, a moving mean) of $n = 17$. The choice of 17 is somewhat arbitrary here; too small an n leaves too much variability, and too large a sample leaves too few observations. The graph of that moving mean is superposed on the data in Fig. 25.9A. The resulting graph still varies up and down in what seems to be random cycles, but a pattern begins to emerge. The prostate volume seems to be increasing with PSA level. Also, we might suspect that the increase is not quite in a straight line but might be slightly curved, concave downward. We could fit an ARMA model to the data, but we will retain the simplicity of regression methods already discussed in Chapter 24. The fit of a first-degree (straight-line) model yields $p < 0.001$; we have strong evidence that volume increases with PSA throughout the entire PSA spectrum. However, $R^2 = 0.185$, indicating that, although PSA level is a significant and important predictor of volume, it is certainly not a decisive predictor. Furthermore, we test the proposal of a curvilinear fit. We fit a second-degree regression curve to it, using the model $y = \beta_0 + \beta_1 x + \beta_2 x^2$ (see Chapter 23 on models). The resulting $p < 0.001$, as before, but R^2 remains at 0.185, indicating that the second-degree model prediction of the smoothed prostate volume by PSA rank adds nothing to the first-degree model. We conclude that prostate volume increases with PSA over the entire PSA range.

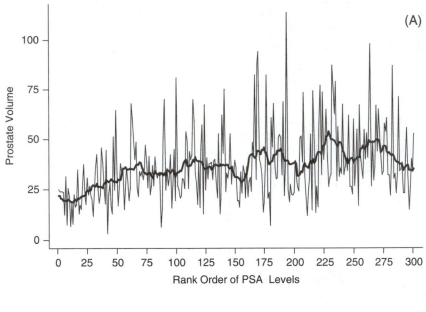

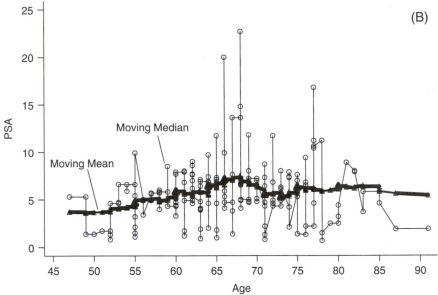

Figure 25.9 (A) Prostate volumes of 301 men depending on increasing levels of prostate-specific antigen (PSA) with a moving mean of 17 superposed. (B) Moving median (of 3) of PSA levels of 301 men depending on age with moving mean of 51 on the values of the moving median superposed.

Relation of Prostate-Specific Antigen Level to Age

A number of large PSA values obscures any pattern that might be present. Some investigators might suggest deleting the large levels, but doing so would no longer leave the same data set; we might even be obscuring the pattern we wish to see. Can we mute the large values without destroying the pattern? We apply a moving median of 3, which is a wide enough moving sample to reduce the most extreme levels, because most large levels do not occur next to each other. Then we apply a moving mean of 51, which is chosen somewhat arbitrarily. The moving median and moving mean are shown superposed in Fig. 25.9B. Note that the vertical axis has a much smaller scale, allowing the local variation to be perceived. Inspection of the shape of the smoothed PSA levels by age reveals that it is not far from a straight line, but slightly curved, opening downward. On fitting an ordinary second-degree regression model, we find the F statistic = 401.93, $p < 0.001$, and $R^2 = 0.730$—a rather good fit. We might interpret the result as PSA level increasing with age until the late 60s, at which time it becomes stable.

Cross-correlation of Prostate-Specific Antigen Level and Prostate Volume as Depending on Age

We noted that smoothed volume appears to depend on PSA level and smoothed PSA appears to depend on age. Let us consider both smoothed PSA level across age (as in Fig. 25.9B) and smoothed volume across age (a newly created moving average of 51). Do they have similar patterns of behavior as age increases? Figure 25.10 shows a plot of the smoothed paths of volume and PSA level plotted against increasing values of age. Both curves seem to increase at first, until age reaches the middle to late 60s, and then they level out. The cross-correlation between these curves is calculated to be 0.727, indicating that the perceived similarity of pattern is real.

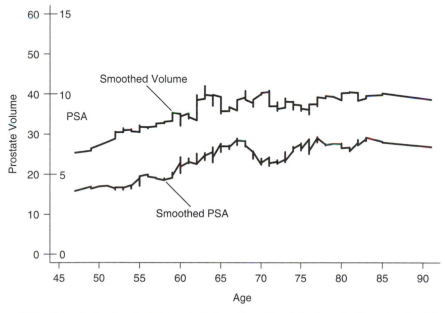

Figure 25.10 Smoothed series of prostate volume (outside axis) and prostate-specific antigen level (inside axis). Both appear to increase until age of middle to late 60s and then level out. Cross-correlation coefficient is 0.73.

ADDITIONAL EXAMPLE: PURITY OF DENTAL WASH WATER

A large dental clinic was concerned about the purity of water being used to wash drilled teeth and small, open wounds. Dental unit water tubing harbors bacteria-laden biofilms, which often contain pathogens. Although technicians are instructed to use purified water and purge the lines with compressed air at scheduled intervals, they do not always follow these procedures with sufficient care. A 12-week study compared eight technician groups.[51] The level of contamination of water ejected from the spray head was examined by a pathologist and rated according to the number of colony-forming bacterial units found. Technician group and date were recorded. Figure 25.11 shows data smoothed by a moving mean of 9 (initial data overlapped too much to perceive a pattern) in the order recorded (see Fig. 25.11A) and further smoothed by a moving mean of 21 (see Fig. 25.11B). Examination of the data in the vicinity of the peaks shows that group seven was using tap water, a discovery that might not have been made from the raw data. In addition, it appears that the curve averages a constant level at first but then begins a steady rise about halfway through the study. Methods to test for a point of change are examined in Section 25.6.

Autocorrelation: Is the Contamination Periodic?

The tap-water peaks look as if they might be periodic rather than haphazard. The autocorrelation coefficient was calculated for various lags. The greatest coefficient, appearing at a lag of about 50, was 0.18. We conclude that tap-water usage is not periodic.

Exercise 25.4. The systolic time ratio (STR) indicates the strength and efficiency of the heart. Does STR relate to heart rate (HR)? Figure 25.12A displays STR graphed against HRs for 228 patients in increasing order.[70] A linear regression shows no relationship, which just indicates that a straight-line fit does not have a significant slope, not that no pattern of relationship exists. Perhaps a smoothing process will demonstrate some relationship. A moving mean of 51 is superposed on the raw data. Figure 25.12B shows the moving mean with the

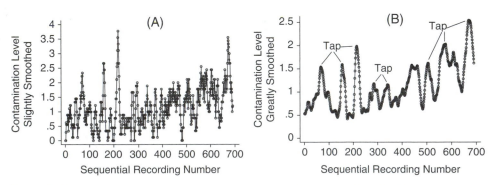

Figure 25.11 Contamination level of water for eight procedures over 12 weeks. Rating ≤2 satisfies American Dental Association standards. (A) Data smoothed by a moving mean of 9. (B) Data further smoothed by a moving mean of 21. A technician group using tap water was discovered.

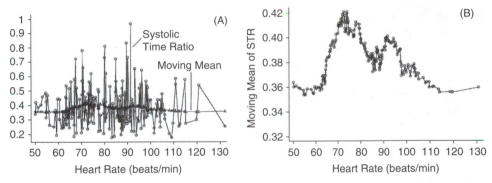

Figure 25.12 (A) Systolic time ratio against heart rate for 228 patients with moving mean of 51 superposed. (B) Moving mean with vertical axis scale enlarged to perceive effect.

vertical axis enlarged so that the pattern can be perceived. What hypothesis for further testing may be posed from Fig. 25.12B?

25.6. TIME-SERIES DATA: TESTING

In time series, testing is primarily concerned with verifying a model form and with identifying change points. Testing the nature of a model has been addressed in sections on regression; this section focuses on testing for a point of change from one model to another.

MOST FREQUENTLY USED METHODS OF CHANGE-POINT IDENTIFICATION

There exists a number of change-point tests, but most are rather complicated. Most change-point methods require a baseline against which to compare evolving data. In baseline-based methods, a statistic triggered by a change from baseline in the longitudinal process must be calculated. In one exception, a method based on current data, a change in the functional form of a smoothed sample must be detected. One convenient categorization of change identification approaches is by sampling schemes from which this triggering statistic is calculated. These may be denoted as (1) periodic samples, (2) accumulating samples, (3) smoothed samples, and (4) moving samples. Each has some advantages and some disadvantages.

(1) Periodic Samples

Although periodic or sporadic sampling usually costs less than complete sampling and allows destructive sampling, it will detect a change, but fails to locate it.

(2) Accumulating Samples

In cumulative sampling, the concept is to accumulate errors until the sum becomes improbably large. This form of analysis has memory: an unusual deviation from the expected path is never forgotten, but neither is the influence of an outlier. An accumulation of no-longer-relevant events, or of outliers, may trigger an indication of change where there is none. The dominant methods of this type are named CUSUM, in which the probability of the observed cumulative sum is found, and EWMA, a control chart scheme based on an exponentially weighted (geometric) moving average. CUSUM appears to be the most frequently applied change-point method, although EWMA weighs data less as they recede into the past, and is thus less limited by memory of irrelevant events. Comparisons show little difference in the decision outcomes of the two methods.

(3) Smoothed Samples

Current-data methods detect patterns that emerge from smoothing by a moving-sample regression fit. Independence from baseline data is the advantage of this method. However, it detects only a limited set of change types, for example, jumps in the moving average but not changes in variance or the nature of the model, and it does not detect outliers.

(4) Moving Samples

Moving samples, or more exactly, a moving sample compared with a baseline sample, are of two distinct types: *Monte Carlo Markov chains* (MCMCs) and *Moving F.*

Monte Carlo Markov Chains

Monte Carlo Markov chains uses a baseline to provide a Bayesian prior probability and a likelihood model. Basically, an appropriate model is posed, tested, and Monte Carlo integrated; then it is carried along the time series by MCMC methods. Monte Carlo Markov chains address the widest variety of change-point issues of all methods and solve a great many problems other than change-point identification. However, it is a method that requires considerable mathematical ability, coupled with the intuition for good model building, a talent more rare than we would wish. It is not a method that could be used by the medical investigator who has a background of only a few statistics courses.

Moving F

By contrast, Moving F is a rather simple test, but one that seems to work well for most change-point purposes. In this test, the variance of a moving sample is divided by the variance of the baseline sample, creating a Moving F statistic. The point at which this Moving F first exceeds a critical value as it moves through time both detects and locates a change point. Moving F will detect any sort of change in a time series, including jumps (trauma has changed a physiologic process), changes in slope (a disease begins), changes in variability (laboratory test results change standard deviation when a disease worsens), and changes in model (a laboratory test result changes from approximately constant to an exponential increase when cancer starts). In addition, outliers are obvious (a decimal point is left out of a laboratory test result).

Because Moving F is the simplest of the change-point identifiers with adequate generality, this method and several examples are given here.

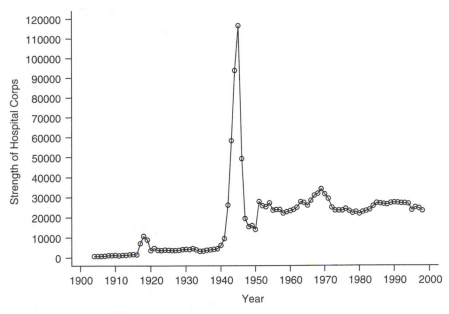

Figure 25.13 Strength of U.S. Navy Hospital Corps.

EXAMPLE POSED: LEVEL OF MEDICAL TECHNICAL SUPPORT

The U.S. Navy Hospital Corps is composed of enlisted technicians who support medical care operations. Corpsmen provide basic medical service aboard ships, to marines in land action, and at ashore bases. The size of the Hospital Corps through time presents an interesting commentary on the politico-military history of the United States. Figure 25.13 shows the strength of the Corps during the twentieth century.[23] Growth spurts can be seen at the times of WWI, WWII, and the Korean War, after each of which the Corps was maintained at greater manning than before. We hardly need tests. However, were the seeming increases at the times of the Vietnam War, the military buildup of the 1980s (what might be thought of as an economic war against the former Soviet Union), and the First Gulf War significant? If not, we would model post–Korean War manning level as a constant with random fluctuations. Let us detect and locate any change points in that period.

METHOD

Moving Sample Designations

A moving sample is composed of a sequence of (mostly overlapping) values, for example, x-values. We need to keep track of which member of the sequence of samples we are dealing with at any moment. As in Section 25.5, let us designate by k the rightmost value in a moving sample of size n so that the sample elements are $x_{k-n+1}, x_{k-n+2}, \ldots, x_k$. The moving mean is $(x_{k-n+1} + x_{k-n+2} + \cdots + x_k)/n = m_k$. The first moving mean is m_1, the second m_2, and so on. m_k is just the mean of n data ending with number k. Similarly, we could think of a moving variance s_i^2 as just the sample variance of the same n data.

513

The Concept of a Simple Moving Test

Suppose we start a time series with a stable baseline of, say, b values, perhaps a critical care blood pressure monitor. This baseline has variance s_b^2 (associating it with its rightmost datum, b). Recall that an F test is just the ratio of two variances (assuming the data are approximately normal). If we start with datum $b + 1$ and calculate the sequence of moving variances s_k^2 of size n, dividing each by s_b^2, we have a Moving F statistic traveling through the time series. Upon comparing this with a critical value of F from Table V (see Tables of Probability Distributions) for $df\,(b - 1, n - 1)$, perhaps for $\alpha = 0.05$, we can see where the variability exceeds baseline variability so much that it is unlikely to have occurred by chance. This is a Moving F test. It will detect a sudden increase in blood pressure. (The monitor example is given to aid conceptualization; it is not suggested that a Moving F test is needed to perceive a clinically crucial change in blood pressure.)

The Reason for Denoting a Moving Sample by Its Leading Member

A moving statistic does not occur at a point on the time axis but rather is calculated from an interval. As we move through the series, the first point at which a new model form has changed the moving sample enough to trigger significance will be taken as the point of change. Thus, if the moving statistic becomes significant at leading point k, k is the change point.

Calculation

The Moving F method is considered relatively simple because it can be handled by users outside the field of statistics applying widely available software. For example, a medical investigator can perform a Moving F in a few steps on Microsoft Excel software or on a statistics package, for example, Stata (Stata Corporation, College Station, TX). (1) Starting with spreadsheet columns of time (or other independent variable) t_i and readings x_i, choose a baseline sample of size b and regress the first b x's on t to find the functional form of the model, say, f, perhaps a constant or a sloped line. Make a column of f-values, f_i, such as by substituting t_i in the regression equation for each i. (2) Calculate the baseline variance about this model, s_b^2. s_b^2 is usually designated as the residual mean square in the software's regression display. (3) Make a column $(x_i - f_i)^2/(n - 1)s_b^2$. (4) Use the software's moving average capability to create the moving average of the last column, which is the Moving F.

Graphing a Moving F Test

On a graph with horizontal time-series axis, t, and F as a vertical axis, plot Moving F and draw a horizontal line at the critical F-value. The time-series values above this line, that is, for which the variability is significantly greater than baseline, will be clearly visible. If Moving F were displayed on a blood pressure monitor, a sudden increase would be visible immediately.

A Two-Tailed Test

We may be concerned not only with an increase in blood pressure but also a decrease. We could split the α, for example, 5%, into two 2.5% areas and identify an upper and a lower critical F, the lower value having 97.5% of the F distribution greater than this value and the upper value having 2.5% greater. (A more complete F table than given in this book would be required.) Then we could detect both significant increases and decreases.

Multiple Causes of Variability Are Possible

The Moving F considered to this point will test for any combination of changes in the process: a change in the shape of the process (the form of the model), a shift in average, and/or a change in the standard deviation. Often, we need to identify which influence is causing the significant F. Consider an electronic monitoring instrument. Its readings may have become only more variable. Alternatively, it could have shifted its setting while maintaining the same variability about the setting. Either event would trigger an increase in F in comparison with the baseline. How are these different sources of variability separated and identified? The numerator of the moving sample's variance is the sum of squares (SS) of deviations of x-values from the baseline mean, m_b for simplicity, or $SS_k = \sigma(x_i - m_b)^2$. We can add and subtract the moving mean value m_k from x_i to obtain $\sigma(x_i - m_k + m_k - m_b)^2$, which can be shown mathematically to be $\Sigma(x_i - m_k)^2 + n(m_k - m_b)^2$, a component due to randomness at the moving sample's position in the time series and a component caused by the difference between the baseline mean and the mean at the moving sample's position. When divided by df, they become variances. Taken in ratio to the baseline variance, we have a Moving F caused by mean shift and one due to random variability. Each may be compared with critical F-values to form a test. (It is also possible to separate out and test a component due to the shape of the process, but that subtlety is not pursued in this chapter.)

EXAMPLE COMPLETED: LEVEL OF MEDICAL TECHNICAL SUPPORT

Let us choose a moving sample of size $n = 7$, which is long enough for smoothing but not so long as to use too much of the time series. The 7-year mean level (from 1955 to 1961) is taken as baseline (6 df). Its mean is $m_b = 23,310$, and variance is $s_b^2 = 457,410.8$ (this value is not much different from the regression's residual variance). We begin a moving sample of 7 (6 df) with 1962, dividing its variance by the residual variance to create a Moving F statistic. From Table V, we see that a critical value of F at $\alpha = 0.05$ and 6,6 df is 4.28. Figure 25.14 shows the Moving F for the years 1962–1998. The increase in manning level becomes significant in 1964, a year after President Johnson assumes the presidency. The buildup and decline of military support for the Vietnam War can be followed year by year. Manning level returns to baseline for 1975 under President Ford and remains there during President Carter's administration. It increases in the 1980s under President Reagan's guidance during the arms race with the former Soviet

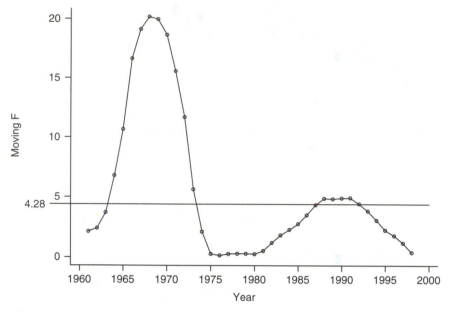

Figure 25.14   Post–Korean War pattern of Moving F calculated from U.S. Navy Hospital Corps strength. Shown is variance of moving sample of 7 in ratio to that of baseline from 1955 to 1961.

Union, becoming significant in 1987. It remains significant during the period of collapse of the former Soviet Union, which begins to overlap with the First Gulf War. Manning level declines at the end of that war, losing significance in 1993, and declines again to baseline during President Clinton's administration.

ADDITIONAL EXAMPLE: HEART RATE BY TYPE OF ANESTHETIC IN TRAUMA

Morphine given intravenously (IV) to a severely injured patient with extensive blood loss impairs the already traumatized vascular system. It was hypothesized that intrathecal (IT) administration of morphine (injection under the nerve-covering sheath) would mitigate that impairment. An experiment[64] used a sample of two pigs, one assigned to IV and the other to IT injection. A number of variables were measured for a 30-minute baseline and a 60-minute simulated hemorrhage. The hemorrhage was stopped and morphine was injected. Measurements were continued for an additional 180 minutes. Figure 25.15A shows HR for the two pigs during this period. The variance of the first 15 prehemorrhage HRs averaged for the 2 pigs was taken as baseline. The moving sample also was taken as size 15. From Table V (interpolated), the 95% critical value of F for 14,14 *df* is 2.48. Figure 25.15B shows the plots of Moving F for IV and IT treatments. Whereas both curves rise above the critical value during hemorrhage, the F for IV morphine rises even further after the end of hemorrhage and the F for IT morphine drops back to the vicinity of baseline, losing significance. The evidence indicates that IT morphine is better for the vascular system of trauma patients after hemorrhage.

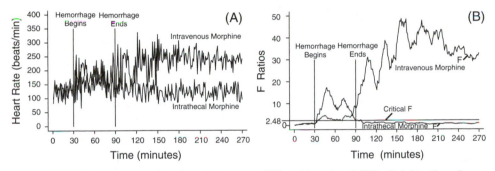

Figure 25.15 (A) Heart rates (HRs) for intravenous (IV) and intrathecal (IT) administration of morphine at baseline, during hemorrhage, and after hemorrhage. (B) Moving F of HRs for IV and IT morphine over same periods. Both subjects show significantly increased HRs during hemorrhage, but IV morphine at the end of hemorrhage exacerbates HR even further, whereas IT morphine allows HR to return to normal.

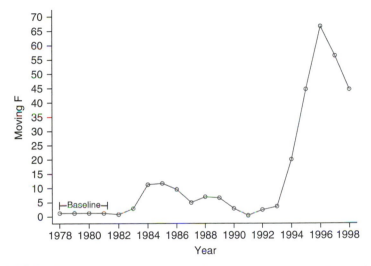

Figure 25.16 Moving F for Dr. Shipman's excess number of death certificates. The baseline used four points and the moving sample three.

Exercise 25.5. Harold Shipman was a rural physician in the United Kingdom who was discovered in 1998 to have murdered a large number of his patients. A recent article[79] provides data on the cumulative number of death certificates by sex that he signed in excess of the local average, totaling 224. Data were taken starting from 1978 with the first excess. Sexes were pooled, and total number of excess certificates were counted by year. The first 4 years (1978–1981) were used as baseline, and a constant baseline was assumed. The baseline variance was used as the denominator of the Moving F. A Moving F of three data, calculated for 1982–1998, is shown in Fig. 25.16. Find the critical F-value and identify where the number of excess death certificates first becomes significant in probability. How many years of murdered patients would have been prevented if Dr. Shipman had been stopped then?

ANSWERS TO EXERCISES

25.1.

Table 25.7

Completion of Table 25.3: Survival Data of 274 Women in Rochester, Minnesota, with Adult-Onset Diabetes Mellitus Who Were Older Than 45 Years at Onset During 1970–1980

Interval (years)	Begin (n_{i-1})	Died	Lost	End (n_i)	S_i (survived)	Confidence intervals
0 (outset)	274	0	0	274	1.0000	
>0–2	274	14	0	260	0.9489	0.9234, 0.9744
>2–4	260	13	0	247	0.9015	0.8672, 0.9358
>4–6	247	14	0	233	0.8504	0.8094, 0.8914
>6–8	233	18	0	215	0.7847	0.7379, 0.8315
	215	0	1	214		
>8–10	214	19	0	195	0.7150	0.6638, 0.7662

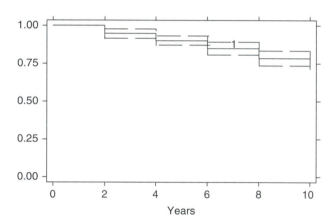

Figure 25.17   Survival curve with confidence limits obtained from Table 26.7.

25.2. From Table III, the critical χ^2 value for $\alpha = 0.05$ for 1 df is 3.84. The calculated 0.63 is much less than 3.84. *Conclusion:* No difference between the survival patterns has been shown. (The actual $p = 0.427$.)

25.3. (1) H_0: population death rate $= 0.052$. H_1: population death rate $= 0.50$. (2) Any difference found will not be used in clinical decisions but will tell the investigator that a different baseline death rate should be calculated for intubated patients. Thus, there is no clinical reason for different α's and β's; they are both chosen as 0.05. From Eq. (25.4), the critical values are $0.05/0.95 = 0.0526$ and $0.95/0.05 = 19$. (3) The first intubated patient is patient 32, who survived. $D_1 = [1 - P(1|H_1)]/[1 - P(1|H_0)] = [1 - 0.50]/[1 - 0.052] = 0.5274$. This will also be the multiplying increment in Eq. (25.5) for each surviving patient, and $[P(1|H_1)]/[P(1|H_0)] = [0.50]/[0.052] = 9.6154$ will be the multiplying increment for each dying patient. (4) Table 25.8 provides a completed version of Table 25.5. The upper critical value is surpassed on observation 8. The test indicates that H_1 should be accepted: The population death rate for intubated patients is 50%.

Table 25.8

Completion of Table 25.5

Patient no.	32	54	65	68	71	83	85	95
k (observation no.)	1	2	3	4	5	6	7	8
Observation outcome	0	0	1	0	0	1	0	1
D_k	0.5274	0.2782	2.6750	1.4108	0.7441	7.1548	3.7734	36.2828

25.4. It appears that the heart exhibits greater efficiency when the HR is in the reference range of about 70–80 beats/min, with inferior efficiency less than 65 beats/min and efficiency diminishing gradually as the HR increases to greater than 80 beats/min. This pattern demonstrated by time-series analysis is unlikely to be detected with regression or other curve-fitting techniques.

25.5. The critical $F_{2,3}(0.95) = 9.55$. Moving F first becomes significant in 1984. Had Dr. Shipman been stopped in 1984, 15 years of patient murders would have been prevented, representing 178 patients.

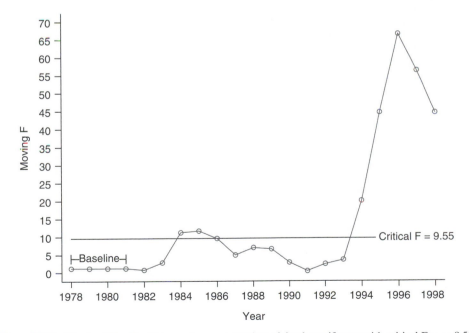

Figure 25.18 Moving F for Dr. Shipman's excess number of death certificates with critical $F_{2,3} = 9.55$. The significant $F_{2,3} = 9.55$ was breached in 1985–1986 and again in 1994–1998.

Chapter 26

Methods You Might Meet, But Not Every Day

26.1. OVERVIEW

Medical investigators often encounter statistical methods with which they are not familiar in journal articles, references from mentors and reviewers, and the like. This book discusses most of the commonly used statistical methods. What of those not commonly met? This chapter lists and provides definitions for some of the less common statistical methods. In cases in which a definition is not adequately descriptive, an example is given of a situation in which it might be applied. This chapter does not intend to present methodology. The purpose is solely orientation.

26.2. ANALYSIS OF VARIANCE ISSUES

FIXED VERSUS RANDOM EFFECTS MODELS IN ANALYSIS OF VARIANCE

Only *fixed effects* analysis of variance (*Model I* ANOVA) is addressed in this book. The effects of the independent variable categories are assumed to be fixed (i.e., constant). The equivalent effects in a random effects model (*Model II* ANOVA, or *components of variance* model) are assumed to be drawn from a probability distribution. The arithmetic is the same in both models for the most common cases, but the interpretation differs. For example, consider testing for patients' prostate-specific antigen level differences in prostate biopsy results classified as benign, suspect, and malignant, where the number of tissue samples collected differs according to the biopsy instruments and the urologist's discretion. Model I assumes that a prostate as a whole is characterized by the diagnosis class. Model II assumes that the biopsy represents only a portion of the prostate and the diagnosis class is itself a probability.

PAIRWISE COMPARISONS AFTER ANALYSIS OF VARIANCE

Section 17.5 addresses the interpretation of a significant ANOVA with three or more categories. Does the significance lie between the means of categories A and B, A and C, B and C, or some combination? Subordinate testing of each pair may be done by multiple comparisons or multiple-range tests, which provide the same sort of interpretation. Table 17.5 lists five multiple comparisons tests. The reader might also meet those of *Dunnett, Gabriel*, or *Hochberg*, or the *multiple-range tests* of *Duncan, Student–Newman–Keuls*, or *Tukey*.

EQUALITY OF VARIANCES ASSUMPTION

Levene's test is a test of the equality of variances assumption in ANOVA.

The *Brown–Forsythe* and *Welch tests* are calculations of ANOVA significance by adjusting results for unequal variances.

The *Box-Cox* transform is designed to make the distributions of residuals closer to normal and their variances more similar.

GENERAL LINEAR MODELS

The methods of ANOVA, analysis of covariance (ANCOVA), and regression presented in this book require a high level of balance and symmetry in design. An investigator does not always have the luxury of data satisfying such requirements. General linear model (GLM) is an umbrella that embraces these three methods and does not require the limiting data balance. Usually, GLM software packages include several options for further analysis, for example, multiple comparisons and transformations of data sets. The additional capability is not "free" but requires more mathematical sophistication of and model specification by the user than do the simpler forms. Some authorities make a distinction between general and generalized linear models.

Nonlinear Models

General linear model is not restricted to linear algebraic models; it may include nonlinear forms, for example, log, exponential, and Gompertz, among others, as well as combinations of algebraic and/or nonlinear forms.

26.3. REGRESSION ISSUES

The *Heckman selection* model provides an adjustment for biased regression variables.

26.4. MULTIVARIATE METHODS

The terms *multiple, multivariate*, and *multivariable* are defined and contrasted in Section 10.6. Usually, *multiple* and *multivariate* refer to methods treating one dependent variable and more than one independent variable. *Multivariable* applies to methods with more than one dependent variable.

HOTELLING'S T^2

A two-sample t test compares means on a continuous measure between two groups. The T^2 test of Harold Hotelling compares means of two or more continuous measures simultaneously for the two groups. For example, it has been suggested that because increased levels of vitamin D have been shown to reduce tooth loss, it may increase the rate of correction of pediatric scoliosis. An investigator measures the difference between left and right thoracic height, Cobb angle, kyphosis, and lordosis on two groups of children with scoliosis. One group was given large vitamin D doses and the other group was not. The investigator uses T^2 to test the difference between the two groups on the basis of the four spinal measurements simultaneously. T^2 will also test a single or paired sample against a hypothesis of zero, as a generalization of the paired t test.

DISCRIMINANT ANALYSIS

Discriminant analysis classifies sets of patients or measures into groups on the basis of multiple measures simultaneously. A line (or plane or hyperplane, depending on number of classifying variables) is constructed between the two groups in a way that minimizes misclassifications. Patients whose results appear on one side are classified as having arisen from one group and those with results on the other side are classified as from the other group. An otolaryngologist wants to classify snoring patients into two groups: one group of patients requires surgery, whereas the other group requires less aggressive treatment. He uses apnea–hypopnea index, maximum decibels (dB) of snoring, and percentage of snoring time at greater than 50 dB as classifying variables. The method produces a plane in three dimensions separating the groups, and the otolaryngologist classifies patients according to their position relative to the plane in the three dimensions. More advanced forms of discriminant analysis address classification into more than two groups.

MAHALANOBIS' DISTANCE

A measure of distance D between the multiple means (centroids) of the two groups in discriminant analysis. D^2 may be used to test the significance of this distance, as does T^2.

MULTIVARIATE ANALYSIS OF VARIANCE AND COVARIANCE

Multivariate ANOVA (MANOVA) and analysis of covariance (ANCOVA) extend those methods to situations having more than one dependent variable. In DB10, we investigate the effects of surgery on hamstrings or quadriceps by (1) time to perform hops and (2) centimeters covered in hops for the operated leg compared with the nonoperated leg. We could analyze either (1) or (2) by the methods of this book. If we wanted to contrast legs by both (1) and (2) simultaneously, we could use Hotelling's T^2 (see earlier). If we had not only legs but also two types of surgery to compare, we would have a 2×2 ANOVA for either (1) or (2) or a *MANOVA* for both simultaneously. If we added the age of the patient (continuous) to leg and surgery type (categorical) and analyzed both (1) and (2) simultaneously, we would have a *multivariate analysis of covariance (MANCOVA)*.

PRINCIPAL COMPONENTS ANALYSIS

Let us start with an example. We have a sample of patients at risk for heart disease with the following measures: cholesterol level; age; socioeconomic level; dietary intakes of saturated fats, carbohydrates, and protein; and mean daily burn off of calories through exercise. Several of the variables are correlated, but cause and effect is not clear. Is there a component of several variables that accounts for their correlation? We do not know what this component might be, but if we find it, we might be able to divine a name or meaning for it by noting the relative contributions from each variable. The search for such a component, and for a second after the first has been removed, and so on, is termed *principal component analysis*. The matrix (array) of variances and covariances is subjected to mathematical procedures that produce "eigenvalues," a measure of "strength" of each component, and "eigenvectors," a list of contributions to the eigenvalue of each variable (but that may not have been adjusted for scale differences). The production of eigenvalues and eigenvectors is mathematical and unambiguous. The trick is to make sense of their meaning.

FACTOR ANALYSIS

Factor analysis is similar in method to principal component analysis, but it uses correlations in the matrix that produce eigenvalues and eigenvectors. The goal is to identify one or a few factors existing in a set of many variables that will explain most of the outcomes. Again, the mathematical results are unambiguous. The challenge is to name and characterize the emerging factors.

CLUSTER ANALYSIS

Suppose you have clinical and laboratory test results on a large sample of patients. If we should group similar sets of results together, perhaps we could see a pattern of disease similarity in a cluster. We might find that diseases we had believed to be quite different have similarities. We might find that we can form a list of potential diseases to be considered when we are given a pattern of test results. Clustering is obtained mathematically by statistics that compare numerical similarities or distances in multivariate space.

26.5. NONPARAMETRIC TESTS

The *Runs Test (Wald–Wolfowitz)* is a simple procedure that tests if there are too many or too few binary observations of the same type in a sequence to be randomly occurring. For example, when allocating patients to two arms of a study, it tests if there are too many of one characteristic, for example, sex or advanced age, in sequence for the sample to be unbiased.

The *Sign test* is another name for a binomial test of proportion (see Section 15.6).

In Section 16.5, we find that ranked outcomes of three or more treatments given to each patient of a sample could be tested by Friedman's test. *Cochran's Q* is a version of Friedman's test in which the outcomes are not ranks but rather are binary outcomes (healed or not, reaction or not, etc.).

26.6. IMPUTATION OF MISSING DATA

Imputation is the term applied to estimating the values of missing data. Missing data may imply bias. For example, the Connecticut Tumor Registry started in 1941 contained data from records back to 1936, from which patients who had died were omitted, leading to an underestimation of mortality. Imputation may not rectify such bias, because imputed data are estimated only from data that are present. For unbiased data, missing data causes little to huge effects, depending on the analysis being used. Many statistical procedures may be conducted without concern for missing data. If it is a concern, the computer user should investigate the software to see how it handles missing data. Some software drops every case (patient) in which there is a single missing datum. A study starting with 1000 patients may result in analysis conducted on 100 because of missing data deletions, and often the user is not alerted to this reduction in sample size. Usually, one or a few missing data may be imputed without devastating the results, but it is not advisable to impute a large number of missing data. Imputation may be done by a variety of methods. The simplest, but not the best, method is to insert the average of that variable for other cases. A more sophisticated method is to use all cases except that with the missing datum to predict the missing-datum variable in a multiple regression of all variables other than that with the missing datum and then use that regression to predict the missing datum. Others methods exist as well.

26.7. RESAMPLING METHODS

THE BOOTSTRAP

Statistical sampling is drawing a set of observations randomly from a population distribution. Often, we do not know the nature of the population distribution, so we cannot use standard formulas to generate estimates of one statistic or another. *If we are willing to assume that a sample distribution adequately reflects the population distribution*, we can resample from the sample distribution to obtain descriptive estimates such as the median, the standard error, a confidence interval, and the like. By repeating the sampling operation a large number of times, perhaps 1000, we decrease the sampling error and increase the quality of the estimates. The assumption strikes many users as a large leap of faith. However, if the sampling has been done carefully, is truly random, and is of a reasonable size, the *bootstrap* provides surprisingly good estimates. Some statistical software packages provide the capability for bootstrap sampling and estimation.

THE JACKKNIFE

A method of resampling from a sample that preceded the bootstrap, the *jackknife* is less efficient than the bootstrap and is more influenced by outliers. However, this influence may be useful in identifying outliers or evaluating how much a particular datum affects the sample. An estimate of a statistic is made *n* times, omitting one datum each time.

26.8. AGREEMENT MEASURES AND CORRELATION

RELIABILITY (INTERNAL AGREEMENT)

Cronbach's α measures reliability, or, more exactly, internal consistency. Technically, α is the square of the expected correlation with a data set of similar format having perfect reliability. As such, it lies between 0 and 1.

AGREEMENT AMONG RATINGS

(Cohen's) Kappa measures interrater agreement in excess of the agreement that is expected to occur by chance. Kappa $= 0$ implies no agreement better than chance, and kappa $= 1$ implies perfect agreement. Different versions allow for two raters, multiple items; multiple raters, two items; and multiple raters, multiple items. The mathematical bases for them are rather different.

AGREEMENT AMONG RANKINGS

Recall that ranks are assigned to items relative to each other, whereas ratings are chosen for each item regardless of ratings for other items. Ratings may all be the same, whereas ranks should all be different except for the occasional tie. Indeed, a forced decision is more akin to real-life clinical practice. If the agreement between only two rankers is at issue, rank correlation methods are available, namely *Spearman's* r_s (see Section 24.4) and *Kendall's* τ. These two methods use different scales and therefore yield slightly different values of the coefficient. Whereas Kendall's method is easier to work with mathematically, Spearman's method is easier to calculate and therefore used more often. A more challenging problem is agreement among multiple rankers. Rank correlation measures agreement among all rankers between members of any pair of categories, for example, between the first and second categories, which is of little help in decision making. It does not address agreement overall or of all rankers about a single category. Two methods for multiple rankers exist, *W* and *A*.

Kendall's W, which he termed a *coefficient of concordance*, provides a single value that measures agreement among all rankers across all categories simultaneously. Suppose we are considering a clinical protocol before an institutional review board (IRB). The choices each board member is asked to rank for this protocol are "Pass," "Revise," "Table," and "Reject." Perhaps a member's first choice would be "Revise" and the second choice "Table." If the protocol cannot be revised or tabled, the board member would select "Pass," reserving "Reject" for a last choice. Kendall's *W* would provide a single value representing how well the board as a whole agrees on what to do with the protocol.

Riffenburgh's A measures interranker agreement among all rankers on each category. Thus, the IRB could see how well they all agreed on passing the protocol, on revising it, and so on, providing guidance for the board's decision. Like Kappa, *A* measures agreement remaining after chance agreement is removed. $A = 0$ implies no agreement, and $A = 1$ implies perfect agreement. Currently awaiting publication, this method may be obtained by contacting the author through Elsevier's Web site (http://books.elsevier.com/companions/0120887703).

AGREEMENT AMONG CONTINGENCY CATEGORIES

The *Phi coefficient* and *Cramér's V* are measures of association among (nonrankable) categories in a contingency table: Phi for a 2×2 table, and V for a larger table. For example, association of frequencies of four diseases among three races could be measured by Cramér's V. Both are functions of chi-square, Phi being the square root of chi-square divided by *n* and V being the square root of chi-square divided by a formula using *n*. Both lie between 0 and 1.

CANONICAL CORRELATION

Canonical correlation is a term for the correlation coefficient among items in two lists (vectors of observations). For example, does a list of laboratory test results correlate with a list of clinical observations on a patient?

26.9. BONFERRONI "CORRECTION"

Bonferroni correction extends the Bonferroni approach to multiple comparisons discussed in Section 17.5 to multiple test results. If *k* significance tests, each with error rate α, are conducted on the same set of data answering aspects of the same data question, the overall error will increase to $1 - (1 - \alpha)^k$. The critical value for concluding significance is adjusted to α/k.

26.10. LOGIT AND PROBIT

Suppose we recorded the survival or death of a pathogen as depending on increasing dosages of an antibiotic 5 days after administration to a sample of ill patients. Logit (for "logistic unit") and probit ("probability unit") are mechanisms to transform the data into a form suitable for regression analysis. Logit analysis is almost the same as logistic regression (see Section 24.10), yielding similar results. Probit analysis is a little different, but again yields similar results in the end. In probit analysis, the residuals (errors) are assumed to follow a standard normal distribution; in logit analysis, the residuals follow a logistic distribution. Both methods give coefficients; logit's coefficients are the natural logs of odds ratios, whereas probit's coefficients are more difficult to interpret.

26.11. ADJUSTING FOR OUTLIERS

COOK'S DISTANCE

In regression, Cook's distance measures the effect on regression outcomes of removing each case (e.g., patient) in turn. Large Cook's distances signal outliers.

SAMPLE (OR MEAN) TRIMMING

Sample (or mean) trimming is using an established policy for removing outliers. A mean estimate much closer to the population mean usually results. However, rank-order methods, which are only slightly or not at all affected by outliers, have taken precedence. Current use of mean trimming is rare.

WINDSORIZED STANDARD DEVIATION

The Windsorized standard deviation is a estimate for the standard deviation of a trimmed sample that was devised by Charlie Windsor.

26.12. CURVE FITTING TO DATA

The ideal fitting procedure occurs when the theoretical (e.g., physiologic) cause of data generation is known and specified mathematically. An example is the probability distribution of a sample mean, which is known to be normal (see Section 2.8). When the theoretical curve is unknown, curves may be fitted to data by various means (e.g., the spline fit).

THEORETICAL FITS

In the theoretical fits orientation, the reader need only appreciate the general shape of various theoretical distributions. They may be classified into the following six patterns, but keep in mind that a certain choice of parameters might make them appear different: (1) bell-shaped curves, typified by the normal and t distributions introduced in Chapter 2, to which we can add the *logistic* and *Cauchy* distributions; (2) skewed bell-shaped curves, typified by the chi-square and F distributions introduced in Chapter 2, to which we can add the *beta, gamma, Rayleigh*, and *Weibull* distributions; (3) curves declining with decreasing slope, tailing toward horizontal in the extreme, characterizing the *negative exponential, log-normal*, and *Pareto* distributions; (4) curves increasing with increasing slope, becoming indefinitely large in the extreme, characterizing the *exponential* distribution; (5) a cusp, pointing upward, characterizing the *Laplace* distribution; and (6) curves increasing at first, then decreasing, often used to represent growth (of an infecting pathogen, a cancer, a population, or the height of a human), characterizing the *Gompertz* distribution.

SPLINE FIT

A *spline* curve is fitted by cubic equation solutions for each segment in a sequence of data segments. It fits the data well, but gives little insight into why it fits.

26.13. TESTS OF NORMALITY

The *Lilliefors* test of normality may be added to the Shapiro–Wilk, one-sample Kolmogorov–Smirnov, and chi-square goodness-of-fit tests addressed in Section 20.2.

Chapter Summaries

Chapter Summaries

SUMMARY OF DATA AND NOTATION FROM CHAPTER 1

STAGES OF SCIENTIFIC KNOWLEDGE

Science usually develops in three stages: *describing* a class of scientific events (using descriptive statistics), *explaining* these events (using statistical testing), and *predicting* their occurrence (statistical modeling and regression). The ability to predict an event implies some level of understanding of the rules of nature governing the event.

PHASE I–IV STUDIES

Phase I is to establish safety. *Phase II* is to demonstrate preliminary effectiveness. *Phase III* is to verify effectiveness on a large sample. *Phase IV* is the ongoing monitoring of effectiveness and safety.

DATA TYPES

Continuous data are positions on a scale, which may be as close to one another as the user can discern and record. Example: blood clotting time. *Discrete data* are a subset of continuous data that are recorded only as distinct values; there is a required distance between adjacent data. Example: age in years.

Rank-order data are indicators of the positions of data that can be positioned according to some ordering characteristic of the subject, such as magnitude of each datum. Example: ordered lengths of surgery on six patients, shortest (rank 1) to longest (rank 6).

Categorical (or nominal) data are indicators of type or category and may be thought of as counts. Example: number of surgical patients with versus without complications.

Ratings may be of any form, depending on the indicator being rated. Ratings are based only on the subject being rated; they are not ranks, which are based on the subject's position relative to other subjects. Example: cancer stage.

String (alphabetic) data are data in word form, not numerical form. Example: male versus female rather than 1 versus 2. String data must be converted to numerical data for analysis.

SYMBOLS

Name symbols behave like family names, indicating to which family the thing or event symbolized belongs. Example: x or μ. *Subscripts* behave like first names, indicating which member of the family is being considered. Example: 1 in x_1 or *postop* in Hct_{postop}.

Operator (or command) symbols represent an instruction to act rather than a thing. Example: $\div$ ("divide what appears before by what appears after") or $\sum$ ("add together the elements that follow").

Relationship symbols express a relationship between two families or two members of a family. Example: $=$ ("what appears before is the same as what appears after") or $>$ ("what appears before is greater than what appears after").

Indicator symbols denote a member of a family without specifying which member. Example: i in x_i could be x_1, x_5, or x_{332}.

TERMS

Population. The entire set of subjects about whom (or which) we want information.

Sample. A subset of the population. The usual goal in statistics is to describe the characteristics of a sample and generalize them to the population from which the sample was drawn.

Representativeness and Bias. The generalization from sample to population is valid only if the sample represents the population in the characteristics being generalized. If the sample does not represent the population, it is termed biased.

Random. The physical or mathematical mode of choosing a datum based solely on probability such that no organized bias may influence the choice.

Control Group. A sample group having all the characteristics of the experimental group except the treatment under study.

Placebo. An apparent treatment, given so that the subject and/or investigator cannot distinguish control subjects from experimental subjects.

Variable. An observation or reading giving information used in answering the study question.

Independent Variable. A variable that, for the purposes of the study question to be answered, occurs independently of other influences.

Dependent Variable. A variable that depends on or is influenced by an independent variable.

SUMMARY OF DISTRIBUTION CONCEPTS FROM CHAPTER 2

Tally. A collection of tick marks, one for each datum marked in its respective interval within the range of possible occurrences of variable values. The shape of a completed tally approximates the shape of the frequency distribution.

Frequency Distribution. The pattern in which the data are distributed along the scale (or axis) of the variable of interest. A sample frequency distribution appears as a sequence of bars of varying height, having the tally intervals as width.

Relative Frequency Distribution. A frequency distribution in which the height of the bar over each respective intervals represents the proportion of the sample falling into that interval.

Probability. The relative frequency with which an event occurs when all events in a population are given equal opportunity to occur. As the sample size increases, tending toward the population size, the relative frequency distribution tends toward the population probability distribution.

Mean. An average; the center of gravity of a distribution, denoted μ in a population's probability distribution and m in a sample's relative frequency distribution.

Median. An average; the value of the variable for which half the sample or population is smaller and half is larger.

Mode. An average; the value of the variable having the greatest relative frequency.

Midrange. An average; the midpoint between the smallest and greatest values of a variable that occur. (A probability distribution with one or both tails extending without bound, e.g., the normal distribution, has no midrange.)

Variance. A measure of variability; the average (mean) of squared deviations from the sample or population mean.

Standard Deviation. A measure of variability; the square root of the variance. The purpose is to put the measure of variability in the same units as the observations and mean.

Skewness. A measure of asymmetry of a distribution. If one tail is stretched out more than the other, the distribution is skewed (e.g., right tail stretched out: right skew).

Inference. Concluding characteristics of a population on the basis of a sample. (This is an oversimplification, but it is the kernel of what is a complicated concept.)

Robustness. A characteristic of a statistic (a summarizing calculation from a sample) in which the statistic is little affected by moderate violation of the assumptions under which the statistic was formed.

Normal Distribution. A family of bell-shaped probability distributions.

Standard Distribution. A probability distribution or relative frequency distribution that has been transformed so that it has mean at 0 and standard deviation of 1. This transformation is achieved by subtracting the mean from each data element and dividing by the standard deviation. The most frequent use is in the standard normal distribution.

Central Limit Theorem (CLT). Data from any distribution (whatever its form) have a mean whose distribution is approximately normal.

t *Distribution.* A bell-shaped distribution, slightly wider than the normal, that arises when the standardizing divisor is the sample standard deviation rather than that of the population. ("Student": the nom de plume of W. S. Gossett, who derived the distribution.)

Degrees of Freedom (df). An integer related to sample size that determines which member of a family of probability distributions is appropriate in a particular case. The standard normal has only one member and does not use *df*, but the *t* has many members and does use *df*.

Chi-square (χ^2) Distribution. A right-skewed family of probability curves arising from the variance. The statistic: sample variance $\times$ *df* $\div$ population variance is distributed χ^2.

F Distribution. A right-skewed family of probability curves arising from the ratio of two sample variances. It depends on two *df* values, one for each sample variance.

Binomial Distribution. A discrete probability distribution of the relative likelihood of outcomes of a two-category event, for example, the heads or tails of a coin flip, survival or death of a patient, or success or failure of a treatment.

Poisson Distribution. A discrete probability distribution of the relative likelihood of outcomes of a two-category event when one of the events is rare, for example, a citizen of a developed nation contracting bubonic plague.

Standard Error of the Mean (SEM). The standard deviation of the sample mean. In most cases, it is the standard deviation of the sample observations divided by the square root of the sample size.

Joint Frequency Distribution. The pattern in which the data are distributed in the two-dimensional plane of two simultaneous scales (or axes) representing the two variables of interest. Each variable has a mean and standard deviation, as it would alone, but there is also a covariance (or correlation: covariance divided by the two standard deviations) between the two variables representing the strength of their relationship.

FORMULAS FOR DESCRIPTIVE STATISTICS FROM CHAPTER 3

Mean. μ (population mean) and *m* (sample mean). (*m* sometimes appears as the variable from which it is calculated with an overbar, as $\bar{x}$.)

$$\mu \text{ or } m = \frac{\sum x}{n} \tag{3.1}$$

Median. Algorithm for the *median* (sometimes *md*) in words rather than in symbols:

- Put the *n* observations in order of size.
- *Median* is the middle observation if *n* is odd.
- *Median* is the halfway between the two middle observations if *n* is even.

Mode. Algorithm for approximate *mode* (sometimes *mo*) in words rather than in symbols (*n* must be large), depending on the bar chart's interval widths and starting point:

- Make a bar chart of the data.
- *Mode* is the center value of the highest bar.

Variance. Population:

$$\sigma^2 = \frac{\sum x^2 - n\mu^2}{n} \tag{3.2}$$

Variance. Sample:

$$s^2 = \frac{\sum x^2 - nm^2}{n - 1} \tag{3.3}$$

Standard Deviation σ or s: square roots of the respective variance.

Standard Error of the Mean (SEM). Population (where σ appears) or sample (where s appears):

$$\sigma_m = \frac{\sigma}{\sqrt{n}} \quad \text{or} \quad s_m = \frac{s}{\sqrt{n}} \tag{3.4}$$

Standard Error of the Mean for Two Samples from the Same Population. Population standard deviation (σ) known:

$$\sigma_\mu = \sigma\sqrt{\frac{1}{n_1} + \frac{1}{n_2}} \tag{3.5}$$

Population standard deviation unknown; sample standard deviations s_1 and s_2:

$$s_p = \sqrt{\frac{(n_1 - 1)s_1^2 + (n_2 - 1)s_2^2}{n_1 + n_2 - 2}} \tag{3.6}$$

and

$$s_m = s_p\sqrt{\frac{1}{n_1} + \frac{1}{n_2}} \tag{3.7}$$

Covariance between x and y, population:

$$\sigma_{xy} = \frac{\sum xy - n\mu_x\mu_y}{n} \tag{3.8}$$

Covariance between x and y, sample:

$$s_{xy} = \frac{\sum xy - nm_xm_y}{n - 1} \tag{3.9}$$

Correlation Coefficient Between x and y. Population (where ρ appears) or sample (where r appears):

$$\rho_{xy} = \frac{\sigma_{xy}}{\sigma_x \sigma_y} \quad \text{or} \quad r_{xy} = \frac{s_{xy}}{s_x s_y} \tag{3.10}$$

SUMMARY OF CONFIDENCE INTERVALS AND PROBABILITY FOR CHAPTER 4

GENERAL FORMS FOR CONFIDENCE INTERVALS

Confidence Interval on an Individual Observation. The probability that a randomly drawn observation from a given probability distribution is contained in a specified interval is given by the area of the distribution under the curve over that interval.

Confidence Interval on a Statistic. The probability that a population statistic from a distribution of estimates of that statistic is contained in a specified interval is given by the area of the distribution over that interval. The common format for a confidence interval on a statistic is $P[$lower critical value $<$ population statistic $<$ upper critical value$] = 1 - \alpha$.

CONFIDENCE INTERVAL ON A MEAN, KNOWN σ

A mean from a normal distribution uses the normal distribution, arising from $z = (x - \mu)/\sigma$.

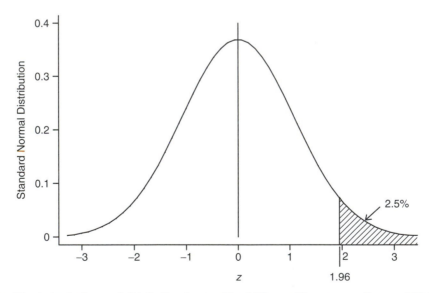

Figure 4.1 A standard normal distribution shown with a 2.5% α and its corresponding $z = 1.96$. The α shown is the area under the curve to the right of a given or calculated z. For a two-tailed computation, α is doubled to include the symmetric opposite tail. A two-tailed $\alpha = 5\%$ lies outside the ± 1.96 interval, leaving 95% of the area under the curve between the tails.

Form of a confidence interval on a mean, known σ:

$$P[m - z_{1-\alpha/2}\sigma_m < \mu < m + z_{1-\alpha/2}\sigma_m] = 1 - \alpha \qquad (4.5)$$

CONFIDENCE INTERVAL ON A MEAN, σ ESTIMATED BY s

A mean with σ estimated by s uses the t distribution, arising from $z = (x - \mu)/s$. Form of a confidence interval on a mean, σ estimated by s:

$$P[m - t_{1-\alpha/2}s_m < \mu < m + t_{1-\alpha/2}s_m] = 1 - \alpha \qquad (4.6)$$

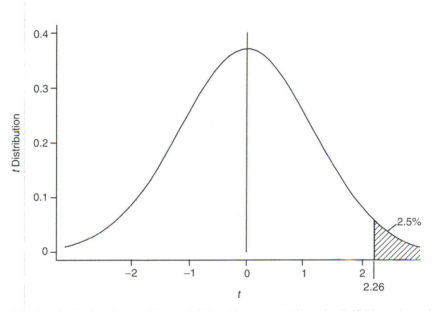

Figure 4.2 A t distribution shown with $\alpha = 2.5\%$ and its corresponding t for 9 df. The α shown is the area under the curve to the right of a given or calculated t. For a two-tailed computation, α is doubled to include the symmetric opposite tail. As pictured for 9 df, 2.5% of the area lies to the right of $t = 2.26$. For two-tailed use, the frequently used $\alpha = 5\%$ lies outside the interval $\pm t$, leaving 95% of the area under the curve between the tails.

CONFIDENCE INTERVAL ON A VARIANCE OR STANDARD DEVIATION

A variance uses the chi-square distribution, arising from $\chi^2 = s^2 \times df/\sigma^2$. Form of a confidence interval on σ^2:

$$P[s^2 \times df/\chi_R^2 < \sigma^2 < s^2 \times df/\chi_L^2] = 1 - \alpha, \qquad (4.7)$$

where χ_R^2 is the right tail critical value (use Table III) and χ_L^2 is the left tail critical value (use Table IV). Form of a confidence interval on σ: same as Eq. (4.7) but with square roots of components within the brackets.

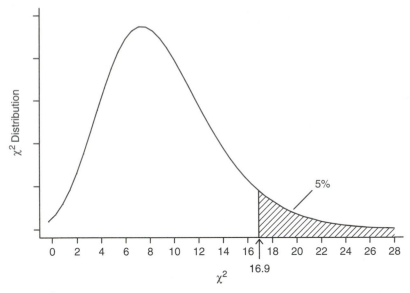

Figure 4.3 The χ^2 (chi-square) distribution for 9 *df* with a 5% α and its corresponding chi-square value of 16.9. The α probability is shown as the shaded area under the curve to the right of a critical chi-square, in this case, representing a 5% probability that a value drawn randomly from the distribution will exceed a critical chi-square of 16.9.

CONCEPTS AND PROCEDURES IN HYPOTHESIS TESTING FROM CHAPTER 5

COMMON ASSUMPTIONS IN HYPOTHESIS TESTING

The Assumption of Independence. All tests assume that "errors" (understood as "deviations from typical") in sample data are independent from one another. This is a major reason for random sampling.

The Assumption of Normality of Errors. Many, but not all, tests also assume that these errors are drawn from a normal distribution. A major exception is rank-order tests.

The Assumption of Equality of Variability. A third assumption frequently made is that the standard deviations (or variances) of the errors are the same.

TYPES AND PROBABILITIES OF ERROR

Here, "error" is used in the more traditional sense: making the wrong decision.

Type I (α) Error and the p-Value. The null hypothesis is rejected when it should not have been, occurring with probability α. After the rejection decision is made, this error may be termed a *false positive* and the data-dependent estimate of α is the *p-value*.

Type II (β) Error and the Power of a Test. The null hypothesis was not rejected when it should have been, occurring with probability β. After the nonrejection decision is made, this error may be termed a *false negative*. $1 - \beta$ is termed the *power* of the test.

Table 5.1

Relationships Among Types of Error and Their Probabilities as Dependent on the Decision and the Truth

		Decision	
		H_0 true	H_0 false
Truth			
	H_0 true	Correct decision True negative (probability $1 - \alpha$)	Type I error False positive (probability α)
	H_0 false	Type II error False negative (probability β)	Correct decision True positive (probability $1 - \beta$)

STEPS IN SETTING UP A TEST

(1) Write down the question you will ask of your data. (Does treating my infected patients with a particular antibiotic make them healthy again?)

(2) Select the variable on which you can obtain data that you believe best to highlight the contrasts in the question. (I think white blood cell [WBC] count will best show the state of health.)

(3) Select the descriptor of the distribution of the variable that will furnish the contrast (e.g., mean, standard deviation). (Mean WBC count will furnish the most telling contrast.)

(4) Write down the null and alternate hypotheses indicated by the contrast. (H_0: $\mu_a = \mu_h$; H_1: $\mu_a \neq \mu_h$, where a represents patients treated with antibiotics and h represents healthy patients.)

(5) Write down a detailed, comprehensive sentence describing the populations measured by the variable involved in that question. (The healthy population is the set of people who have no current or chronic infections affecting their WBC. The treated population is the set of people who have the infection being treated and no other current or chronic infection affecting their WBC.)

(6) Write down a detailed, comprehensive sentence describing the sample(s) from which your data on the variable will be drawn. (My sample is randomly selected from patients presenting to my clinic who passed my exclusion screen [e.g., other infections].)

(7) Ask yourself what biases might emerge from any distinctions between the makeup of the sample(s) and the population(s). Could this infection be worse for age, sex, cultural origin, and so on, of one patient than another? Are the samples representative of the populations with respect to the variable being recorded? (I have searched and found studies that show that mortality and recovery rates, and therefore probably WBC counts, are the same for different sexes and cultural groups. One might infer that the elderly have begun to compromise their immune systems, so I stratify my sample to ensure that my sample reflects the age distribution at large.)

(8) Recycle steps 1–7 until you are satisfied that all steps are fully consistent with one another.

(9) In terms of the variable descriptors and hypotheses being used, choose the most appropriate statistical test and select the α level you will accept.

(10) Verify that you have an adequate sample size to answer the question.

(11) At this point, you are ready to begin collecting your data.

CONCEPTS AND PROCEDURES IN HYPOTHESIS TESTING FROM CHAPTER 6

CONTINGENCY TESTS ON 2 × 2 TABLES

Contingency Table of Prediction Versus Truth Showing n Symbols for Cell Counts, Marginal Totals, and Grand Total

		Prediction		Totals
		Yes	No	
Truth	Yes	n_{11}	n_{12}	$n_{1.}$
	No	n_{21}	n_{22}	$n_{2.}$
	Totals	$n_{.1}$	$n_{.2}$	n (or $n_{..}$)

Chi-square Test of Contingency to Test Independence. The expected value of a cell is

$$e_{ij} = \frac{n_{i.}n_{.j}}{n}.$$ (6.1)

The chi-square statistic with 1 degree of freedom (df) is calculated by

$$\chi^2 = \sum_i^2 \sum_j^2 \frac{(n_{ij} - e_{ij})^2}{e_{ij}}.$$ (6.2)

The critical value of χ^2 is found for the chosen α from the $df = 1$ line of Table III (see Tables of Probability Distributions). If χ^2 is larger than the critical value, reject H_0. To be valid, no cell count should be <5 or expected value <1.

RISKS AND ODDS IN MEDICAL DECISIONS

Table 6.6

Truth Table Showing Counts of the Prediction of Presence or Absence of a Malady as Related to the Truth of That Presence or Absence

		Prediction (exposure or test result)		
		Have disease	Do not have disease	
TRUTH	Have disease	n_{11} true positive	n_{12} false negative	$n_{1.}$(Yes)
	Do not have	n_{21} false positive	n_{22} true negative	$n_{2.}$(No)
		$n_{.1}$ (Predict Yes)	$n_{.2}$ (Predict No)	n (or $n_{..}$)

Table 6.8

Concepts of False Positive, False Negative, Sensitivity, Specificity, Accuracy, and Odds Ratio Based on Sample Error Rates Arising from the Format of Table 6.6

Values found from a sample truth table		Population definitions of probabilities and odds of values being estimated	
Names of estimates	Sample computing formulas	Definitions	Relationships
False-positive sample rate (p-value)	$n_{21}/n_{2.}$	Probability of a false positive (α)	$P(\text{predict yes} \mid \text{no})$
False-negative sample rate	$n_{12}/n_{1.}$	Probability of a false negative (β)	$P(\text{predict no} \mid \text{yes})$
Sensitivity	$n_{11}/n_{1.}$	Probability of a true positive: *power* $(1 - \beta)$	$P(\text{predict yes} \mid \text{yes})$
Specificity	$n_{22}/n_{2.}$	Probability of a true negative $(1 - \alpha)$	$P(\text{predict no} \mid \text{no})$
Accuracy	$(n_{11} + n_{22})/n_{..}$	Overall probability of a correct decision	$P(\text{predict no} \mid \text{no or yes} \mid \text{yes})$
Odds ratio	$\dfrac{n_{11}/n_{21}}{n_{12}/n_{22}}$ or $n_{11}n_{22}/n_{12}n_{21}$	Odds of a disease when predicted in ratio to odds of the disease when not predicted	$\dfrac{\text{ratio(yes/no} \mid \text{predicted yes})}{\text{ratio(yes/no} \mid \text{predicted no})}$

RANK-SUM TEST TO COMPARE TWO SAMPLES

Steps in Rank-Sum Test Using Table IX (see Tables of Probability Distributions):

1. Samples sizes are n_1 and n_2; n_1 is smaller. If $n_2 > 8$, use the normal approximation.
2. Rank combined data, keeping track of the sample from which each datum arose.
3. Add up the ranks of the data from the smaller sample and name it T.
4. Calculate $U = n_1 n_2 + n_1(n_1 + 1)/2 - T$.
5. Look up p-value from Table IX and use it to accept or reject the null hypothesis.

Normal Approximation to Rank-Sum Test:

1. Follow steps 1–3 just listed.
2. Calculate $\mu = n_1(n_1 + n_2 + 1)/2$, $\sigma^2 = n_1 n_2(n_1 + n_2 + 1)/12$, and $z = (T - \mu)/\sigma$.

3. Find the normal critical value from Table I. If z is larger than the critical value, reject H_0.

NORMAL OR t TEST TO COMPARE TWO SAMPLE MEANS

If σ is known or estimated from a large sample, use the normal test. If σ is estimated by s from a small sample, use the t test with $n_1 + n_2 - 2$ df. Calculate

$$z = (m_1 - m_2)/\sigma_d,$$

where

$$\sigma_d = \sigma \sqrt{\frac{1}{n_1} + \frac{1}{n_2}},$$

or

$$t = (m_1 - m_2)/s_d,$$

where

$$s_d = \sqrt{\left(\frac{1}{n_1} + \frac{1}{n_2}\right)\left[\frac{(n_1 - 1)s_1^2 + (n_2 - 1)s_2^2}{n_1 + n_2 - 2}\right]}.$$

Reject H_0 if z or t is larger than the critical z from Table I or the critical t from Table II.

CONCEPTS AND FORMULAS FOR SAMPLE SIZE FROM CHAPTER 7

CONCEPT OF CHOOSING SAMPLE SIZE FOR A MEANS TEST

See Fig. 7.1 on p. 545.

FORMULA TO CALCULATE MINIMUM REQUIRED SAMPLE SIZE FOR A TEST OF MEANS

For one mean tested against a theoretical mean μ, substitute error risks α and β, estimated standard deviation σ^2, and clinically important deviation d of sample mean m from theoretical mean μ, in the following equation to solve for n:

$$n = \frac{\left(z_{1-\alpha/2} + z_{1-\beta}\right)^2 \sigma^2}{d^2}$$

For two means, substitute α, β, the two standard deviation estimates, and $d = m_1 - m_2$ in the following formula:

$$n_1 = n_2 = \frac{\left(z_{1-\alpha/2} + z_{1-\beta}\right)^2 \left(\sigma_1^2 + \sigma_2^2\right)}{d^2}$$

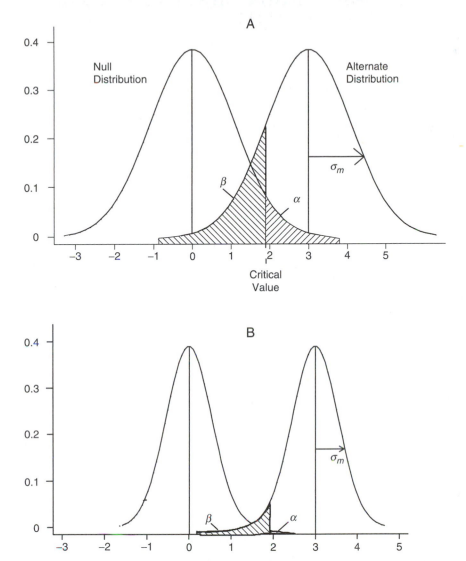

Figure 7.1 Distributions for null and alternate hypotheses in a test of means (A) with sizes of associated error probabilities indicated by *shaded areas* (see Section 5.2). If the sample size is quadrupled, $\sigma_m = \frac{\sigma}{\sqrt{n}}$ shrinks to half, yielding an equivalent diagram (B) with more slender curves. Note how much β has shrunk for a fixed α. The method of estimating a minimum sample size to yield specified error sizes for detecting a given distance between means is to start with those error sizes and that distance and "back-solve" the relationships to find the associated n.

FORMULAS AND RELATIONSHIPS OF STATISTICAL PREDICTION FROM CHAPTER 8

REGRESSION

The purpose is to predict a dependent variable y by an independent variable x. From analytic geometry, the point–slope form for a point (x_1, y_1) and slope b_1 is $y - y_1 = b_1(x - x_1)$. We choose the joint mean (m_x, m_y) as the point, and the best fit slope is shown to be

$$b_1 = \frac{s_{xy}}{s_x^2} = \frac{\text{cov}(x, y)}{\text{var}(x)}.$$

This yields a straight-line regression fit as

$$y - m_y = \frac{s_{xy}}{s_x^2}(x - m_x).$$

Assumptions: Errors in y are independent from one another; x is measured without error, and the standard deviation in y is the same for all x.

CORRELATION AS RELATED TO REGRESSION

Because the sample correlation coefficient r is

$$r = \frac{s_{xy}}{s_x s_y} = \frac{\text{cov}(x, y)}{sd(x)sd(y)},$$

the relationship between the slope of the regression line and the correlation coefficient is

$$r = b_1 \frac{s_x}{s_y}.$$

The implication is as follows. The closer the slope of the regression line is to $45°$, the greater is the relationship between x and y. The tighter is the envelope of data points about the plot of the correlation, the greater is the relationship between x and y.

Also, the square of the correlation coefficient, r^2, indicates the proportion of the possible causal influence on y by x; the closer r^2 is to 1.0, the better x is as a predictor of y. This statistic is termed the *coefficient of determination*. It usually is written R^2 because it largely is used with multiple independent variables.

OUTCOMES ANALYSIS

The concept of examining the total impact of a treatment on a patient (i.e., Is the patient's quality of life improved?) rather than a single clinical indicator (e.g., Is platelet count increased?) is termed *outcomes analysis*. The approach is to find a *measure of effectiveness* (MOE), usually a combination of indicators, that represents a total influence on the patient.

DEFINITIONS AND FORMULAS FOR EPIDEMIOLOGY FROM CHAPTER 9

Incidence Rate of a Disease. The rate at which new cases of the disease occur.

Prevalence Rate. the proportion of the population having that disease at a point in time.

Mortality Rate. The rate at which the population is dying.

If n denotes the number of people in the epidemiologic population, n_{new} denotes the number of new cases in a specified interval, $n_{present}$ denotes the number of cases present at any one point in time, and n_{dying} denotes the number dying during the specified interval, then:

Incidence rate is

$$I = 1000 \times \frac{n_{new}}{n}.$$

Prevalence rate is

$$P = 1000 \times \frac{n_{present}}{n}.$$

Mortality rate is

$$M = 1000 \times \frac{n_{dying}}{n}.$$

$$\text{Odds ratio (OR)} = \text{odds of a disease given a characteristic} \div$$
$$\text{odds of the disease without the characteristic}$$

Life tables provide the proportion surviving at end of associated interval. (Also, the probability that any randomly chosen member alive at the outset will survive longer than to the end of that interval.)

The following first six lines of the infant malaria morbidity table are shown to illustrate the format:

Interval (weeks)	Begin	Died	Lost	End	S (survival)
0 (outset)	155	0	0	155	1.0000
>0–13	155	14	0	141	0.9097
>13–26	141	23	0	118	0.7613
	118	0	3	115	
>26–39	115	17	0	98	0.6488

A survival curve shows survival and lost to follow-up (censored) by interval. A survival curve for infant malaria morbidity data is shown in the following figure to illustrate the format.

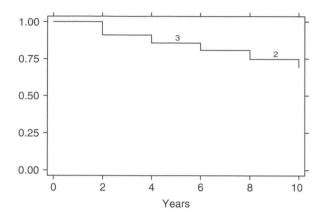

Criteria of evidence supporting causality:

(1) *Size of Effect.* Relative risk > 2.0.
(2) *Strength of Association.* $p < 0.05$.
(3) *Consistency of Association.* Reproducible effect.
(4) *Specificity of Association.* Effect from a single cause.
(5) *Temporality.* Cause precedes effect.
(6) *Biological Gradient.* Evidence of dose–response effect.
(7) *Biological Plausibility.* Reasonable explanatory model.

HINTS ON READING MEDICAL ARTICLES FROM CHAPTER 10

WAYS TO IMPROVE EFFICIENCY IN READING MEDICAL ARTICLES

(1) Allow enough time to *think* about the article.
(2) Identify the central question and read with it in mind.
(3) Think about how you would have attacked the problem posed.
(4) Verify that the author has specified goals; selected variables and descriptive statistics that satisfy the goals; chosen null and alternate hypotheses and error risks; described population and sample, and how biases were avoided; demonstrated that sample size was adequate; used appropriate statistical methods; and passed logically from goals through data and analyses to conclusions.
(5) Read the article repeatedly.
(6) Plan a revision to the study that would avoid any flaws you find.

STUDY TYPES

(1) *Registry.* An accumulation of data from an uncontrolled sample.
(2) *Case–Control.* Subjects selected for outcome (e.g., disease) and studied for risk factor.
(3) *Cohort.* Subjects selected for risk factor and studied for outcome.
(4) *Randomized Controlled Trial.* Subjects selected randomly, with risk factor and outcome being allowed to develop. A study b*linded* to subjects and investigators is best.

SOURCES OF BIAS

Sources of bias are myriad. Section 10.4 describes some of the more common sources.

SOME BIOSTATISTICAL ASPECTS OF ARTICLES TO WATCH FOR

(1) Confusion of statistical and clinical significance.
(2) Violation of assumptions upon which statistical methods are based.
(3) Generalization from a biased or poorly behaved sample to the population.
(4) Failure to define data formats, symbols, or statistical terms not obvious from context.
(5) Use of multiple related tests leading to a cumulative p-value.

META-ANALYSIS

A Meta-analysis Should Be Developed as Follows:

(1) Define the inclusion/exclusion criteria for admitting articles.
(2) Search exhaustively and locate all articles addressing the issue.
(3) Assess the articles against the criteria.
(4) Quantify the admitted variables on common scales.
(5) Aggregate the admitted databases through an organized scheme.

Biases That May Infiltrate an Integrative Literature Review:

(1) Data on which conclusions are based are not given or even summarized.
(2) Some of the studies were carried out with insufficient scientific rigor.
(3) Negative-result findings are absent, leading to overestimation of the success of an approach.

Criteria for an Acceptable Meta-analysis:

(1) The study objectives were clearly identified.
(2) Inclusion criteria of articles and data were established before selection.
(3) An active effort was made to find and include all relevant articles.
(4) An assessment of publication bias was made.
(5) Specific data used were identified.
(6) Assessment of article comparability (controls, circumstances, etc.) was made.
(7) The meta-analysis was reported in enough detail to allow replication.

CHAPTER 11

Chapter 11 has no Summary.

HINTS ON PLANNING MEDICAL STUDIES FROM CHAPTER 12

EVIDENCE-BASED MEDICINE

(1) Acquire evidence (medical history, clinical picture, test results, and relevant published studies).

(2) Prune for credibility, weighting and prioritizing that remaining according to its importance.

(3) Integrate evidence of different types and from different sources.

(4) Add nonmedical aspects (cost considerations, patient cooperation, and likelihood of follow-up).

(5) Embed the integrated totality of evidence into a decision model.

SAMPLING SCHEMES

(1) *Simple Random Sampling.* Every member of population is equally likely to be sampled.

(2) *Systematic Sampling.* For sample of n, divide population into k equal portions and randomly sample n/k from each.

(3) *Stratified Sampling.* Divide population into unequal proportions and randomly sample proportionally from each.

(4) *Cluster Sampling.* Randomly select a cluster and sample it 100%.

RANDOMIZATION

Random sampling selects subjects only on the basis of probability, for example, a number from a random number generator, not haphazard appearance. Too often haphazard sampling is designated incorrectly as random. A mechanism that produces a number solely by chance is a random number generator. Such numbers may be generated mechanically, electronically, or drawn from prepared tables.

PLANNING A STUDY

(1) Start with objectives. Do not start with the abstract.

(2) Develop the background from prior studies and relevance. (What will it contribute?)

(3) Plan your materials, methods, and data, and their integration in conducting the study.

(4) Define your population. Verify representative sampling. Ensure adequate sample size.

(5) Anticipate which statistical analysis will yield results that will satisfy your objectives.

(6) Plan the bridge from results to conclusions (the discussion).

(7) Anticipate the form in which your conclusions will be expressed.

(8) Now write the abstract, summarizing all the foregoing in half a page to a page.

(9) After drafting this terse summary, review steps (1) through (8) and revise as required.

SOME MECHANISMS TO FACILITATE STUDY PLANNING

(1) *Work backward through the logical process.* (a) What conclusions are needed to answer these questions? (b) What data results and how many data will I need to reach these conclusions? (c) What statistical methods will I need to obtain these results? (d) What is the nature and format of the data I need to apply these statistical methods? (e) What is the design and conduct of the study I need to obtain these data? (f) And, finally, what is the ambiance in the literature that leads to the need for this study in general and this design in particular?

(2) *Analyze dummy data.* Make up representative numbers and analyze them.

(3) *Play the role of devil's advocate.* Criticize your work as if you were a reviewer.

DATA MANAGEMENT

(1) Plan data usage start to finish. Note what data will be needed in final analysis.

(2) Ensure that raw data are all present and sufficiently accurate.

(3) Reformat raw data in a form, including entirely numerical, consistent with a statistical software package that will answer the study's questions.

(4) Verify the correctness of the data at each stage.

A FIRST-STEP GUIDE TO CHOOSING STATISTICAL TESTS

Table 12.1 also appears on the inside back cover. It provides "first aid" in selecting a statistical test for a given set of conditions. The method of selection is explained in the legend for this table.

ETHICS IN STUDY DESIGN

(1) *Keep the Patient in Mind.* The design of studies affects not only scientific knowledge but also patients. A study affects two groups of patients: those used in the study and the future patient who will be affected by the study results.

(2) *What Affects the Patient?* Factors that affect the patient include (1) the choice of the clinically important difference δ between a sample and a population characteristic (e.g., a mean) or between that of two samples; (2) the choice of α, the risk for a false positive; (3) the choice of β, the risk for a false negative; (4) the $\alpha:\beta$ ratio; (5) the sidedness of the test chosen; and (f) the sample size chosen.

(3) *Verify That the Study Will Not Affect Patients in Unanticipated Ways.* After the study is planned to satisfy scientific and statistical design criteria, review the study design with both current and future patients in mind. If the design will affect

either group in an undesirable way, revise the design as necessary to eliminate undesirable aspects.

(4) *Ethical Considerations Beyond the Statistical.* Various international, national, and state governments have instituted extensive patient protection procedures to which the investigator must conform. Be familiar with these in the early stages of any study.

FINDING PROBABILITIES OF ERROR GIVEN IN TABLES FROM CHAPTER 13

The distributions from which Tables I–VII are obtained appear on p. 553. (Tables VIII and IX are specialty tables from Chapter 16 on rank-order methods.) The tables provide probabilities (areas under portions of the curves) for left, right, or both tails for different members of the curve families as affected by sample size.

FORMULAS FOR CONFIDENCE INTERVALS FROM CHAPTER 14

General Form of Confidence Interval on a Statistic:

The probability

that a population statistic

from a distribution of estimates of that statistic

is contained in a specified interval

is given by

the area of the distribution within that interval.   (14.1)

or

P[lower critical value < population statistic < upper critical value] $= 1 - \alpha$.
                    (14.2)

1 − α Confidence Interval on a Mean, Known (Population) Standard Deviation σ:

$$P[m - z_{1-\alpha/2}\sigma_m < \mu < m + z_{1-\alpha/2}\sigma_m] = 1 - \alpha \qquad (14.3)$$

Above Interval for 95% Confidence:

$$P[m - 1.96 \times \sigma_m < \mu < m + 1.96 \times \sigma_m] = 0.95 \qquad (14.4)$$

1 − α Confidence Interval on a Mean, Estimated (Sample) Standard Deviations:

$$P[m - t_{1-\alpha/2}s_m < \mu < m + t_{1-\alpha/2}s_m] = 1 - \alpha \qquad (14.7)$$

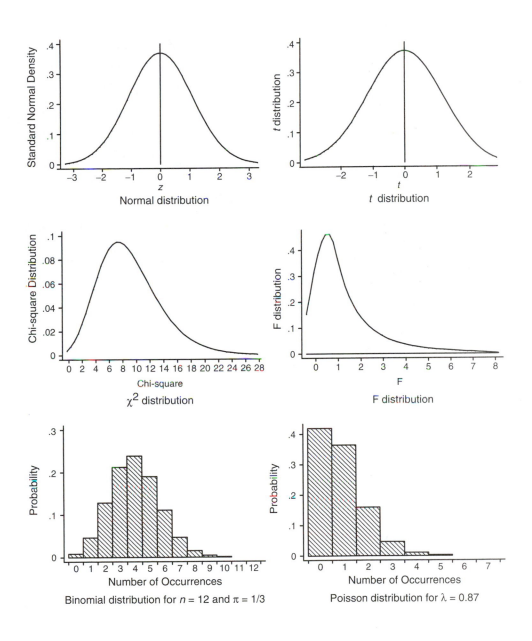

1 − α Confidence Interval on a Variance or Standard Deviation:

$$P[s^2 \times df/\chi_R^2 < \sigma^2 < s^2 \times df/\chi_L^2] = 1 - \alpha, \tag{14.8}$$

where χ_R^2 is Table III (see Tables of Probability Distributions) area under right tail of chi-square (χ^2) and χ_L^2 is Table IV (see Tables of Probability Distributions) area under left tail. For a confidence interval on the standard deviation, take the square root of each component within the brackets in Eq. (14.8).

Confidence Interval on a Proportion. If π, the unknown population proportion, is not near 0 or 1, calculate the sample proportion p and $\sigma = \sqrt{p(1-p)/n}$ and substitute to obtain

$$P[p - 1.96 \times \sigma - 1/2n < \pi < p + 1.96 \times \sigma + 1/2n] = 0.95. \tag{14.9}$$

If π is near 0 or 1, substitute p and $\sigma = \sqrt{p/n}$ to obtain

$$P[p - 1.96 \times \sigma < \pi < p + 1.96 \times \sigma] = 0.95. \tag{14.10}$$

If confidence other than 95% is desired, replace 1.96 as appropriate from Table I (see Tables of Probability Distributions).

Confidence Interval on a Correlation Coefficient. The normal-transformed correlation coefficient has mean

$$m = \frac{1}{2} \ln \frac{1+r}{1-r} \tag{14.11}$$

and standard deviation

$$\sigma = \frac{1}{\sqrt{n-3}}. \tag{14.12}$$

A $1 - \alpha$ (e.g., 95%) confidence interval is

$$P\left[\frac{1+r-(1-r)e^{\frac{2z_{1-\alpha/2}}{\sqrt{n-3}}}}{1+r+(1-r)e^{\frac{2z_{1-\alpha/2}}{\sqrt{n-3}}}} < \rho < \frac{1+r-(1-r)e^{\frac{2z_{1-\alpha/2}}{\sqrt{n-3}}}}{1+r+(1-r)e^{\frac{2z_{1-\alpha/2}}{\sqrt{n-3}}}}\right] = 1 - \alpha. \tag{14.13}$$

FORMULAS FOR TESTS OF CATEGORIES FROM CHAPTER 15

TRUTH TABLE, SHOWING FREQUENCIES n_{ij}

		Prediction		
		Yes	No	
Truth	Yes	n_{11} (correct decision)	n_{12} (Type II error)	$n_{1.}$ (truly yes)
	No	n_{21} (Type I error)	n_{22} (correct decision)	$n_{2.}$ (truly no)
		$n_{.1}$ (predict yes)	$n_{.2}$ (predict no)	n (or $n_{..}$)

CONTINGENCY TABLES: $r \times c$ (FISHER'S EXACT AND CHI-SQUARE TESTS)

A 2×2 table is a special case in which $r = c = 2$. H_0: Row variable independent of column variable. Fisher's exact test (FET) and its p-value are calculated by a statistical software package for small r and c. If $r + c > 7$, every $e_{ij} \geq 1$, and every $n_{ij} \geq 5$, Table III (see Tables of Probability Distributions) gives the p-value (looked up as if it were α) from a chi-square statistic with $(r - 1)(c - 1)$ degrees of freedom (df), where $e_{ij} = n_{i.} \times n_{.j} \div n$, using

$$\chi^2 = \sum_{i}^{r} \sum_{j}^{c} \frac{(n_{ij} - e_{ij})^2}{e_{ij}}. \tag{15.3}$$

RISKS AND ODDS IN MEDICAL DECISIONS

Names of sample estimates	Sample formulas	Population definitions of parameters being estimated	Relationships
False-positive rate (p-value)	n_{21}/n_2.	P(false positive): α	P(predict Y | N)
False-negative rate	n_{12}/n_1.	P(false negative): β	P(predict N | Y)
Sensitivity	n_{11}/n_1.	P(true positive): $1 - \beta$	P(predict Y | Y)
Specificity	n_{22}/n_2.	P(true negative): $1 - \alpha$	P(predict N | N)
Accuracy	$(n_{11} + n_{22})/n$	Overall P(correct decision)	P(predict N | N or Y | Y)
Positive predictive value	$n_{11}/n_{.1}$	P(positive prediction is correct)	P(Y | predicted Y)
Negative predictive value	$n_{22}/n_{.2}$	P(negative prediction is correct)	P(N | predicted N)
Relative risk	$n_{11}n_{.2}/n_{12}n_{.1}$	P(disease predicted) in ratio to P(disease not)	P(Y | predicted Y) P(Y | predicted N)
Odds ratio	$n_{11}n_{22}/n_{12}n_{21}$	Odds of disease predicted in ratio to odds of disease not predicted	Ratio(Y/N | predicted Y) Ratio(Y/N | predicted N)
Likelihood ratio	$n_{11}n_{2.}/n_{21}n_{1.}$	P(correctly predicting disease) in ratio to P(incorrectly predicting)	P(predict Y | Y) P(predict Y | N)
Attributable risk	$\dfrac{n_{11}}{n_{.1}} - \dfrac{n_{12}}{n_{.2}}$	Amount of incidence rate attributable to exposure factor	P(Y | predicted Y) $-$ P(Y | predicted N)

N, no; Y, yes.

Receiver Operating Characteristic Curves. A receiver operating characteristic (ROC) curve displays rate of true positive (vertical axis) against rate of false positive (horizontal axis). It can help choose the best (weighted if desired) critical value for a medical decision (sensitivity-to-specificity trade-off). It can help choose the better predictor of two risk factors.

TESTS OF ASSOCIATION

Log Odds Ratio (L) *Test.* H_0: L = a hypothesized log odds ratio λ ($\lambda = 0$ if the category types are independent) tested by

$$\chi^2 = \left(\frac{L - \lambda}{SEL}\right)^2, \tag{15.6}$$

where

$$L = \ln\left[\frac{(n_{11} + 0.5) \times (n_{22} + 0.5)}{(n_{12} + 0.5) \times (n_{21} + 0.5)}\right], \tag{15.4}$$

with standard error

$$SEL = \sqrt{\left(\frac{1}{n_{11} + 0.5}\right) + \left(\frac{1}{n_{12} + 0.5}\right) + \left(\frac{1}{n_{21} + 0.5}\right) + \left(\frac{1}{n_{22} + 0.5}\right)}. \tag{15.5}$$

Table III using 1 *df* gives the *p*-value (looked up as if it were α).

TEST OF A PROPORTION π NOT CLOSE TO ZERO

π is estimated by the sample proportion $p_s = n_s/n$.

Small Sample (Binomial Calculation). H_0: $p_s = \pi$. Table VI (see Tables of Probability Distributions) gives the *p*-value, the probability that $p_s > \pi$ by chance alone.

Large Sample (Normal Approximation). H_0: $p_s = \pi$, and $n \geq 5/\pi$. Table I (see Tables of Probability Distributions) gives the two-tailed *p*-value (looked up as if it were α) for

$$z = \frac{p_s - \pi}{\sqrt{\pi(1 - \pi)/n}}. \tag{15.8}$$

TEST OF TWO PROPORTIONS

A normal statistic to test H_0: $\pi_1 = \pi_2$ is calculated by

$$z = \frac{p_1 - p_2}{\sqrt{p_c(1 - p_c)\left(\frac{1}{n_1} + \frac{1}{n_2}\right)}}. \tag{15.11}$$

TEST OF A PROPORTION π NEAR ZERO

π is estimated by the sample proportion $p_s = n_s/n$.

Small Sample (Poisson Approximation). H_0: $p_s = \pi$, π is near 0, and $n \gg n\pi \gg \pi$. $\lambda = n\pi$. Table VII (see Tables of Probability Distributions) gives the probability that $p_s > \pi$ by chance alone (*p*-value), given n_s and λ.

Large Sample (Normal Approximation). $\lambda > 9$. H_0: $p_s = \pi$, π is near 0, and $n \gg n\pi \gg \pi$. $\lambda = n\pi$. Table I gives *p*-value (looked up as if it were α) for

$$z = (p_s - \pi)\sqrt{n/\pi}. \tag{15.13}$$

MATCHED PAIR SAMPLE (MCNEMAR'S TEST)

To test a certain characteristic for association with a disease, pair n patients with disease and n control patients. b = number of cases with disease yes but control no, and c = number of cases with disease no but control yes. Calculate chi-square with 1 *df*. Table III gives the *p*-value (looked up as if it were α).

$$\chi^2_{1df} = \frac{(|b - c| - 1)^2}{b + c} \tag{15.14}$$

FORMULAS FOR TESTS OF RANKS FROM CHAPTER 16

SINGLE OR PAIRED SMALL SAMPLES: THE SIGNED-RANK TEST

Hypothesis: median difference = 0. It may test (1) observations deviating from a hypothetical common value or (2) pairs on the same individuals, such as before and after data.

1. Calculate the differences of the observations as in (1) or (2).
2. Rank the magnitudes (i.e., the differences without signs).
3. Reattach the signs to the ranks.
4. Add up the positive and negative ranks.
5. Denote by T the unsigned value of the smaller sum of ranks.
6. Look up the *p*-value for the test in Table VIII (see Tables of Probability Distributions). If $n > 12$, use the normal approximation.

TWO SMALL SAMPLES: THE RANK-SUM TEST

Think of the rank-sum test informally as testing if the two distributions have the same median.

1. Name the sizes of the two samples n_1 and n_2; n_1 is the smaller sample.
2. Combine the data, keeping track of the sample from which each datum arose.
3. Rank the data.
4. Add up the ranks of the data from the smaller sample and name it T.
5. Calculate $U = n_1 n_2 + n_1(n_1 + 1)/2 - T$.

6. Look up the p-value from Table IX (see Tables of Probability Distributions) and use it to decide whether or not to reject the null hypothesis. (Table IX gives two-tailed error probabilities.)

THREE OR MORE INDEPENDENT SAMPLES: THE KRUSKAL–WALLIS TEST

The Kruskal–Wallis test is just the rank-sum test extended to more than two samples. Think of it informally as testing if the distributions have the same median. The chi-square (χ^2) approximation requires five or more members per sample.

1. Name the number of samples m $(3, 4, \ldots)$.
2. Name the sizes of the several samples $n_1, n_2, \ldots, n_m$; n is the grand total.
3. Combine the data, keeping track of the sample from which each datum arose.
4. Rank the data.
5. Add up the ranks of the data from each sample separately.
6. Name the sums $T_1, T_2, \ldots, T_m$.
7. Calculate the Kruskal–Wallis H statistic, which is distributed as chi-square, by

$$H = \frac{12}{n(n+1)} \left(\frac{T_1^2}{n_1} + \frac{T_2^2}{n_2} + \cdots + \frac{T_k^2}{n_k} \right) - 3(n+1). \qquad (16.1)$$

Obtain the p-value (as if it were α) from Table III (χ^2 right tail) (see Tables of Probability Distributions) for $m - 1$ degrees of freedom (df).

THREE OR MORE "PAIRED" SAMPLES: THE FRIEDMAN TEST

The Friedman test is an extension of the paired-data concept. In the example, a dermatologist applied three skin patches to test for an allergy to each of eight patients. Hypothesis: the several treatments have the same distributions. This test requires five or more patients.

1. Name the number of treatments k $(3, 4, \ldots)$ and of blocks (e.g., patients) n.
2. Rank the data within each block (e.g., rank the treatment outcomes for each patient).
3. Add the ranks for each treatment separately; name the sums $T_1, T_2, \ldots, T_k$.
4. Calculate the Friedman F_r statistic, which is distributed as chi-square, by

$$F_r = \frac{12}{nk(k+1)} \left(T_1^2 + T_2^2 + \cdots + T_k^2 \right) - 3n(k+1). \qquad (16.2)$$

5. Obtain the p-value (as if it were α) from Table III (χ^2 right tail) for $k - 1$ df.

SINGLE LARGE SAMPLES: THE NORMAL APPROXIMATION TO THE SIGNED-RANK TEST

Hypothesis: median difference = 0. It may test (1) observations deviating from a hypothetical common value or (2) pairs on the same individuals, such as before and after data.

1. Calculate the differences of the observations as in (1) or (2).
2. Rank the magnitudes (i.e., the differences without signs).
3. Reattach the signs to the ranks.
4. Add up the positive and negative ranks.
5. Denote by T the unsigned value of the smaller sum of ranks; n is sample size (number of ranks).
6. Calculate $\mu = n(n + 1)/4$, $\sigma^2 = (2n + 1)\mu/6$, and then $z = (T - \mu)/\sigma$.
7. Obtain the p-value (as if it were α) from Table I (see Tables of Probability Distributions) for a two- or one-tailed test as appropriate.

TWO LARGE SAMPLES: THE NORMAL APPROXIMATION TO THE RANK-SUM TEST

Think of the rank-sum test for large samples informally as testing if the two distributions have the same median.

1. Name the sizes of the two samples n_1 and n_2; n_1 is the smaller sample.
2. Combine the data, keeping track of the sample from which each datum arose.
3. Rank the data.
4. Add up the ranks of the data from each sample separately.
5. Denote by T the sum associated with n_1.
6. Calculate $\mu = n_1(n_1 + n_2 + 1)/2$; $\sigma^2 = n_1 n_2(n_1 + n_2 + 1)/12$; and $z = (T - \mu)/\sigma$.
7. Obtain the p-value (as if it were α) from Table I (see Tables of Probability Distributions) for a two- or one-tailed test as appropriate.

FORMULAS FOR TESTS OF MEANS FROM CHAPTER 17

SINGLE OR PAIRED NEAR-NORMAL SAMPLES: NORMAL AND t TESTS

Paired data: change to single by subtraction (e.g., before − after a treatment). Assume that data form a normal distribution. H_0: $\mu_0 = \mu$. Choose H_1: $\mu_0 \neq \mu$, $\mu_0 < \mu$, or $\mu_0 > \mu$. Choose α. If σ_m is known or sample is large, calculate z and use Table I (see Tables of Probability Distributions).

$$z = \frac{m - \mu}{\sigma_m} = \frac{m - \mu}{\sigma/\sqrt{n}} \tag{17.1}$$

If n is small and σ is estimated by s, calculate t and use Table II (see Tables of Probability Distributions) with $n - 1$ df.

$$t = \frac{m - \mu}{s_m} = \frac{m - \mu}{s/\sqrt{n}} \tag{17.2}$$

TWO NEAR-NORMAL SAMPLES: NORMAL AND t TESTS

Assume that data form a normal distribution. H_0: $\mu_1 = \mu_2$. Choose H_1: $\mu_1 \neq \mu_2$, H_1: $\mu_1 < \mu_2$, or H_1: $\mu_1 > \mu_2$. Choose α. Method depends on variances equal or not equal, according to the following table:

Total sample size	Subgroup sample size	Variances about the same	Variances somewhat different	Variances extremely different
Large size *or* σ's known	About equal	Normal (z) test, equal variances	Normal (z) test, equal variances	Rank-sum test
	Very different		Normal (z) test, unequal variances	Rank-sum test
Small size, s estimates σ	About equal	t test, equal variances	t test, equal variances	Rank-sum test
	Very different		t test, unequal variances	Rank-sum test

If σ's composing σ_d are known or samples are large, calculate z and use Table I.

$$z = (m_1 - m_2)/\sigma_d, \tag{17.5}$$

where

$$\sigma_d = \sigma\sqrt{\frac{1}{n_1} + \frac{1}{n_2}} \tag{17.6}$$

if variances are equal, and

$$\sigma_d = \sqrt{\frac{\sigma_1^2}{n_1} + \frac{\sigma_2^2}{n_2}} \tag{17.7}$$

if variances are unequal.

If n's are small and s's estimate σ's, calculate t and use Table II.

$$t = (m_1 - m_2)/s_d, \tag{17.8}$$

with $df = n_1 + n_2 - 1$, where

$$s_d = \sqrt{\left(\frac{1}{n_1} + \frac{1}{n_2}\right)\left[\frac{(n_1 - 1)s_1^2 + (n_2 - 1)s_2^2}{n_1 + n_2 - 2}\right]} \tag{17.9}$$

if variances are equal, and

$$s_d = \sqrt{\frac{s_1^2}{n_1} + \frac{s_2^2}{n_2}}$$ (17.10)

with *df* rounded to the integer nearest to

$$\text{approx. } (df) = \frac{\left(\dfrac{s_1^2}{n_1} + \dfrac{s_2^2}{n_2}\right)^2}{\dfrac{\left(\dfrac{s_1^2}{n_1}\right)^2}{n_1 - 1} + \dfrac{\left(\dfrac{s_2^2}{n_2}\right)^2}{n_2 - 1}}$$ (17.11)

if variances are unequal.

THREE OR MORE MEANS: ONE-WAY ANALYSIS OF VARIANCE

Assume: (1) distributed approximately normal and (2) variances approximately equal. H_0: No differences among means. H_1: One or more differences among means. Choose α and find the associated critical value of F in Table V (see Tables of Probability Distributions) for $k - 1$ and $n - k$ *df*. Total sample has n observations, mean m, and variance s^2, divided into k groups having n_1, $n_2, \ldots, n_k$ observations with group means $m_1, m_2, \ldots, m_k$.

Source of variability	Sum of squares			Variances (mean squares)	
	Designation	Formula	*df*	Designation	Formula
Mean	SSM	$\sum n_i(m_i - m)^2$	$k - 1$	s_m^2 (or MSM)	SSM/$(k - 1)$
Error	SSE	SST $-$ SSM	$n - k$	s_e^2 (or MSE)	SSE/$(n - k)$
Total	SST	$(n - 1)s^2$	$n - 1$	s^2 (or MST)	SST/$(n - 1)$

MSE, mean square of error; MSM, mean square of means; MST, mean square of the total; SSE, sum of squares for error; SSM, sum of squares for means; SST, sum of squares for the total.

Calculate $F = \text{MSM/MSE} = s_m^2/s_e^2$ and compare with the critical value from Table V.

Multiple comparisons using statistical software: Outcomes are *p*-values adjusted so that each may be compared with chosen overall α while retaining this overall α. If statistical software is not available, use Bonferroni's approach: Make two-sample *t* tests on each pair but choose the critical *t* from Table II using $2\alpha/k(k - 1)$ rather than α, interpolating as required.

FORMULAS FOR MULTIFACTOR TESTS ON MEANS FROM CHAPTER 18

For two-factor analysis of variance (ANOVA), with the following definitions,

$$i = 1, 2, \ldots, r$$
$$j = 1, 2, \ldots, c$$
$$k = 1, 2, \ldots, w$$
$$m_{ij} = \sum_{k=1}^{w} x_{ijk}/w$$

use the following formulas of Table 18.2 to complete the ANOVA display in Table 18.3.

Table 18.2

Formulas for Components in a Two-Factor Analysis of Variance[a]

$$A = rcw \times m_{..}^2$$
$$SST = \sum_i^r \sum_j^c \sum_k^w x_{ijk}^2 - A$$
$$SSR = cw \sum_i^r m_{i.}^2 - A$$
$$SSC = rw \sum_j^c m_{.j}^2 - A$$
$$SSI = w \sum_i^r \sum_j^c m_{ij}^2 - A - SSR - SSC$$
$$SSE = SST - SSR - SSC - SSI$$

[a]The symbolism $\sum_i^r$ implies "sum over i from 1 to r," and the equivalent is true for other indicator symbols.
SSC, sum of squares for columns; SSE, sum of squares for error (or residual); SSI, sum of squares for interaction (row-by-column); SSR, sum of squares for rows; SST, sum of squares for the total.

Table 18.3

Two-Factor Analysis of Variance

Source	Sums of squares	df	Mean squares	F	p
Rows	SSR from Table 18.2	$r - 1$	MSR = SSR/df	MSR/MSE	
Columns	SSC from Table 18.2	$c - 1$	MSC = SSC/df	MSC/MSE	
Interaction (r × c)	SSI from Table 18.2	$(r - 1)(c - 1)$	MSI = SSI/df	MSI/MSE	
Error (residual)	SSE from Table 18.2	$rc(w - 1)$	MSE = SSE/df		
Total	SST from Table 18.2	$rcw - 1$			

df, degrees of freedom; MSC, mean square of columns; MSE, mean square of error; MSI, mean squares for interaction (row-by-column); MSR, mean squares for rows; SSC, sum of squares for columns; SSE, sum of squares for error; SSI, sum of squares for interaction (row-by-column); SSR, sum of squares for rows; SST, sum of squares for the total.

For repeated measures two-factor ANOVA, add the supplemental formulas Table 18.4 to complete the ANOVA table (see Table 18.5).

For three-factor or more ANOVA and for analysis of covariance (ANCOVA), use statistical software as guided by the text in Chapter 18.

Table 18.4

Formulas Supplemental to Those in Table 18.2 for Components in a Repeated-Measures Analysis of Variance

$$\text{SSAcross} = \sum_j^c m_{i \cdot k}^2 - A$$

$$\text{SSE(B)} = \text{SSAcross} - \text{SSC}$$

$$\text{SSE(W)} = \text{SST} - \text{SSAcross} - \text{SSR} - \text{SSI}$$

SSAcross, sum of squares across repeated measures; SSC, sum of squares for columns; SSE(B), between-repeated-measures sum of squares for error; SSE(W), within-repeated-measures sum of squares for error; SSI, sum of squares for interaction (row-by-column); SSR, sum of squares for rows; SST, sum of squares for the total.

Table 18.5

Repeated-Measures (Two-Factor) Analysis of Variance

Source	Sums of squares	df	Mean squares	F	p
Columns (groups)	SSC (Table 18.2)	$c - 1$	MSC = SSC/df	MSC/MSE(B)	
Error between	SSE(B) (Table 18.4)	$c(w - 1)$	MSE(B) = SSE(B)/df		
Rows (repeated)	SSR (Table 18.2)	$r - 1$	MSR = SSR/df	MSR/MSE(W)	
Interaction	SSI (Table 18.2)	$(r - 1)(c - 1)$	MSI = SSI/df	MSI/MSE(W)	
Error within	SSE(W) (Table 18.4)	$c(r - 1)(w - 1)$	MSE(W) = SSE(W)/df		
Total	SST (Table 18.2)	$rcw - 1$			

df, degrees of freedom; MSC, mean square of columns; MSE(B), between-repeated-measures mean square of error; MSE(W), within-repeated-measures mean square of error; MSI, mean squares for interaction (row-by-column); MSR, mean squares for rows; SSAcross, sum of squares across repeated measures; SSC, sum of squares for columns; SSE(B), between-repeated-measures sum of squares for error; SSE(W), within-repeated-measures sum of squares for error; SSI, sum of squares for interaction (row-by-column); SSR, sum of squares for rows; SST, sum of squares for the total.

FORMULAS FOR TESTS OF VARIANCES FROM CHAPTER 19

SINGLE SAMPLES

Assume that data form a normal distribution. Identify the theoretical σ. Choose α. $H_0: \sigma_0^2 = \sigma^2$. $H_1: \sigma_0^2 > \sigma^2$ (see Table III in Tables of Probability Distributions), $H_1: \sigma_0^2 < \sigma^2$ (see Table IV in Tables of Probability Distributions), or $H_1: \sigma_0^2 \neq \sigma^2$ (see Tables III and IV).

$$\chi^2 = \frac{df \times s^2}{\sigma^2} \tag{19.1}$$

Use the critical value(s) of χ^2 (with $n - 1\,df$) to decide whether or not to reject H_0.

TWO SAMPLES

Assume normality. s_1^2 is the larger variance. H_0: $\sigma_1^2 = \sigma_2^2$. H_1: $\sigma_1^2 > \sigma_2^2$. Choose α.

$$F = \frac{s_1^2}{s_2^2} \tag{19.2}$$

Use the critical value of $F(n_1 - 1,\ n_2 - 1\ df)$ from Table V (see Tables of Probability Distributions) to decide whether or not to reject H_0.

THREE OR MORE SAMPLES

Assume normality in all samples. H_0: $\sigma_1^2 = \sigma_2^2 = \ldots = \sigma_k^2$. H_1: Not H_0. Choose α. i denotes sample number, 1 to k. n is total number. n_i is number of data x_i in ith sample. Find usual variance s_i^2 for each sample:

$$s_i^2 = \frac{\sum x_i^2 - n_i m_i^2}{n_i - 1} \tag{19.3}$$

Pool the k sample variances to find the overall variance s^2:

$$s^2 = \frac{\sum (n_i - 1)s_i^2}{n - k} \tag{19.4}$$

The test statistic, Bartlett's M, is given by

$$M = \frac{(n - k)\ln(s^2) - \sum (n_i - 1)\ln(s_i^2)}{1 + \dfrac{1}{3(k - 1)}\left(\sum \dfrac{1}{n_i - 1} - \dfrac{k}{n - k}\right)}. \tag{19.5}$$

Use critical value of χ^2 (with $k - 1\,df$) from Table III to decide whether or not to reject H_0.

FORMULAS FOR TESTS OF DISTRIBUTION SHAPE FROM CHAPTER 20

TESTS OF NORMALITY OF A DISTRIBUTION

Details of the Kolmogorov–Smirnov (KS) and chi-square goodness-of-fit methods are given next. (Use the Shapiro–Wilk test only with a statistical software package.)

Table 20.1

Partial Guide to Selecting a Test of Normality of a Distribution

	Prefer less conservative test	Prefer more conservative test
Small sample (5–50)	Shapiro–Wilk test	Kolmogorov–Smirnov test (one-sample form)
Medium to large sample (>50)	Shapiro–Wilk test	Chi-square goodness-of-fit test

KOLMOGOROV–SMIRNOV TEST (ONE-SAMPLE FORM)

Format of Table Providing Calculations Required for the Kolmogorov–Smirnov Test of Normality

Data	x	k	$F_n(x)$	z	$F_e(x)$	$|F_n(x) - F_e(x)|$
⋮	⋮	⋮	⋮	⋮	⋮	⋮
⋮	⋮	⋮	⋮	⋮	⋮	⋮

Hypotheses are H_0: Sample distribution not different from specified normal distribution; and H_1: Sample distribution is different. Choose α.

1. Arrange the n sample values in ascending order.
2. Let x denote the sample value each time it changes. Write x-values in a column in order.
3. Let k denote the number of sample members less than x. Write k-values next to x-values.
4. List each $F_n(x) = k/n$ in the next column corresponding to the associated x.
5. List $z = (x - \mu)/\sigma$ for each x to test against an *a priori* distribution or $z = (x - m)/s$ for each x to test for a normal shape but not a particular normal.
6. For each z, list an expected $F_e(n)$ as the area under the normal distribution to the left of z. (This area can be found from software, from a very complete normal table, or by interpolation from Table I.)
7. List next to these the differences $|F_n(x) - F_e(x)|$.
8. The test statistic is the largest of these differences, say, L.
9. Calculate the critical value. For a 5% α, use $(1.36/\sqrt{n}) - (1/4.5n)$. Critical values for $\alpha = 1\%$ and 10% are given in the text. If L is greater than the critical value, reject H_0; otherwise, do not reject H_0.

LARGE-SAMPLE TEST OF NORMALITY OF A DISTRIBUTION: CHI-SQUARE GOODNESS-OF-FIT TEST

1. Choose α. Define k data intervals as in forming a histogram of the data.
2. Standardize interval ends by subtracting mean and dividing by standard deviation.

Table 20.4

Format for Table of Values Required to Compute the Chi-square Goodness-of-Fit Statistic

Interval	Standard normal z to end of interval	P	Expected frequencies (e_i)	Observed frequencies (n_i)
⋮	⋮	⋮	⋮	⋮
⋮	⋮	⋮	⋮	⋮

3. Find the area under the normal curve to the end of an interval from a table of normal probabilities and subtract the area to the end of the preceding interval.
4. Multiply normal probabilities by n to find expected frequencies e_i.
5. Tally the number of data n_i in each interval.
6. Calculate χ^2 value: $\chi^2 = \sum \frac{n_i^2}{e_i} - n$.
7. Find critical χ^2 from Table III (see Tables of Probability Distributions). If calculated χ^2 is greater than critical χ^2, reject H$_0$; otherwise, do not reject H$_0$.

TEST OF EQUALITY OF TWO DISTRIBUTIONS (KOLMOGOROV–SMIRNOV TEST)

Table 20.8

Format for Table of Data and Calculations Required for the Two-Sample Kolmogorov–Smirnov Test of Equality of Distributions

Ordered data	k_1	k_2	F_1	F_2	$|F_1 - F_2|$
⋮	⋮	⋮	⋮	⋮	⋮
⋮	⋮	⋮	⋮	⋮	⋮

H$_0$: The population distributions from which the samples arose are not different; H$_1$: They are different. The sample sizes are n_1 and n_2; n_1 is the larger sample. Choose α.

1. Combine the two data sets and list in ascending order.
2. For each sample 1 datum different from the datum above it, list for k_1 the number of data in sample 1 preceding it. Repeat the process for sample 2, listing entries for k_2.
3. For every k_1, list $F_1 = k_1/n_1$. Repeat for sample 2. In all blanks, list F from line above.
4. List $|F_1 - F_2|$ for every datum.
5. The test statistic is the largest of these differences; call it L.
6. Calculate critical value. For 5% α, it is $1.36\sqrt{\frac{n_1+n_2}{n_1 n_2}}$; for 1% and 10%, see text.

If L is greater than the critical value, reject H_0; otherwise, do not reject H_0.

FORMULAS FOR EQUIVALENCE TESTING FROM CHAPTER 21

To test if d is not more than a critical value distance from Δ in standardized units (i.e., *nonsuperiority testing*), the test statistics are

$$z = \frac{\Delta - d}{\sigma_d} \qquad (21.1)$$

or

$$t = \frac{\Delta - d}{s_d}. \qquad (21.2)$$

If the test statistic z or t is as large as z_{1-a} or t_{1-a}, respectively, reject H_0, which implies accepting nonsuperiority.

Table 21.1

Test Cases and Standard Deviations with Corresponding Standard Errors and Probability Tables from Which to Find Critical Values[a,b]

Case	Standard deviation(s)	s_d (standard error)	Probability table[c]
$d = m - \mu$	σ	$\dfrac{\sigma}{\sqrt{n}}$	z table (Table I)
$d = p - \pi$	Assumed[d]	$\sqrt{\dfrac{\pi(1-\pi)}{n}}$	z table (Table I)
$d = m - \mu$	s $df = n - 1$	$\dfrac{s}{\sqrt{n}}$	t table (Table II)
$d = m_1 - m_2$	σ_1, σ_2 $\sigma_1 = \sigma_2 = \sigma$	$\sigma\sqrt{\dfrac{1}{n_1} + \dfrac{1}{n_2}}$	z table (Table I)
$d = m_1 - m_2$	σ_1, σ_2 $\sigma_1 \neq \sigma_2$	$\sqrt{\dfrac{\sigma_1^2}{n_1} + \dfrac{\sigma_2^2}{n_2}}$	z table (Table I)
$d = p_1 - p_2$	Assumed[d]	$\sqrt{\left(\dfrac{1}{n_1} + \dfrac{1}{n_2}\right)\left(\dfrac{n_1 p_1 + n_2 p_2}{n_1 + n_2}\right)\left(1 - \dfrac{n_1 p_1 + n_2 p_2}{n_1 + n_2}\right)}$	z table (Table I)
$d = m_1 - m_2$	s_1, s_2 $df = n_1 + n_2 - 2$	$\sqrt{\left(\dfrac{1}{n_1} + \dfrac{1}{n_2}\right)\left[\dfrac{(n_1 - 1)s_1^2 + (n_2 - 1)s_2^2}{n_1 + n_2 - 2}\right]}$	t table (Table II)

[a]Subscripts correspond to sample number.
[b]Expressions for proportions assume moderate to large sample sizes.
[c]See Tables of Probability Distributions.
[d]Standard deviations for proportions are assumed from binomial approximation.

To test if d is neither more nor less than a critical value distance from Δ in standardized units (i.e., *equivalence testing*), the test uses two one-sided tests (TOST), and the test statistics are

$$z_L = \frac{d - (-\Delta)}{\sigma_d}, \quad z_R = \frac{\Delta - d}{\sigma_d} \tag{21.3}$$

or

$$t_L = \frac{d - (-\Delta)}{s_d}, \quad t_R = \frac{\Delta - d}{s_d}, \tag{21.4}$$

where L and R denote left and right. Replace Eq. (21.1) by Eq. (21.3) and Eq. (21.2) by Eq. (21.4) in Table 21.1. For Eq. (21.3), if both $z_L \geq z_{1-\alpha}$ and $z_R \geq z_{1-\alpha}$, reject H_0. For Eq. (21.4), if both $t_L \geq t_{1-\alpha}$ and $t_R \geq t_{1-\alpha}$, reject H_0.

The use of a confidence interval instead of a test statistic is diagrammed in Fig. 21.1.

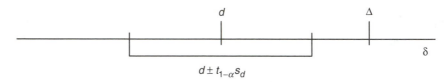

Figure 21.1 Diagram of the δ-axis illustrating d, its confidence interval when using a critical t-value, and a possible location of Δ. H_0 is rejected if Δ falls outside the confidence interval.

FORMULAS FOR REQUIRED SAMPLE SIZES FROM CHAPTER 22

SAMPLE SIZE FOR TESTS ON MEANS

Case 1: One Mean, Normal Distribution (σ Known or from Large Sample)

Is m different from μ? Choose clinically meaningful difference $d = m - \mu$ to answer this question. Look up z-values from Table I (see Tables of Probability Distributions) for $z_{1-\alpha/2}$ and $z_{1-\beta}$. Substitute in Eq. (22.1) to find n, the minimum sample size required.

$$n = \frac{(z_{1-\alpha/2} + z_{1-\beta})^2 \sigma^2}{d^2} \tag{22.1}$$

Case 2: Two Means, Normal Distributions

Is μ_1 (estimated by m_1) different from μ_2 (estimated by m_2)? Choose a clinically meaningful difference $d = m_1 - m_2$ to answer this question. Find σ_1^2 and σ_2^2. Look up z-values from Table I for $z_{1-\alpha/2}$ and $z_{1-\beta}$. Substitute in Eq. (22.2) to find $n_1 (= n_2)$, the minimum sample size required in *each* sample.

$$n_1 = n_2 = \frac{(z_{1-\alpha/2} + z_{1-\beta})^2 (\sigma_1^2 + \sigma_2^2)}{d^2} \qquad (22.2)$$

Case 3: Poorly Behaved or Unknown Distributions (σ Known or from Large Sample)

The Sample Size Is a Conservative Overestimate, Usually More Than Required. Is m different from μ? Choose clinically meaningful difference k as the number of standard deviations m is away from μ. Choose α. Substitute in Eq. (22.4) to find the minimum sample size required.

$$n = \frac{\sigma^2}{\alpha k^2} \qquad (22.4)$$

Case 4: No Objective Prior Data

For your variable, guess the smallest and largest values you have noticed. σ is estimated as 0.25(largest − smallest). Use this value of σ in case 1. Clearly, this "desperation" estimate is of extremely low confidence, but it may be better than picking a number out of the air.

SAMPLE SIZE FOR CONFIDENCE INTERVALS ON MEANS

For a $1 - \alpha$ confidence interval on a theoretical mean μ, find $z_{1-\alpha/2}$ from Table I, choose the desired difference $d = m - \mu$ between the observed and theoretical means, estimate σ from other sources, and substitute these values in

$$n = \frac{(z_{1-\alpha/2})^2 \sigma^2}{d^2}. \qquad (22.5)$$

SAMPLE SIZE FOR TESTS ON RATES (PROPORTIONS)

Case 1: Sample Rate Against a Theoretical Rate

Theoretical rate (proportion) is π. Sample rate is p. Look up z-values in Table I for $z_{1-\alpha/2}$ and $z_{1-\beta}$. Substitute in Eq. (22.8) for π not near 0 or 1 (binomial) or in Eq. (22.9) for π near 0 or 1 (Poisson) to find n, the minimum sample size required.

$$n = \left[\frac{z_{1-\alpha/2}\sqrt{\pi(1-\pi)} + z_{1-\beta}\sqrt{p(1-p)}}{p - \pi} \right]^2 \qquad (22.8)$$

$$n = \left[\frac{z_{1-\alpha/2}\sqrt{\pi} + z_{1-\beta}\sqrt{p}}{p - \pi} \right]^2 \qquad (22.9)$$

Case 2: Two Rates

Sample rates (proportions) are p_1 and p_2; $p_m = (p_1 + p_2)/2$. Look up z-values in Table I for $z_{1-\alpha/2}$ and $z_{1-\beta}$. Substitute in Eq. (22.10) for p_m not near 0 or 1 (binomial) or in Eq. (22.11) for p_m near 0 or 1 (Poisson) to find n, the minimum sample size required.

$$n_1 = n_2 = \left[\frac{z_{1-\alpha/2}\sqrt{2p_m(1-p_m)} + z_{1-\beta}\sqrt{p_1(1-p_1) + p_2(1-p_2)}}{p_1 - p_2} \right]^2 \quad (22.10)$$

$$n_1 = n_2 = \left[\frac{(z_{1-\alpha/2} + z_{1-\beta})\sqrt{p_1 + p_2}}{p_1 - p_2} \right]^2 \quad (22.11)$$

Case 3: Contingency Tables

Use case 2, calculating proportions from cell entries.

CONFIDENCE INTERVAL ON A RATE (PROPORTION)

For a $1 - \alpha$ confidence interval on a theoretical rate (proportion) π, find $z_{1-\alpha/2}$ from Table I, choose the desired interval width $|p - \pi|$ between the observed and theoretical rates, and estimate p from other sources. If π is not near 0 or 1, substitute these values in Eq. (22.13). If π is near 0 or 1, substitute in Eq. (22.14).

$$n = \frac{z_{1-\alpha/2}^2 p(1-p)}{(p - \pi)^2} \quad (22.13)$$

$$n = \frac{z_{1-\alpha/2}^2 p}{(p - \pi)^2} \quad (22.14)$$

CONFIDENCE INTERVAL ON A CORRELATION COEFFICIENT

Sample sizes for 95% confidence intervals on a theoretical correlation coefficient ρ for various values of a sample correlation coefficient r are given in Table 22.1. Some representative values are as follows:

r	0.05	0.10	0.15	0.20	0.25	0.30	0.35	0.40	0.45	0.50
n	1538	385	172	97	63	44	32	25	20	16

FORMULAS AND STEPS FOR MODELING AND DECISION FROM CHAPTER 23

A MODEL IS A QUANTITATIVE OR GEOMETRIC REPRESENTATION OF A RELATIONSHIP OR PROCESS

(1) Algebraic Models

Algebraic models have one or more equations. The dependent variable, representing a biological event or process, is a *function* (of any shape) of one or more independent variables.

Straight-Line Models. Slope-intercept: $y = \beta_0 + \beta_1 x$. Slope-mean point: $y - m_y = \beta_1(x - m_x)$.

Curved Models. Adding a squared term provides a (portion of a) parabola: $y = \beta_0 + \beta_1 x + \beta_2 x^2$.

Commonly seen curves are illustrated in Fig. 23.3. These include the upward opening parabola (see Fig. 23.3A), a third-degree curve (adding an x^3 term to a parabolic model; see Fig. 23.3B), a logarithmic curve (see Fig. 23.3C), an exponential curve (see Fig. 23.3D), a biological growth curve (see Fig. 23.3E), and a sine wave (see Fig. 23.3F).

Constants of Fit for Any Model. Four constants specify a particular member of a family of curves by choosing the four constants of scale and position, as shown in Table 23.1. For a family $y = f(x)$, the constants and their location in the equation are $y = d + cf(ax + b)$.

Table 23.1

Effects of Fit Constants (Parameters) That Change the Function $y = f(x)$ to $y = d + cf(ax + b)$ and Effects of the Values They May Take On

Fit constant	Along which axis	General effect	Detailed effect
a	x	Stretches–shrinks curve	$a < 1$: stretches $a > 1$: shrinks
b	x	Slides curve	$b > 0$: to left $b < 0$: to right
c	y	Stretches–shrinks curve	$c > 1$: stretches $c < 1$: shrinks
d	y	Slides curve	$d > 0$: up $d < 0$: down

Multivariable Models. The three-dimensional case of y depending on two x's can be visualized geometrically, as with the expression $y = \beta_0 + \beta_1 x_1 + \beta_2 x_2$ (see Fig. 23.11). A sample point is ($y = 3.5$, $x_1 = 2.5$, $x_2 = 2$).

A curved surface in three dimensions is possible (perhaps $y = \beta_0 + \beta_1 x_1 + \beta_2 x_2 + \beta_3 x_2^2$), as is one in more than three dimensions. Add as many variables in whatever powers as desired. An algebraic expression of a four-dimensional plane would be $y = \beta_0 + \beta_1 x_1 + \beta_2 x_2 + \beta_3 x_3$.

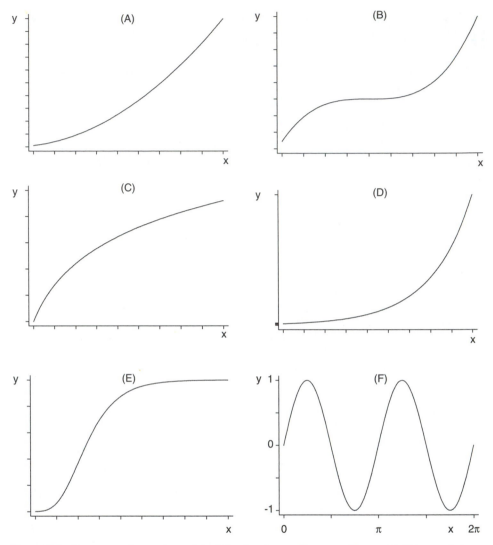

Figure 23.3 Representative portions of the following classes of curve are illustrated: (A) second-degree curve (parabola) opening upward; (B) third-degree curve; (C) logarithmic curve; (D) exponential curve; (E) biological growth curve; and (F) sine wave.

(2) Clinical Decision Based on Recursive Partitioning Models

Steps to Develop the Decision Process. Sort the possible outcomes (diagnoses or result of treatment), listing those alike in adjoining rows. Identify the related indicator (symptom, sign, or condition) most consistently related to the outcome, and then the next, and so on. Draw a decision box with arrows extending out, depending on initial partitioning. Draw the next decision boxes with appropriate arrows extending to them and more arrows extending out, depending on the next round of partitions. Continue until reaching the final outcomes. The boxes should be ordered such that no looping back can occur and no box is repeated. There is usually some trial and error.

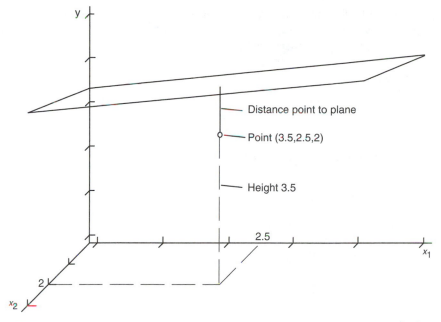

Figure 23.11 Conceptual geometry of a three-dimensional model showing a sample point ($y = 3.5$, $x_1 = 2.5$, and $x_2 = 2$) and a plane of fit to a set of points in the space.

(3) Logical Models: Number Needed to Treat

The number needed to treat (NNT) is an estimate of the number of people that must be screened to detect one case to treat. If we denote by n_s the number of subjects in the population or catchment screened and by n_d the number of disease cases discovered by screening, the ratio $p_d = n_d/n_s$ is the proportion of disease cases discovered by screening. If all disease cases are found by screening, p_d also estimates the disease prevalence. The NNT is the reciprocal of the proportion discovered, or

$$\text{NNT} = \frac{1}{p_d} = \frac{n_s}{n_d}. \tag{23.4}$$

In most screening programs, some of the disease cases would be discovered through ordinary medical practice, for example, p_m. Then $p_d - p_m$ is the proportion of additional detections due to the screening and

$$\text{NNT} = \frac{1}{p_d - p_m}. \tag{23.5}$$

If c_s denotes the cost for the screening program, that is, the cost to screen n_s people, then the cost per subject screened is c_s/n_s and the cost per detection c_d is

$$c_d = \frac{c_s}{n_s} \times \text{NNT}. \tag{23.6}$$

If NNT is the number needed per additional detection, then c_d is the cost per additional detection.

Is the benefit of a new treatment or mode of diagnosis worth its cost? If we denote as p_1 the proportion of successful outcomes using the established treatment or mode of diagnosis and by p_2 that of the new treatment or mode, then $d = p_2 - p_1$ is the difference in success rates and the number of patients so treated or diagnosed required to benefit (NNB) from the change is

$$\text{NNB} = \frac{1}{d}. \tag{23.7}$$

(4) Clinical Decision Based on Measure of Effectiveness Models: Outcomes Analysis

Steps in Measure of Effectiveness Development:

(1) Clearly and unequivocally define the goal.
(2) Identify the (quantifiable) components of the measure of effectiveness (MOE) that will satisfy the goal.
(3) Weight each component to adjust for different units systems.
(4) Weight the components of the MOE (the variables) by their relative importance.

FORMULAS FOR REGRESSION AND CORRELATION FROM CHAPTER 24

Assumptions Underlying Regression:

(1) The errors (i.e., the deviations from average) are independent from one another.
(2) The regression model used in the fit is appropriate.
(3) The independent (x) readings are measured as exact values (without randomness).
(4) The variance of y is the same for all values of x.
(5) The distribution of y is approximately normal for all values of x.

Assumptions Underlying Correlation:

(1) The errors in data values are independent from one another.
(2) Correlation always requires the assumption of a straight-line relationship.
(3) x and y are assumed to follow a bivariate normal distribution. (Rank correlation assumes only the first and second assumptions.)

Simple Regression Equation. $y - m_y = b_1(x - m_x)$ or $y = b_0 + b_1 x$, where m_x and m_y are the respective means, $b_0 = m_y - b_1 m_x$, and $b_1 = s_{xy}/s_x^2$.

Correlation Coefficients.
Continuous variables:

$$r = \frac{s_{xy}}{s_x s_y} \left(= b_1 \frac{s_x}{s_y} \right) \tag{24.3}$$

Rank-order variables:

$$r_s = 1 - \frac{6 \sum d_i^2}{n(n^2 - 1)} \tag{24.4}$$

Tests and Confidence Intervals on Regression Parameters.
Standard error of residuals:

$$s_e = \sqrt{\frac{n-1}{n-2} \left(s_y^2 - b_1^2 s_x^2\right)} = s_y \sqrt{\frac{n-1}{n-2} \left(1 - R^2\right)} \tag{24.5}$$

Standard error of slope:

$$s_b = \frac{s_e}{\sqrt{n-1} s_x} \tag{24.6}$$

t test on slope, H_0: $\beta = 0$:

$$t_{(n-2)df} = \frac{b_1}{s_b} \tag{24.7}$$

t test on slope, H_0: $\beta = \beta_1$:

$$t_{(n-2)df} = \frac{b_1 - \beta_1}{s_b} \tag{24.8}$$

Confidence interval on slope:

$$P[b_1 - t_{1-\alpha/2} s_b < \beta_1 < b_1 + t_{1-\alpha/2} s_b] = 1 - \alpha \tag{24.10}$$

Standard error of mean prediction:

$$s_{m|x} = s_e \sqrt{\frac{1}{n} + \frac{(x - m_x)^2}{(n-1) s_x^2}} \tag{24.11}$$

Confidence interval on mean prediction:

$$P\left[m|x - t_{1-\alpha/2} s_{m|x} < \mu|x < m|x + t_{1-\alpha/2} s_{m|x}\right] = 1 - \alpha \tag{24.12}$$

Standard error of individual prediction:

$$s_{y|x} = s_e \sqrt{1 + \frac{1}{n} + \frac{(x - m_x)^2}{(n-1) s_x^2}} \tag{24.13}$$

Confidence interval on individual prediction:

$$P\left[y|x - t_{1-\alpha/2} s_{y|x} < E(y)|x < y|x + t_{1-\alpha/2} s_{y|x}\right] = 1 - \alpha \tag{24.14}$$

Standard error on slope difference:

$$s_{b:1-2} = \sqrt{\left(\frac{(n_1 - 2)s_{e:1}^2 + (n_2 - 2)s_{e:2}^2}{n_1 + n_2 - 4} \right) \left(\frac{1}{(n_1 - 1)s_1^2} + \frac{1}{(n_2 - 1)s_2^2} \right)} \qquad (24.15)$$

t test on slope difference:

$$t = \frac{b_{1:1} - b_{1:2}}{s_{b:1-2}} \qquad (24.16)$$

Tests and Confidence Intervals on Correlation Coefficients.
t test on correlation, H_0: $\rho = 0$:

$$t_{(n-2)df} = \sqrt{\frac{(n - 2)r^2}{1 - r^2}} \qquad (24.17)$$

z test on correlation, H_0: $\rho = \rho_0$:

$$z = \frac{m - \mu}{\sigma}, \qquad (24.21)$$

where

$$m = \frac{1}{2} \ln \left(\frac{1 + r}{1 - r} \right), \ \mu = \frac{1}{2} \ln \left(\frac{1 + \rho_0}{1 - \rho_0} \right), \ \sigma = \frac{1}{\sqrt{n - 3}}.$$

z test on two coefficients:

$$z = \frac{m_1 - m_2}{\sqrt{\sigma_1^2 + \sigma_2^2}} \qquad (24.22)$$

A confidence interval on ρ:

$$P \left[\frac{1 + r - (1 - r)e^{\frac{2z_{1-\alpha/2}}{\sqrt{n - 3}}}}{1 + r + (1 - r)e^{\frac{2z_{1-\alpha/2}}{\sqrt{n - 3}}}} < \rho < \frac{1 + r - (1 - r)e^{\frac{2z_{1-\alpha/2}}{\sqrt{n - 3}}}}{1 + r + (1 - r)e^{\frac{2z_{1-\alpha/2}}{\sqrt{n - 3}}}} \right] = 1 - \alpha \quad (24.23)$$

Logistic Regression:
Model:

$$\ln \left(\frac{p_m}{1 - p_m} \right) = \beta_0 + \beta_1 x \qquad (24.24)$$

Prediction of proportion "successes":

$$P_m = \frac{e^{b_0 + b_1 x}}{1 + e^{b_0 + b_1 x}} \qquad (24.25)$$

Choosing the Type of Regression to Use:

Table 24.3

Classes of Regression Model by Type of Outcome Variable They Predict

Nature of outcome (dependent) variable	Name of regression type
Continuous	(Ordinary) regression
Ranks	Ordered regression
Categorical: two categories	Logistic regression
Categorical: several categories	Multinomial regression
Counts (number of occurrences)	Poisson regression
Survival (independent variable: time)	Cox (proportional hazards)

METHODS FOR SURVIVAL AND TIME-SERIES ANALYSIS FROM CHAPTER 25

LIFE TABLES AND SURVIVAL CURVES

See section for Chapter 9 earlier in this summary for discussion of life tables and survival curves.

CONFIDENCE INTERVALS ON SURVIVAL CURVES

The 95% confidence interval on the true survival proportion for each time period, where S_i is the estimated survival proportion for that period, is

$$S_i \pm 1.96 \times S_i \sqrt{\frac{1 - S_i}{n_{i-1}}}. \tag{25.2}$$

LOG-RANK TEST OF TWO OR MORE SURVIVAL CURVES

Use a statistical software package on a computer to perform a log-rank test of two or more survival curves.

SEQUENTIAL ANALYSIS

Definitions: x_k is the value of the kth observation in the sequence, $P(x_k|H_0)$ is the probability of x_k occurring if H_0 is true, and $P(x_k|H_1)$ is the probability of x_k occurring if H_1 is true.
Steps:

(1) Pose the null and alternate hypotheses.
(2) Assign error risk values to α and β and calculate the critical values.
(3) Find the probability ratio $D_1 = P(x_k|H_1)/P(x_k|H_0)$.

(4) Fill in Table 25.5, calculating $D_k = D_{k-1} \times P(x_k|H_1)/P(x_k|H_0)$.

For a test of proportions: If the observation is binary, x_k can be written as either 0 or 1. The calculation of D_k will be $D_k = D_{k-1}P(1|H_1)/P(1|H_0)$ if event 1 occurs, or $D_k = D_{k-1}[1 - P(1|H_1)]/[1 - P(1|H_0)]$ if event 1 does not occur. For a test of means: If the mean under the null hypothesis is μ_0, that under the alternate is μ_1, and the variance is σ^2, the multiplying increment is

$$\exp\left\{-\frac{1}{2\sigma^2}\left[(x_k - \mu_1)^2 - (x_k - \mu_0)^2\right]\right\}. \tag{25.6}$$

(5) After each observation in the sequence, make a decision using the following criteria:

$$\text{Accept } H_0 \text{ when } \quad D_k \leq \frac{\beta}{1 - \alpha}.$$

$$\text{Reject } H_0 \text{ when } \quad D_k \geq \frac{1 - \beta}{\alpha}. \tag{25.4}$$

Continue sampling otherwise.

If H_0 is accepted or rejected, stop; if not, sample another patient and repeat steps (4) and (5).

MOVING SAMPLES IN TIME SERIES: ESTIMATION

Consider a moving sample of readings x of width k with rightmost element i, that is, $(x_{i-k+1}, x_{i-k+2}, \ldots, x_i)$. A moving mean, median, or variance is just the sequence of ordinary means, medians, or variances of the k elements of the moving sample. Because the sample drops the leftmost reading and adds one to the right end to create the movement, a moving statistic needs to be adjusted for only the two readings changed.

SERIAL CORRELATION IN TIME SERIES

A serial correlation coefficient is just the ordinary correlation coefficient of two columns of data, except that the data are ordered in sequence through time. A serial correlation may have a lag in time, which is equivalent to sliding one column down the number of elements in the lag, and then calculating the correlation coefficient on the column pairs.

CROSS-CORRELATION IN TIME SERIES

A cross-correlation is a serial correlation in which the two columns of data arise from two different variables. A cross-correlation may or may not be lagged.

AUTOCORRELATION IN TIME SERIES

An autocorrelation coefficient is a lagged serial correlation coefficient in which the second column of data is a repetition of the first.

MOVING SAMPLES IN TIME SERIES: TESTING

Consider a moving sample in which the beginning sample represents a baseline period (e.g., measurements before treatment or onset of disease). A moving variance in ratio to the baseline variance provides a Moving F statistic that can be compared for significance with a critical F-value. The baseline variance is an estimate of preevent random variability. The Moving F ratio tests if succeeding moving variances are increased by a change in the mean, by larger random variability, by a change in the pattern (model) of the path through time, or by some combination of these factors. The Moving F may be graphed against the time values to identify the points of change and the periods during which differences persist.

Other methods to test for changes in a time-series pattern exist but are either more difficult to use or fail to detect and locate some types of change. These methods include quality or process control, cumulative sum (CUSUM), exponentially weighted moving average (EWMA), significance of zero crossing (SIZER), and Monte Carlo Markov chains (MCMC).

LISTING OF METHODS LESS FREQUENTLY ENCOUNTERED FROM CHAPTER 26

PURPOSE

Medical investigators sometimes encounter references to statistical methods less frequently used than those delineated in this book. Chapter 26 defines a number of these methods so that the investigator can discern whether or not they apply to the issue at hand and search more advanced texts.

LIST OF TERMS AND METHODS NOTED IN CHAPTER 26

Analysis of Variance (ANOVA) and Covariance (ANCOVA):

Random effects models
Pairwise comparison methods other than those addressed in this book
Levene's test for equality of variances
Tests when variances are unequal
Box-Cox transform to enhance assumptions
General or generalized linear models (GLM)

Regression:

Heckman selection model to reduce bias

Multivariate Methods:

Hotelling's T^2
Discriminant analysis
Mahalanobis' distance
Multivariate analysis of variance (MANOVA) and covariance (MANCOVA)

Principal components analysis
Factor analysis
Cluster analysis

Nonparametric Tests:

Runs test
Sign test
Cochran's Q

Imputation:

Estimation of missing data

Resampling Methods:

Bootstrap
Jackknife

Agreement Measures and Correlation:

Cronbach's α
(Cohen's) Kappa
Kendall's τ
Kendall's W
Riffenburgh's A
Phi coefficient
Cramér's V
Canonical correlation

Bonferroni "Correction":

Controlling error in multiple significance tests

Logit and Probit:

Transformation of binary outcomes for regression analysis

Outliers:

Cook's distance
Sample (or mean) trimming
Windsorizing

Curve Fitting:

Theoretical fits: logistic, Cauchy, beta, gamma, Rayleigh, Weibull, exponential, negative exponential, log-normal, Pareto, Laplace, and Gompertz distributions
Spline fits

Test of Normality:

Lilliefors test

References and Data Sources

The data used for examples in this book arose from real medical studies. In many cases, data were edited to better illustrate the statistical method, so the treatment in this book may not represent the true study results or the intent of the investigator. I advised on most of the studies, some of which have been published and are so referenced and some of which are still in progress or in press. Other data arose from studies published in the medical literature and are so referenced.

The terms, prose, and interpretations are mine and are not necessarily attributable to the investigator referenced. In cases of misstatement, I beg the reader's indulgence. We both will be better served if you look to the statistical method being illustrated rather than to the medical nuance.

1. Antevil JL, Buckley RG, Johnson AS, et al. Treatment of suspected symptomatic cholelithiasis with glycopyrrolate: A prospective, randomized clinical trial. *Ann Emerg Med* 2005;45:172–176.
2. Bailar JC, Mosteller F, eds. *Medical Uses of Statistics,* 2nd ed. Boston: NEJM Books, 1992.
3. Battaglia, Michael J, MD (CDR, MC, USN), Orthopedics Department, Naval Medical Center, San Diego, CA.
4. Blanton, Christopher L, MD, Ophthalmologist, Inland Eye Institute, Glendale, CA.
5. Blanton, CL, Schallhorn S, Tidwell J. Radial keratotomy learning curve using the American technique. *J Cataract Refractive Surg* 1998;24:471–476.
6. Choplin, Neil, MD, Ophthalmologist, San Diego, CA.
7. Crain, Donald, MD (LCDR, MC, USN), Urology Department, Naval Medical Center, San Diego, CA.
8. Crum, Nancy F, MD (LCDR, MC, USN), Infectious Diseases Department, Naval Medical Center, San Diego, CA.
9. Cupp, Craig, MD (CAPT, MC, USN), Otolaryngology Department, Naval Medical Center, San Diego, CA.
10. Curtis KM, Savitz DA, Arbuckle TE. Effects of cigarette smoking, caffeine consumption, and alcohol intake on fecundability. *Am J Epidemiol* 1997;146:32–41.
11. Daly, Karen A, MD (CAPT, MC, USN), Mental Health Department, Naval Hospital Bremerton, WA.
12. Devereaux, Asha, MD, Critical Care Practice, Coronado, CA.
13. Elsas T, Johnsen H. Long-term efficacy of primary laser trabeculoplasty. *Br J Ophthalmol* 1991;75:34–37.
14. Engle AT, Laurent JM, Schallhorn SC, et al. Masked comparison of silicone hydrogel lotrafilcon A and etafilcon A extended-wear bandage contact lenses after photorefractive keratectomy. *J Cataract Refract Surg* 2005;31:681–686.
15. Farkas T, Thornton SA, Wilton N, et al. Homologous versus heterologous immune responses to Norwalk-like viruses among crew members after acute gastroenteritis outbreaks on 2 US Navy vessels. *J Infect Dis* 2003;187:187–193.
16. Freilich, Daniel, MD (CDR, MC, USN), Naval Medical Research Center, Silver Spring, MD.
17. Gilmore DM, Frieden TR. Universal radiographic screening for tuberculosis among inmates upon admission to jail. *Am J Public Health* 1997;87:1335–1337.

18. Goldberg MA. Erythropoiesis, erythropoietin, and iron metabolism in elective surgery: Preoperative strategies for avoiding allogeneic blood exposure. *Am J Surg* 1995; 170:37S–43S.

19. Gordon, Richard L, DPharm, Director, Coumadin Clinic, Naval Medical Center, San Diego, CA.

20. Grau, Kriste J, RN (CDR, NC, USN), Head, Pediatric Nursing, Naval Medical Center, San Diego, CA.

21. Greason, Kevin, MD (CDR, MC, USN), Cardiothoracic Surgery, Naval Medical Center, San Diego, CA.

22. Grossman, Ira, PhD, Supervisor, Psychology Services, Sharp Mesa Vista Hospital, San Diego, CA.

23. Hacala, MT. U.S. Navy Hospital Corps: A century of tradition, valor, and sacrifice. *Navy Medicine* 1998;89:12–26.

24. Hedges LV, Olkin I. *Statistical Methods for Meta-analysis*. San Diego: Academic Press, 1985.

25. Hennekens CH, Buring JE. *Epidemiology in Medicine*. Boston: Little, Brown and Company, 1987.

26. Hennrikus W, Mapes R, Lyons P, Lapoint J. Outcomes of the Chrisman-Snook and Modified Bostrom procedures for chronic lateral ankle instability. *Am J Sports Med* 1996;24:400–404.

27. Hennrikus W, Shin A, Klingelberger C. Self-administered nitrous-oxide and a hematoma block for analgesia in the outpatient reduction of fractures in children. *J Bone Joint Surg* 1995;77A:335–339.

28. Hennrikus W, Simpson B, Klingelberger C, Reis M. Self-administered nitrous-oxide analgesia for pediatric fracture reductions. *J Pediatr Orthop* 1994;14:538–542.

29. Hoffer, Michael E, MD (CDR, MC, USN), Co-Director, Defense Spatial Orientation Center, Naval Medical Center, San Diego, CA.

30. Hong WK, Endicott J, Itri LM, et al. 13-*cis*-Retinoic acid in the treatment of oral leukoplakia. *N Engl J Med* 1986;315:1501–1505.

31. Howe, Steven C, MD, Private surgical practice, Towson, MD.

32. Huff, D. *How to Lie with Statistics*. New York: WW Norton, 1954.

33. Huisman, Thomas, MD, Urologist, Chiaramonte and Associates Urology, Clinton, MD.

34. Johnstone, Peter AS, MD, Professor of Radiation Oncology, Emory University, Atlanta, GA.

35. Johnstone PA, Norton MS, Riffenburgh RH. Survival of patients with untreated breast cancer. *J Surg Oncol* 2000;73:273–277.

36. Johnstone PAS, Powell CR, Riffenburgh RH, et al. The fate of 10-year clinically recurrence-free survivors after definitive radiotherapy for $T_{1-3}N_0M_0$ prostate cancer. *Radiat Oncol Invest* 1998;6:103–108.

37. Johnstone PAS, Sindela WS. Radical reoperation for advanced pancreas carcinoma. *J Surg Oncol* 1996;61:7–13.

38. Kelso, John M, MD (CAPT, MC, USN), Allergy-Immunology Department, Naval Medical Center, San Diego, CA.

39. Klein HG. Allogeneic transfusion risks in the surgical patient. *Am J Surg* 1995;170: 21S–26S.

40. Last JM, Ed. *A Dictionary of Epidemiology,* 2nd ed. New York: Oxford University Press, 1988.

41. Le Hesran JY, Cot M, Personne P, et al. Maternal placental infection with *Plasmodium falciparum* and malaria morbidity during the first 2 years of life. *Am J Epidemiol* 1997;146:826–831.

42. Leibson CL, O'Brien PC, Atkinson E, et al. Relative contributions of incidence and survival to increasing prevalence of adult-onset diabetes mellitus: A population-based study. *Am J Epidemiol* 1997;146:12–22.

43. Leivers, David, MD (CAPT, MC, USN), Anesthesiology Department, Naval Medical Center, San Diego, CA.

44. Lilienfeld DE, Stolley PD. *Foundations of Epidemiology*, 3rd ed. New York: Oxford University Press, 1994. (This is a revision of the original book by A. M. Lilienfeld.)

45. Lyons AS, Petrucelli RJ. *Medicine, An Illustrated History.* New York: Harry N. Abrams, 1978.

46. Mahon, Richard, MD (CDR, MC, USN), Undersea Medical Research Division, Naval Medical Research Center, Silver Spring, MD.

47. Mausner JS, Kramer S. *Mauser & Bahn Epidemiology—An Introductory Text.* Philadelphia: WB Saunders Company, 1985.

48. McNulty PA. Prevalence and contributing factors of eating disorder behaviors in a population of female Navy nurses. *Mil Med* 1997;162:703–706.

49. Meier P. *Introductory Lecture Notes on Statistical Methods in Medicine and Biology.* Chicago: University of Chicago, 1974.

50. Missing sources. The sources of a few examples could not be found despite a strong effort to locate them. Such data that could not be referenced were slightly altered so as not to reflect on any investigator later appearing.

51. Mitchell, Jan, DDS (CAPT, DC, USN), Branch Director, U.S. Marine Corps, San Diego, CA.

52. Mitchell, John D, MD (LCDR, MC, USN), Cardiothoracic Surgeon, University of Colorado Health Science Center, Denver, CO.

53. Mitts, Kevin G, MD, Orthopaedic practice, Pittsfield, MA.

54. Muldoon, Michael, MD (CDR, MC, USN), Orthopedics Department, Naval Medical Center, San Diego, CA.

55. Murray, James, MD, Head, Vascular Surgery, Kaiser Medical Center, Baldwin Park, CA.

56. O'Leary, Michael J, MD, Neurotology/Skull Base Surgery, Senta Medical Clinic, San Diego, CA.

57. Panos, Reed G, MD, Plastic Surgeon, Champagne, IL.

58. Pekarske, William, MD, Department of Anesthesiology, Northwest Medical Center, Tucson, AZ.

59. Peto R, Pike MC, Armitage P, et al. Design and analysis of randomized clinical trials requiring prolonged observation of each patient. II. Analysis and examples. *Br J Cancer* 1977;35:1–39.

60. Poggi, Matthew, MD (CDR, MC, USN), Radiation Oncology Department, Naval Medical Center, Bethesda, MD.

61. Rao PM, Rhea JT, Novelline RA, et al. Effect of computed tomography of the appendix on treatment of patients and use of hospital resources. *N Engl J Med* 1998;338:141–146.

62. Riffenburgh, Ralph S, MD, Ophthalmologic Surgeon, Pasadena, CA.

63. Riffenburgh, Robert H, PhD, Clinical Investigation Department, Naval Medical Center, San Diego, CA.

64. Riffenburgh RH. Detecting a point of any change in a time series. *J Appl Statist Sci* 1994;1:487–488.

65. Riffenburgh RH. Reverse gullibility and scientific evidence. *Arch Otolaryngol Head Neck Surg* 1996;122:600–601.

66. Riffenburgh RH, Johnstone PA. Survival patterns of cancer patients. *Cancer* 2001;91:2469–2475.

67. Riffenburgh RH, Olson PE, Johnstone PA. Association of schistosomiasis with cervical cancer: Detecting bias in clinical studies. *East Afr Med J* 1997;74:14–16.

68. Ross, E Victor, MD (CAPT, MC, USN), Dermatology Department, Naval Medical Center, San Diego, CA.

69. Rothman KJ. *Modern Epidemiology.* Boston: Little, Brown and Company, 1986.

70. Sageman, W Scott, MD, Pulmonary Critical Care Medicine, Monterey, CA.

71. Schallhorn, Stephen, MD (CAPT, MC, USN), Ophthalmology Department, Naval Medical Center, San Diego, CA.

72. Schellenberg G. Apo-E4: A risk factor, not a diagnostic test. *Sci News* 1994;145:10.

73. Schnepf, Glenn A, MD (CAPT, MC, USN), Navy Central HIV Department, Naval Medical Center, Bethesda, MD.

74. Scott CB. Validation and predictive power of Radiation Therapy Oncology Group RTOG recursive partitioning analysis classes for malignant glioma patients: A report using RTOG 90-06. *Int J Radiat Oncol Biol Phys* 1998;40:51–55.

75. Shimoda O, Ikuta Y, Sakamoto M, Terasaki H. Skin vasomotor reflex predicts circulatory responses to laryngoscopy and intubation. *Anesthesiology* 1998;88:297–304.

76. Silverman MK, Kopf AW, Grin CM, et al. Recurrence rates of treated basal cell carcinomas, part 2: Curettage-Electrodessication. *J Dermatol Surg Oncol* 1991;17:720–726.

77. Smith, Stacy, MD, Dermatologist, San Diego, CA.

78. Snedecor GW, Cochran WG. *Statistical Methods,* 6th ed. Ames, IA: Iowa State University Press, 1967.

79. Spiegelhalter DJ, Best N. Shipman's statistical legacy. *Significance* 2004;1:10–12.

80. Student [pseud.]. The probable error of a mean. *Biometrika* 1908;6:1–25.

81. Sutherland JC. Cancer in a mission hospital in South Africa. With emphasis on cancer of the cervix uteri, liver and urinary bladder. *Cancer* 1968;22:372–378. (Child:adult ratios were inferred from other sources for larger African areas.)

82. Tanen, David A, MD (CDR, MC, USN), Naval Medical Center, San Diego, CA.

83. Tanen DA, Danish DC, Grice G, et al. Fasciotomy worsens the amount of myonecrosis in a porcine model of crotaline envenomation. *Ann Emerg Med* 2004;44:99–104.

84. Tanen DA, Miller S, French T, Riffenburgh RH. Intravenous sodium valproate vs. prochlorperazine for the emergency department treatment of acute migraine headaches. *Ann Emerg Med* 2003;41(6):847–853.

85. Tasker SA, O'Brien WA, Treanor JJ, et al. Effects of influenza vaccination in HIV-infected adults: A double-blinded placebo-controlled trial. *Vaccine* 1998;16:1039–1042.

86. Tockman MS. Progress in the early detection of lung cancer. In: Aisner J, Arriagada R, Green MR, Martini N, Perry MC, eds. *Comprehensive Textbook of Thoracic Oncology.* Baltimore: Williams & Wilkins, 1996:105–111.

87. Tufte E. *Visual Display of Quantitative Information,* 2nd ed. Cheshire, CT: Graphics Press, 2001.

88. Wald A. *Sequential Analysis.* New York: John Wiley & Sons, 1947.

89. Wilde J. A 10-year follow-up of semi-annual screening for lung cancer in Erfurt County, GDR. *Eur Respir J* 1989;2:656–662.

90. Wright L. Postural hypotension in late pregnancy. *Br Med J* 1962;17:760–762.

91. Zilberfarb J, Hennrikus W, Reis M, et al. Treatment of acute septic bursitis. *Surgical Forum XLIII* 1992:577–579.

Tables of Probability Distributions

Table I
Normal Distribution[a]

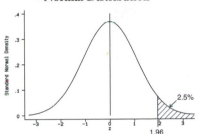

z (no. std. deviations to right of mean)	One-tailed applications		Two-tailed applications	
	One-tailed α (area in right tail)	$1 - \alpha$ (area except right tail)	Two-tailed α (area in both tails)	$1 - \alpha$ (area except both tails)
0	.500	.500	1.000	.000
.10	.460	.540	.920	.080
.20	.421	.579	.842	.158
.30	.382	.618	.764	.236
.40	.345	.655	.690	.310
.50	.308	.692	.619	.381
.60	.274	.726	.548	.452
.70	.242	.758	.484	.516
.80	.212	.788	.424	.576
.90	.184	.816	.368	.632
1.00	.159	.841	.318	.682
1.10	.136	.864	.272	.728
1.20	.115	.885	.230	.770
1.281	*.100*	*.900*	*.200*	*.800*
1.30	.097	.903	.194	.806
1.40	.081	.919	.162	.838
1.50	.067	.933	.134	.866
1.60	.055	.945	.110	.890
1.645	*.050*	*.950*	*.100*	*.900*
1.70	.045	.955	.090	.910
1.80	.036	.964	.072	.928
1.90	.029	.971	.054	.946
1.960	*.025*	*.975*	*.050*	*.950*
2.00	.023	.977	.046	.954
2.10	.018	.982	.036	.964
2.20	.014	.986	.028	.972
2.30	.011	.989	.022	.978
2.326	*.010*	*.990*	*.020*	*.980*
2.40	.008	.992	.016	.984
2.50	.006	.994	.012	.988
2.576	*.005*	*.995*	*.010*	*.990*
2.60	.0047	.9953	.0094	.9906
2.70	.0035	.9965	.0070	.9930
2.80	.0026	.9974	.0052	.9948
2.90	.0019	.9981	.0038	.9962
3.00	.0013	.9987	.0026	.9974
3.100	*.0010*	*.9990*	*.0020*	*.9980*
3.20	.0007	.9996	.0014	.9986
3.300	*.0005*	*.9995*	*.0010*	*.9990*
3.40	.0003	.9997	.0006	.9994
3.50	.0002	.9998	.0004	.9996

[a]For selected distances (z) to the right of the mean are given (1) one-tailed α, the area under the curve in the positive tail; (2) one-tailed $1 - \alpha$, the area under all except the tail; (3) two-tailed α, the areas combined for both positive and negative tails; and (4) two-tailed $1 - \alpha$, the area under all except the two tails. Entries for the most commonly used areas are in italics.

Table II

t Distribution[a]

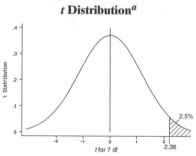

One-tailed α (right tail area)	.10	.05	.025	.01	.005	.001	.0005
One-tailed $1 - \alpha$ (except right tail)	.90	.95	.975	.99	.995	.999	.9995
Two-tailed α (area both tails)	.20	.10	.05	.02	.01	.002	.001
Two-tailed $1 - \alpha$ (except both tails)	.80	.90	.95	.98	.99	.998	.999
up df = 2	1.886	2.920	4.303	6.965	9.925	22.327	31.598
3	1.638	2.353	3.182	4.541	5.841	10.215	12.941
4	1.533	2.132	2.776	3.747	4.604	7.173	8.610
5	1.476	2.015	2.571	3.365	4.032	5.893	6.859
6	1.440	1.943	2.447	3.143	3.707	5.208	5.959
7	1.415	1.895	2.365	2.998	3.499	4.785	5.405
8	1.397	1.860	2.306	2.896	3.355	4.501	5.041
9	1.383	1.833	2.262	2.821	3.250	4.297	4.781
10	1.372	1.812	2.228	2.764	3.169	4.144	4.587
11	1.363	1.796	2.201	2.718	3.106	4.025	4.437
12	1.356	1.782	2.179	2.681	3.055	3.930	4.318
13	1.350	1.771	2.160	2.650	3.012	3.852	4.221
14	1.345	1.761	2.145	2.624	2.977	3.787	4.140
15	1.341	1.753	2.131	2.602	2.947	3.733	4.073
16	1.337	1.746	2.120	2.583	2.921	3.686	4.015
17	1.333	1.740	2.110	2.567	2.898	3.646	3.965
18	1.330	1.734	2.101	2.552	2.878	3.610	3.922
19	1.328	1.729	2.093	2.539	2.861	3.579	3.883
20	1.325	1.725	2.086	2.528	2.845	3.552	3.850
21	1.323	1.721	2.080	2.518	2.831	3.527	3.819
22	1.321	1.717	2.074	2.508	2.819	3.505	3.792
23	1.319	1.714	2.069	2.500	2.807	3.485	3.767
24	1.318	1.711	2.064	2.492	2.797	3.467	3.745
25	1.316	1.708	2.060	2.485	2.787	3.450	3.725
26	1.315	1.706	2.056	2.479	2.779	3.435	3.707
27	1.314	1.703	2.052	2.473	2.771	3.421	3.690
28	1.313	1.701	2.048	2.467	2.763	3.408	3.674
29	1.311	1.699	2.045	2.462	2.756	3.396	3.659
30	1.310	1.697	2.042	2.457	2.750	3.385	3.646
40	1.303	1.684	2.021	2.423	2.704	3.307	3.551
60	1.296	1.671	2.000	2.390	2.660	3.232	3.460
100	1.290	1.660	1.984	2.364	2.626	3.174	3.390
∞	1.282	1.645	1.960	2.326	2.576	3.090	3.291

[a]Selected distances (*t*) to the right of the mean are given for various degrees of freedom (*df*) and for (1) one-tailed α, area under the curve in the positive tail; (2) one-tailed $1 - \alpha$, area under all except the tail; (3) two-tailed α, areas combined for both positive and negative tails; and (4) two-tailed $1 - \alpha$, area under all except the two tails.

Table III

Chi-square Distribution, Right Tail[a]

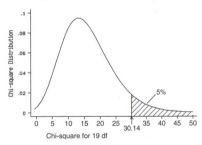

α (area in right tail)	.10	.05	.025	.01	.005	.001	.0005
$1 - \alpha$ (except right tail)	.90	.95	.975	.99	.995	.999	.9995
$df = 1$	2.71	3.84	5.02	6.63	7.88	10.81	12.13
2	4.61	5.99	7.38	9.21	10.60	13.80	15.21
3	6.25	7.81	9.35	11.34	12.84	16.26	17.75
4	7.78	9.49	11.14	13.28	14.86	18.46	20.04
5	9.24	11.07	12.83	15.08	16.75	20.52	22.15
6	10.64	12.59	14.45	16.81	18.54	22.46	24.08
7	12.02	14.07	16.01	18.47	20.28	24.35	26.02
8	13.36	15.51	17.53	20.09	21.95	26.10	27.86
9	14.68	16.92	19.02	21.67	23.59	27.86	29.71
10	15.99	18.31	20.48	23.21	25.19	29.58	31.46
11	17.28	19.68	21.92	24.72	26.75	31.29	33.13
12	18.55	21.03	23.34	26.22	28.30	32.92	34.80
13	19.81	22.36	24.74	27.69	29.82	34.54	36.47
14	21.06	23.69	26.12	29.14	31.32	36.12	38.14
15	22.31	25.00	27.49	30.57	32.81	37.71	39.73
16	23.54	26.30	28.84	32.00	34.27	39.24	41.31
17	24.77	27.59	30.19	33.41	35.72	40.78	42.89
18	25.99	28.87	31.53	34.80	37.16	42.32	44.47
19	27.20	30.14	32.85	36.19	38.58	43.81	45.97
20	28.41	31.41	34.17	37.57	39.99	45.31	47.46
21	29.62	32.67	35.48	38.94	41.40	46.80	49.04
22	30.81	33.92	36.78	40.29	42.80	48.25	50.45
23	32.01	35.17	38.08	41.64	44.19	49.75	52.03
24	33.20	36.41	39.36	42.97	45.56	51.15	53.44
25	34.38	37.65	40.65	44.31	46.93	52.65	51.93
26	35.56	38.88	41.92	45.64	48.30	54.05	56.43
27	36.74	40.11	43.20	46.97	49.65	55.46	57.83
28	37.92	41.34	44.46	48.28	51.00	56.87	59.24
29	39.09	42.56	45.72	49.59	52.34	58.27	60.73
30	40.26	43.77	46.98	50.89	53.68	59.68	62.23
35	46.06	49.80	53.20	57.34	60.27	66.62	69.26
40	51.80	55.76	59.34	63.69	66.76	73.39	76.11
50	63.17	67.51	71.42	76.16	79.50	86.66	89.56
60	74.40	79.08	83.30	88.38	91.96	99.58	102.66
70	85.53	90.53	95.02	100.43	104.22	112.32	115.66
80	96.58	101.88	106.63	112.32	116.32	124.80	128.32
100	118.50	124.34	129.56	135.81	140.16	149.41	153.11

[a]Selected χ^2 values (distances above zero) are given for various degrees of freedom (df) and for (1) α, the area under the curve in the right tail, and (2) $1 - \alpha$, the area under all except the right tail.

Table IV

Chi-square Distribution, Left Tail[a]

α (area in left tail)	.0005	.001	.005	.01	.025	.05	.10
1 − α (area except left tail)	.9995	.999	.995	.99	.975	.95	.90
$df = 1$	.0000004	.0000016	.000039	.00016	.00098	.0039	.016
2	.00099	.0020	.010	.020	.051	.10	.21
3	.015	.024	.072	.12	.22	.35	.58
4	.065	.091	.21	.30	.48	.71	1.06
5	.16	.21	.41	.55	.83	1.15	1.61
6	.30	.38	.68	.87	1.24	1.64	2.20
7	.48	.60	.99	1.24	1.69	2.17	2.83
8	.71	.86	1.34	1.65	2.18	2.73	3.49
9	.97	1.15	1.73	2.09	2.70	3.33	4.17
10	1.26	1.48	2.16	2.56	3.25	3.94	4.87
11	1.58	1.83	2.60	3.05	3.82	4.57	5.58
12	1.93	2.21	3.07	3.57	4.40	5.23	6.30
13	2.31	2.61	3.57	4.11	5.01	5.89	7.04
14	2.70	3.04	4.07	4.66	5.63	6.57	7.79
15	3.10	3.48	4.60	5.23	6.26	7.26	8.55
16	3.54	3.94	5.14	5.81	6.91	7.96	9.31
17	3.98	4.42	5.70	6.41	7.56	8.67	10.09
18	4.44	4.90	6.26	7.01	8.23	9.39	10.86
19	4.92	5.41	6.84	7.63	8.91	10.12	11.65
20	5.41	5.92	7.43	8.26	9.59	10.85	12.44
21	5.89	6.45	8.03	8.90	10.28	11.59	13.24
22	6.42	6.99	8.64	9.54	10.98	12.34	14.04
23	6.92	7.54	9.26	10.20	11.69	13.09	14.85
24	7.45	8.09	9.89	10.86	12.40	13.85	15.66
25	8.00	8.66	10.52	11.52	13.12	14.61	16.47
26	8.53	9.23	11.16	12.20	13.84	15.38	17.29
27	9.10	9.80	11.81	12.88	14.57	16.15	18.11
28	9.68	10.39	12.46	13.57	15.31	16.93	18.94
29	10.24	10.99	13.12	14.25	16.05	17.71	19.77
30	10.81	11.58	13.79	14.95	16.79	18.49	20.60
35	13.80	14.68	17.19	18.51	20.57	22.46	24.80
40	16.92	17.93	20.71	22.16	24.43	26.51	29.05
50	23.47	24.68	27.99	29.71	32.36	34.76	37.69
60	30.32	31.73	35.53	37.49	40.48	43.19	46.46
70	37.44	39.02	43.28	45.44	48.76	51.74	55.33
80	44.82	46.49	51.17	53.54	57.15	60.39	64.28
100	59.94	61.92	67.32	70.07	74.22	77.93	82.36

[a]Selected χ^2 values (distances above zero) are given for various degrees of freedom (df) and for (1) α, the area under the curve in the left tail and (2) $1 - \alpha$, the area under all except the left tail.

Table V

F Distribution[a]

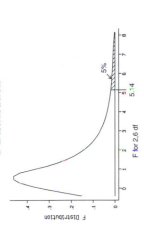

F Distribution

5%

5.14

F for 2,6 df

Denominator df	Numerator df																		
	1	2	3	4	5	6	7	8	9	10	12	15	20	25	30	40	60	100	∞
2	18.51	19.00	19.16	19.25	19.30	19.33	19.35	19.37	19.38	19.40	19.41	19.43	19.45	19.46	19.46	19.47	19.48	19.49	19.50
3	10.13	9.55	9.28	9.12	9.01	8.94	8.89	8.85	8.81	8.79	8.74	8.70	8.66	8.63	8.62	8.59	8.57	8.55	8.53
4	7.71	6.94	6.59	6.39	6.26	6.16	6.09	6.04	6.00	5.96	5.91	5.86	5.80	5.77	5.75	5.72	5.69	5.66	5.63
5	6.61	5.79	5.41	5.19	5.05	4.95	4.88	4.82	4.77	4.74	4.68	4.62	4.56	4.52	4.50	4.46	4.43	4.41	4.36
6	5.99	5.14	4.76	4.53	4.39	4.28	4.21	4.15	4.10	4.06	4.00	3.94	3.87	3.83	3.81	3.77	3.74	3.71	3.67
7	5.59	4.74	4.35	4.12	3.97	3.87	3.79	3.73	3.68	3.64	3.57	3.51	3.44	3.40	3.38	3.34	3.30	3.27	3.23
8	5.32	4.46	4.07	3.84	3.69	3.58	3.50	3.44	3.39	3.35	3.28	3.22	3.15	3.11	3.08	3.04	3.01	2.97	2.93
9	5.12	4.26	3.86	3.63	3.48	3.37	3.29	3.23	3.18	3.14	3.07	3.01	2.94	2.89	2.86	2.83	2.79	2.76	2.71
10	4.96	4.10	3.71	3.48	3.33	3.22	3.14	3.07	3.02	2.98	2.91	2.85	2.77	2.73	2.70	2.66	2.62	2.59	2.54
11	4.84	3.98	3.59	3.36	3.20	3.09	3.01	2.95	2.90	2.85	2.79	2.72	2.65	2.60	2.57	2.53	2.49	2.46	2.40
12	4.75	3.89	3.49	3.26	3.11	3.00	2.91	2.85	2.80	2.75	2.69	2.62	2.54	2.50	2.47	2.43	2.38	2.35	2.30
13	4.67	3.81	3.41	3.18	3.03	2.92	2.83	2.77	2.71	2.67	2.60	2.53	2.46	2.41	2.38	2.34	2.30	2.26	2.21
14	4.60	3.74	3.34	3.11	2.96	2.85	2.76	2.70	2.65	2.60	2.53	2.46	2.39	2.34	2.31	2.27	2.22	2.19	2.13
15	4.54	3.68	3.29	3.06	2.90	2.79	2.71	2.64	2.59	2.54	2.48	2.40	2.33	2.28	2.25	2.20	2.16	2.12	2.07
16	4.49	3.63	3.24	3.01	2.85	2.74	2.66	2.59	2.54	2.49	2.42	2.35	2.28	2.23	2.19	2.15	2.11	2.07	2.01
17	4.45	3.59	3.20	2.96	2.81	2.70	2.61	2.55	2.49	2.45	2.38	2.31	2.23	2.18	2.15	2.10	2.06	2.02	1.96
18	4.41	3.55	3.16	2.93	2.77	2.66	2.58	2.51	2.46	2.41	2.34	2.27	2.19	2.14	2.11	2.06	2.02	1.98	1.92
19	4.38	3.52	3.13	2.90	2.74	2.63	2.54	2.48	2.42	2.38	2.31	2.23	2.16	2.11	2.07	2.03	1.98	1.94	1.88
20	4.35	3.49	3.10	2.87	2.71	2.60	2.51	2.45	2.39	2.35	2.28	2.20	2.12	2.07	2.04	1.99	1.95	1.91	1.84
22	4.30	3.44	3.05	2.82	2.66	2.55	2.46	2.40	2.34	2.30	2.23	2.15	2.07	2.02	1.98	1.94	1.89	1.85	1.78
25	4.24	3.39	2.99	2.76	2.60	2.49	2.40	2.34	2.28	2.24	2.16	2.09	2.01	1.96	1.92	1.87	1.82	1.78	1.71
30	4.17	3.32	2.92	2.69	2.53	2.42	2.33	2.27	2.21	2.16	2.09	2.01	1.93	1.88	1.84	1.79	1.74	1.70	1.62
40	4.08	3.23	2.84	2.61	2.45	2.34	2.25	2.18	2.12	2.08	2.00	1.92	1.84	1.78	1.74	1.69	1.64	1.59	1.51
50	4.03	3.18	2.79	2.56	2.40	2.29	2.20	2.13	2.07	2.03	1.95	1.87	1.78	1.73	1.69	1.63	1.57	1.52	1.44
60	4.00	3.15	2.76	2.53	2.37	2.25	2.17	2.10	2.04	1.99	1.92	1.84	1.75	1.69	1.65	1.59	1.53	1.48	1.39
80	3.96	3.11	2.72	2.49	2.33	2.21	2.13	2.06	2.00	1.95	1.88	1.79	1.70	1.64	1.60	1.54	1.48	1.43	1.32
100	3.94	3.09	2.70	2.46	2.31	2.19	2.10	2.03	1.97	1.93	1.85	1.77	1.68	1.62	1.57	1.52	1.45	1.39	1.28
150	3.90	3.06	2.66	2.43	2.27	2.16	2.07	2.00	1.94	1.89	1.82	1.73	1.64	1.58	1.54	1.48	1.41	1.34	1.22
∞	3.84	3.00	2.60	2.37	2.21	2.10	2.01	1.94	1.88	1.83	1.75	1.67	1.57	1.51	1.46	1.39	1.32	1.24	1.00

[a] Selected distances (F) are given for various degrees of freedom (df) for $\alpha = 5\%$, the area under the curve in the positive tail, and $1 - \alpha = 95\%$, the area under all except the tail. Numerator df appears in column headings, denominator df in row headings, and F at the intersection of row and column in the table body.

Table VI
Binomial Distribution[a]

n	n_o	.05	.10	.15	.20	.25	.30	.35	.40	.45	.50
2	1	.098	.190	.278	.360	.438	.510	.578	.640	.698	.750
	2	.003	.010	.023	.040	.063	.090	.123	.160	.203	.250
3	1	.143	.271	.386	.488	.578	.657	.725	.784	.834	.875
	2	.007	.028	.061	.104	.156	.216	.282	.352	.425	.500
	3	.000	.001	.003	.008	.016	.027	.043	.064	.091	.125
4	1	.186	.344	.478	.590	.684	.760	.822	.870	.909	.938
	2	.014	.052	.110	.181	.262	.348	.437	.525	.609	.688
	3	.001	.004	.012	.027	.051	.084	.127	.179	.242	.313
	4	.000	.000	.001	.002	.004	.008	.015	.026	.041	.063
5	1	.226	.410	.556	.672	.763	.832	.884	.922	.950	.969
	2	.023	.082	.165	.263	.367	.472	.572	.663	.744	.813
	3	.001	.009	.027	.058	.104	.163	.235	.317	.407	.500
	4	.000	.001	.002	.007	.016	.031	.054	.087	.131	.188
	5		.000	.000	.000	.001	.002	.005	.010	.019	.031
6	1	.265	.469	.629	.738	.822	.882	.925	.953	.972	.984
	2	.033	.114	.224	.345	.466	.580	.681	.676	.836	.891
	3	.002	.016	.047	.099	.169	.256	.353	.456	.559	.656
	4	.000	.001	.006	.017	.038	.071	.117	.179	.255	.344
	5		.000	.000	.002	.005	.011	.022	.041	.069	.109
	6				.000	.000	.001	.002	.004	.008	.016
7	1	.302	.522	.679	.790	.867	.918	.951	.972	.985	.992
	2	.044	.150	.283	.423	.555	.671	.766	.841	.898	.938
	3	.004	.026	.074	.148	.244	.353	.468	.580	.684	.773
	4	.000	.003	.012	.033	.071	.126	.200	.290	.392	.500
	5		.000	.001	.005	.013	.029	.056	.096	.153	.227
	6			.000	.000	.001	.004	.009	.019	.036	.063
	7					.000	.000	.001	.002	.004	.008
8	1	.337	.570	.728	.832	.900	.942	.968	.983	.992	.996
	2	.057	.187	.343	.497	.633	.745	.831	.894	.937	.965
	3	.006	.038	.105	.203	.322	.448	.572	.685	.780	.856
	4	.000	.005	.021	.056	.114	.194	.294	.406	.523	.637
	5		.000	.003	.010	.027	.058	.106	.174	.260	.363
	6			.000	.001	.004	.011	.025	.050	.089	.145
	7				.000	.000	.001	.004	.009	.018	.035
	8						.000	.000	.001	.002	.004

[a]Values of cumulative binomial distribution, depending on π (theoretical proportion of occurrences in a random trial), n (sample size), and n_o (number of occurrences observed). Given π, n, and n_o, the corresponding entry in the table body represents the probability that n_o or more occurrences (or alternatively that n_o/n proportion observed occurrences) would have been observed by chance alone.

Table VI

(Continued)

n	n_o	.05	.10	.15	.20	.25	.30	.35	.40	.45	.50
							π				
9	1	.370	.613	.768	.866	.925	.960	.979	.990	.995	.998
	2	.071	.225	.401	.564	.700	.804	.879	.930	.962	.981
	3	.008	.053	.141	.262	.399	.537	.663	.768	.851	.910
	4	.001	.008	.034	.086	.166	.270	.391	.517	.639	.746
	5	.000	.001	.006	.020	.045	.099	.172	.267	.379	.500
	6		.000	.001	.003	.010	.025	.054	.099	.166	.254
	7			.000	.000	.001	.004	.011	.025	.050	.090
	8					.000	.000	.001	.004	.009	.020
	9							.000	.000	.001	.002
10	1	.401	.651	.803	.893	.944	.972	.987	.994	.998	.999
	2	.086	.264	.456	.624	.756	.851	.914	.954	.977	.989
	3	.012	.070	.180	.322	.474	.617	.738	.833	.900	.945
	4	.001	.013	.050	.121	.224	.350	.486	.618	.734	.828
	5	.000	.002	.010	.033	.078	.150	.249	.367	.496	.623
	6		.000	.001	.006	.020	.047	.095	.166	.262	.377
	7			.000	.001	.004	.011	.026	.055	.102	.172
	8				.000	.000	.002	.005	.012	.027	.055
	9						.000	.001	.002	.005	.011
	10							.000	.000	.000	.001
11	1	.431	.686	.833	.914	.958	.980	.991	.996	.999	.999
	2	.102	.303	.508	.678	.803	.887	.939	.970	.986	.994
	3	.015	.090	.221	.383	.545	.687	.800	.881	.935	.967
	4	.002	.019	.069	.161	.287	.430	.574	.704	.809	.887
	5	.000	.003	.016	.050	.115	.210	.332	.467	.603	.726
	6		.000	.003	.012	.034	.078	.149	.247	.367	.500
	7			.000	.002	.008	.022	.050	.099	.174	.274
	8				.000	.001	.004	.012	.029	.061	.113
	9					.000	.001	.002	.006	.015	.033
	10						.000	.000	.001	.002	.006
	11								.000	.000	.001
12	1	.460	.718	.858	.931	.968	.986	.994	.998	.999	1.000
	2	.118	.341	.557	.725	.842	.915	.958	.980	.992	.997
	3	.020	.111	.264	.442	.609	.747	.849	.917	.958	.981
	4	.002	.026	.092	.205	.351	.508	.653	.775	.866	.927
	5	.000	.004	.024	.073	.158	.276	.417	.562	.696	.806
	6		.001	.005	.019	.054	.118	.213	.335	.473	.613
	7		.000	.001	.004	.014	.039	.085	.158	.261	.387
	8			.000	.001	.003	.010	.026	.057	.112	.194
	9				.000	.000	.002	.006	.015	.036	.073

(Continued)

Table VI
(*Continued*)

n	n_o	.05	.10	.15	.20	.25	.30	.35	.40	.45	.50
	10						.000	.001	.003	.008	.019
	11							.000	.000	.001	.003
	12									.000	.000
13	1	.487	.746	.879	.945	.976	.990	.996	.999	1.000	1.000
	2	.135	.379	.602	.766	.873	.936	.970	.987	.995	.999
	3	.025	.134	.270	.498	.667	.798	.887	.942	.973	.989
	4	.003	.034	.097	.253	.416	.579	.722	.831	.907	.954
	5	.000	.007	.026	.009	.206	.346	.500	.647	.772	.867
	6		.001	.005	.030	.080	.165	.284	.426	.573	.710
	7		.000	.001	.007	.024	.062	.130	.229	.356	.500
	8			.000	.001	.006	.018	.046	.098	.179	.291
	9				.000	.001	.004	.013	.032	.070	.133
	10					.000	.001	.003	.008	.020	.046
	11						.000	.000	.001	.004	.011
	12								.000	.001	.002
	13									.000	.000
14	1	.512	.771	.897	.956	.982	.993	.998	.999	1.000	1.000
	2	.153	.415	.643	.802	.899	.953	.980	.992	.997	.999
	3	.030	.158	.352	.552	.719	.839	.916	.960	.983	.993
	4	.004	.044	.147	.302	.479	.645	.780	.876	.937	.971
	5	.000	.009	.047	.130	.259	.416	.577	.721	.833	.910
	6		.002	.012	.044	.112	.220	.360	.514	.663	.788
	7		.000	.002	.012	.038	.093	.184	.308	.454	.605
	8			.000	.002	.010	.032	.075	.150	.259	.395
	9				.000	.002	.008	.024	.058	.119	.212
	10					.000	.002	.006	.018	.043	.090
	11						.000	.001	.004	.011	.029
	12							.000	.001	.002	.007
	13								.000	.000	.001
	14										.000
15	1	.537	.794	.913	.965	.987	.995	.998	1.000	1.000	1.000
	2	.171	.451	.681	.833	.920	.965	.986	.995	.998	1.000
	3	.036	.184	.396	.602	.764	.873	.938	.973	.989	.996
	4	.006	.056	.177	.352	.539	.703	.827	.910	.958	.982
	5	.001	.013	.062	.164	.314	.485	.648	.783	.880	.941
	6	.000	.002	.017	.061	.148	.278	.436	.597	.739	.849
	7		.000	.004	.018	.057	.131	.245	.390	.548	.696
	8			.001	.004	.017	.050	.113	.213	.347	.500
	9			.000	.001	.004	.015	.042	.095	.182	.304

Table VI
(*Continued*)

n	n_o	.05	.10	.15	.20	.25	.30	.35	.40	.45	.50
								π			
	10				.000	.001	.004	.012	.034	.077	.151
	11					.000	.001	.003	.009	.026	.059
	12						.000	.001	.002	.006	.018
	13							.000	.000	.001	.004
	14									.000	.001
	15										.000
16	1	.560	.815	.926	.972	.990	.997	.999	1.000	1.000	1.000
	2	.189	.485	.716	.859	.937	.974	.990	.997	.999	1.000
	3	.043	.211	.439	.648	.803	.901	.955	.982	.993	.998
	4	.007	.068	.210	.402	.595	.754	.866	.935	.972	.989
	5	.001	.017	.079	.202	.370	.550	.711	.833	.915	.962
	6	.000	.003	.024	.082	.190	.340	.510	.671	.802	.895
	7		.001	.006	.027	.080	.175	.312	.473	.634	.723
	8		.000	.001	.007	.027	.074	.159	.284	.437	.598
	9			.000	.002	.008	.026	.067	.142	.256	.402
	10				.000	.002	.007	.023	.058	.124	.227
	11					.000	.002	.006	.019	.049	.105
	12						.000	.001	.005	.015	.038
	13							.000	.001	.004	.011
	14								.000	.001	.002
	15									.000	.000

Table VII

Poisson Distribution[a]

n_o	.1	.2	.3	.4	.5	.6	.7	.8	.9	1.0	1.1	1.2	1.3	1.4	1.5
						$\lambda(= n\pi)$									
1	.095	.181	.259	.330	.394	.451	.503	.551	.593	.632	.667	.699	.728	.753	.777
2	.005	.018	.037	.062	.090	.122	.159	.191	.228	.264	.301	.337	.373	.408	.442
3	.000	.001	.004	.008	.014	.023	.034	.047	.063	.080	.100	.121	.143	.167	.191
4		.000	.000	.001	.002	.003	.006	.009	.014	.019	.026	.034	.043	.054	.066
5				.000	.000	.000	.001	.001	.002	.004	.005	.008	.011	.014	.019
6							.000	.000	.000	.001	.001	.002	.002	.003	.005
7										.000	.000	.000	.000	.001	.001

n_o	1.6	1.7	1.8	1.9	2.0	2.1	2.2	2.3	2.4	2.5	2.6	2.7	2.8	2.9	3.0
						$\lambda(= n\pi)$									
1	.798	.817	.835	.850	.865	.878	.889	.900	.909	.918	.926	.933	.939	.945	.950
2	.475	.507	.537	.566	.594	.620	.645	.669	.692	.713	.733	.751	.769	.785	.801
3	.217	.243	.269	.296	.323	.350	.377	.404	.430	.456	.482	.506	.531	.554	.577
4	.079	.093	.109	.125	.143	.161	.181	.201	.221	.242	.264	.286	.308	.330	.353
5	.024	.030	.036	.044	.053	.062	.073	.084	.096	.109	.123	.137	.152	.168	.185
6	.006	.008	.010	.013	.017	.020	.025	.030	.036	.042	.049	.057	.065	.074	.084
7	.001	.002	.003	.003	.005	.006	.008	.009	.012	.014	.017	.021	.024	.029	.034
8	.000	.000	.001	.001	.001	.002	.002	.003	.003	.004	.005	.007	.008	.010	.012
9		.000	.000	.000	.000	.000	.001	.001	.001	.001	.002	.002	.002	.003	.004
10							.000	.000	.000	.000	.000	.001	.001	.001	.001

n_o	3.1	3.2	3.3	3.4	3.5	3.6	3.7	3.8	3.9	4.0	4.1	4.2	4.3	4.4	4.5
						$\lambda(= n\pi)$									
1	.955	.959	.963	.967	.970	.973	.975	.978	.980	.982	.983	.985	.986	.988	.989
2	.815	.829	.841	.853	.864	.874	.884	.893	.901	.908	.916	.922	.928	.934	.939
3	.599	.620	.641	.660	.679	.697	.715	.731	.747	.762	.776	.790	.803	.815	.826
4	.375	.398	.420	.442	.463	.485	.506	.527	.547	.567	.586	.605	.623	.641	.658
5	.202	.219	.237	.256	.275	.294	.313	.332	.352	.371	.391	.410	.430	.449	.468
6	.094	.105	.117	.130	.142	.156	.170	.184	.199	.215	.231	.247	.263	.280	.297
7	.039	.045	.051	.058	.065	.073	.082	.091	.101	.111	.121	.133	.144	.156	.169
8	.014	.017	.020	.023	.027	.031	.035	.040	.045	.051	.057	.064	.071	.079	.087
9	.005	.006	.007	.008	.010	.012	.014	.016	.019	.021	.025	.028	.032	.036	.040
10	.001	.002	.002	.003	.003	.004	.005	.006	.007	.008	.010	.011	.013	.015	.017
11	.000	.001	.001	.001	.001	.001	.002	.002	.002	.003	.003	.004	.005	.006	.007
12		.000	.000	.000	.000	.000	.001	.001	.001	.001	.001	.001	.002	.002	.002
13							.000	.000	.000	.000	.000	.000	.001	.001	.001

[a]Values of the cumulative Poisson distribution, depending on $\lambda = n\pi$ (sample size × theoretical proportion of occurrences in random trials) and n_o (number of occurrences observed). Given λ and n_o, the corresponding entry in the table body represents the probability that n_o or more occurrences (or alternatively that n_o/n proportion observed occurrences) would have been observed by chance alone. (Probability is always 1.000 for $n_o = 0$.)

Table VII

(*Continued*)

n_O	4.6	4.7	4.8	4.9	5.0	5.1	5.2	5.3	5.4	5.5	5.6	5.7	5.8	5.9	6.0
							$\lambda(= n\pi)$								
1	.990	.991	.992	.993	.993	.994	.995	.995	.996	.996	.996	.997	.997	.997	.998
2	.944	.948	.952	.956	.960	.963	.966	.969	.971	.973	.976	.978	.979	.981	.983
3	.837	.848	.858	.867	.875	.884	.891	.898	.905	.912	.918	.923	.929	.933	.938
4	.674	.690	.706	.721	.735	.749	.762	.775	.787	.798	.809	.820	.830	.840	.849
5	.487	.505	.524	.542	.560	.577	.594	.611	.627	.643	.658	.673	.687	.701	.715
6	.314	.332	.349	.367	.384	.402	.419	.437	.454	.471	.488	.505	.522	.538	.554
7	.182	.195	.209	.223	.238	.253	.268	.283	.298	.314	.330	.346	.362	.378	.394
8	.095	.104	.113	.123	.133	.144	.155	.167	.178	.191	.203	.216	.229	.242	.256
9	.045	.050	.156	.062	.068	.075	.082	.089	.097	.106	.114	.123	.133	.143	.153
10	.020	.022	.025	.028	.032	.036	.040	.044	.049	.054	.059	.065	.071	.077	.084
11	.008	.009	.010	.012	.014	.016	.018	.020	.023	.025	.028	.031	.035	.039	.043
12	.003	.003	.004	.005	.006	.006	.007	.008	.010	.011	.013	.014	.016	.018	.020
13	.001	.001	.001	.002	.002	.002	.003	.003	.004	.005	.005	.006	.007	.008	.009
14	.000	.000	.001	.001	.001	.001	.001	.001	.001	.002	.002	.002	.003	.003	.004
15			.000	.000	.000	.000	.000	.000	.000	.001	.001	.001	.001	.001	.001

n_O	6.1	6.2	6.3	6.4	6.5	6.6	6.7	6.8	6.9	7.0	7.1	7.2	7.3	7.4	7.5
							$\lambda(= n\pi)$								
1	.998	.998	.998	.998	.999	.999	.999	.999	.999	.999	.999	.999	.999	.999	.999
2	.984	.985	.987	.988	.989	.990	.991	.991	.992	.993	.993	.994	.994	.995	.995
3	.942	.946	.950	.954	.957	.960	.963	.966	.968	.970	.973	.975	.976	.978	.980
4	.858	.866	.874	.881	.888	.895	.901	.907	.913	.918	.923	.928	.933	.937	.941
5	.728	.741	.753	.765	.776	.787	.798	.808	.818	.827	.836	.845	.853	.861	.868
6	.570	.586	.601	.616	.631	.645	.659	.673	.686	.699	.712	.724	.736	.747	.759
7	.410	.426	.442	.458	.474	.489	.505	.520	.535	.550	.565	.580	.594	.608	.622
8	.270	.284	.298	.313	.327	.342	.357	.372	.386	.401	.416	.431	.446	.461	.475
9	.163	.174	.185	.197	.208	.220	.233	.245	.258	.271	.284	.297	.311	.324	.338
10	.091	.098	.106	.114	.123	.131	.140	.150	.151	.170	.180	.190	.201	.212	.223
11	.047	.051	.056	.061	.067	.073	.079	.085	.092	.099	.106	.113	.121	.129	.138
12	.022	.025	.028	.031	.034	.037	.041	.045	.049	.053	.058	.063	.068	.074	.079
13	.010	.011	.013	.014	.016	.018	.020	.022	.025	.027	.030	.033	.036	.039	.043
14	.004	.005	.006	.006	.007	.008	.009	.010	.012	.013	.014	.016	.018	.020	.022
15	.002	.002	.002	.003	.003	.003	.004	.004	.005	.006	.007	.007	.008	.009	.010
16	.001	.001	.001	.001	.001	.001	.002	.002	.002	.002	.003	.003	.004	.004	.005
17	.000	.000	.000	.000	.000	.001	.001	.001	.001	.001	.001	.001	.002	.002	.002
18						.000	.000	.000	.000	.000	.000	.001	.001	.001	.001

(*Continued*)

Table VII

(*Continued*)

n_o	7.6	7.7	7.8	7.9	8.0	8.1	8.2	8.3	8.4	8.5	8.6	8.7	8.8	8.9	9.0
							$\lambda(=n\pi)$								
1	.999	1.00	1.00	1.00	1.00	1.00	1.00	1.00	1.00	1.00	1.00	1.00	1.00	1.00	1.00
2	.996	.996	.996	.997	.997	.997	.998	.998	.998	.998	.998	.998	.999	.999	.999
3	.981	.983	.984	.985	.986	.987	.988	.989	.990	.991	.991	.992	.993	.993	.994
4	.945	.948	.952	.955	.958	.960	.963	.965	.968	.970	.972	.974	.976	.977	.979
5	.875	.882	.888	.895	.900	.906	.911	.916	.921	.926	.930	.934	.938	.942	.945
6	.769	.780	.790	.799	.809	.818	.826	.835	.843	.850	.858	.865	.872	.878	.884
7	.635	.649	.662	.674	.687	.699	.710	.722	.733	.744	.754	.765	.774	.784	.793
8	.490	.504	.519	.533	.547	.561	.575	.588	.601	.614	.627	.640	.652	.664	.676
9	.352	.366	.380	.394	.408	.421	.435	.449	.463	.477	.491	.504	.518	.531	.544
10	.235	.247	.259	.271	.283	.296	.309	.321	.334	.347	.360	.373	.386	.399	.413
11	.147	.156	.165	.174	.184	.194	.204	.215	.226	.237	.248	.259	.271	.282	.294
12	.085	.092	.098	.105	.112	.119	.127	.135	.143	.151	.160	.169	.178	.187	.197
13	.046	.050	.055	.059	.064	.069	.074	.079	.085	.091	.097	.104	.110	.117	.124
14	.024	.026	.029	.031	.034	.037	.041	.044	.048	.051	.056	.060	.064	.069	.074
15	.011	.013	.014	.016	.017	.019	.021	.023	.025	.027	.030	.033	.035	.038	.042
16	.005	.006	.007	.007	.008	.009	.010	.011	.013	.014	.015	.017	.018	.020	.022
17	.002	.003	.003	.003	.004	.004	.005	.005	.006	.007	.007	.008	.009	.010	.011
18	.001	.001	.001	.001	.002	.002	.002	.002	.003	.003	.003	.004	.004	.005	.005
19	.000	.000	.000	.001	.001	.001	.001	.001	.001	.001	.002	.002	.002	.002	.002
20				.000	.000	.000	.000	.000	.001	.001	.001	.001	.001	.001	.001

Table VIII

Signed-Rank Probabilities[a]

	Sample Size n								
T	4	5	6	7	8	9	10	11	12
1	.250	.125	.063	.031	.016	.008	.004	.002	.001
2	.375	.188	.094	.047	.024	.012	.006	.003	.002
3	.625	.313	.156	.078	.039	.020	.010	.005	.003
4	.875	.438	.219	.109	.055	.027	.014	.007	.004
5		.625	.313	.156	.078	.039	.020	.010	.005
6		.801	.438	.219	.109	.055	.027	.014	.007
7		1.000	.563	.297	.149	.074	.037	.019	.009
8			.688	.375	.195	.098	.049	.025	.012
9			.844	.469	.250	.129	.065	.032	.016
10			1.000	.578	.313	.164	.084	.042	.021
11				.688	.383	.203	.106	.054	.027
12				.813	.461	.250	.131	.067	.034
13				.938	.547	.301	.160	.083	.043
14					.641	.359	.193	.102	.052
15					.742	.426	.233	.1123	.064
16					.844	.496	.275	.148	.077
17					.945	.570	.322	.175	.092
18						.652	.375	.206	.110
19						.7344	.432	.240	.129
20						.820	.492	.278	.151
21						.910	.557	.320	.176
22						1.000	.625	.365	.204
23							.695	.413	.233
24							.770	.465	.266
25							.846	.520	.301
26							.922	.577	.339
27							1.000	.638	.380
28								.700	.424
29								.765	.470
30								.831	.519
31								.899	.569
32								.966	.622
33									.677
34									.733
35									.791
36									.850
37									.910
38									.970

[a]Two-tailed probabilities for the distribution of T, the signed-rank statistic. For a sample size n and value of T, the entry gives the p-value. If a one-tailed test is appropriate, halve the entry. If $n > 12$, go to Section 16.6.

Tables of Probability Distributions

Table IX

Rank-Sum *U* Probabilities[a]

	n_2: 3			n_2: 4				n_2: 5				
n_1:	1	2	3	1	2	3	4	1	2	3	4	5
U 0	.500	.200	.100	.400	.133	.056	.028	.333	.094	.036	.016	.008
1	1.00	.400	.200	.800	.267	.114	.058	.337	.190	.072	.032	.016
2		.800	.400		.534	.228	.114	1.00	.380	.142	.064	.032
3			.700		.800	.400	.200		.572	.250	.112	.056
4			1.00			.628	.342		.858	.392	.190	.096
5						.858	.486			.572	.286	.150
6							.686			.786	.412	.222
7							.886			1.00	.556	.310
8											.730	.420
9											.904	.548
10												.690
11												.842
12												1.00

	n_2: 6						n_2: 7						
n_1:	1	2	3	4	5	6	1	2	3	4	5	6	7
U 0	.286	.072	.024	.010	.004	.002	.250	.054	.016	.006	.002	.002	.001
1	.572	.142	.048	.020	.008	.004	.500	.112	.034	.012	.006	.002	.001
2	.856	.286	.096	.038	.018	.008	.750	.222	.066	.024	.010	.004	.002
3		.428	.166	.066	.030	.016	1.00	.334	.116	.042	.018	.008	.004
4		.642	.262	.114	.052	.026		.500	.184	.072	.030	.014	.006
5		.858	.380	.172	.082	.042		.666	.233	.110	.048	.022	.012
6			.548	.258	.126	.064		.888	.384	.164	.074	.034	.018
7			.714	.352	.178	.094			.516	.230	.106	.052	.026
8			.904	.476	.246	.132			.666	.316	.148	.074	.038
9				.610	.330	.180			.834	.412	.202	.102	.054
10				.762	.428	.240			1.00	.528	.268	.138	.072
11				.914	.536	.310				.648	.344	.180	.098
12					.662	.394				.788	.432	.234	.128
13					.792	.484				.928	.530	.296	.164
14					.930	.588					.638	.366	.208
15						.700					.756	.446	.260
16						.818					.876	.534	.318
17						.938					1.00	.628	.382
18												.730	.456
19												.836	.534
20												.946	.620
21													.710
22													.804
23													.902
24													1.00

Table IX

(*Continued*)

	n_2:					8			
	n_1:	1	2	3	4	5	6	7	8
U	0	.222	.044	.012	.004	.002	.001	.000	.000
	1	.444	.088	.024	.008	.004	.002	.001	.001
	2	.666	.178	.048	.016	.006	.003	.002	.001
	3	.888	.266	.082	.028	.010	.004	.003	.002
	4		.400	.134	.048	.018	.008	.004	.002
	5		.534	.194	.072	.030	.012	.006	.003
	6		.712	.278	.110	.046	.020	.010	.004
	7		.888	.376	.154	.066	.030	.014	.006
	8			.496	.214	.094	.042	.020	.010
	9			.630	.282	.138	.060	.028	.014
	10			.774	.368	.170	.082	.040	.020
	11			.922	.460	.222	.108	.052	.028
	12				.570	.284	.142	.072	.038
	13				.682	.354	.182	.094	.050
	14				.808	.434	.228	.120	.064
	15				.934	.524	.282	.152	.082
	16					.622	.344	.190	.104
	17					.724	.414	.232	.130
	18					.832	.490	.280	.160
	19					.944	.572	.336	.194
	20						.662	.396	.234
	21						.754	.464	.278
	22						.852	.536	.328
	23						.950	.612	.382
	24							.694	.442
	25							.778	.506
	26							.866	.574
	27							.956	.626
	28								.720
	29								.798
	30								.878
	31								.960

[a]Two-tailed probabilities for the distribution of U, the rank-sum statistic. For two samples of size n_1 and n_2 ($n_2 > n_1$) and the value of U, the entry gives the p-value. If a one-tailed test is appropriate, halve the entry. If $n_2 > 8$, go to Section 16.7. It can be seen from inspection that, if $n_1 + n_2 < 7$, a rejection of H_o at $\alpha = 0.05$ is not possible.

Symbol Index

Classes of Symbols

Subscript	Designates member of family indicated by symbol to which it is attached
Superscript	Power of symbol to which it is attached
Superscript (in parentheses)	Acts as subscript when subscript position is preempted
Greek letters	Population parameters
Roman letters	Corresponding to a Greek letter, the corresponding sample parameter

Mathematical Symbols

$+$	Plus (add what appears before it to what appears after it)
$-$	Minus (subtract what appears after it from what appears before it)
$\pm$	Plus or minus, according to context
$\times$	(1) Times (multiply what appears before it by what appears after it); (2) by (indicating two dimensions of an entity, as a 2×2 table)
$\div$ or $/$	Division (divide what appears before it by what appears after it)
$=$	Equals (what appears before it is the same as what appears after it)
$\neq$	Unequal (what appears before it is different from what appears after it)
$<$	Less than (what appears before it is less than what appears after it)
$\leq$	Less than or equal to
$>$	Greater than (what appears before it is greater than what appears after it)
$\geq$	Greater than or equal to
$\approx$	Approximately equal to
$\equiv$	Identical to (always equal to)
$\sqrt{}$	Square root (taken of what comes under the bar)
$'$	In position of a superscript, sometimes indicates a ranked observation
$\circ$	In position of a superscript, degrees of an angle
$\%$	Percent (proportion of 100 units)
$\mid$	Given (what comes before it is conditional upon what comes after it)

|...|	Absolute (what appears within bars has minus sign, if any, dropped)
e	Natural number (approximately 2.7183); base of exponential distribution
$f(..)$	Function of what appears within parentheses
$\log(..)$	Logarithm of what appears within parentheses, often to base 10
$\ln(..)$	Natural logarithm of what appears within parentheses to base e

Statistical Symbols

$\sim$	Is distributed as [e.g., $x \sim N(0,1)$]
ANOVA	Analysis of variance
ANCOVA	Analysis of covariance
AR	Attributable risk
α (alpha)	(1) Predata probability that a false hypothesis is accepted; (2) name of a particular probability distribution
β (beta)	(1) Predata probability that a true hypothesis is rejected; (2) coefficient (multiplier) of a variable in a regression or other equation, often theoretical or modeled equation; name of a particular probability distribution
b	Often, a sample estimate of a β coefficient
χ^2 (chi-square)	(1) A particular probability distribution; (2) a value that obeys that distribution
c	Sometimes, number of columns of a data table
$c.i.$	Confidence interval
$\mathrm{cov}(x, y)$	Covariance of x with y (or whatever variables are indicated)
δ	A theoretical or putative difference
Δ	A clinically relevant or important difference
d	A sample difference between two defined quantities; an estimate of δ
df	Degrees of freedom
e	Expected value; also see under Mathematical Symbols
F	(1) A particular probability distribution; (2) a value that obeys that distribution
FET	Fisher's exact test of contingency
H	Hypothesis designator, as H_0: null hypothesis, H_1 or H_a: alternate
i or j	Usually as subscript, indicator of member of set of variables
k	(1) As i or j; (2) as letter or superscript, indicator of position, number, or category
KS test	Kolmogorov–Smirnov test
λ	Occurrence rate in a Poisson distribution
LR	Likelihood ratio
μ (mu)	Mean of a population or a probability distribution
m	Mean of a sample
md	Median
MS	Mean square, as MSE: MS error, MSM: MS means, MST: MS Total

mo	Mode
MOE	Measure of effectiveness
n	Number of members of a sample
NNB	Number needed to benefit
NNT	Number needed to treat
NPV	Negative predictive value
OR	Odds ratio
π (pi)	(1) A theoretical or population probability or proportion; (2) a product of following terms
p	Among many meanings, (1) a probability; (2) a proportion; (3) as a "*p*-value," the postdata estimate of α; (4) as a subscript, symbol to which it is attached is pooled over two or more samples
PPV	Positive predictive value
ρ (rho)	A theoretical or population correlation coefficient
r	(1) A sample correlation coefficient; (2) sometimes, number of rows of a data table
R^2	(1) Coefficient of determination; (2) if a single variable, r^2
ROC	Receiver operating characteristic
RR	Relative risk, risk ratio
$\sum$	(1) Sum of values appearing after it; (2) sometimes, dispersion (variance–covariance) matrix in multivariate analysis; (3) as otherwise defined
σ	Population or theoretical standard deviation (square root of variance)
σ^2	Population or theoretical variance (square of standard deviation)
s	Sample standard deviation
s^2	Sample variance
SED	Standard error of the difference
SEE	Standard error of the estimate
SEM	Standard error of the mean
SS	Sum of squares, as SSE: SS error, SSM: SS means, SST: SS Total
t	(1) A particular probability distribution; (2) a value that obeys that distribution
T	A sum of ranks in rank-order tests
TOST	Two one-sided tests (in equivalence testing)
U	The statistic of the Mann–Whitney form of the rank-sum test
x or *y*	A variable; a value of that variable
$\bar{x}$	Alternative notation for m_x, the mean of variable *x*; also $\bar{y}$, etc.
z	(1) Often, the normal (or Gaussian) distribution; (2) a value that obeys that distribution

Subject Index